Foundations of Analysis

David E. Belding
Hobart & Williams Smith Coleges

Kevin J. Mitchell
Hobart & Williams Smith Coleges

Prentice Hall, Englewood Cliffs, New Jersey 07632

Library of Congress Cataloging-in-Publication Data

Belding, David F.
Foundations of analysis / David F. Belding, Kevin J. Mitchell.
p. cm.
Include bibliographical references (p.) and index.
ISBN 0-13-326679-6
1. Mathematical analysis. 2. Mitchell, Kevin J. II. Title.
QA300.b452 1991
515--dc20 90-20163
CIP

Editorial/production supervision: *bookworks*
Cover design: Bruce Kenselaar
Manufacturing buyers: Paula Massenaro and Lori Bulwin
Acquisition editor: Steve Conmy

Printed in the United States of America

10 9 8 7 6 5 4 3 2 1

ISBN 0-13-326679-6

Prentice-Hall International (UK) Limited, *London*
Prentice-Hall of Australia Pty. Limited, *Sydney*
Prentice-Hall Canada Inc., *Toronto*
Prentice-Hall Hispanoamericana, S.A., *Mexico*
Prentice-Hall of India Private Limited, *New Delhi*
Prentice-Hall of Japan, Inc., *Tokyo*
Simon & Schuster Asia Pte. Ltd., *Singapore*
Editora Prentice-Hall do Brasil, Ltda., *Rio de Janeiro*

For

Susan Brand Belding

and

Ellen Duncan Mitchell

Contents

Preface

This text is an introduction to basic analysis. It presents a careful development of the real number system and the theory of calculus on the real line followed by extensions of the theory to $\mathbf{R}^2$ and the complex plane. The text was designed to be a first encounter with rigorous, formal mathematics for serious mathematics students with a year of calculus. Early mathematical education up through calculus seems to focus on learning *how to* carry out various mathematical operations. But the heart of mathematics does not lie in reproducing computational procedures or memorizing mathematical facts, but in seeing *why* the facts are true, in grasping the arguments which show *why* the computation works, in understanding the connections which weave the individual pieces of mathematics into a theory. The discovery of this "heart" of mathematics begins by cultivating a deep curiousity about what underlies the methods and conclusions of mathematics. A desire to investigate how calculus is put together and what makes it tick is the primary prerequisite for reading this book.

In writing this text we had three major goals in mind.

- First, we wished to help students develop some understanding and control of the language of mathematics: definitions, theorems, and proofs.
- Secondly, we wanted to present the central concepts and theory of calculus thoroughly and clearly enough to provide a secure foundation for more advanced mathematics courses and enable students to grasp some of the unity and beauty of the subject.

- Finally, we wished to foster an appreciation for the living, human nature of mathematics, a sense of how mathematics is created and evolves.

The transition to formal mathematics based on rigorous proofs is a necessary but difficult step. Reading and writing proofs requires a new level of precision and attention to detail. Rigorous definitions force us to rework our vague intuitive understanding of concepts such as limits or integrals. In this text we try to provide the kind of demanding but supportive environment needed to make this transition successful. We supply a context for the more difficult definitions and proofs by tracing the history behind the concepts. Hidden problems which motivate the design of a proof or wording of a definition are pointed out and discussed. The proofs have been crafted with the student in mind. Thus, many details are included which draw attention to points that may seem obvious or repetitive to more veteran mathematicians. Nearly all problems require a proof of some kind. Some are simply intended to guide the review of material. These often require only slight modification of earlier arguments or immediate reflection on the meaning of a definition or theorem. However, written answers to such straightforward questions serve to reveal serious misconceptions early on. When an involved argument is required, the problem is frequently broken into steps to point toward fruitful questions and to reveal the structure and level of detail needed for a thorough proof.

A firm understanding of basic analysis requires more than a single exposure to the important ideas. This text is designed to present the central concepts of calculus several times in a parallel fashion in different contexts. For example, limits, derivatives and integrals are developed in three settings: functions of one variable, functions of two variables, and functions of a complex variable. Axioms systems are introduced for the real number system in Chapter 1, encountered again in Chapter 5 with vector spaces, and a third time in Chapter 7 with complex numbers. Prepartion for the difficult concept of uniformity starts in Chapter 1 with the Heine-Borel theorem. It is fully introduced in the definition of uniform continuity in Chapter 2 and used at various points thereafter. The related idea of uniform convergence of functions is an important theme in Chapter 4. Green's theorem is developed as an extension of the fundamental theorem of calculus and then employed to prove Cauchy's theorem in Chapter 7. This integrated approach featuring repeated encounters from different perspectives is designed to reinforce and enrich the readers understanding of the fundamental concepts of analysis. It is also intended to demonstrate the unity and power of analysis and afford a more secure and meaningful preparation for later analysis courses.

The awareness of mathematics as a living subject is frequently lost in formal presentations. In this text we have tried to keep that spirit alive. The evolution of the central ideas are discussed from the perspective of mathematicians struggling to forge important concepts and definitions. Several historically important applications are presented which highlight the roots and connections of analysis with other disciplines. Our intention thoughout the text is to engage the reader in the process of creating mathematics, in thinking through arguments.

This text does not pretend to be a comprehensive treatment of basic analysis and instructors are likely to find various favorite topics missing. Instead, we have tried to present the main current of analysis faithfully and with enough spirit to inspire the student to explore further.

An index of symbols is provided for convenience. Problems which are of special importance to later work are marked with the symbol •. Those marked with a ⋆ are somewhat more challenging. Statements upon which the theory is built, such as definitions, theorems, and corollaries, are numbered consecutively within each section. For example, Theorem 3.7.4 refers to the fourth statement in Section 7 of Chapter 3. The conclusion of proofs is marked by the symbol ▮. As examples do not play a direct role in the theory, they are numbered separately and conclude with □.

We wish to express special appreciation to our families, students, colleagues and friends who have sustained us with their help and encouragement thoughout this project. In particular we thank David Eck for his invaluable help in negotiating the subtleties of TEX. We also thank the following reviewers for their useful comments and suggestions: Professors Michael E. Mays, Richard Chandler, John Konvalina, Roger H. Marty, Daniel Kocan, R. B. Burckel, Eli Passow, Charles N. Friedman and Peter Colwell.

Kevin Mitchell

David Belding

Geneva

1

The Real Number System

This chapter introduces the basic mathematical object that underlies all of analysis: the real number system. No doubt you are already familiar with real numbers from your previous study of calculus, at least in an intuitive working sense. Here we will explore the real numbers more deeply. Our study begins with the Greeks' discovery of irrational quantities. The early Greek conception of number was based on whole numbers and quantities derived directly as the ratio of whole numbers. Since irrational quantities cannot be described in this concrete way, they were not regarded as proper numbers by the Greeks. For a long time irrationals, as their name indicates, occupied an anomalous status as phantom quantities scattered among the true "rational" numbers. It was not until the nineteenth century that a precise and logically sound meaning for the irrationals as numbers was developed by *constructing* the set of real numbers (rationals and irrationals) directly from the rational numbers.

In this construction the real numbers are endowed with a special property called completeness, which is the basis for all limit processes in calculus. In Section 1.2 we explore this construction and some of the consequences of this special completeness property. Following this, we examine the set of axioms that underlie the real number system and see how the common operations with real numbers are consequences of these axioms. Finally, in Section 1.4 the completeness of the real numbers is used to prove the important *Heine-Borel theorem*, which forms the foundation for several key results in the theory of calculus.

Mathematics evolves by abstracting and generalizing the properties and structures that it uncovers in the course of investigating particular mathematical objects. Perhaps no other object has inspired more mathematics in this way than the real number system. Each of the major branches of mathematics can trace its roots back to the fundamental example of the real numbers. In **analysis** the completeness property of real numbers and the attendant notions of limit are the primary motivating ideas. In **algebra** the stucture of addition and multiplication of real numbers forms the central paradigm. The fundamental concept of *orders of infinity* in **logic** arose from questions about the size of the set of real numbers. Finally, the branch of **topology** originated in the study of open and closed sets of real numbers whose significance was highlighted by the Heine-Borel theorem. Thus this single object, the real number system, has been the primary inspiration and guiding hand in the development of a large part of mathematics. A thorough study of the real numbers is indispensable for all serious mathematics students.

Foundational problems such as providing clear and precise definitions and rules for operation are often misunderstood as merely pedantic exercises. Students tend to rely on their original intuitions of the subject, ignoring the difficult definitions and constructions. But good definitions need to be taken seriously, for they are actually meant to reshape our intuitions. It is with these newly reshaped tools that we are able to resolve the subtle confusions in our earlier understanding. Furthermore, these definitions and constructions form the foundation for exploration in areas where our earlier notions would have had no meaning. At first, the new definitions and constructions seem abstract and contrived. However, with work they become more concrete and new, very keen intuitions develop. At its best, a thorough foundational study can be a truly enlightening experience. It can profoundly simplify and unify a subject, opening the way to powerful generalizations. In mathematics there is perhaps no better or more important foundational study than that of the real number system.

1.1 Irrational Numbers

The Greeks were the first civilization to view mathematics as an abstract deductive system. Earlier cultures had considered mathematics primarily as a tool kit of techniques to aid practical computations in activities such as navigation, construction, and commerce. The Greeks' interest in mathematics went beyond these practical applications to a deep concern for the logical integrity and consistency of arguments. This interest was rooted in a deep, almost religious belief in the power of pure rational thought to enlighten the mind. By training the mind logically, particularly through the study of mathematics, mortals could glimpse the underlying rational design of the universe and learn to lead a moral and just life. The following words of Proclus, a chronicler of early Greek mathematics, epitomize the lofty role of mathematics in Greek culture.

> This, therefore is mathematics; she reminds you of the invisible form of the soul; she gives life to her own discoveries; she awakens the mind and purifies the intellect; she brings light to our intrinsic ideas; she abolishes oblivion and ignorance which are ours by birth.[1]

1.1.1 The Pythagoreans

The elevation of mathematics to this preeminent status was the work of a famous school of philosophers known as the Pythagoreans, which arose in Greece about the sixth century B.C. under the leadership of the mathematician Pythagoras. They lived as a secretive communal society, sharing their worldly goods and following strict dietary codes. As scholars the Pythagoreans were chiefly concerned with their studies, which they organized into four main branches of learning: *arithmetica* (number theory), *harmonia* (music), *geometria* (geometry), and *astrologia* (astronomy).

The Pythagoreans were motivated by the belief that through pure rational thought they could uncover and understand the design of the universe and how it worked. A basic tenet of this belief system was that numbers, by which they meant positive whole numbers, formed the key to understanding in all fields of knowledge. In the words of a famous follower, Philolaus:

> All things which can be known have number; for it is not possible that without number anything can be either conceived or known.[2]

The central role assigned to numbers by the Pythagoreans was mingled with mystical tendencies and led to a great deal of numerology. It was believed that objects could be represented by numbers and that the relationships between the numbers revealed truths about the relationships between the corresponding objects.

> The number one is the generator of numbers, the number of reason; two is the first even or female number, the number of opinion; three is the first true male number, the number of harmony, being composed

[1] Morris Kline, *Mathematical Thought from Ancient to Modern Times* (New York: Oxford University Press, 1972), p. 24.

[2] Carl B. Boyer, *A History of Mathematics* (Princeton, N.J.: Princeton University Press, 1985), p. 60.

of unity and diversity; four is the number of justice or retribution, indicating the squaring of accounts; five is the number of marriage, the union of male and female numbers; and six is the number of creation.[3]

For the Pythagoreans ratios of integers were the most important tool in the study of these relationships and thus formed the unifying thread connecting all the areas of learning. In music, fundamental integral ratios underlying the theory of harmony were studied. These ideas were applied to *astrologia* in which it was believed that the motions of heavenly bodies were governed by the same ratios, and students ostensibly learned to hear "the music of the spheres."

In the same way it was believed that the basic truths of geometry could be revealed and understood through numbers and their ratios. A major tenet of this system was that any two line segments were **commensurable** (evenly measurable using a common unit). That is, for any two line segments it was believed possible to find a unit small enough so that each segment would be an exact integral multiple of that unit.

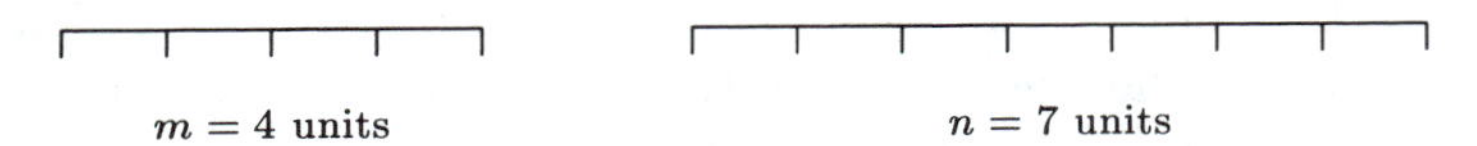

Figure 1.1.1. Two commensurable line segments.

Ironically, the Pythagoreans themselves discovered the fallacy in this belief. Certain familiar lengths such as the side of a square and its diagonal are inherently **incommensurable** and hence cannot be evenly measured by a common unit no matter how small. The existence of incommensurable lengths was a mysterious and incomprehensible fact to the Pythagoreans. Since the ratios between such lengths could not be described by whole numbers, they were called *alogos* meaning "without word" or inexpressible. This discovery implied that whole numbers and their ratios were in some way inadequate or too incomplete to describe geometric lengths. Obviously this was a major setback for a philosophy that had postulated number as the central key to all understanding. The problem was so embarrassing that according to legend the discoverer was thrown overboard at sea and members were forbidden to reveal the secret of the inexpressible quantities.[4]

Today we may find some of the Pythagoreans' beliefs naïve and superstitious, but we owe a great deal to these philosophers. They inspired

---

[3] Boyer, p. 57.

[4] Ernest Sondheimer and Alan Rogerson, *Numbers and Infinity: A Historical Account of Mathematical Concepts* (Cambridge: Cambridge University Press, 1981), p. 43.

some of the basic principles of western culture: that nature is susceptible to systematic rational understanding, and that moral conduct and justice should rest on logical reasoning. The four areas of learning that the Pythagoreans outlined were the original four "liberal arts" known as the *Quadrivium*. In the Middle Ages three more subjects, logic, rhetoric, and grammar, called the *Trivium*, were added. In the West these seven liberal arts have for centuries been considered the foundation of all learning. Notice the central role of mathematics in these liberal studies. Even the word *mathematics*, which means "that which is learned", is attributed to Pythagoras.

1.1.2 Irrational Numbers

Let's examine how the problem of incommensurable lengths arose. Suppose for a moment that the side of a square and its diagonal *were* commensurable. This would mean there exists a unit small enough so that the side would be some integer multiple of this unit, say N, and the diagonal would be another whole multiple of the same unit, say M.

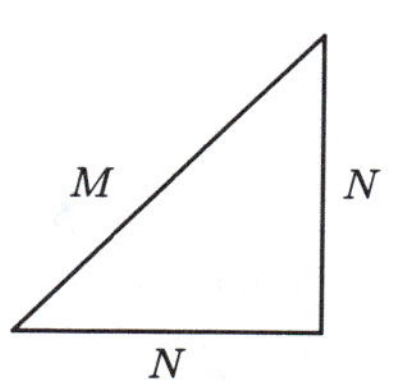

Figure 1.1.2

By the Pythagorean theorem these numbers must satisfy

$$\begin{aligned} N^2 + N^2 &= M^2 \\ 2N^2 &= M^2 \\ 2 &= M^2/N^2. \end{aligned}$$

This last equality would then imply that $\sqrt{2} = M/N$ for integers M and N. Any number that can be expressed as the quotient of two integers is called a **rational number**. Hence using the Pythagorean theorem we see that if the diagonal and side of a square are commensurable, then $\sqrt{2}$ must be rational. But the following argument shows that $\sqrt{2}$ is *not* rational.

THEOREM 1.1.1 *There are no integers M and N such that $\sqrt{2} = M/N$.*

PROOF (By contradiction.) Suppose that $\sqrt{2} = M/N$ for some integers M and N. Since any fraction can be expressed in lowest terms by dividing out common factors, *we may further assume that M and N have no common divisors.* We will obtain a contradiction by showing that M and N must both be divisible by 2.

If $\sqrt{2} = M/N$, we may write $2 = M^2/N^2$ and

$$2N^2 = M^2.$$

Since N^2 is an integer, this implies that M^2 is divisible by 2 (i.e., that it is an even integer). What about M? If M were odd, M^2 would also be odd; thus M must be even. We conclude that *if M^2 is divisible by 2, M must be divisible by 2.* So we may write $M = 2P$ for some integer P and substitute for M in the foregoing equation:

$$2N^2 = (2P)^2, \qquad \text{or equivalently,} \qquad N^2 = 2P^2.$$

The last equality says that N^2 is divisible by 2; hence N is also divisible by 2. We have shown that M and N *have a common factor*, 2, contrary to our original choice of M and N. Therefore there can be no integers M and N with $\sqrt{2} = M/N$. ∎

The Greeks were masters at discovering inconsistencies buried in statements, as is beautifully illustrated in the dialogues of Socrates. They honed the technique of proof by contradiction, known as *reductio ad absurdum*, to a fine art. The preceding argument shows that if a rational expression for $\sqrt{2}$ existed, it could not have been in lowest terms. Since any ratio of integers can always be reduced to lowest terms, this must mean that no such expression can exist. It is considered one of the most beautiful examples of proof by contradiction in elementary mathematics. In the remainder of this section we look at several other famous proofs by contradiction.

1.1.3 The Fundamental Theorem of Arithmetic

The heart of the proof that $\sqrt{2}$ is irrational lies in the fact that if 2 divides M^2, it must also divide M. The same is true if 3 divides M^2, and an analogous argument shows that $\sqrt{3}$ is also irrational (see problem 2). In fact, it is true of any **prime number**, that is, any integer greater than 1 that is evenly divisible only by itself and 1. At this point you might guess that the irrationality of square roots is linked to questions of divisibility. The most formidable tool for the investigation of divisibility is the fact that any integer can be written in only one way as the product of prime numbers. The proof of this fact, known as the fundamental theorem of arithmetic, provides another example of proof by contradiction. We begin with a lemma.

LEMMA 1.1.2 *Any natural number $N > 1$ can be written as a product of prime numbers.*

PROOF Suppose that this were not true. Let N be the *first* number that cannot be written as a product of one or more primes (thus all smaller numbers can be expressed as products of primes).[5] We may conclude that N itself is not a prime (since then it would already be a product of primes). Hence N is divisible by some number $1 < K < N$ and we may write $N = KM$. Since both K and M are smaller than N they can be written as products of primes: $K = p_1p_2\dots p_n$ and $M = q_1q_2\dots q_m$. But this allows us to express N itself as a product of primes, contradicting our assumption:

$$N = KM = p_1p_2\dots p_nq_1q_2\dots q_m. \qquad \blacksquare$$

Next we prove that each number has only *one* such representation as a product of primes.

THEOREM 1.1.3 ***(The Fundamental Theorem of Arithmetic)*** *For any natural number, $N > 1$, there is a unique factorization in terms of prime numbers. That is, N can be written in one and only one way as*

$$N = p_1p_2p_3\cdots p_r,$$

where the p's are all prime and $p_1 \le p_2 \le p_3 \le \cdots \le p_r$.

Note that several of the p's may be equal when numbers have multiple factors of a prime; for example, $540 = 2\cdot2\cdot3\cdot3\cdot3\cdot5$. The inequalities serve only to put the factorization in a standard order. The real content of the theorem is that the prime factors and their multiplicities are *unique* for each integer.

PROOF From the lemma we know that all numbers can be factored into primes. Now suppose there exist numbers that may be factored into primes in at least two distinct ways. If any such numbers exist, there must be a smallest one, which we will designate by N. Our contradiction will rest on finding another number smaller than N that also has two factorizations.

---

[5] The fact that any nonempty subset of natural numbers has a first element is known as the **well-ordering property** of the natural numbers.

Let the two distinct prime factorizations of N be given as

$$N = p_1 p_2 p_3 \cdots p_k \qquad \text{and} \qquad N = q_1 q_2 q_3 \cdots q_l. \tag{1}$$

First observe that if there were any common primes in these factorizations, we could divide them out and obtain two different factorizations of an even smaller number. This is impossible since N is assumed to be the *smallest* number with two distinct factorizations. *Hence these two factorizations have no primes in common.*

Now assume that $p_1 > q_1$ and consider the number

$$M = N - q_1 p_2 p_3 \cdots p_k = (p_1 - q_1) p_2 p_3 \cdots p_k \, .$$

Clearly M is a smaller number than N, and our proof will be complete if we can find two distinct factorizations of M. To obtain the first prime factorization for M expand $(p_1 - q_1)$ in primes $r_1 r_2 \cdots r_i$ and then write M as the following product of primes:

$$M = r_1 r_2 \cdots r_i p_2 p_3 \cdots p_k. \tag{2}$$

The second factorization of M is derived using the other expression for N given in equation (1) above.

$$\begin{aligned} M = N - q_1 p_2 p_3 \cdots p_k &= (q_1 q_2 \cdots q_l) - (q_1 p_2 p_3 \cdots p_k) \\ &= q_1 (q_2 \cdots q_l - p_2 p_3 \cdots p_k) \\ &= q_1 s_1 s_2 \cdots s_j, \end{aligned} \tag{3}$$

where $s_1 s_2 \cdots s_j$ is the prime factorization of $(q_2 \cdots q_l - p_2 p_3 \cdots p_k)$. To show that the factorizations of M in equations (2) and (3) are distinct, first observe that q_1 *does* appear in the factorization given in (3). We now show that it does not appear in the factorization in (2). To see this first recall that q_1 is not equal to any of the p's in (2). Furthermore, q_1 does not divide $(p_1 - q_1)$, since if it did we could write $(p_1 - q_1) = kq_1$ or $p_1 = kq_1 + q_1 = (k+1)q_1$, which contradicts the fact that p_1 is prime. Therefore it cannot appear among the r's. Thus these two prime factorizations of M are *distinct.* This implies that N was not the smallest such number, contrary to our choice of N. ∎

The contradiction here showed that it is impossible to find a *smallest* number with two prime factorizations. But if there were any such numbers, there would have to be a smallest one; therefore there must be none.

This theorem provides the key step needed to prove that $\sqrt{p}$ is irrational for any prime p. Again the proof is by contradiction.

LEMMA 1.1.4 *If a prime p divides M^2, then p divides M.*

PROOF We will use the fact that *a prime p divides a number M if and only if p appears in the prime factorization of M.* This is a consequence of the uniqueness of prime factorizations, and its proof is left as an exercise.

Now assume that p divides M^2. If p does not divide M, by the foregoing statement its prime factorization $M = r_1 r_2 r_3 \cdots r_s$ could not include p among any of the r's. But from this factorization we can obtain the unique prime factorization of M^2,

$$M^2 = r_1 r_1 r_2 r_2 r_3 r_3 \cdots r_s r_s .$$

Since p does not appear in this factorization, p could not divide M^2. But this contradicts our assumption; hence p must divide M. ∎

Now the way is paved to show, exactly as in Theorem 1.1.1, that the square root of any prime is irrational. The proof is a straightforward exercise.

THEOREM 1.1.5 *For any prime, p, $\sqrt{p}$ is irrational.*

The inexpressibility of irrationals in terms of whole numbers forces us to use proofs by contradiction. In the same way, many facts about infinity can only be understood indirectly. We conclude with the famous proof by contradiction attributed to Euclid, which shows that there are an infinite number of primes.

THEOREM 1.1.6 ***(Infinitude of Primes)*** *There are an infinite number of primes.*

PROOF Suppose that there were only a finite number of primes, say n of them:

$$p_1, \; p_2, \; p_3, \ldots, \; p_n \, .$$

We will construct a number that is not factorable into these primes. This contradicts the fundamental theorem of arithmetic which states that all numbers are amenable to prime factorization. Consider the number

$$K = p_1 p_2 p_3 \cdots p_n + 1 \, .$$

By the fundamental theorem of arithmetic K is factorable into a product of these n primes. But this is impossible, since none of the primes p_i evenly divides K. To see this, note that any prime p_i will divide the first part evenly, leaving the 1 as a remainder. ∎

Problems for Section 1.1

1.1.1. Prove the Pythagorean theorem. (Compute the total area of Figure 1.1.3 and subtract the area of the four triangles.)

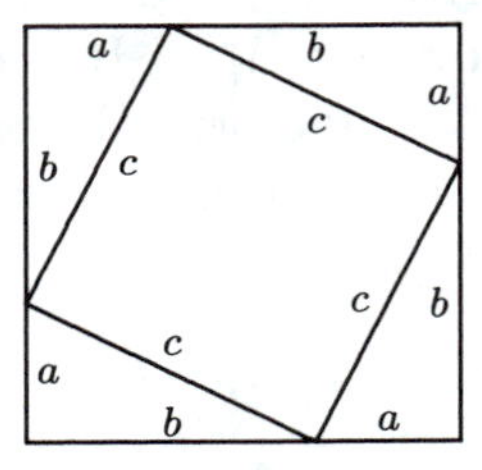

Figure 1.1.3

1.1.2. Prove that $\sqrt{3}$ is irrational. (First you must show that if M^2 is divisible by 3, then M is divisible by 3. To get started note that any whole number M can be written as $3n$, $3n+1$, or $3n+2$. Now show that only in the case where $M = 3n$ is M^2 divisible by 3. Now complete the argument by mimicking the proof of Theorem 1.1.1.)

1.1.3. Use Lemma 1.1.4 to prove Theorem 1.1.5.

1.1.4. Carefully explain where the argument of Theorem 1.1.1 would fail if we were to try to use it to prove that $\sqrt{4}$ is irrational.

1.1.5. Prove that $\sqrt[3]{2}$ is irrational.

1.1.6. Prove that a prime p divides an integer M if and only if p appears in the prime factorization of M.

1.1.7. Mimic the proof of Lemma 1.1.4 to prove that if a prime p divides M^k for some positive integer k, then p divides M.

1.1.8. Prove that $\sqrt[k]{p}$ is irrational where p is a prime.

1.1.9. One nice property of integers is that when added or multiplied together, these sums or products are again integers. We say that the integers are **closed** under addition and multiplication.

a) Prove that the rationals are closed under multiplication and addition. That is, when two rationals are added or multiplied the result is again a rational.

b) Find counterexamples to show that the irrationals are *not* closed under either addition or multiplication.

1.1.10. **a)** Prove by contradiction that if a is irrational, $a + r$ is also irrational for any rational number r.

b) Show that if a is irrational and r and s are rational with $s \neq 0$, then $r + sa$ is irrational.

1.1.11. Show that if a is irrational, then $-a$ and $1/a$ are both irrational.

1.1.12. Prove that if a is irrational, then $\sqrt{a}$ is irrational.

1.1.13. We have seen that $\sqrt{p}$ is irrational for every prime. It is natural to ask for which integers will the square root be rational and for which will it be irrational. Prove that $\sqrt{k}$ is rational if and only if all of its prime factors occur with even multiplicity. (Hint: Show for any integer M that the prime factors of M^2 will occur with even multiplicity. Then, assuming that $\sqrt{k} = M/N$ or $kN^2 = M^2$, show that each prime in the factorization of k must occur an even number of times.)

1.1.14. Complete the following steps to show that $\sqrt{2} + \sqrt{3}$ is irrational. Begin by assuming that $\sqrt{2} + \sqrt{3}$ is rational and proceed to a contradiction.

a) Show (by contradiction) that if $\sqrt{2} + \sqrt{3}$ is rational, $\sqrt{2} - \sqrt{3}$ must also be rational. (Consider the product of both numbers.)

b) Now show that both numbers in **(a)** cannot be rational. (Consider their sum.)

1.1.15. Show that an irrational raised to an irrational power can be rational. (Consider $(\sqrt{3})^{\sqrt{2}}$. If it is rational, you are done. If it is irrational, try raising it to $\sqrt{2}$.)

1.1.16. Logarithms and trigonometric functions also give rise to irrational numbers.

a) Prove by contradiction that $\log_{10} 2$ is irrational. Suppose that $\log_{10} 2 = p/q$ for some integers p and $q \neq 0$. Show that this would imply that $2^q = 10^p$. Carefully explain why this violates the fundamental theorem of arithmetic.

b) Repeat the argument in **(a)** to show that $\log_2 6$ is irrational.

c) Use the trigonometric identity $\cos 2\theta = 2\cos^2\theta - 1$ to prove that if $\cos 2\theta$ is irrational, then $\cos\theta$ is also irrational.

1.1.17. In high school you learned about decimal expressions for rational numbers. The important facts are that any repeating decimal can be expressed as the quotient of two integers, and any quotient of integers can be expressed as either a terminating or repeating decimal. Represent each of these repeating decimals as the quotient of integers.

a) $.\overline{64}$ **b)** $.3\overline{21}$ **c)** $.\overline{a_1a_2a_3}$ **d)** $.\overline{9}$

1.1.18. The general expression for a repeating decimal is $0.b_1b_2\dots b_n\overline{a_1a_2\dots a_m}$. Represent it as the quotient of integers.

1.1.19. Express the following fractions as repeating decimals:

a) $1/4$ **b)** $4/11$ **c)** $2/7$

⋆1.1.20. Carefully explain why the decimal expansion of p/q must repeat. (First consider what must happen in the division process when some remainder recurs for the second time. For example, what do you know must happen when the remainder 4 crops up for the second time in $3/7$? Next consider how many possible remainders can occur when dividing out p/q.)

1.1.21. The equivalence between repeating decimals and rationals gives us a way to generate decimal expressions for irrational numbers. All we have to do is invent a pattern that is sure never to repeat. For example, the number

$$a = .51511511151111511111\ldots$$

is generated by making each successive segment the same as before but with an additional 1 added on. This precludes the repetition of any finite sequence (why?); thus the number cannot be rational. Invent several other irrational numbers using these ideas.

1.1.22. Let a, b, and c be rationals with $a \neq 0$. Suppose that the polynomial ax^2+bx+c has two distinct roots. Show that if one of the roots is rational, so is the other.

⋆1.1.23. The three consecutive odd integers 3, 5, and 7 are all prime. Prove that these are the only **triple primes**. (Note that there are many **twin primes**—that is, consecutive odd integers that are both primes—such as 11 and 13, 17 and 19, and 59 and 61. It is still unknown whether there are a finite number of twin primes.)

1.2 Constructing the Real Numbers

1.2.1 The Struggle to Understand Irrationals

With the discovery of incommensurable line segments the Greeks confronted the inherent incompleteness of the rational number system. In some mysterious way certain continuous geometric magnitudes seemed to transcend numerical description. In the Pythagorean philosophy the essence of number was *whole number*. Rational numbers could be accepted as ratios of whole numbers, but there was no clear way to give a similar meaning in terms of whole numbers to irrational quantities like $\sqrt{2}$. Today we work routinely with numbers like $\sqrt{2}$ but the Greeks could accept such quantities only as geometrical magnitudes and not as numbers.

This problem ended the hopes of the Pythagoreans to subsume all knowledge under the concept of number. Furthermore, it undermined their extensive work in geometry, which had rested on arguments involving numerical ratios and proportions. How is it possible to compare line segments saying "length a is to length b" if no common unit can be found with which to measure the two lines? The whole basis of their elaborate edifice of geometry was threatened.

Eudoxus, a Greek mathematician of the fourth century B.C., settled this crisis in the foundations of geometry by creating a definition of proportionality that applied to *all* geometrical lengths. We will examine Eudoxus' solution, since it foreshadows the modern construction of real numbers. His definition rested on two ingredients: the ability to compare two lengths, saying which is larger; and the ability to construct multiples of any length. Eudoxus then compared lengths as follows:

> Length a is to length b as length c is to length d if for any whole numbers N and M the length of the multiple Na compares to Mb in the same way as Nc compares to Md. That is, either
>
> 1. both $Na = Mb$ and $Nc = Md$ or,
> 2. both $Na > Mb$ and $Nc > Md$ or,
> 3. both $Na < Mb$ and $Nc < Md$.

To better understand the significance of Eudoxus' definition we translate it from a statement about proportional line segments into a statement about the numerical ratios a/b and c/d. Taking each of the three statements in turn and dividing by Nb in the first part and by Nd in the second converts the definition to

> The numbers a/b and c/d are equal if and only if for *any rational number* M/N one of the following cases holds:
>
> 1. $a/b = M/N$ and $c/d = M/N$ or,
> 2. $a/b < M/N$ and $c/d < M/N$ or,
> 3. $a/b > M/N$ and $c/d > M/N$.

Any time case 1 holds for some pair of integers N and M, the number a/b is rational, since it equals M/N. This is equivalent to the line segments a and b being commensurable. Now if a/b is not rational, it will split all the rationals into two separate groups: those rationals that are larger than a/b and those that are smaller. Cases 2 and 3 state that two quantities c/d and a/b are the same if and only if there is no rational number M/N that is simultaneously larger than one and smaller than the other (that is, no rational lies between the two). This means *an irrational quantity is completely determined by the way it divides the rationals.* The idea of splitting the rationals is at the heart of how real numbers are constructed.

With his definition Eudoxus had restored integrity to the foundations of Greek geometry. Since their number system had proved to be inadequate, the Greeks looked to geometry to provide the foundation for their mathematics. All quantities and operations in mathematics were given geometric interpretations. For example, "x squared" and "x cubed" were not considered algebraic operations carried out with a number x, but instead referred to the actual construction of a square (or cube) from a given line segment x. The Greeks developed many ingenious methods for constructing various lengths and areas from given geometrical figures. Today these constructions are recognized as the geometric counterpart to solving algebraic equations. As the reader can imagine, solving equations by geometric constructions is much more awkward than by our current powerful algebraic methods. Furthermore, there are inherent limitations in working with geometric constructions. The famous problem of "doubling the cube," that is, of constructing the side of a cube that would have twice the volume of a given cube, translates into solving the equation

$$x^3 = 2\,.$$

Algebraically the solution is trivial if we accept irrationals like $\sqrt[3]{2}$. However, it turns out to be *impossible* to construct the length $\sqrt[3]{2}$ from a unit length using only straightedge and compass. This problem together with squaring the circle and trisecting an arbitrary angle are the most famous impossible geometrical contructions. They serve to illustrate the inherent limitations of a mathematics based solely on geometry.

The Hindu and Arab mathematicians who followed the Greeks were less rigidly concerned with the logical foundations of their mathematics and took a different approach toward irrationals. When irrationals arose in their algebraic work as roots to equations they were treated as if they obeyed the same operations as rational numbers. Facts such as $\sqrt{a}\sqrt{c} = \sqrt{ac}$ were reasoned out based on commutativity and the meaning of a square root (i.e., that $\sqrt{a^2} = a$ and that $\sqrt{a}\sqrt{a} = a$).

$$\begin{aligned}\sqrt{a}\sqrt{c} = \sqrt{\left(\sqrt{a}\sqrt{c}\right)^2} &= \sqrt{\left(\sqrt{a}\sqrt{c}\right)\left(\sqrt{a}\sqrt{c}\right)} \\ &= \sqrt{\left(\sqrt{a}\sqrt{a}\right)\left(\sqrt{c}\sqrt{c}\right)} = \sqrt{ac}.\end{aligned}$$

The willingness to work with irrationals as numbers paved the way for many important developments in algebra, such as general solutions to quadratic and cubic equations. Yet the meaning of these operations raised difficult questions. Whereas the product of an integer n with a quantity a can be given a definite interpretation as

$$na \equiv \underbrace{a + a + a + \cdots + a}_{n \text{ times}},$$

what similar "meaning" can be assigned to a product like $\sqrt{2} \cdot \sqrt{3}$? The dichotomy between the apparent usefulness of irrationals and their unclear meaning continued for centuries. An early text on arithmetic reveals the confusion.

> Since in proving geometrical figures, when rational numbers fail us irrational numbers take their place and prove exactly those things which rational numbers could not prove ... we are moved and compelled to assert that they truly are numbers, compelled that is, by

> the results which follow from their use—results which we perceive to be real, certain, and constant. On the other hand, other considerations compel us to deny that irrational numbers are numbers at all. To wit, when we seek to subject them to numeration [decimal representation] ... we find that they flee away perpetually, so that not one of them can be apprehended precisely in itself .... Now that cannot be called a true number which is of such a nature that it lacks precision .... Therefore, just as an infinite number is not a number, so an irrational number is not a true number but lies hidden in a kind of cloud of infinity.[1]

The development of calculus introduced more elusive quantities such as the famous infinitesimals. The usefulness and power of these methods to describe continuous motion was apparent to all. Yet no one could adequately describe their basic meaning. As with the Pythagoreans before them, mathematicians again confronted the inability of the number system to deal with continuous quantities.

1.2.2 The Definition of Real Numbers

In the late nineteenth century mathematicians were deeply involved with the project of forming a solid and coherent foundation for calculus. At that time Richard Dedekind recognized that the construction of a complete and rigorous number system that included irrationals was a necessary part of this project and addressed the problem in his book *Continuity and Irrational Numbers.*

> The domain of rational numbers is insufficient and it becomes absolutely necessary that ... the (rational numbers) be essentially improved by the creation of new numbers such that the domain of numbers shall gain the same completeness, or as we may say at once, the same continuity, as the straight line.[2]

The whole problem hinges on continuity. But what exactly is the nature of continuity? Dedekind examined this question.

> The comparison of the domain **Q** of rational numbers with a straight line has led to the recognition of the existence of gaps, of a certain incompleteness or discontinuity of the former, while we ascribe to the straight line completeness, absence of gaps, or continuity. In what then does this continuity consist? Everything must depend on the answer to this question, and only through it shall we obtain a

---

[1] Kline, *Mathematical Thought*, p. 251.

[2] David Eugene Smith, *A Source Book in Mathematics* (New York: Dover Publications, Inc., 1984), p. 36.

> scientific basis for the investigation of all continuous domains .... For a long time I pondered over this in vain, but I finally found what I was seeking. It consists of the following ... the fact that every point p of the straight line produces a separation of the same into two portions such that every point of one portion lies to the left of every point of the other. I find the essence of continuity in the converse, i.e., in the following principle:
>
> If all points of the straight line fall into two classes such that every point of the first class lies to the left of every point of the second class, then there exists one and only one point which produces this division of all points into two classes, this severing of the straight line into two portions.[3]

As we saw earlier, Eudoxus had implicitly used this severing property to make sense of the ratios of continuous lengths. In this way Eudoxus had brought arithmetic to continuous geometric lengths. Now, 2300 years later, Dedekind used this same property to extend the number system and bring continuity to the arithmetic of numbers. Dedekind's construction of the new "real" numbers rested on the property that any point on the number line severed or split the rational numbers in a unique way. By defining numbers at every possible splitting, the mysterious gaps that had existed in the rationals were filled in and the real number system was endowed with continuity. Dedekind defined the new "real" numbers as special subsets of rational numbers called cuts.

DEFINITION 1.2.1 *A **Dedekind cut** is a subset α of the rational numbers $\mathbf{Q}$ with the following properties:*

i) α *is not empty and* $\alpha \neq \mathbf{Q}$;

ii) if $p \in \alpha$ *and* $q < p$, *then* $q \in \alpha$;

iii) if $p \in \alpha$, *then there is some* $r \in \alpha$ *such that* $r > p$.

*The collection of all Dedekind cuts is called the set of **real numbers**. Two real numbers α and β are **equal** if and only if both cuts are the same subset of $\mathbf{Q}$.*

Notice that just as rational numbers are built from pairs of integers, real numbers are now constructed as subsets of rational numbers. To avoid confusion in this section we will denote the rationals used in constructing cuts by roman letters (p, q, r, etc.) and reserve Greek letters ($\alpha, \beta, \lambda, \gamma$, etc.) to denote the cuts or subsets themselves. By property (ii), whenever a rational is in one of these cuts, all other rationals that are smaller (i.e., lie to the left) are also included. Therefore a cut may be visualized geometrically as all rationals on a left half-line (see

[3] Smith, *Source Book in Mathematics*, pp. 36, 37.

Figure 1.2.1). Lastly, by property (iii), the set has no largest element; therefore the cut for a particular real number will include all the rationals up to that number *but not including it.*

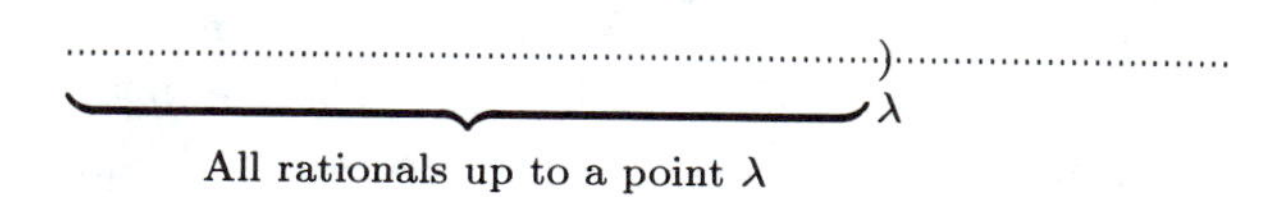

Figure 1.2.1. A cut determining the real number λ.

The number 1 is represented as a cut by the set of all rational numbers including negative rationals that are strictly less than 1. To find the cut representing a particular irrational, simply expand the number as an infinite decimal (easier said than done). For instance,

$$\pi = 3.1415926\ldots$$

The cut is formed by first taking all the rationals less than 3.1, including all the rationals up to 3.14, including all rationals to 3.141, and so on. Continuing this process indefinitely we obtain a particular subset of the rationals that uniquely defines the number π.

Real numbers are usually thought of as points on the real number line. This picture serves us well for intuitive purposes. If needed, the Dedekind cut can be visualized as the set of rationals to the left of the point. However, formally we do not rely on the vague notions of "point" and "line." Instead, each real number is defined as a concrete subset of rational numbers. This is how the real numbers may be *constructed* out of the rationals.

1.2.3 The Least Upper Bound Property

The purpose of Dedekind's definition was to bring continuity and completeness to the number system. This fundamental characteristic is embodied in the least upper bound property.

To begin we must define the relationship of **less than** for real numbers. Considered as points on a line, one number is smaller than another if it lies to the left. In terms of their Dedekind cuts, the cut for the smaller number is included in the cut for the larger, as in Figure 1.2.2.

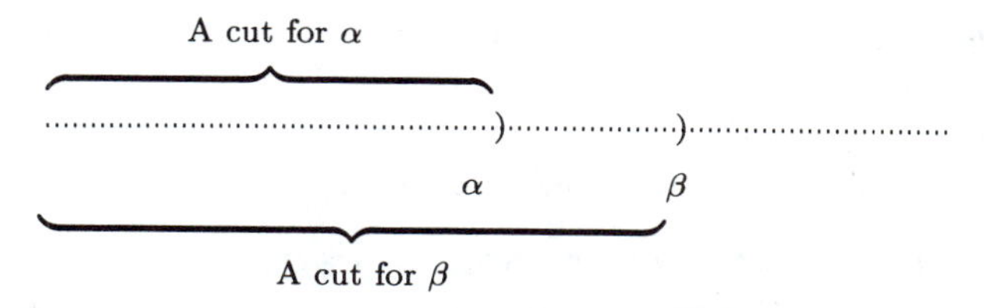

Figure 1.2.2. We say $\alpha < \beta$ since the cut for α is contained in β.

Definition 1.2.2 *Given two real numbers (cuts) α and β we say that $\alpha < \beta$ ($\alpha \leq \beta$) whenever $\alpha \subset \beta$ ($\alpha \subseteq \beta$).*

Definition 1.2.3 *Let α, λ, and γ represent real numbers (cuts). A nonempty set S of real numbers has an* **upper bound** *γ, if for any number $\alpha \in S$ we have $\alpha \leq \gamma$. An upper bound λ is called the* **least upper bound** *(l.u.b.) for the set S if for any other upper bound γ we have $\lambda \leq \gamma$.*

To develop a sense for upper bounds think of the set S as a collection of points on the real number line. An upper bound for S is any point to the right of all points in S. The least upper bound is the leftmost of these upper bounds (see Figure 1.2.3). As our intuition suggests, there is at most a single least upper bound for a set (see problem 2 of this section). A least upper bound may or may not belong to the set, depending on whether S has a maximum element.

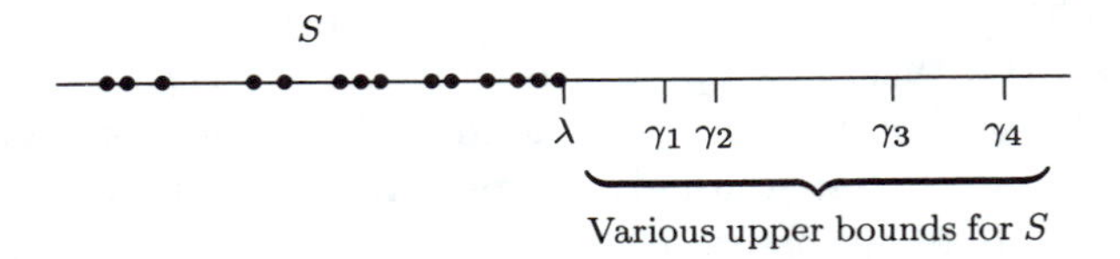

Figure 1.2.3. λ is the least upper bound for S.

Example 1.2.1

a) If $A = \{-1, -\frac{1}{2}, -\frac{1}{3}, \ldots\}$, the l.u.b. for the set A is 0. Note that it is not in A.

b) If $B = \{1, 2, 3, \ldots\}$, there are no upper bounds at all for B; therefore there can be no l.u.b. for B.

c) If $C = \{1, 1.4, 1.41, 1.414, 1.4142, \ldots\}$, where the dots indicate the continued decimal expansion of $\sqrt{2}$, the l.u.b. of C is $\sqrt{2}$. □

What kinds of sets have least upper bounds? To explore this question let's momentarily return to the rational number system. The set C in the preceding example is a set in this system, since it consists entirely of rational numbers. This set has upper bounds, for example, 2, 4, and 1.415. Yet in the rational number system C has no *least* upper bound. This is because any rational upper bound must exceed $\sqrt{2}$. But for any fixed rational number greater than $\sqrt{2}$ there will always be another rational even closer to $\sqrt{2}$. *Thus in the rational number system there are bounded sets that have no least upper bound.* The problem is that rational numbers can converge to points that are not rational. With the real numbers these gaps have been filled and every bounded set will have a least upper bound among the reals.

Theorem 1.2.4 ***(The Least Upper Bound Property)*** *Every nonempty set of real numbers that is bounded above has a least upper bound.*

PROOF Let A be a nonempty subset of the real numbers, and suppose that γ is an upper bound for A. Keep in mind that γ and all elements of A are cuts, that is, *subsets* of rationals. The l.u.b. for A, (call it λ), is formed by taking the *union* of the cuts $\alpha \in A$. That is, $\lambda = \bigcup\{\alpha \mid \alpha \in A\}$. To show that λ is the least upper bound for A we must first prove that λ actually defines a number. That is, we must show that as a set λ satisfies the three properties of Definition 1.2.1.

1. λ is nonempty, since it is the nonempty union of nonempty sets. Next we must show that $\lambda \neq \mathbf{Q}$. This is done by finding some rational that is not in λ. To do this observe that since γ itself is a cut there is some rational $q \notin \gamma$. We will show $q \notin \lambda$. First note that, since γ is an upper bound for A, for all $\alpha \in A$ we have $\alpha \leq \gamma$, or equivalently, $\alpha \subseteq \gamma$. Therefore λ, which is the union of the α's, is also a subset of γ. This implies that the rational q could not belong to λ.

2. Suppose that $p \in \lambda$. We must show that if $q < p$, then $q \in \lambda$. Since λ is the union of the α's, we must have $p \in \alpha$ for some α. Now because α itself is a cut, we have $q \in \alpha$. This implies that $q \in \lambda$, since λ contains α.

3. Given any $p \in \lambda$, we must exhibit a rational $r > p$ that is also in λ. But as in step 2 we again know that $p \in \alpha$ for some α, and since α itself is a cut, there is an $r \in \alpha$ with $r > p$. Hence $r \in \lambda$, since λ contains α.

At this point we have established that λ is a cut and thus qualifies as a real number. We must now show that λ is the l.u.b. of A.

First observe that for any $\alpha \in A$, we have $\alpha \subseteq \lambda$, that is, $\alpha \leq \lambda$. So by Definition 1.2.3, λ is an upper bound for A. Next suppose that γ is any other upper bound for A. By Definition 1.2.3, for every $\alpha \in A$, we have $\alpha \leq \gamma$, that is, $\alpha \subseteq \gamma$. This implies that λ, which is the union of the α's, must also be a subset of γ. That is, $\lambda \leq \gamma$. Since γ was an arbitrary upper bound, this shows that λ is the least upper bound. ∎

The least upper bound property is one version of what is called the **completeness** of the real numbers. It is this property that distinguishes the rational number system from the reals and provides the link between continuity and the number system. Many important theorems in succeeding chapters will use this property at some crucial point. This is not surprising, for calculus is eminently the mathematics of continuity.

1.2.4 Addition and Multiplication of Real Numbers

If we are to do arithmetic with real numbers, we must learn how to add and multiply these subsets called cuts. This is a lengthy and technical process, and we do not carry it out in full here. For details see Rudin's book *Principles of Mathematical Analysis.* However, to glimpse a little of what is involved, let's examine the definition for the addition of two cuts.

DEFINITION 1.2.5 *Let α and β be two real numbers (i.e., Dedekind cuts). Then the* **sum** *of α and β is defined by*

$$\alpha + \beta \equiv \{p + r \mid p \in \alpha \text{ and } r \in \beta\}.$$

First it must be verified that this sum is a legitimate number. To do this we must show that it possesses the properties of a Dedekind cut.

THEOREM 1.2.6 *If α and β are Dedekind cuts, then $\alpha + \beta$ is a Dedekind cut.*

PROOF We must check the three properties of Definition 1.2.1 for the subset $\alpha + \beta$.

1. Since α and β are themselves cuts, neither α nor β is empty; therefore $\alpha + \beta$ is not empty. To verify that $\alpha + \beta$ is not all of $\mathbf{Q}$, we must exhibit at least one rational q that is not in $\alpha + \beta$. Again, since α and β are cuts there are rationals $q_1 \notin \alpha$ and $q_2 \notin \beta$. But if $q_1 \notin \alpha$, then $q_1 > p$ for all $p \in \alpha$ (since if $q_1 < p$ for some p, then q_1 would be in α by property ii for a cut). Similarly $q_2 > r$ for all $r \in \beta$. This implies $q_1 + q_2 > p + r$ for all of the sums $p + r$; therefore $q_1 + q_2 \notin \alpha + \beta$. Hence $\alpha + \beta \neq \mathbf{Q}$.
2. We need to show that if $(p + r) \in \alpha + \beta$ and $q < (p + r)$, then $q \in \alpha + \beta$. To do this we must show that q can be written as the sum of two rationals, one from each cut. This can be done by writing $q = p + (q - p)$. Then $p \in \alpha$. Furthermore, since $q < (p + r)$ we know that $(q - p) < r$ and therefore that $(q - p) \in \beta$.
3. It must be shown that for any $(p + r) \in \alpha + \beta$, there is some $q + s > p + r$ that is also in $\alpha + \beta$. This is left as an exercise. ∎

To define multiplication, a similar process is repeated: define the product of cuts and verify that the product of cuts is again a cut. This is technically much harder, since we cannot simply define $\alpha \cdot \beta \equiv \{p \cdot r \mid p \in \alpha \text{ and } r \in \beta\}$. (Can you see where the problem arises?) For a treatment of the multiplication of cuts we refer the reader to Rudin's book.

These subsets are extremely clumsy to work with, and we would like to avoid using them whenever possible. The systematic way to accomplish this is to prove certain basic properties about the addition and multiplication of cuts. Then in our routine operations with real numbers we may appeal to these properties instead of returning to the definition. For example, two frequently used properties for addition are the following:

THEOREM 1.2.7 *If α, β, and γ are cuts, then*

a) $\alpha + \beta = \beta + \alpha$;

b) $(\alpha + \beta) + \gamma = \alpha + (\beta + \gamma)$.

PROOF **a)** This first property is called **commutativity**. From the definition, $\alpha + \beta \equiv \{p + r \mid p \in \alpha \text{ and } r \in \beta\}$; therefore $\beta + \alpha \equiv \{r + p \mid r \in \beta \text{ and } p \in \alpha\}$. But since for all rational numbers $p + r = r + p$, these two subsets are equal. Hence the two numbers $\alpha + \beta$ and $\beta + \alpha$ are equal.

b) The second property is called **associativity**. Its proof is an exercise. ∎

Exactly which basic properties are needed? In the next section we study a small set of key properties called the **axioms for the real number system** from which all the laws of arithmetic can be derived. If we could prove that these axioms held for the addition and multiplication of cuts, all other arithmetical operations and properties would follow and we could operate freely without reference to the definition of real numbers. A small part of this work is done in the problems, but the complete demonstration is a long, involved exercise. (For details, the reader is again referred to Rudin). If carried out in full this work shows that *the addition and multiplication of real numbers obey all the same rules of operation as the rational numbers.*

1.2.5 The Density of Rational and Irrational Numbers

The new numbers constructed by the process of Dedekind cuts are scattered *everywhere* over the real line. In fact, between any two real numbers there will always be a rational number. We call this property the **density** of the rational numbers as a subset of the reals. The irrationals are also dense, since there is always an irrational between any two real numbers. In a sense rationals and irrationals are thoroughly shuffled. To prove this density property we begin with a consequence of the least upper bound property called the Archimedean property.

THEOREM 1.2.8 ***(The Archimedean Property for the Real Numbers)*** *If α and β are positive real numbers, then there is some positive integer n such that $n\alpha > \beta$.*

In the proofs in this section we will assume that real numbers obey the rules of arithmetic for inequalities.

PROOF (By contradiction.) Suppose that $n\alpha > \beta$ did not hold for any integer n, that is, for all integers n we have $n\alpha \leq \beta$. Then the set $A = \{n\alpha \mid n \in N\}$ is bounded above by β. By Theorem 1.2.4 there is a l.u.b. for A, which we will call λ.

Now since α is positive, $\lambda - \alpha < \lambda$; therefore $\lambda - \alpha$ is not an upper bound for the set A. Thus for some integer m we must have $\lambda - \alpha < m\alpha$, or equivalently, $\lambda < (m+1)\alpha$. But this contradicts the choice of λ as an upper bound for A. ∎

THEOREM 1.2.9 ***(Density of the Rational Numbers)*** *If α and β are real numbers with $\alpha < \beta$, then there is a rational number r such that $\alpha < r < \beta$.*

PROOF Both 1 and $\beta - \alpha$ are positive numbers, so we may use the Archimedean property to conclude that for some integer m we have $1 < m(\beta - \alpha)$, or equivalently,

$$m\alpha + 1 < m\beta\,.$$

Let n be the *largest* integer such that $n \leq m\alpha$. Adding 1 to both sides gives

$$n + 1 \leq m\alpha + 1 < m\beta\,.$$

But since n is the *largest* integer less than or equal to $m\alpha$, we know that $m\alpha < n + 1$ and therefore that

$$m\alpha < n + 1 < m\beta, \quad \text{or equivalently,} \quad \alpha < \frac{n+1}{m} < \beta.$$

Thus we have constructed a rational number between α and β. ∎

A proof of the following analogous theorem is outlined in the exercises.

THEOREM 1.2.10 ***(Density of the Irrational Numbers)*** *If α and β are real numbers with $\alpha < \beta$, then there is an irrational number γ such that $\alpha < \gamma < \beta$.*

If you find this construction of real numbers difficult, you are not alone. Morris Kline describes the student reaction to a similarly rigorous construction of the integers by Peano:

> He used [his construction] in his lectures and the students rebelled. He tried to satisfy them by passing all of them but that did not work and he was obliged to resign his professorship.[4]

We hasten to note that, despite student feelings, both Dedekind's and Peano's constructions have become the accepted standard. However, you may take comfort in the fact that all subsequent work uses only the characteristic properties of the real numbers, such as the least upper bound property and the axioms described in the next section. Thus there is no further direct need for the actual definition of real numbers as Dedekind cuts, and with some caution we may return to thinking of numbers as points on the number line.

Problems for Section 1.2

1.2.1. Let $\mathbf{N} = \{1, 2, 3, \ldots\}$ be the set of natural numbers. Let

$$A = \{(-1)^n \mid n \in \mathbf{N}\}, \quad B = \left\{ \frac{(-1)^n}{n} \,\middle|\, n \in \mathbf{N} \right\}, \quad C = \left\{ \frac{n-1}{n} \,\middle|\, n \in \mathbf{N} \right\},$$

$$D = \left\{1,\ 1 + \frac{1}{2},\ 1 + \frac{1}{2} + \frac{1}{3},\ 1 + \frac{1}{2} + \frac{1}{3} + \frac{1}{4}, \ldots \right\}, \quad E = \{\cos n \mid n \in \mathbf{N}\},$$

$$F = \{100n - n^2 \mid n \in \mathbf{N}\}.$$

Determine which of these sets have upper bounds. For each set with an upper bound, find three different upper bounds and, where possible, find the least upper bound.

•1.2.2. Show that if a set of real numbers S has a least upper bound λ, this least upper bound is *unique*. That is, if β is also a least upper bound for S, then $\lambda = \beta$.

•1.2.3. Suppose that $\lambda = \text{l.u.b.}\ A$. Let $B = \{ka \mid a \in A\}$, where $k > 0$.

- **a)** Show that $k\lambda$ is an upper bound for the set B.
- **b)** Show that $k\lambda$ is the least upper bound for B. (Hint: Suppose that γ were another upper bound for B that was less than $k\lambda$. Use γ to construct an upper bound for A that is smaller than λ.)
- **c)** What can happen if $k < 0$?

1.2.4. Suppose that you are given two sequences of real numbers, $A = \{a_1, a_2, a_3, \ldots\}$ and $B = \{b_1, b_2, b_3, \ldots\}$. Define $C = \{a_1 + b_1, a_2 + b_2, a_3 + b_3, \ldots\}$.

- **a)** Show that if A and B have least upper bounds λ_1 and λ_2, respectively, then $\lambda_1 + \lambda_2$ is an upper bound for C.
- **b)** Find an example showing that $\lambda_1 + \lambda_2$ need not be the *least* upper bound for C.

---

[4] Kline, *Mathematical Thought*, p. 988.

•1.2.5. **Theorem:** *λ is the least upper bound of S if and only if (i) $\lambda \geq \alpha$ for all $\alpha \in S$; and (ii) for any real number $\beta < \lambda$, there is a real number $\alpha \in S$ such that $\beta < \alpha$.* Prove this as follows: First assume that λ is a l.u.b. for S.

a) Show that (i) holds because λ is an upper bound for S.

b) Show that (ii) holds because λ is the *least* upper bound for S. (Hint: Use a proof by contradiction. Assume that there is some number $\beta < \lambda$ and that there is *no* element α in S such that $\beta < \alpha$. What property of λ does this contradict?)

Now assume that λ satisfies (i) and (ii). Show that λ is the least upper bound for S.

c) Begin by showing that λ is an upper bound for S.

d) Let γ be another upper bound for S. If $\gamma < \lambda$, use (ii) to derive a contradiction.

1.2.6. Let T and V be sets of real numbers with least upper bounds λ_1 and λ_2, respectively. Consider the set $S = \{t + v \mid t \in T,\ v \in V\}$.

a) Show that $\lambda_1 + \lambda_2$ is an upper bound for S.

b) Now show that any number γ such that $\gamma < \lambda_1 + \lambda_2$ is not an upper bound for S. (Hint: Let $d = \lambda_1 + \lambda_2 - \gamma$. Why is there an element $t' \in T$ so that $t' > \lambda_1 - \frac{d}{2}$? Why is there an element $v' \in V$ so that $v' > \lambda_2 - \frac{d}{2}$? Show that $t' + v' > \gamma$ and use this to complete the proof.)

•1.2.7. Prove that the irrationals are dense (Theorem 1.2.10). One possible proof would be to construct an irrational between α and β based on $\sqrt{2}$ by following an argument analogous to Theorem 1.2.9. Start by reviewing this proof.

a) Prove that if r and $s \neq 0$ are rational, then $r + s\sqrt{2}$ is irrational.

b) Show that $\sqrt{2} < m(\beta - \alpha)$ for some positive integer m.

c) Let n be the largest integer less than $m\alpha$. Show that $m\alpha < n + \sqrt{2} < m\beta$.

d) Locate an irrational number between α and β.

•1.2.8. Let α, β, and γ be real numbers (cuts). We say that a nonempty set S of real numbers has a **lower bound** β, if for any number $\alpha \in S$ we have $\alpha \geq \beta$. A lower bound β is called the **greatest lower bound** (g.l.b.) for the set S if for any other lower bound γ we have $\beta \geq \gamma$. Intuitively, a lower bound for S is any point to the left of all points in S. The greatest lower bound is the rightmost of these lower bounds.

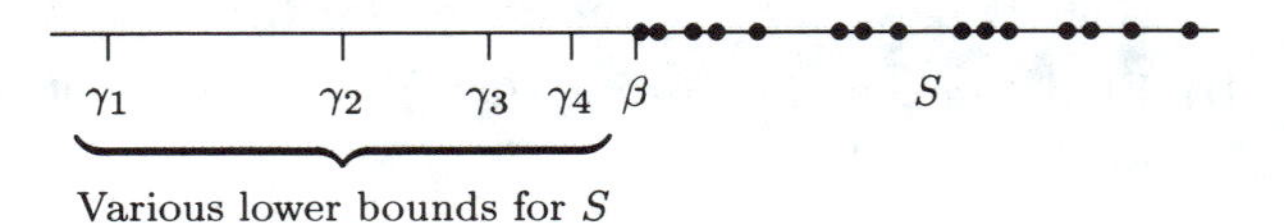

Figure 1.2.4. β is the greatest lower bound for S.

Examples:

1. If $A = \{-1, -\frac{1}{2}, -\frac{1}{3}, \ldots\}$, the g.l.b. of A is -1. Note that it is in A.
2. If $B = \{-1, -2, -3, \ldots\}$, there are no lower bounds for B.
3. $C = \{1, \frac{1}{2}, \frac{1}{3}, \frac{1}{4}, \ldots\}$ has a greatest lower bound of 0 (which is not in C).

The Greatest Lower Bound Property: *Every nonempty set of real numbers that has a lower bound has a greatest lower bound.* One way to prove this result is to mimic the steps used in the least upper bound property. An alternative approach is to "invert" the set and use the least upper bound property in the following manner:

a) Let S be a nonempty set of real numbers and assume that γ is a lower bound for S. Define A to be the set of additive inverses of elements in S, that is, $A = \{-s \mid s \in S\}$. Show that $-\gamma$ is an *upper bound* for A.

b) Why does A have a *least upper bound*? Call this least upper bound λ.

c) Show that $-\lambda$ is a *lower bound* for S.

d) Show that $-\lambda$ is the *greatest lower bound* for S by showing that $\gamma \leq -\lambda$.

1.2.9. Determine which sets in problem 1 have lower bounds; determine greatest lower bounds where possible.

1.2.10. Use the ideas in problem 5 to prove that β is the greatest lower bound of S if and only if both (i) $\beta \leq \alpha$ for all $\alpha \in S$; and (ii) for any real number $\tau > \beta$, there is a real number $\alpha \in S$ such that $\tau > \alpha$.

1.2.11. Let $f(x) = x$ and let $g(x) = 1 - x^2$.

a) Let $A = \{f(x) \mid x \in [-1, 1]\}$. Find l.u.b. A. At which x does this l.u.b. occur?

b) Let $B = \{g(x) \mid x \in [-1, 1]\}$. Find l.u.b. B. At which x does this l.u.b. occur?

c) Let $C = \{f(x) + g(x) \mid x \in [-1, 1]\}$. Find l.u.b. C.

d) Is l.u.b C = l.u.b. A + l.u.b. B? Why?

1.2.12. Show that the set $\mathbf{N}$ of natural numbers is not bounded above, that is, has no upper bound. (Hint: This seems obvious but who knows what happens "way out there." Suppose there were some upper bound γ. Apply the Archimedean property with $\alpha = 1$ and $\beta = \gamma$.)

•1.2.13. **a)** Prove that given any $\epsilon > 0$ there is some natural number n such that $\frac{1}{n} < \epsilon$. (Hint: Apply the Archimedean property with $\alpha = \epsilon$ and $\beta = 1$.)

b) Prove the result in **(a)** in another way using the density of the rationals.

1.2.14. Let C_0 = {all negative rational numbers}. Prove that C_0 satisfies the three properties for a cut. What number is represented by this cut?

1.2.15. Prove that $\alpha + \beta$ in Theorem 1.2.6 satisfies the third property of Dedekind cuts.

1.2.16. Prove **(b)** of Theorem 1.2.7.

1.2.17. The first part of the proof of Theorem 1.2.4 shows that the union of cuts is again a cut (provided that they are bounded above). Show that the *intersection* of two cuts is again a cut. If C_1 and C_2 are cuts, show that $C_1 \cap C_2$ is also a cut. What cut is it?

1.2.18. The argument in the preceding problem works for any finite number of cuts, but it can fail for an infinite number of cuts. Let $S_{1/n}$ be the cut for $\frac{1}{n}$. What is the intersection of these cuts for all positive integers n? Why is this set not a cut?

1.2.19. The intersection in the preceding problem satisfies properties (i) and (ii) for a cut but fails to satisfy property (iii). Such sets fail only because they have a maximum element, but they can be modified as follows: Suppose that S is a set that satisfies properties (i) and (ii) of Definition 1.2.1. Then define $S^- = \{r \mid r \text{ rational and } r < p \text{ for some } p \in S\}$. Thus any rational number qualifies for S^- as long as it is strictly less than some element in S. Prove that S^- is a cut and that if S is a cut, then $S^- = S$.

⋆1.2.20. In this problem we will develop the notion of the negative of a given cut C. Some gymnastics are necessary, since the obvious attempts fail.

a) As first attempts to define the negative of C, consider the set $C_- = \{-r \mid r \in C\}$ or the set $C^c = \{r \mid r \text{ is rational and } r \notin C\}$. Sketch diagrams of C_- and C^c (relative to a cut C on the number line) and show that neither is a cut.

b) Take the cut for a positive number and sketch the set $(C_-)^c$. Repeat for a negative number.

c) Show that $(C_-)^c$ satisfies properties (i) and (ii) of Definition 1.2.1. Thus we can make this into a cut using the technique of the preceding problem. We call this cut $-C = ((C_-)^c)^-$.

d) If this is all to work out correctly we must show that the real number $-C$ obeys the properties that the negative or additive inverse of C should obey. In particular, show that $C + -C = C_0$, where C_0 is the cut for 0 (see problem 15).

1.3 An Axiom System for the Real Numbers

An axiom system consists of a small but complete set properties from which all other laws and facts about the subject can be derived. In an axiomatic approach we are not concerned with the particular interpretation assigned to the elements of the system but rather with the basic rules that the elements obey and the logical consequences of those rules. For example, in an axiomatic treatment of geometry the meaning of basic terms such as *point, line,* and *between* is left open to any consistent interpretation. Instead the emphasis is on the fundamental properties and *relationships* the terms satisfy, such as the fact that every two points determine a unique line, and on the theorems which can be deduced from these properties. The strength of the axiomatic approach resides in this freedom of interpretation, since it allows the entire theory to be transported to any system whose basic elements satisfy the axioms. In this way axiom systems have the power to reveal deep underlying similarities between apparently different systems. David Hilbert, one of the central architects of twentieth century mathematics, described the importance of the axiomatic approach.

> Everything that can be the object of mathematical thinking as soon as the erection of a theory is ripe, falls into the axiomatic method and thereby directly into mathematics. By pressing to ever deeper layers of axioms... we can obtain deeper insights into scientific thinking and learn the unity of our knowledge. Especially by virtue of the axiomatic method mathematics appears called upon to play a leading role in all knowledge.... The axiomatic method is and remains the one suitable and indispensable aid to the spirit of every exact investigation no matter in what domain; it is logically unassailable and at the same time fruitful; it guarantees thereby complete freedom of investigation.[1]

In this section we present an axiom system for the real numbers. Our interest is no longer focused on the nature and definition of real numbers as Dedekind cuts, but rather on the properties that these numbers obey. These fundamental properties are divided into three groups: the field axioms, the order axioms, and the completeness axiom. Many other important mathematical structures obey a subset of these axioms (e.g., the complex numbers satisfy the field axioms). However, it is remarkable that *only* the real numbers (as defined by Dedekind cuts or some equivalent construction) satisfy the full set of axioms.[2] Thus this set of properties uniquely *characterizes* the real numbers.

1.3.1 The Field Axioms

DEFINITION 1.3.1 *A **field** is a set together with two binary operations, called **addition** and **multiplication**, that satisfies the following eleven axioms.*

Since our main interest in this chapter is the field of real numbers, we denote the set for our field by **R** and call the elements of our set **numbers**.

[1] Kline, *Mathematical Thought*, p. 1027.

[2] For a beautiful exposition of this fact, see Michael Spivak, *Calculus*, 2nd ed., (Berkeley, Calif.: Publish or Perish, Inc., 1980), pp. 547–572.

I. The Field Axioms

Axioms for Addition: for all a, b, and c in $\mathbf{R}$

1. Closure: There is a unique number $a + b \in \mathbf{R}$.
2. Commutativity: $a + b = b + a$.
3. Associativity: $a + (b + c) = (a + b) + c$.
4. Identity: There is a number 0 such that $a + 0 = a$.
5. Inverses: There is a number $-a \in \mathbf{R}$ such that $a + (-a) = 0$.

Axioms for Multiplication: for all a, b, and c in $\mathbf{R}$

6. Closure: There is a unique number $a \cdot b$ (sometimes written ab) in $\mathbf{R}$.
7. Commutativity: $a \cdot b = b \cdot a$.
8. Associativity: $a \cdot (b \cdot c) = (a \cdot b) \cdot c$.
9. Identity: There is a number called 1, which does not equal 0, such that $a \cdot 1 = a$.
10. Inverses: If $a \neq 0$, there is a number $a^{-1} \in \mathbf{R}$ such that $a \cdot (a^{-1}) = 1$.

Distributive Axiom

11. If a, b, and c are in $\mathbf{R}$, then $a \cdot (b + c) = a \cdot b + a \cdot c$.

We begin exploring the consequences of these axioms with a question of uniqueness. The axioms assert the existence of identity elements and inverses with special properties. Could several elements in our number system have these same properties or are identities and inverses unique?

Theorem 1.3.2 ***(Uniqueness of Identities and Inverses)*** *The additive and multiplicative identities are unique. That is:*

a) *if $a + b = a$, then $b = 0$;*

b) *if $a \cdot b = a$ and $a \neq 0$, then $b = 1$.*

For each element the additive and multiplicative inverses are unique. That is:

c) *If $a + b = 0$, then $b = -a$.*

d) *If $a \cdot b = 1$, then $b = a^{-1}$.*

PROOF We will prove only **(a)** and **(c)**. The other parts are proved similarly and are left as exercises.

a) Suppose that $a + b = a$. Then by Axiom 5 there is an inverse for a and we may add it to both sides of the equation.

$$\begin{aligned} (a+b)+(-a) &= a+(-a) && \text{(Axiom 5)} \\ (b+a)+(-a) &= 0 && \text{(Axioms 2 and 5)} \\ b+(a+(-a)) &= 0 && \text{(Axiom 3)} \\ b+0 &= 0 && \text{(Axiom 5)} \\ b &= 0 && \text{(Axiom 4)} \end{aligned}$$

c) Suppose that $a + b = 0$. To show that $b = -a$ simply add the inverse of a to each side and mimic the foregoing proof. ∎

Uniqueness is a powerful principle. It implies that *any time an element acts like an identity or a particular inverse, it must be that identity or inverse.* For example, the equality $-a + a = 0$ shows that a acts like the inverse for $-a$. We may conclude by part c of the preceding theorem that it must be the inverse, that is, $a = -(-a)$. Of course, the same argument works for multiplication.

COROLLARY 1.3.3 *If $a \in \mathbf{R}$, then $-(-a) = a$. If $a \neq 0$, then $(a^{-1})^{-1} = a$.*

We have noted that the additive identity, 0, is always excluded when using multiplicative inverses. In light of the striking symmetry in the properties of addition and multiplication, it is natural to wonder why it is necessary in Axiom 10 to exclude 0. The reason for this exclusion follows from how the additive identity interacts with multiplication.

THEOREM 1.3.4 *If $a \in \mathbf{R}$, then $a \cdot 0 = 0$.*

PROOF To prove this we begin with the fact that $a \cdot 1 = a$ and then expand 1 and use distributivity.

$$\begin{aligned} a \cdot 1 &= a && \text{(Axiom 9)} \\ a \cdot (1+0) &= a && \text{(Axiom 4)} \\ a \cdot 1 + a \cdot 0 &= a && \text{(Axiom 11)} \\ a + a \cdot 0 &= a && \text{(Axiom 5)} \\ a \cdot 0 &= 0 && \text{(Theorem 1.3.2a)} \end{aligned}$$ ∎

Now if 0 had a multiplicative inverse, 0^{-1}, then for any element a we would have

$$a = a \cdot 1 = a \cdot (0^{-1} \cdot 0) = (a \cdot 0^{-1}) \cdot 0 = 0.$$

Thus the whole system would necessarily collapse to a single element 0. This trivial case of a field consisting of only one element is excluded by Axiom 9. The problems with division by 0 run throughout calculus and require a careful treatment with the concept of limit.

We now prove some other familiar properties of numbers.

THEOREM 1.3.5 *If $a \cdot b = 0$, then either $a = 0$ or $b = 0$.*

PROOF Suppose that $a \cdot b = 0$ and $a \neq 0$; then we must show that $b = 0$.

$$\begin{aligned} a^{-1} \cdot (a \cdot b) &= a^{-1} \cdot 0 && \text{(Axiom 10)} \\ (a^{-1} \cdot a) \cdot b &= a^{-1} \cdot 0 && \text{(Axiom 8)} \\ 1 \cdot b &= 0 && \text{(Axiom 10 and Theorem 1.3.4)} \\ b &= 0\,. && \text{(Axiom 9)} \ \blacksquare \end{aligned}$$

This property of numbers is invoked in finding roots to algebraic expressions by factoring. For example, if x satisfies

$$x^2 + 6x + 5 = (x+5)(x+1) = 0,$$

then by Theorem 1.3.5 either $(x+5) = 0$ or $(x+1) = 0$.

Another consequence of these axioms is the way minus signs cancel under multiplication.

THEOREM 1.3.6 *Let a and b be real numbers. Then*

a) *$(-a) \cdot b = -(a \cdot b)$ and, in particular, $(-1) \cdot b = -b$;*

b) *$(-a) \cdot (-b) = a \cdot b$.*

PROOF **a)** All we need to prove is that $(a \cdot b) + ((-a) \cdot b) = 0$, since this shows that $(-a) \cdot b$ acts like an additive inverse for $a \cdot b$; therefore by Theorem 1.3.2 it must equal the inverse, $-(a \cdot b)$. We begin by using the distributive law to add $a \cdot b$ and $(-a) \cdot b$; we have

$$\begin{aligned} (a \cdot b) + ((-a) \cdot b) &= (b \cdot a) + (b \cdot (-a)) && \text{(Axiom 7)} \\ &= b \cdot (a + (-a)) && \text{(Axiom 11)} \\ &= b \cdot 0 && \text{(Axiom 5)} \\ &= 0 && \text{(Theorem 1.3.4)} \end{aligned}$$

b) This result is now used to prove the second statement.

$$
\begin{aligned}
(-a)\cdot(-b) &= -(a\cdot(-b)) && \text{(part a))}\\
&= -((-b)\cdot a) && \text{(Axiom 7)}\\
&= -(-(b\cdot a)) && \text{(part a))}\\
&= (b\cdot a) && \text{(Corollary 1.3.3)}\\
&= a\cdot b && \text{(Axiom 7)} \blacksquare
\end{aligned}
$$

The important point to keep in mind is that all the laws of arithmetic are direct consequences of these basic axioms. Furthermore, any other field (i.e., any system for which these axioms hold) also obeys these same laws. Here are two examples of other fields. Remember that we are free to interpret multiplication and addition in any way desired provided the axioms are satisfied.

EXAMPLE 1.3.1 This field has only three elements, a, b, and c. Addition, $\oplus$, and multiplication, $\odot$, are defined by the following tables.

$\oplus$	a	b	c
a	a	b	c
b	b	c	a
c	c	a	b

$\odot$	a	b	c
a	a	a	a
b	a	b	c
c	a	c	b

Observe that a acts as the additive identity and b is the multiplicative identity. The symmetry across the diagonals of each table shows that both operations are commutative. The other properties can be checked on a case by case basis. $\square$

EXAMPLE 1.3.2 Let **C** be the set of all ordered pairs (a, b), where a and b are real numbers. Addition and multiplication are defined as follows:

$$(a, b) \oplus (c, d) = (a + b, c + d)$$

and

$$(a, b) \odot (c, d) = (a \cdot c - b \cdot d, a \cdot d + b \cdot c).$$

This is the field of complex numbers. If you think of the ordered pair (a, b) as $a+bi$ and (c, d) as $c+di$, the rules for addition and multiplication of two of these numbers follow from the fact that $i^2 = -1$. We will prove that this system is a field in Chapter 7 when we study the complex numbers more thoroughly. $\square$

1.3.2 The Order Axioms

The axioms for a field alone are not strong enough to guarantee that the elements can be ordered according to natural concepts of *greater than* and *less than.* For instance, we can show that the field in Example 1.3.1 has no such ordering. If there were an ordering, then either $a < b$ or $b < a$. Suppose first that $a < b$; then using the addition table we would have

$$a + b < b + b \qquad \text{or} \qquad b < c.$$

Together $a < b$ and $b < c$ would then imply that $a < c$. Finally, adding b to both sides of $a < c$ gives

$$a + b < c + b \qquad \text{or} \qquad b < a.$$

Thus starting with $a < b$ we have concluded $b < a$. The same contradiction occurs if we begin with the assumption $b < a$. A consistent ordering obeying familiar properties (i.e., the ability to add the same quantity to both sides) is impossible to construct for this field. In fact, it is true that any field with a finite number of elements *cannot be ordered* because the same problems are encountered as in this three element field. Apparently a stronger axiom system is required to guarantee that the elements of our field can be consistently ordered. The key lies in having a set of "positive" elements that behave properly under addition and multiplication. We therefore add the following three axioms to our system of real numbers.

II. The Order Axioms

12. Trichotomy: There exists a subset P of $\mathbf{R}$ such that for any number, a, exactly one of the following holds:
    i) $a = 0$;
    ii) $a \in P$, in which case we say $a > 0$;
    iii) $-a \in P$, in which case we say $a < 0$.

    P is called the set of ***positive elements*** of $\mathbf{R}$.
13. Additive Closure of P: If $a > 0$ and $b > 0$, then $a + b > 0$.
14. Multiplicative Closure of P: If $a > 0$ and $b > 0$, then $a \cdot b > 0$.

Axiom 12 provides for a set P of positive elements that actually *defines* the meaning of $a > 0$ and $a < 0$. The next definition extends this relationship of "greater than" and "less than" to all pairs of elements by considering the difference of the elements. (Note that $a - b$ is defined as $a + (-b)$.)

DEFINITION 1.3.7 *The element a is* **greater than** *b (denoted $a > b$) if and only if $a - b > 0$. We say a is* **less than** *b ($a < b$) if and only if $b > a$. (Similar definitions hold for $\geq$ and $\leq$.)*

We are now able to prove some standard properties of inequalities.

THEOREM 1.3.8 *The following properties of inequalities hold:*

a) *If $a > b$ and $b > c$, then $a > c$ (**transitivity**).*

b) *If $a > b$ and $c \geq d$, then $a + c > b + d$.*

PROOF **a)** We leave the proof of the transitivity property as an exercise.

b) By Definition 1.3.7 $a > b$ means that $a - b > 0$. So we must prove that if $a - b > 0$ and $c - d \geq 0$, then $(a + c) - (b + d) > 0$.

$$\begin{aligned} (a-b)+(c-d) &> 0 && \text{(Axiom 13)} \\ (a+c)+(-b)+(-d) &> 0 && \text{(Axioms 2 and 3)} \\ (a+c)+((-1)\cdot b+(-1)\cdot d) &> 0 && \text{(Theorem 1.3.6a)} \\ (a+c)-(b+d) &> 0 && \text{(Axiom 11, Theorem 1.3.6a)} \ \blacksquare \end{aligned}$$

The analogous statements of Theorem 1.3.8 using $<$, $\leq$, and $\geq$ also hold true. Multiplication of inequalities requires some care and is a common cause of errors.

THEOREM 1.3.9 *The following properties of inequalities under multiplication hold.*

a) *If $a > b$ and $c > 0$, then $a \cdot c > b \cdot c$.*

b) *If $a > b$ and $c < 0$, then $a \cdot c < b \cdot c$.*

PROOF Part **(a)** is a straightforward exercise. We prove **(b)**. From Definition 1.3.7 we must show that if $a - b > 0$ and $c < 0$, then $ac - bc < 0$. By Axiom 12, $-c > 0$ and

$$\begin{aligned} (a-b)\cdot(-c) &> 0 && \text{(Axiom 14)} \\ -((a-b)\cdot c) &> 0 && \text{(Axiom 7 and Theorem 1.3.6)} \\ -(ac-bc) &> 0 && \text{(Axiom 11)} \\ (ac-bc) &< 0 && \text{(Axiom 12)} \ \blacksquare \end{aligned}$$

The order axioms enable us to define a measure of distance between elements in our set. This will be important for our work, since the notion of distance or closeness is at the heart of calculus. We begin by defining absolute value.

DEFINITION 1.3.10 *The* ***absolute value*** *of a real number a is denoted by $|a|$ and is defined as*

$$|a| = \begin{cases} a, & \text{if } a \geq 0 \\ -a, & \text{if } a < 0. \end{cases}$$

DEFINITION 1.3.11 *The* ***distance*** *from a to b is $|b - a|$.*

The concept of a distance measure or **metric** on a set in mathematics has itself been axiomatized by the following three properties:

1. The distance from a to b is always a positive number except in the case that $a = b$ when the distance is 0.
2. The distance from a to b is equal to the distance from b to a. (symmetry)
3. The distance from a to b plus the distance from b to c is always greater than or equal to the distance from a to c. (triangle inequality)

It is straightforward to check that our definition of distance based on absolute values satisfies properties 1 and 2. To prove property 3 we first develop a very useful method for handling inequalities involving absolute values.

THEOREM 1.3.12 *The statement $|a| \leq b$ is equivalent to $-b \leq a \leq b$.*

PROOF We prove $|a| \leq b \Rightarrow -b \leq a \leq b$, leaving the converse as an exercise. Thus we assume $|a| \leq b$, and to prove $-b \leq a \leq a$ we must show both $a \leq b$ and $-b \leq a$. From the definition of absolute value there are two cases to consider: $a \geq 0$ and $a < 0$.

Suppose first that $a \geq 0$; so by definition $|a| = a$. Then $|a| \leq b$ implies $a \leq b$. Hence we must only show $-b \leq a$.

$$\begin{aligned} 0 &\leq b && \text{(Transitivity from } 0 \leq a \text{ and } a \leq b) \\ 0 + (-b) &\leq b + (-b) && \text{(Theorem 1.3.8b)} \\ -b &\leq 0 && \text{(Axioms 4 and 5)} \\ -b &\leq a && \text{(Transitivity from } -b \leq 0 \text{ and } 0 \leq a) \end{aligned}$$

Next suppose that $a < 0$, so that $|a| = -a$. Then $|a| \leq b$ implies $-a \leq b$. Adding b to both sides of this inequality (Theorem 1.3.8b) gives $-b - a \leq 0$, which by Definition 1.3.7 is $-b \leq a$. Thus we must only show that $a \leq b$.

$$\begin{aligned} 0 &\leq -a && \text{(Trichotomy, since } a < 0) \\ 0 &\leq b && \text{(Transitivity from } 0 \leq -a \text{ and } -a \leq b) \\ a &\leq b && \text{(Transitivity from } a \leq 0 \text{ and } 0 \leq b) \quad \blacksquare \end{aligned}$$

For our distance measure property 3 states that $|b-a|+|c-b| \geq |c-a|$. We will use the previous theorem to prove a more useful form of the triangle inequality from which property 3 can be immediately obtained.

THEOREM 1.3.13 ***(The Triangle Inequality)*** *For any real numbers a and b,*

$$|a|+|b| \geq |a+b|.$$

PROOF Clearly $|a| \leq |a|$, and by Theorem 1.3.12 this implies that $-|a| \leq a \leq |a|$. Similarly, $-|b| \leq b \leq |b|$. Adding these two inequalities (Theorem 1.3.8b) gives $-|a|+(-|b|) \leq a+b \leq |a|+|b|$. By distributivity we have

$$-(|a|+|b|) \leq a+b \leq |a|+|b|$$

Employing Theorem 1.3.12 in the reverse direction, we have shown that $|a+b| \leq |a|+|b|$. ∎

1.3.3 The Completeness Axiom

The field axioms together with the order axioms still do not account for all properties of the real number system. This is evident since the rationals alone satisfy all 14 of these axioms. Therefore anything that could be proven using only these axioms would necessarily hold for both the real numbers and the rationals. This implies that the important continuity properties of the real number system that were explored in Section 1.2 are not consequences of these axioms. Therefore we augment our axiom system with the following familiar property.

III. The Least Upper Bound Axiom (Completeness Axiom)

15. For every nonempty set of real numbers that has an upper bound, there exists a least upper bound.

DEFINITION 1.3.14 *A system of elements with two operations obeying Axioms 1 through 15 is called a* ***complete ordered field.***

With our earlier sets of axioms there have been different systems for which the axioms held. In the case of fields we saw that the reals, the rationals, and the systems in Examples 1.3.1 and 1.3.2 all satisfy Axioms 1 to 11. Both **Q** and **R** satisfy the first 14 axioms. Thus to say that we have a field or an ordered field does not determine the particular mathematical object under discussion. But in contrast, it can be shown that the real numbers are the only system that satisfies all 15 axioms. This axiom system actually *characterizes* the real numbers. The real number system can be extended, but only at the expense of some of the properties. If we extend the real numbers to the complex numbers, the order Axioms 12 to 14 are forfeited. Logicians have extended **R** by adding "infinitesimals" in between all of the reals. This preserves the ordering, but completeness is sacrificed. Thus the real numbers constructed in Section 1.2 constitute the *only* complete ordered field.

Problems for Section 1.3

•1.3.1. Finsh the proof of Corollary 1.3.3 by showing that if $a \neq 0$, then $(a^{-1})^{-1} = a$.

•1.3.2. Prove that if $a = -a$, then $a = 0$.

•1.3.3. Prove **(b)** and **(d)** of Theorem 1.3.2.

1.3.4. Under what conditions on real numbers a, b, and c can we conclude that $a \cdot b = c \cdot b$ implies that $a = c$? Prove your answer.

1.3.5. **a)** Which of the field axioms are satisfied by the set of natural numbers, **N**?

b) Which of the field axioms are satisfied by the set of integers, **Z**?

1.3.6. **a)** Determine whether the field axioms are satisfied by the two element set $F_2 = \{a, b\}$ where the following tables define the operations of addition and multiplication.

$\oplus$	a	b
a	a	b
b	b	a

$\odot$	a	b
a	a	a
b	a	b

b) Show that the order axioms are *not* satisfied by F_2 no matter how one choses the set P.

c) What happens if we switch the roles of the two operations? That is, if "$\odot$" is addition and "$\oplus$" is multiplication, are the field axioms satisfied?

1.3.7. In this problem we *extend* the rational numbers by *adjoining* $\sqrt{2}$ to them as follows. Let

$$\mathbf{Q}[\sqrt{2}] = \left\{ r + s\sqrt{2} \mid r,\, s \in \mathbf{Q} \right\}.$$

Notice, of course, that $\mathbf{Q}[\sqrt{2}]$ is a subset of **R**, so we can define the operations of addition and multiplication on $\mathbf{Q}[\sqrt{2}]$ to be the usual "real" addition and "real" multiplication. Show that $\mathbf{Q}[\sqrt{2}]$ satisfies all the field axioms.

1.3.8. Let M_{22} denote the set of 2×2 matrices with real entries. Using the standard operations of matrix addition and multiplication, which of the field axioms does M_{22} satisfy? Use any results you know from linear algebra.

•1.3.9. In analogy with Definition 1.3.7, define $a \geq b$ and $a \leq b$.

1.3.10. Prove **(a)** of Theorem 1.3.9, that is, if $a > b$ and $c > 0$, then $a \cdot c > b \cdot c$.

•1.3.11. Use Definition 1.3.7 and Axiom 13 to show the transitive property of inequalities stated in Theorem 1.3.8a.

1.3.12. Prove that if $a \neq 0$, then $a^2 > 0$. Conclude that $1 > 0$.

1.3.13.
a) Our axiom system only explicitly gives the existence of two numbers: 1 and 0. Use these to find a reasonable definition of the other numbers $2, 3, \ldots$

b) Using your definition for 2 and 4 prove that $2 \cdot 2 = 4$.

c) Show that 2 is really a distinct number by proving that it doesn't equal 0 or 1.

1.3.14. Complete the proof of Theorem 1.3.12.

1.3.15. The third property for a measure of distance, or metric, is that the distance from x to y plus the distance from y to z should be greater than or equal to the distance from x to z. Use the definition of distance to restate this property in terms of absolute values; then prove it using the triangle inequality. (Hint: Let $a = y - x$ and let $b = z - y$.)

•1.3.16. Several useful inequalities involving absolute values can be derived from the triangle inequality (Theorem 1.3.13) by a judicious choice for a and b.

a) $|x| - |y| \leq |x - y|$. (Hint: Let $a = x - y$ and $b = y$.)

b) $|x| - |y| \leq |x + y|$. (Hint: Let $a = x + y$ and $b = -y$.)

c) $|x| + |y| \geq |x - y|$.

d) $||x| - |y|| \leq |x - y|$.

1.3.17. Prove that $|x| = |-x|$

1.3.18. Show that $|xy| = |x||y|$.

•1.3.19. Several inequalities, which we will use often in the text, follow by looking at differences of two inequalities. Prove each of them. Theorem 1.3.8 may help.

a) If $a \leq b$ and $c \leq d$, then $c - b \leq d - a$.

b) If $a \leq b \leq d$ and $a \leq c \leq d$, then $c - b \leq d - a$.

c) If $a \leq b \leq d$ and $a \leq c \leq d$, then $b - c \leq d - a$.

d) If $a \leq b \leq d$ and $a \leq c \leq d$, then $|c - b| \leq d - a$.

e) If $a \leq b \leq c \leq d$, then $c - b \leq d - a$.

•1.3.20.
a) Show that if $|a - m| < 1$, then $|a| < |m| + 1$.

b) Show that if $|a - m| < |m|/2$, then $|m|/2 < |a|$.

•1.3.21.
a) Let $\delta > 0$. Use absolute values to describe all real numbers x that lie within the distance δ of the real number a.

b) Let $\delta > 0$. Use absolute values to describe all real numbers $x \neq a$ that lie within the distance δ of the real number a.

1.4 The Heine-Borel Theorem

The Heine-Borel theorem provides an important link between our study of the real numbers and the theory of calculus. Its proof depends on the least upper bound property; therefore this theorem is rooted in the continuity and completeness properties of the real numbers. In turn, the Heine-Borel theorem plays a key role in the proofs of several central results concerning the differentiation and integration of functions. Some of these connections are traced in the following diagram.

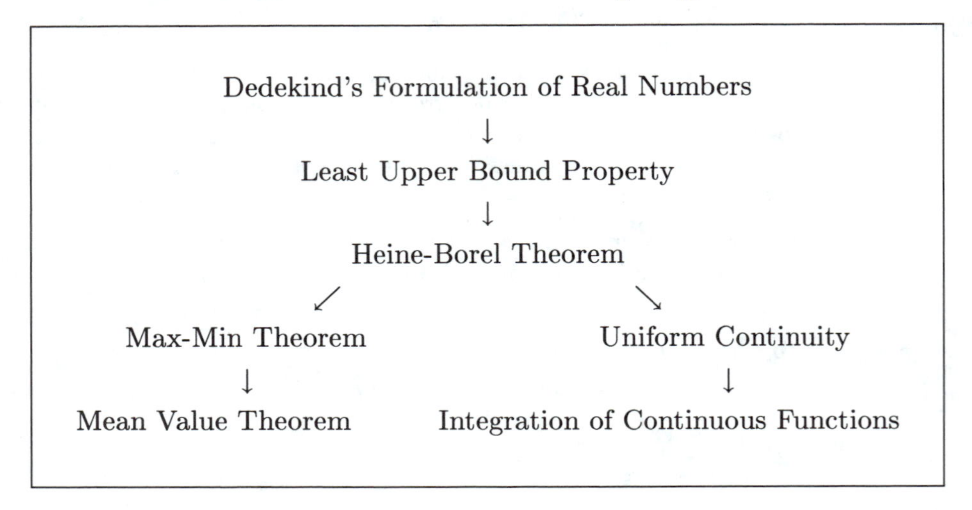

The central idea in the Heine-Borel theorem arose from trying to find uniform limits on the behavior of a function over all points in a set. Such limits can often be found for some small open interval (*locally*) about any point in a set, and the question is, can these local limits somehow be combined to form *global* uniform limits over the entire set? We might try this by covering all the points in the set with little overlapping open intervals of local control. However, if it requires an infinite number of open intervals to cover the set, we cannot be assured of global, *uniform* limits, since the behavior can get worse with each successive open interval. On the other hand, if we knew that the set could be covered by a finite number of the open intervals, there would be a limit to how bad the behavior could be and uniform control over all points in the set would be possible. The Heine-Borel theorem speaks to this problem by showing that if the set in question is a closed and bounded interval, then within any collection of open intervals that covers the set there will be a finite subcollection that also covers the set.

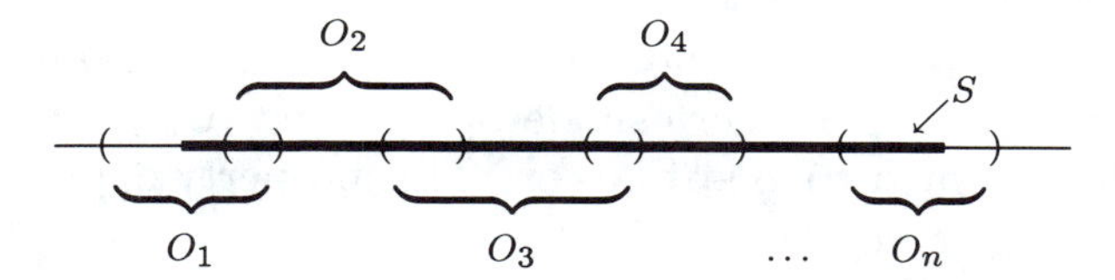

Figure 1.4.1. An open cover for S.

On the surface this result about open coverings of sets would seem of only technical interest. However, as we can gather from the preceding chart, this theorem is the basis for important results in calculus. Often in mathematics an apparently minor technical fact is the key to important results. Later the fact is abstracted and becomes the basis for a new area of study. In the case of the Heine-Borel theorem these technical ideas about open coverings of sets stimulated a study of the nature of *open* and *closed* sets that developed into a major branch of mathematics called (point-set) *topology*. Sets for which any open covering can be reduced to a finite cover have a special importance in topology and are known as **compact sets**. In this section we first prove the Heine-Borel theorem and then briefly investigate one of its consequences concerning bounded sets that have an infinite number of elements.

1.4.1 Open Covers and the Heine-Borel Theorem

Our work will employ collections of open intervals, that is, intervals of the form (a, b). These collections will be denoted by $\{O_\alpha \mid \alpha \in A\}$, where each O_α is an open interval and A is a (possibly infinite) index set. This notation is sometimes abbreviated as $\{O_\alpha\}$.

DEFINITION 1.4.1 *Let S be a set of real numbers. An* ***open cover*** *for S is a collection of open intervals, $\{O_\alpha \mid \alpha \in A\}$, such that each point in S is contained in at least one of these intervals. That is, for each $s \in S$ we have $s \in O_\beta$ for some $\beta \in A$. A* ***finite subcover*** *for an open cover $\{O_\alpha\}$ consists of a finite number of the open intervals O_α, which together form an open cover for S.*

Sets can have a multitude of different open covers.

EXAMPLE 1.4.1 One open cover for the set $[0, 1)$ is given by the collection $\{O_\alpha\} = \{(-.2, .2), (0, .4), (.2, .6), (.4, .8), (.6, 1)\}$. But it is also covered by many other collections such as $\{U_n\} = \{(-1, 1 - \frac{1}{n}) \mid n \in \mathbf{N}\}$ and $\{V_1\} = \{\mathbf{R}\}$. $\square$

In the preceding example the collection $\{V_1\}$ shows that *any* set of numbers has an open cover; hence the existence of an open cover is not an issue. Instead, in the Heine-Borel theorem we are interested in knowing when *every* single open cover $\{O_\alpha\}$ for a set must necessarily have some **finite subcover**. This property depends on the nature of the set being covered.

EXAMPLE 1.4.2 Each of the following three sets illustrates how a covering of a set can fail to have a finite subcover.

a) The set $[1, \infty)$ is covered by the collection of intervals $\{O_n\} = \{(0, n) \mid n \in \mathbf{N}\}$, since each number in the set is in some interval $(0, n)$ for a large enough n. However, no finite number of these sets will cover all of $[1, \infty)$. Thus this set has an open cover with no finite subcover.

b) The collection of open intervals $\{W_n\} = \{(1/n, 1 - 1/n) \mid n \in \mathbf{N}\}$ forms an open cover for the set $(0, 1)$. But it has no finite subcover.

c) The set consisting of the points $\{\frac{1}{n} \mid n \in \mathbf{N}\}$ is covered by the open intervals $\{O_n\} = \{(\frac{1}{n}, 2) \mid n \in \mathbf{N}\}$. Again, there is no finite subcover. □

Notice that $[1, \infty)$ in **(a)** fails to have a finite subcover because the set is not bounded. A set is **bounded** if it is contained in the interval $[-n, n]$ for some sufficiently large $n \in \mathbf{N}$. Intuitively, this means that the points in a bounded set cannot travel off to infinity in either direction. Now if a set is *not* bounded, the collection $\{(-n, n) \mid n \in \mathbf{N}\}$ will form an open cover for the set (since it covers all of $\mathbf{R}$); yet because of unboundedness there will be no finite subcover. Similarly, **(b)** illustrates how an open interval can have an open bounded cover with no finite subcover. But the Heine-Borel theorem states that if the set is a *closed and bounded* interval, *any* open cover $\{O_\alpha\}$ will necessarily have a finite subcover.

THEOREM 1.4.2 ***(The Heine-Borel Theorem)*** *Given any open cover $\{O_\alpha\}$ for the closed and bounded interval $[a, b]$, there is a finite subcover (i.e., a finite subcollection of sets in $\{O_\alpha\}$ will include all points of the interval $[a, b]$).*

PROOF We start by covering subintervals of $[a, b]$ with finite subcollections of the sets in $\{O_\alpha\}$, beginning at a and moving to the right. Let P be the set

$$P = \{p \in [a, b] \mid [a, p] \text{ has a finite subcover in } \{O_\alpha\}\}.$$

In order to be in the set P a point p must first be in $[a, b]$ and then there must exist some finite subcover of the interval $[a, p]$. The proof depends on a clear understanding of this set P. There are three parts to the proof.

STEP 1 *P has a least upper bound λ.* P is not empty. Since $\{O_\alpha\}$ covers the entire interval, a itself must be in some open interval O_α, say $a \in (c,d)$. If $d > b$, then the entire interval $[a,b]$ is contained in (c,d). So the single set (c,d) is a finite subcover and we are done. If $d \leq b$, at least the interval $[a, \frac{d+a}{2}]$ has a finite covering by (c,d). So by the definition of P we have $\frac{d+a}{2} \in P$ (Figure 1.4.2).

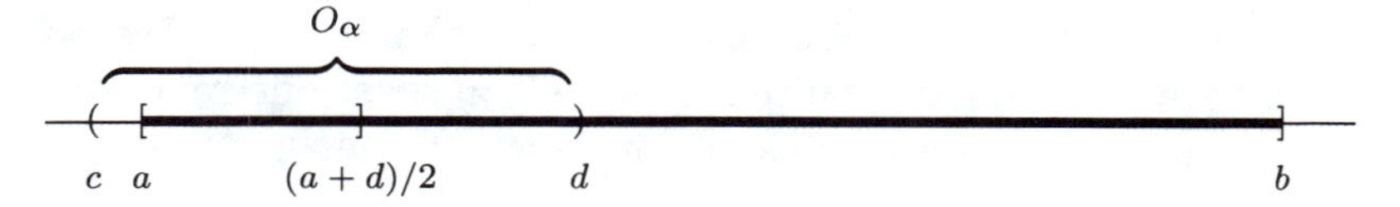

Figure 1.4.2. $[a,(a+d)/2]$ has a finite cover consisting of one set, O_α.

Now P is bounded above by b, since P is a subset of $[a,b]$. Therefore by Axiom 15 the set P has a least upper bound, which we denote by λ.

STEP 2 *The least upper bound λ equals b.* This is proved by contradiction. First observe that $\lambda \leq b$. (Why?) Suppose now that $\lambda < b$. Since $\{O_\alpha\}$ covers all of $[a,b]$ there is some interval (r,s) in the collection $\{O_\alpha\}$ such that $r < \lambda < s$. Since $r < \lambda$ and λ is the least upper bound, there is some point $p > r$ with $p \in P$. (If this were not the case, then r, which is smaller than λ, would be an upper bound for P, contradicting the fact that λ is the *least* upper bound.)

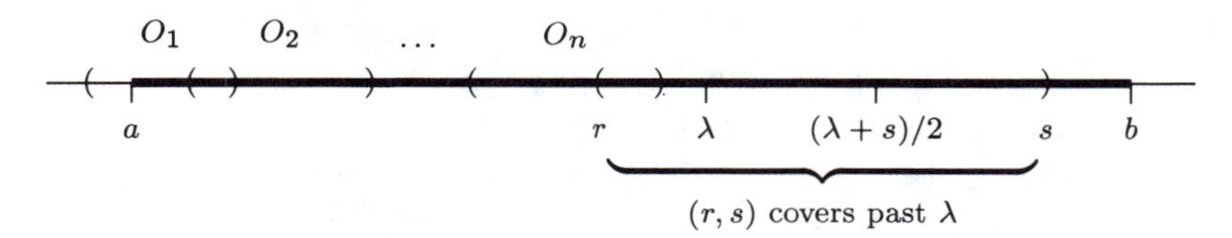

Figure 1.4.3. A finite open cover that extends past λ. Contradiction!

By the definition of P this means that $[a,p]$ has a finite subcover which we denote by $O_1, O_2, \ldots, O_n$. Adding the single set (r,s) to this cover forms a finite collection, $\{O_1, O_2, \ldots, O_n, (r,s)\}$, *which covers past λ* (Figure 1.4.3). In particular, the interval $[a,p']$, for $p' = \frac{\lambda+s}{2} > \lambda$, is covered by this finite collection. *But this contradicts the fact that λ is an upper bound for P!* Therefore our assumption that $\lambda < b$ cannot be true and we must have $\lambda = b$.

STEP 3 *There is a finite subcover of $[a,b]$.* We argue as in Step 2. That is, since $\{O_\alpha\}$ covers b there must be an interval (u,v) with $u < b < v$. But because b is the l.u.b. of P there must be a point $\tilde{p} \in P$ with $\tilde{p} > u$ such that $[a,\tilde{p}]$ has a finite cover. Adding the single set (u,v) to this cover gives a finite cover for the whole interval. ∎

1.4.2 The Bolzano-Weierstrass Theorem

We have discussed the completeness of the real number system in two ways, informally with Dedekind cuts and axiomatically with the least upper bound property. A first consequence of the completeness of the real number system was the Heine-Borel theorem, which dealt with infinite open covers of closed, bounded intervals. A second consequence involves infinite bounded sets of real numbers. We will show that for every bounded infinite set there is at least one real number about which points of the set "cluster" or "accumulate."

DEFINITION 1.4.3 *Let S be a subset of* $\mathbf{R}$. *A point x in* $\mathbf{R}$ *is an* ***accumulation point****, or* ***cluster point****, of S if for every $\epsilon > 0$ there is a point s in S such that* $0 < |s - x| < \epsilon$.

The condition that for every $\epsilon > 0$ there is a point s in S such that $0 < |s - x| < \epsilon$ means that there are points of S that are not equal to x but which are arbitrarily close to x.

EXAMPLE 1.4.3 Let $S = \{1/n \mid n \in \mathbf{N}\}$. Then 0 is an accumulation point of S because for any $\epsilon > 0$, we can choose some sufficiently large $n \in \mathbf{N}$ such that $0 < |1/n - 0| = 1/n < \epsilon$. This example also shows that an accumulation point of a set need not be a member of that set. □

EXAMPLE 1.4.4 The set of all integers, $\mathbf{Z}$, has no accumulation points. For any real number x that is not an integer, let ϵ_x denote the distance from x to the nearest integer. Then for any $s \in \mathbf{Z}$, we have $|n - x| \geq \epsilon_x$; therefore x cannot be an accumulation point of $\mathbf{Z}$. If x is an integer, then for any $s \in \mathbf{Z}$, if $0 < |s - x|$, then $s \neq x$; therefore $|s - x| \geq 1$. So no integer is an accumulation point of $\mathbf{Z}$ either. □

THEOREM 1.4.4 *(**The Bolzano-Weierstrass Theorem**) Every bounded infinite set of real numbers has an accumulation point.*

PROOF The proof is similar to the proof of the Heine-Borel theorem. Let S be an infinite set that is bounded above by b and below by a. Define the set P as follows: *a real number p is in P if and only if there are at most a finite number of elements of S that are less than p.* Observe that if $p \in P$ and $q < p$, then $q \in P$ because any element of S less than q is less than p.

STEP 1 *P has an upper bound.* Notice that $a \in P$ because no element of S can be less than a lower bound of S. Thus S is not empty. Further, P is bounded above by b, since the infinite number of elements of S are all less than or equal to an upper bound of S. By Axiom 15, P has a least upper bound λ.

STEP 2 *λ is an accumulation point for S.* Given any $\epsilon > 0$, there are two cases to consider: either $\lambda \in P$ or $\lambda \notin P$. If $\lambda \in P$, then $\lambda + \epsilon$ is not in P, since λ is the least upper bound for P. This means that there are an infinite number of elements of S that are less than $\lambda+\epsilon$. But only a finite number of these same elements are less than or equal to λ, since λ is in P. So there must be at least one element s of S between λ and $\lambda + \epsilon$. Consequently $0 < |s - \lambda| < \epsilon$, which shows that λ is an accumulation point of S.

Assume now that $\lambda \notin P$. Since λ is the least upper bound for P, there is some element $p \in P$ such that $\lambda - \epsilon \leq p < \lambda$ (else $\lambda - \epsilon$ would be a smaller upper bound for P than λ). Since $p \in P$, there are only a finite number of elements of S that are less than p. On the other hand $\lambda \notin P$; therefore λ is greater than or equal to an infinite number of elements of S. So there must be at least one element s of S between p and λ. That is, $\lambda - \epsilon \leq p < s < \lambda$, so $0 < |s - \lambda| < \epsilon$, which shows that λ is an accumulation point of S. ∎

Observe that in the preceding proof we showed that λ was an accumulation point of a set S by showing that there were an infinite number of points of S within ϵ of λ, though the definition of an accumulation point requires only a single point s (not equal to x) of S to be within ϵ of λ. It is left as an exercise to show that this behavior is true of all accumulation points.

LEMMA 1.4.5 *If x is an accumulation point of S and $\epsilon > 0$, then there are an infinite number of points of S within ϵ of x.*

Problems for Section 1.4

1.4.1. For the interval $[1, 10]$ find an open cover using open intervals of length less than 1.

1.4.2. For each of the following sets find an an open cover that has no finite subcover.

a) The set $\mathbf{N}$ of natural numbers.

b) The interval $[0, 10)$.

c) The set $A = \{\frac{1}{2^n} \mid n \in \mathbf{N}\}$.

d) The rationals between 0 and $\sqrt{2}$ (including 0).

1.4.3. Suppose that the collection $\{O_\alpha\}$ is an open cover for the interval $[0, 1)$. Suppose further that $1 \in \cup\{O_\alpha\}$. Show that $[0, 1)$ has a finite subcover from among the collection $\{O_\alpha\}$. (Hint: Think about the interval $[0, 1]$ with respect to the open cover $\{O_\alpha\}$. What does the Heine-Borel theorem say about this interval?)

1.4.4. **a)** Show that if $\{O_\alpha\}$ is an open cover of the interval $[a, b]$, then for any subset $C \subset [a, b]$ there is a finite subcover of C from the sets $\{O_\alpha\}$.

b) Why doesn't **(a)** prove that we can find a finite subcover for any open cover of any bounded set?

1.4.5. Suppose that $\{O_\alpha\}$ is an open cover of $[0, \infty)$. If the open interval $(1000, \infty)$ is one of the sets in the collection $\{O_\alpha\}$, show that there is a *finite* subcover for $[0, \infty)$. (Hint: What can you say about $[0, 1000]$ with respect to the open cover $\{O_\alpha\}$?)

1.4.6. Show that if $I_n = (0, \frac{1}{n})$, the intersection of I_n over all $n \in \mathbf{N}$ is empty (i.e., $\bigcap_{n=1}^{\infty} I_n = \emptyset$). (Hint: Suppose that a number α is in the intersection. Show that α must be greater than 0 and then use the Archimedean principle to show that there must be an integer n such that $\frac{1}{n} < \alpha$.)

1.4.7. A collection $\{I_n\}$ of intervals are said to be **nested** if $I_{n+1} \subset I_n$. Prove the following:

The Closed Nested Interval Theorem: *Let $\{I_n\} = \{[a_n, b_n]\}$ be a collection of closed nested intervals. Then $\bigcap_{n=1}^{\infty} I_n \neq \emptyset$, that is, the intersection of all the I_n is not empty.*

a) Let $B = \{b_n \mid n \in \mathbf{N}\}$. Show that B is bounded below and so must have a greatest lower bound, b.

b) Show that $b \geq a_n$ for any n.

c) Show that b must be in $\bigcap_{n=1}^{\infty} I_n$.

d) Repeat this type of argument using the set $A = \{a_n \mid n \in \mathbf{N}\}$ in place of B. What exactly is the intersection $\bigcap_{n=1}^{\infty} I_n$?

1.4.8. Suppose that a set S of numbers has the property that if a and b are in S, then for any $c \in (a, b)$ we have $c \in S$. Show that S must be an interval of one of these forms: (g, h), $[g, h)$, $(g, h]$, or $[g, h]$, where g and h are either finite numbers or $\pm\infty$.

1.4.9. Prove Lemma 1.4.5. (Suppose that for some $\epsilon > 0$, there were only a finite number of points, $s_1, s_2, \ldots, s_n$ of S within ϵ of x. Let $\epsilon' = \min\{|s_1 - x|, |s_2 - x|, \ldots, |s_n - x|\}$. Now show that no $s \in S$ satisfies $0 < |s - x| < \epsilon'$.)

1.4.10. Show that if S is a finite set, it has no accumulation points.

1.4.11. What are the accumulation points of the following sets?

a) $B = [0, 1) \cup \{2\}$ **b)** $C = \{n/3 \mid n \in \mathbf{N}\}$

c) $D = \{3/n \mid n \in \mathbf{N}\}$

1.4.12. **a)** Suppose that λ is the least upper bound of a set S and that λ is *not* in S. Show that λ is an accumulation point of S. (Hint: For any $\epsilon > 0$, show that there is a point $s \in S$ such that $\lambda - \epsilon < s < \lambda$. Now use the definition of accumulation point to finish the proof.)

b) Give an example of a set S with a least upper bound λ that is *not* an accumulation point of S.

1.4.13. **a)** Show that any real number x is an accumulation point of $\mathbf{Q}$. (Hint: x is an accumulation point of $\mathbf{Q}$ if and only if for any $\epsilon > 0$, there is a rational number $r \neq x$ that lies in the interval $(x-\epsilon, x+\epsilon)$. That is, r lies in either $(x-\epsilon, x)$ or $(x, x+\epsilon)$. Is there always a rational in such an interval?)

b) What are the accumulation points of the irrationals?

1.4.14. A set S is said to be **dense** in the set of real numbers if for any two real numbers α and β with $\alpha < \beta$, there is an element $s \in S$ such that $\alpha < s < \beta$. If S is dense, show that any real number x is an accumulation point of S.

1.4.15. Suppose that λ is *both* an upper bound and an accumulation point for a set S. Show that λ is actually the *least* upper bound for S. (Use a proof by contradiction.)

a) Assume that β is an upper bound of S and that $\beta < \lambda$. Let $\epsilon = \lambda - \beta$ (which is positive). Show that since λ is an accumulation point for S, there must exist $s \in S$ such that $-\epsilon < s - \lambda$.

b) Show that $\beta < s$ and complete the proof.

2

Functions, Limits, and Continuity

One of the most fundamental and pervasive concepts in mathematics is the idea of a function. From earlier courses you have developed a strong intuitive sense of what a function is: a rule that associates to each "input" a unique "output." Given a graph in the plane you can tell whether it is the graph of a function or not by using the "vertical line test." You may not be able to rigorously define "function," but you know one when you see it.

Mathematicians themselves struggled with the definition well into the nineteenth century. This is not so startling until one recalls that the calculus of Newton and Leibniz was constructed in the last half of the seventeenth century. With the rise of calculus, the "analytic" function concept was ushered in and used extensively throughout the eighteenth century. Such functions had the same rule or analytic expression throughout their entire domains. In 1748 Euler wrote:

> A function of a variable quantity is an analytical expression composed in any way from this variable quantity and from numbers or constant quantities.[1]

Euler did not bother to define what he meant by the term "analytic

---

[1] C. H. Edwards, Jr., *The Historical Development of the Calculus* (New York: Springer-Verlag, Inc., 1979), p. 271.

expression," though he probably had algebraic and simple transcendental operations in mind. The notion that Lacroix used in 1797 sounds even more vague than our intuitive definition:

> Every quantity whose value depends on one or several others is called a function of the latter, whether one knows or does not know by what operations it is necessary to go from the latter to the first quantity.[2]

Lagrange's definition of function in the 1813 edition of a text he wrote on analysis was not too different.

> The word *function* has been used by the first analysts to denote in general the powers of one quantity. Since then, the meaning of this word has been extended to any quantity formed in any manner from another quantity.[3]

With the nineteenth century came an increasing interest in rigor as a wider variety of functions were studied. In 1821 Cauchy defined a function this way.

> When variable quantities are so joined between themselves so that, the value of one of these being given, one may determine the values of all the others, one ordinarily conceives these diverse quantities expressed by means of the one among them, which then takes the name independent variable; and the other quantities expressed by means of the independent variable are those which one calls functions of this variable.[4]

Finally, in 1837 Dirichlet gave a definition of function that is quite similar in spirit to the modern one, that is, y is a function of x when to each value of x in an interval there corresponds a unique value of y.[5]

2.1 Functions

In hindsight, we see that mathematicians were struggling to produce a definition that was precise enough to be useful while still general enough to be applicable in a wide variety of settings. A function specifies a special type of correspondence between two sets. Such general functions should include among other things linear transformations of vector spaces, determinants, and the more familiar calculus functions such as

[2] Morris Kline, *Mathematical Thought from Ancient to Modern Times* (New York: Oxford University Press, 1972), p. 949.

[3] I. Grattan-Guinness, "The Emergence of Mathematical Analysis and its Foundational Progress, 1780–1880," in *From the Calculus to the Set Theory, 1630–1910*, ed. I. Grattan-Guinness (London: Gerald Duckworth & Co. Ltd., 1980), p. 100.

[4] Kline, *Mathematical Thought*, p. 950.

[5] Kline, p. 950.

polynomials, logarithms, and exponentials. We will use the following modern formulation of Dirichlet's idea.

Let a and b be elements of the sets A and B, respectively. Recall that the **ordered pair** of elements (a, b) is uniquely determined by the elements a and b and the order in which they are given. That is,

$$(a, b) = (a', b') \iff a = a' \quad \text{and} \quad b = b'.$$

Definition 2.1.1 *A **function** f is a collection of ordered pairs of elements such that if (a, b) and (a, c) are both in the collection, then $b = c$. That is, the collection cannot contain two different ordered pairs with the same first element.*

Notice that the definition does not specify that the elements of the ordered pairs need to be real numbers. A function may relate any two sets of objects. For the moment, we will be concerned with **real-valued** functions of a **real variable**. That is, both elements of the ordered pairs (a, b) in the function will be real numbers. With this assumption, the **domain** of a function f is the set of all real numbers a for which there is some real number b such that the pair (a, b) is in f. When the pair (a, b) belongs to f, b is called the **value** of f at a and is denoted by $b = f(a)$. From the definition of f, the meaning of $f(a)$ is clear because both (a, b) and (a, c) cannot belong to f unless $b = c$. Finally, the **range** of f is the set of all numbers b for which there is a number a such that (a, b) belongs to f.

Example 2.1.1 The collection f of all pairs of real numbers (x, y) for which $y = 2x$ is clearly a function. For if (a, b) and (a, c) are both in f, then $b = 2a$ and $c = 2a$, and $b = c$. In contrast, the collection g of all pairs of real numbers (x, y) for which $y^2 = x$ is not a function. If (a, b) and (a, c) are both in f, then $b^2 = a$ and $c^2 = a$. So $b^2 = c^2$ and $b = \pm c$. That is, g contains ordered pairs of the form (b^2, b) and $(b^2, -b)$ and so is not a function. Finally, the collection h of all pairs of real numbers (x, y) for which $x^2 = y + 1$ is a function. For if (a, b) and (a, c) are both in f, then $a^2 = b + 1$ and $a^2 = c + 1$, and $b = c = a^2 - 1$. □

The definition of a function makes it impossible for both (a, b) and (a, c) to belong to f unless $b = c$. Thus the second member of any ordered pair that belongs to f is completely determined by the first member of the pair. For this reason, the second element y of any ordered pair (x, y) in f is denoted by $f(x)$. For example, the functions f and h described in Example 2.1.1 may be denoted by

$$f(x) = 2x \quad \text{and} \quad h(x) = x^2 - 1.$$

We will also adopt the convention that when no explicit description of the domain of a function is given, the domain is assumed to be the largest set of real numbers for which the correspondence makes sense. For example, the domain of the function

$$q(x) = \frac{\sqrt{x}}{x-1}$$

is assumed to be all $x \geq 0$ except $x = 1$.

Occasionally it will be important to restrict the domain of a function to some smaller set. For example, if we are interested in finding the area of a square that has sides of length x, the area function is

$$A(x) = x^2 \qquad (x \geq 0).$$

Here the domain has been restricted because lengths are nonnegative. The function A is not the same as the function $S(x) = x^2$, whose domain is all real numbers. The two functions consist of different collections of ordered pairs. For example, S contains the ordered pair $(-2, 4)$, whereas A does not. More generally, since two functions are equal if and only if they consist of exactly the same set of ordered pairs and since the second element in each ordered pair is uniquely determined by the first, we see that *two functions f and g are equal if and only if they have the same domain D and $f(x) = g(x)$ for all x in D.*

Many functions encountered in a first calculus course are defined by a single equation, such as $f(x) = 2x^3 - x^2 + 1$. But not all functions are so easily described. The most familiar example of a function that has more than one "rule,' is the absolute value function, which is defined by

$$|x| = \begin{cases} x, & \text{if } x \geq 0 \\ -x, & \text{if } x < 0. \end{cases}$$

Among the more exotic functions of this type is the following one, which was first considered by Dirichlet in 1829:

$$D(x) = \begin{cases} 1, & \text{if } x \text{ is rational} \\ 0, & \text{if } x \text{ is irrational.} \end{cases}$$

Sometimes describing a function "in words" is more convenient than using a formula or a rule. For example, consider the **the nearest integer function** $N(x)$ determined by

$N(x)$ is the distance from x to the nearest integer.

The domain of $N(x)$ consists of all real numbers, and the range is the closed interval $[0, 1/2]$. A formal description of $N(x)$ is more complicated. First we define $N(x)$ on the interval $[0, 1)$ by

$$N(x) = \begin{cases} x, & \text{if } 0 \leq x \leq 1/2 \\ 1 - x, & \text{if } 1/2 < x < 1. \end{cases}$$

Next, any real number may be written uniquely as $n + x$, where $n \in \mathbf{Z}$ and $x \in [0, 1)$. We extend the definition of N to the entire real line by setting

$$N(n + x) = N(x) \qquad (n \in \mathbf{Z},\ x \in [0, 1)).$$

For example, $N(.7) = .3$ and $N(-1.7) = .3$.

One way to get a feel for a function is to make a graphic representation of it. Every function we have considered so far has been a collection of ordered pairs of real numbers (a, b). Such a collection is a subset of the plane, $\mathbf{R}^2$. Using the usual rectangular coordinate system in the plane, the graph of a function f may be visualized as the set of all points in $\mathbf{R}^2$ that correspond to the pairs $(x, f(x))$ that comprise f. The graph of the nearest integer function, $N(x)$, is given in Figure 2.1.1.

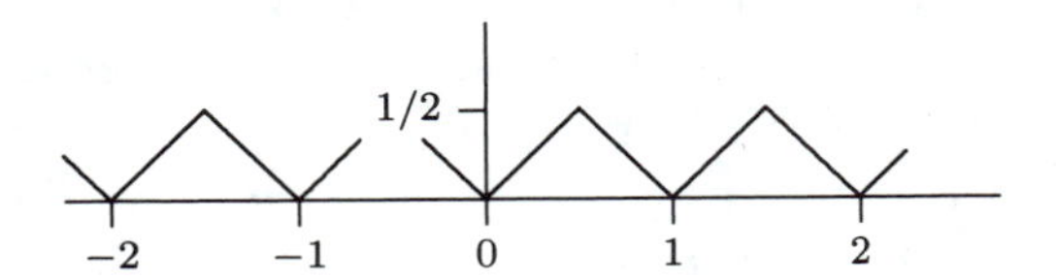

Figure 2.1.1

There are a few common ways to construct new functions from old ones. If f and g are any two functions, their **sum** is the function $f + g$ defined by

$$(f + g)(x) = f(x) + g(x).$$

The natural domain of $f + g$ is the set of those x which lie in both the domain of f and in the domain of g. That is,

$$\text{domain } (f + g) = \text{domain } f \cap \text{domain } g.$$

Similarly the **product** $f \cdot g$ of f and g is defined by

$$(f \cdot g)(x) = f(x) \cdot g(x),$$

and its domain is again domain $f \cap$ domain g. The **quotient** f/g is the function defined by

$$\left(\frac{f}{g}\right)(x) = \frac{f(x)}{g(x)}.$$

Here we cannot allow $g(x)$ to be 0. So the domain of f/g must be a subset of $\{x \mid g(x) \neq 0\}$. The domain can be written as the intersection of two sets:

$$\text{domain}\left(\frac{f}{g}\right) = \text{domain } f \cap \{x \in \text{domain } g \mid g(x) \neq 0\}.$$

For example, if $f(x) = x^2 - 1$ and $g(x) = x + 1$, then the function f/g is

$$\left(\frac{f}{g}\right)(x) = \frac{x^2 - 1}{x + 1}$$

and its domain consists of all real numbers except -1. (Calculate the appropriate intersection to see this.) Notice that f/g is not the same as the function $h(x) = x - 1$ obtained from f/g by canceling common factors because the domain of h consists of all real x.

For any two functions f and g, we define the **composite** function $f \circ g$ by

$$(f \circ g)(x) = f(g(x)).$$

If x is to be in the domain of $f \circ g$, then x must be in the domain of g and $g(x)$ must itself be in the domain of f. Therefore

$$\text{domain } f \circ g = \{x \mid x \in \text{domain } g \text{ and } g(x) \in \text{domain } f\}.$$

For example, let $f(x) = 1/x$ and let $g(x) = x^2 - 1$. Then

$$(f \circ g)(x) = f(g(x)) = \frac{1}{g(x)} = \frac{1}{x^2 - 1}.$$

Since $g(x) = 0$ for $x = \pm 1$,

$$\text{domain } (f \circ g) = (-\infty, -1) \cup (-1, 1) \cup (1, \infty).$$

In this chapter and the next we will concentrate on developing the theory of real-valued functions of a single real variable. In later chapters we will generalize this notion. Instead of both elements of an ordered pair (a, b) of a function f being real numbers, we will allow a or b or both to be ordered pairs of numbers, or vectors, or complex numbers. For example, consider the following real-valued function of two real variables: f is the collection of all ordered pairs $((x, y), z)$, where $z = x^2 + y^2$, that is, $f(x, y) = x^2 + y^2$. The first element of the ordered pair is itself an ordered pair of real numbers (x, y), and the second element of the pair is a single real number z.

By examining the same concept in a variety of settings, one begins to understand the power of careful definitions and obtains a sense of the structure of mathematics.

Problems for Section 2.1

2.1.1. Decide which of the following collections are functions by determining in each case whether the collection contains two pairs (a, b) and (a, c) with $b = c$.

a) The collection f of all pairs of real numbers (x, y) for which $2y = x + 4$.

b) The collection g of all pairs of real numbers (x, y) for which $x = |y|$.

c) The collection h of all pairs of real numbers (x, y) for which $x^2 + y^2 = 1$.

d) The collection F of all pairs of real numbers (x, y) for which $\sqrt{y} = \sqrt{x}$.

e) The collection G of all pairs of real numbers (x, y) for which $y^2 = x^2$.

2.1.2. Determine which of these functions are equal.

a) $f(x) = |x|$ **b)** $g(x) = \left(\sqrt{x}\right)^2$ **c)** $h(x) = \sqrt{x^2}$

2.1.3. What are the problems involved in making an accurate graph of the Dirichlet function

$$D(x) = \begin{cases} 1, & \text{if } x \text{ is rational} \\ 0, & \text{if } x \text{ is irrational?} \end{cases}$$

2.1.4. The **greatest integer function**, denoted by $[x]$, is defined by: $[x]$ is the greatest integer less than or equal to x. For example, $[2] = 2$, $[4.1] = 4$, and $[-4.1] = -5$.

a) Show that $s(x) = -[-x]$ is the smallest integer greater than or equal to x.

b) Draw graphs for both $[x]$ and $-[-x]$ and determine those values of x for which $[x] = -[-x]$.

c) The **fractional part** of x is the function defined by $\{x\} = x - [x]$. What is the range of $\{x\}$? Draw a graph of $\{x\}$.

•2.1.5. **a)** A function f is called a **constant function** if for all x, $f(x) = c$, where c is a real constant. Let g be any function. Use constant functions and the definition of a product of two functions to give a definition of the function $(c \cdot g)$ and give a description of its domain.

b) Define the **difference** of two functions f and g. Be sure to describe its domain.

•2.1.6. A function p is a **real polynomial** if there are real numbers $a_0, a_1, \ldots, a_n$ such that for all x

$$p(x) = a_n x^n + a_{n-1} x^{n-1} + \cdots + a_1 x + a_0.$$

When a polynomial is written this way it is presumed that $a_n \neq 0$. The highest power of x that appears in the polynomial is called the **degree** of p. Rational functions are constructed from polynomials. A function f is **rational** if f can be expressed as the ratio of two polynomials p/q, where q is not the constant function 0. Find the domains of the following rational functions:

a) $f(x) = \dfrac{3x^3 + 7}{x^3 - x}$ **b)** $h(x) = \dfrac{x^4 - x^2 + 2}{x^2 - 6}$

c) $r(x) = \dfrac{-17x + 11}{x^4 + x^2 + 2}$ **d)** $s(x) = \dfrac{x^5 - x^3 + 2x}{x^3 - 6x}$

2.1.7. Let f and g be functions defined on a common domain D. Then the **maximum** and **minimun functions** are defined as follows: for any $x \in D$,

$$\max(f,g)(x) = \begin{cases} f(x), & \text{if } f(x) \geq g(x) \\ g(x), & \text{if } g(x) > f(x) \end{cases}$$

and

$$\min(f,g)(x) = \begin{cases} f(x), & \text{if } f(x) \leq g(x) \\ g(x), & \text{if } g(x) < f(x). \end{cases}$$

a) Draw the graphs of $\max(x, x^2)$, $\min(x, x^2)$, and $\max(\sin x, \cos x)$.

b) Show that $|x| = \max(x, -x)$ and that $|f| = \max(f, -f)$ for any function f.

c) Show that $\max(f,g) = \frac{1}{2}(|f-g|+f+g)$. Similarly, show that $\min(f,g) = \frac{1}{2}(-|f-g|+f+g)$.

2.2 Limits

Next we consider the behavior of functions. In particular, we are interested in what happens to the values of a function $f(x)$ as x approaches some specified value a in the domain of f. The intuitive idea is this: $f(x)$ approaches the "limit" L when x approaches a if $f(x)$ can be made close to L by taking x sufficiently near to a. For example, if $f(x) = x + 4$, we expect the limit of $f(x)$ as x approaches 1 to be 5 because we can make $x + 4$ close to 5 by taking x correspondingly close to 1.

2.2.1 Background

Limits such as the preceding one really pose no problem at all, but others, as we shall see, can be much more difficult to calculate. Historically, the concern with limits arose in the eighteenth century as mathematicians attempted to make precise the ideas of Newton's and Leibniz's calculus. Their first problem was to make (good) sense out of the notion of a derivative or an instantaneous rate of change (a "fluxion" in Newton's terminology). The following example illustrates the problem.

Let $y = f(t)$ represent some quantity that is a function of time t in the time interval $a \leq t \leq b$. The **average rate of change in** f with respect to t over the time interval t_0 to t_1 is the difference quotient:

$$\text{average rate of change} = \frac{\text{change in } f}{\text{change in } t} = \frac{\Delta y}{\Delta t} = \frac{f(t_1) - f(t_0)}{t_1 - t_0}.$$

But this is just the slope of the secant line of f through the points $(t_0, f(t_0))$ and $(t_1, f(t_1))$. Letting t_1 approach t_0, average rates of change over shorter time intervals are obtained, which motivates the following idea: To obtain the instantaneous rate of change in f at t_0, allow the time interval to become "infinitesimally" short. That is, make the difference between t_1 and t_0 "infinitesimally" small or zero. Of course, if this is the case, then $t_0 = t_1$, and the difference quotient becomes

$$\frac{\Delta y}{\Delta t} = \frac{f(t_1) - f(t_0)}{t_1 - t_0} = \frac{0}{0}.$$

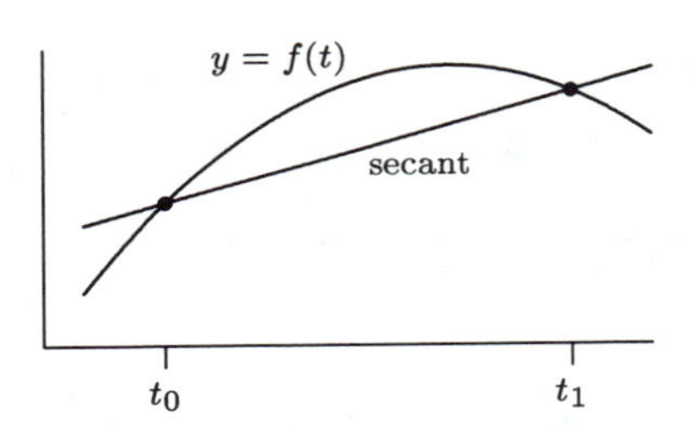

Figure 2.2.1

What to do? Not only is this last expression undefined algebraically (see the discussion after Theorem 1.3.4), but how can there be any (rate of) change if no time has passed? Instead, if we say that Δt gets closer to but never quite reaches 0, how can the limit of the difference quotient ever be reached? The problem is how to maintain an interval of time over which to calculate a rate of change and simultaneously to treat the time interval as if it were 0. Here is how Newton in Book 1 of his *Principia* replied to these questions.

> It is objected that there is no ultimate proportion [limit] of evanescent quantities; because the proportion before the quantities have vanished, is not ultimate; and when they have vanished is none. But by the same argument, it might well be maintained, that there is no ultimate velocity of a body arriving at a certain place, when its motion is ended: because the velocity, before the body arrives at the place, is not its ultimate velocity; when it has arrived, is none. But the answer is easy: for by the ultimate velocity is meant that, with which the body is moved, neither before it arrives at its last place,

> when the motion ceases, nor after; but at the very instant when it arrives; that is, the very velocity with which the body arrives at its last place, when the motion ceases. And in like manner, by the ultimate ratio of evanescent quantities is to be understood the ratio of the quantities, not before they vanish, nor after, but that with which they vanish.[1]

While we may understand the point of Newton's analogy, it hardly helps us with a particular calculation and surely cannot serve as a definition. In fact the notion of "the very velocity with which the body arrives at its last place" is exactly an instantaneous rate of change, which requires us to *already* understand the idea of a limit. What Newton has done is to supply us with a purely physical concept, instantaneous velocity, that seems to lead to a paradox when naïvely defined.

Many mathematicians during the eighteenth century saw no easy answers to such problems. They sought to eliminate the "infinitesimal" from the foundations of calculus and tried to find purely algebraic ways of obtaining (defining) the derivative. Michel Rolle (1652–1719) was among those who completely rejected the idea of infinitesimals, feeling that it was false and led to errors.[2] In fact, "Rolle at one point taught that calculus was a collection of ingenious fallacies."[3]

When definitions of limit were attempted, the results were often vague, misleading, or worse. Typical of such definitions are the following. In 1765 D'Alembert wrote:

> One magnitude is said to be the *limit* of another magnitude when the second may approach the first within any given magnitude, however small, though the first magnitude may never exceed the magnitude it approaches.[4]

Simon L'Huilier (1750–1840) introduced the modern limit notation[5] in an essay on limits that won a prize from the Berlin Academy. He stated,

> Given a variable quantity always smaller or greater than a proposed constant quantity; but which can differ from the latter by less than any proposed quantity however small; this constant quantity is called

---

[1] Florian Cajori, *A History of the Conceptions of Limits and Fluxions in Great Britain from Newton to Woodhouse* (Chicago: Open Court Press, 1919), p. 9.

[2] D. J. Struik, ed., *A Source Book in Mathematics, 1200–1800* (Princeton, N.J.: Princeton University Press, 1986), p. 342.

[3] Kline, *Mathematical Thought*, p. 427.

[4] H. J. M. Bos, "Newton, Leibniz, and the Leibnizian Tradition," in *From the Calculus to the Set Theory, 1630–1910*, ed. I. Grattan-Guinness (London, Gerald Duckworth & Co. Ltd., 1980), p. 91.

[5] Grattan-Guinness, "The Emergence of Mathematical Analysis," p. 101.

> the limit in greatness or smallness of the variable quantity.[6]

Both definitions completely ignore the possibility that the quantity may oscillate in value about the limit. (This is the case with the function $x\sin(\frac{1}{x})$ when x is near 0; see Example 2.2.3.) L'Huilier deserves some credit, however. In a revision of his essay he did permit oscillation in certain specific cases.[7] Augustin-Louis Cauchy (1789–1857) in the early part of the nineteenth century gave a somewhat clearer definition.

> When successive values attributed to a variable approach indefinitely a fixed value so as to end by differing from it by as little as one wishes, this last is called the limit of all the others.[8]

A problem remains, however: who gets to decide how little is little enough? How does one actually verify that some number L is (or is not) the limit of a function at a specified point? No formal conditions had been given that a limit had to satisfy.

It was Karl Weierstrass (1815–1897) who finally gave such formal conditions. He removed the vagueness of such phrases as "approach indefinitely" and "differing by as little as one wishes" from the definition of limit by using the notion of distance.

2.2.2 The Limit Definition

Recall that the set of all numbers x within a distance δ of a is represented by the inequality $|x-a|<\delta$. Similarly, to say that $f(x)$ is within a distance ϵ of L means that $|f(x)-L|<\epsilon$.

Suppose that $f(x)=\frac{2x^2+x}{x}$, for $x\neq 0$. Even though f is not defined at 0, we can ask what happens to f as x gets close to 0. Dividing by the common factor of x, f can be expressed as

$$f(x)=\frac{2x^2+x}{x}=\begin{cases} 2x+1, & \text{if } x\neq 0 \\ \text{undefined}, & \text{if } x=0. \end{cases}$$

As x gets close (but is not equal) to 0, we expect $f(x)$ to be close to 1. For example, can values of x be chosen sufficiently close to 0 to force $f(x)$ to be within .01 of 0? Absolute values are helpful here. We want

$$|f(x)-1|<.01 \qquad (x\neq 0),$$

[6] Carl B. Boyer, *The History of the Calculus and Its Conceptual Development* (New York: Dover Publications, Inc., 1959), p. 256.

[7] Judith V. Grabiner, *The Origins of Cauchy's Rigorous Calculus* (Cambridge, Mass.: MIT Press, 1981), p. 84.

[8] Boyer, *The History of the Calculus*, p. 272.

which for our function is $|(2x+1)-1| < .01$, where $x \neq 0$. Simplifying, we see that we need

$$|2x| < .01 \qquad (x \neq 0).$$

Can x be chosen close enough to 0 to make $|2x| < .01$? In other words, is there a number $\delta > 0$ such that if the distance between x and 0 is less than δ then $|2x| < .01$? We must find $\delta > 0$ such that

$$\text{if } \ |x-0| < \delta, \quad \text{then } \ |2x| < .01\,.$$

But,

$$|2x| < .01 \iff 2|x| < .01 \iff |x| < .005 \iff |x-0| < .005\,.$$

If we choose δ to be any positive number less than or equal to .005, then

$$|x-0| < \delta \Longrightarrow |x-0| < .005 \Longrightarrow |2x| < .01 \Longrightarrow |f(x)-1| < .01\,.$$

Except that $x = 0$ satisfies the first inequality, since $|0-0| < \delta$, but it does not satisfy the last inequality ($|f(0)-1| < .01$ is meaningless because $f(0)$ is not even defined). We must omit 0 from consideration. This is conveniently done by writing $0 < |x-0| < \delta$ ($x = 0$ does not satisfy this inequality, but every other value of x within δ of 0 does). Now our argument works without exception:

$$\text{if } \ 0 < |x-0| < \delta, \quad \text{then } \ |f(x)-1| < .01\,.$$

There is nothing special about .01 here. Choosing an arbitrary positive distance ϵ, an argument entirely analogous to the preceding one shows that we can make $|f(x)-1| < \epsilon$ by choosing x such that $0 < |x-0| < \epsilon/2$. (Try it!) In Cauchy's terms, we have made $f(x)$ as close as we wish (within ϵ) to 1. Weierstrass's definition makes the ideas in this argument precise.

DEFINITION 2.2.1 *Let f be a function defined at each point in some open interval containing a, except possibly at a itself. Then a number L is the* ***limit of f at*** *a if for every number $\epsilon > 0$, there is a number $\delta > 0$ such that*

$$\textit{if } \ 0 < |x-a| < \delta, \quad \textit{then } \ |f(x)-L| < \epsilon\,.$$

This is denoted by writing $\lim_{x\to a} f(x) = L$.

Two points are worth noting. First, since x must satisfy $0 < |x-a|$, we never consider $x = a$ in the limit calculation. This means that the value of f at a (if f is even defined there) is irrelevant. Remember that it was "$x = a$" (that is, $x = 0$ and $t_1 = t_0$) that got us into trouble in the preceding two examples. Second, showing that a limit exists depends on our ability to find an appropriate δ. That is, $\epsilon > 0$ is arbitrary but assumed to be given. Can we (you!) find a $\delta > 0$ such that if $0 < |x-a| < \delta$, then $|f(x)-L| < \epsilon$?

2.2.3 Examples

EXAMPLE 2.2.1 Show that $\lim_{x\to 2} 3x - 1 = 5$.

SOLUTION Here $f(x) = 3x - 1$, $a = 2$, and $L = 5$. We must show that for every $\epsilon > 0$, there is a $\delta > 0$ such that

$$\text{if } \ 0 < |x-2| < \delta, \ \text{ then } \ |(3x-1) - 5| < \epsilon .$$

Finding δ is most easily accomplished by working backward as we did earlier. Manipulate the second inequality until it contains a term of the form $|x - 2|$ as in the first inequality. This is easy here. First

$$|(3x-1) - 5| = |3x - 6| = 3|x-2|,$$

so

$$|(3x-1)-5| < \epsilon \iff 3|x-2| < \epsilon \iff |x-2| < \epsilon/3.$$

We should choose $\delta = \epsilon/3$. Notice that the requirement that δ be greater than 0 is met, since $\epsilon > 0$.

We can check that our choice of δ works: given $\epsilon > 0$, let $\delta = \epsilon/3$. Then

$$0 < |x-2| < \delta \Longrightarrow |x-2| < \epsilon/3 \Longrightarrow |3x-6| < \epsilon \Longrightarrow |(3x-1)-5| < \epsilon. \ \square$$

Here is an obvious but often overlooked point. Suppose in the preceding example that we had chosen a number δ' such that $0 < \delta' < \delta = \epsilon/3$ (δ' is positive but strictly smaller than $\epsilon/3$). Then δ' also "works." If $0 < |x-2| < \delta'$, we still have $|x-2| < \epsilon/3$, and the argument proceeds exactly as before. That is, *if one value of δ can be found, any smaller positive number δ' will also satisfy the requirements of the limit definition.*

EXAMPLE 2.2.2 Show that $\lim_{x\to -1} 2 - 4x = 6$.

SOLUTION The idea is the same as in Example 2.2.1, but here we have to keep track of signs in the various inequalities. This time, given $\epsilon > 0$, find $\delta > 0$ such that

$$\text{if } \ 0 < |x-(-1)| < \delta, \ \text{ then } \ |(2-4x) - 6| < \epsilon,$$

or equivalently,

$$\text{if } \ 0 < |x+1| < \delta, \ \text{ then } \ |-4x-4| < \epsilon .$$

Working backward,

$$|-4x-4| = |-4||x+1| = 4|x+1|,$$

so

$$|-4x-4| < \epsilon \iff 4|x+1| < \epsilon \iff |x+1| < \epsilon/4.$$

Choose $\delta = \epsilon/4$. If you incorrectly factored out -4 instead of $|-4|$ at the second stage, you would have produced $\delta = -\epsilon/4$, which is nonsense; δ represents a distance and must be positive. As a check, given $\epsilon > 0$, let $\delta = \epsilon/4$. Then

$$0 < |x+1| < \delta \Longrightarrow |x+1| < \epsilon/4 \Longrightarrow |4x+4| < \epsilon \Longrightarrow |-4x-4| < \epsilon. \quad \square$$

EXAMPLE 2.2.3 Show that $\lim_{x\to 0} x\sin(\frac{1}{x}) = 0$.

SOLUTION First notice that $f(x) = x\sin(\frac{1}{x})$ is not even defined at 0. Second, $f(x)$ oscillates an infinite number of times in any open interval containing the origin. To see this, let n be any integer. Then

$$f\left(\frac{1}{n\pi+\pi/2}\right) = \frac{1}{n\pi+\pi/2}\sin(n\pi+\pi/2) = \begin{cases} \frac{1}{n\pi+\pi/2}, & \text{if } n \text{ is even} \\ \frac{-1}{n\pi+\pi/2}, & \text{if } n \text{ is odd.} \end{cases}$$

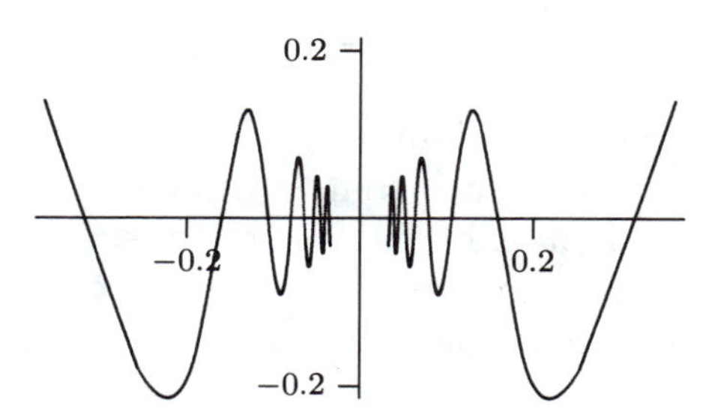

Figure 2.2.2. The graph of $f(x) = x\sin(1/x)$.

L'Huilier's notion of limit would not apply to this type of oscillating function. But it is quite easy to work with the modern definition of limit here. Given $\epsilon > 0$, find $\delta > 0$ such that

$$\text{if } 0 < |x-0| < \delta, \text{ then } \left|x\sin(\tfrac{1}{x}) - 0\right| < \epsilon,$$

or equivalently,

$$\text{if } 0 < |x| < \delta, \text{ then } \left|x\sin(\tfrac{1}{x})\right| < \epsilon.$$

Working backward doesn't get us very far. If we could eliminate the factor of $\sin(\frac{1}{x})$, we would be left with just $|x|$ and we could proceed as in the preceding two examples. This is easily done because the sine function is *bounded.* For any $x \neq 0$, $-1 \leq \sin(\frac{1}{x}) \leq 1$, so $\left|\sin(\frac{1}{x})\right| \leq 1$. Therefore

$$\left|x \sin(\tfrac{1}{x})\right| = |x| \left|\sin(\tfrac{1}{x})\right| \leq |x|(1) = |x| .$$

If we now force this last term to be less than ϵ, we have

$$|x| \left|\sin(\tfrac{1}{x})\right| \leq |x| < \epsilon \Longrightarrow |x| < \epsilon.$$

So we can choose $\delta = \epsilon$.

Check: given $\epsilon > 0$, let $\delta = \epsilon$. Then $0 < |x - 0| < \delta \Longrightarrow |x| < \epsilon$. This implies that

$$\left|x \sin(\tfrac{1}{x}) - 0\right| = |x| \left|\sin(\tfrac{1}{x})\right| \leq |x|(1) = |x| < \epsilon. \quad \square$$

The crucial step in the proof was being able to replace the complicated factor $\sin(\frac{1}{x})$ which was unrelated to the desired factor $|x - 0|$ by a numerical bound. We continue with several more such examples.

EXAMPLE 2.2.4 Show that if $a > 0$, then $\lim_{x \to a} \sqrt{x} = \sqrt{a}$.

SOLUTION Given $\epsilon > 0$, find $\delta > 0$ such that

$$\text{if} \quad 0 < |x - a| < \delta, \quad \text{then} \quad |\sqrt{x} - \sqrt{a}| < \epsilon.$$

Notice that the domain of $f(x) = \sqrt{x}$ consists of all $x \geq 0$. We must be sure to choose δ such that any value of x that satisfies $0 < |x - a| < \delta$ lies in the domain of f. But $|x - a| < \delta$ means $a - \delta < x < a + \delta$. Since we need $0 \leq x$, we must have $0 \leq a - \delta$, or $\delta < a$. That is, any δ we end up choosing must be less than or equal to a. Working backward now

$$|\sqrt{x} - \sqrt{a}| = |\sqrt{x} - \sqrt{a}| \left| \frac{\sqrt{x} + \sqrt{a}}{\sqrt{x} + \sqrt{a}} \right| = \frac{|x - a|}{|\sqrt{x} + \sqrt{a}|} .$$

We used conjugation to get the desired factor of $|x - a|$, but as in Example 2.2.3, there is an unwanted factor: $\frac{1}{|\sqrt{x} + \sqrt{a}|}$. Since $x \geq 0$,

$$\frac{1}{|\sqrt{x} + \sqrt{a}|} \leq \frac{1}{\sqrt{a}} .$$

This implies that

$$\frac{|x - a|}{|\sqrt{x} + \sqrt{a}|} \leq \frac{|x - a|}{\sqrt{a}} .$$

We can now force $\frac{|x-a|}{|\sqrt{x}+\sqrt{a}|}$ to be less than ϵ by forcing $\frac{|x-a|}{\sqrt{a}} < \epsilon$. This is equivalent to taking

$$|x-a| < \epsilon\sqrt{a}\,.$$

We can choose $\delta = \epsilon\sqrt{a}$, but remember that δ must also be less than a. So we should choose δ to be the minimum of the two numbers a and $\epsilon\sqrt{a}$. This is denoted by $\delta = \min\{a, \epsilon\sqrt{a}\}$.

Check: given $\epsilon > 0$, let $\delta = \min\{a, \epsilon\sqrt{a}\}$. Then $0 < |x-a| < \delta$ implies that

$$|\sqrt{x}-\sqrt{a}| = \frac{|x-a|}{|\sqrt{x}+\sqrt{a}|} \leq \frac{|x-a|}{\sqrt{a}} < \frac{\delta}{\sqrt{a}} \leq \frac{\epsilon\sqrt{a}}{\sqrt{a}} = \epsilon\,.$$

We used both restrictions on δ: $\delta \leq a$ means $\sqrt{x}$ will always be defined and $\delta \leq \epsilon\sqrt{a}$ provides the needed inequality at the end of the argument. This trick of letting δ be the minimum of a set of values will be used repeatedly.

Finally, remember that the expression $\lim_{x\to a} f(x) = L$ requires the function f to be defined on both sides of a. Consequently, even though $\sqrt{0} = 0$, $\lim_{x\to 0}\sqrt{x}$ *does not exist.* For any $\delta > 0$, there are always negative values of x that satisfy $0 < |x-0| < \delta$ and are not in the domain of the square root function. We will return to this problem in Section 2.4. ☐

EXAMPLE 2.2.5 Show that $\lim_{x\to -2} x^2+1 = 5\,.$

SOLUTION There are no problems with the domain this time. So given $\epsilon > 0$, find $\delta > 0$ such that

$$\text{if } \; 0 < |x-(-2)| < \delta, \;\; \text{then } \; |(x^2+1)-5| < \epsilon,$$

or equivalently,

$$\text{if } \; 0 < |x+2| < \delta, \;\; \text{then } \; |x^2-4| < \epsilon.$$

Work backward to get a factor of $|x+2|$:

$$|x^2-4| = |x-2||x+2|.$$

Again there is an unwanted factor, $|x-2|$, that must be bounded. When x is close to -2, $|x-2|$ is nearly 4. This needs to be made precise. If we make certain that $\delta < 1$ (so that x is within a unit of -2), then

$$|x+2| < \delta \Longrightarrow |x+2| < 1 \Longrightarrow -1 < x+2 < 1\,.$$

We manipulate the last inequality by subtracting 4 to get a factor of $x-2$:

$$-5 < x-2 < -3 \Longrightarrow 3 < |x-2| < 5\,.$$

In other words, the constraint $|x+2| < 1$ simultaneously bounds $|x-2|$ by 5. So if $\delta < 1$, then

$$|x^2-4| = |x-2||x+2| < 5|x+2|\,.$$

Thus $|x^2-4|$ can be made smaller than ϵ by taking $5|x+2| < \epsilon$, or

$$|x+2| < \epsilon/5\,.$$

Select $\delta = \min\{1,\, \epsilon/5\}$.

You should check that this value of δ works as we have done in preceding examples. Other forms of this answer are possible. Suppose that you had chosen $\delta < 2$ to start with instead of $\delta < 1$. Show that you would end up with $\delta = \min\{2,\, \epsilon/6\}$. ▯

EXAMPLE 2.2.6 Show that $\lim_{x\to 2} \dfrac{1}{2x+1} = \dfrac{1}{5}$.

SOLUTION This time we must not allow $x = -1/2$. Since the distance between 2 and $-1/2$ is $5/2$, we must constrain δ to be smaller than $5/2$. If we take $\delta < 2$ to start with, then

$$\left|\frac{1}{2x+1} - \frac{1}{5}\right| = \left|\frac{5-(2x+1)}{5(2x+1)}\right| = \left|\frac{4-2x}{5(2x+1)}\right| = \frac{|-2|}{5}\left|\frac{x-2}{2x+1}\right|.$$

We need to bound the unwanted factor of $\frac{1}{|2x+1|}$. But

$$|x-2| < \delta < 2 \Longrightarrow -2 < x-2 < 2\,.$$

To get a factor of $2x+1$, we manipulate this last inequality:

$$2 < x-2 < 2 \Longrightarrow 0 < x < 4 \Longrightarrow 0 < 2x < 8 \Longrightarrow 1 < 2x+1 < 9.$$

The last inequality yields

$$\frac{1}{9} < \frac{1}{2x+1} < 1\,.$$

The factor is bounded above by 1. Now we can finish the argument:

$$\left|\frac{1}{2x+1} - \frac{1}{5}\right| = \frac{2}{5}\left|\frac{x-2}{2x+1}\right| \le \frac{2}{5}|x-2|(1) = \frac{2|x-2|}{5}\,.$$

Clearly we need $|x-2| < 5\epsilon/2$. Choose $\delta = \min\{2,\, 5\epsilon/2\}$. ▯

To show that a function f *does not have any limit at a*, we often use a proof by contradiction.

EXAMPLE 2.2.7 Show that $\lim_{x \to 0} f(x)$ does not exist, where

$$f(x) = \begin{cases} 1, & \text{if } x \geq 0 \\ -1, & \text{if } x < 0. \end{cases}$$

SOLUTION Notice in Figure 2.2.3 that the two "halves" of the graph f are separated by 2 units (vertically) near 0. That is, the values of $f(x)$ do not approach a common value as x approaches 0.

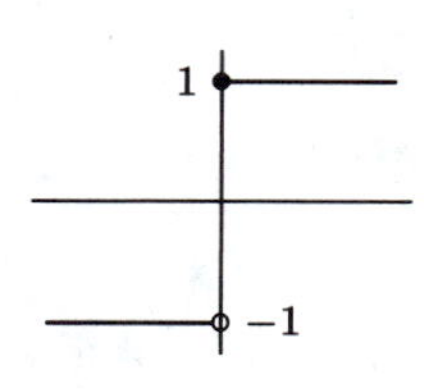

Figure 2.2.3

To make this precise, assume that $\lim_{x \to 0} f(x) = L$. Choose $\epsilon = 1$ (any positive ϵ less than 1 will also do). By the definition of limit, for the given ϵ there is a $\delta > 0$ such that if $0 < |x - 0| < \delta$, then $|f(x) - L| < \epsilon$. Let $x_1 = \delta/2$ and $x_2 = -\delta/2$. Both x_1 and x_2 satisfy $0 < |x - 0| < \delta$, so $|f(x_1) - L| < \epsilon$ and $|f(x_2) - L| < \epsilon$. Further, $f(x_1) = 1$ and $f(x_2) = -1$, so $f(x_1) - f(x_2) = 1 - (-1) = 2$. Combining these facts and using the triangle inequality yields

$$\begin{aligned} 2 = |f(x_1) - f(x_2)| &= |f(x_1) - L + L - f(x_2)| \\ &\leq |f(x_1) - L| + |L - f(x_2)| \\ &< \epsilon + \epsilon = 1 + 1 = 2. \end{aligned}$$

But this gives the contradiction that $2 < 2$. So no limit at 0 can exist. ◻

EXAMPLE 2.2.8 Show that $\lim_{x \to 0} \left|\frac{1}{x}\right|$ does not exist.

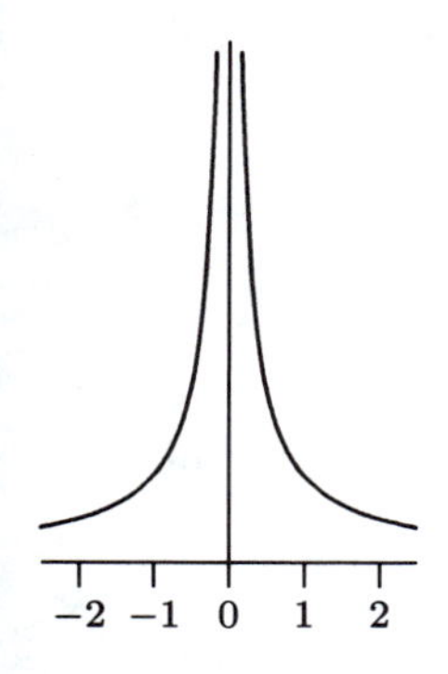

Figure 2.2.4

Figure 2.2.4 indicates that $\left|\frac{1}{x}\right|$ increases without bound as x gets close to 0. Further, small changes in x near 0 produce large changes in $f(x)$. We will use this observation to construct a proof by contradiction. Assume that $\lim_{x \to 0} \left|\frac{1}{x}\right| = L$. Choose $\epsilon = 1$. Then for this ϵ there is a $\delta > 0$ such that if $0 < |x - a| < \delta$, then $|f(x) - L| < \epsilon$. By the Archimedean property of $\mathbf{R}$, there is a positive integer N such that $\frac{1}{N} < \delta$. Now let $x_1 = \frac{1}{N}$ and $x_2 = \frac{1}{N+2}$. Both x_1 and x_2 satisfy $0 < |x - 0| < \delta$, so $|f(x_1) - L| < \epsilon$ and $|f(x_2) - L| < \epsilon$. Further, $f(x_1) = N$ and $f(x_2) = N + 2$, so $f(x_1) - f(x_2) = -2$. Using these facts and the triangle inequality yields

$$\begin{aligned} 2 = |f(x_1) - f(x_2)| &= |f(x_1) - L + L - f(x_2)| \\ &\leq |f(x_1) - L| + |L - f(x_2)| < \epsilon + \epsilon = 1 + 1 = 2. \end{aligned}$$

Again we have the contradiction that $2 < 2$. No limit at 0 can exist. ◻

To show just how badly functions can behave, we prove that the Dirichlet function has no limit at any point even though it is defined for all real numbers.

EXAMPLE 2.2.9 Show that for any number a, $\lim_{x\to a} D(x)$ does not exist where

$$D(x) = \begin{cases} 1, & \text{if } x \text{ is rational} \\ 0, & \text{if } x \text{ is irrational.} \end{cases}$$

SOLUTION The ideas involved are similar to those in the preceding two examples. This time the two 'halves' of the function are separated by a single unit. That is, the value of $D(x)$ is either 0 or 1, so no limit value L can be within 1/2 of both at the same time.

Assume that $\lim_{x\to a} D(x) = L$. Choose $\epsilon = 1/2$. Then for this ϵ there is a $\delta > 0$ such that if $0 < |x - a| < \delta$ then $|D(x) - L| < \epsilon$. But for *any* $\delta > 0$, there is a rational number x_1 between a and $a + \delta$. Similarly there is an irrational number x_2 between a and $a + \delta$. Both x_1 and x_2 satisfy $0 < |x - a| < \delta$, so $|f(x_1) - L| < \epsilon$ and $|f(x_2) - L| < \epsilon$. Further, $f(x_1) = 1$ and $f(x_2) = 0$, so $f(x_1) - f(x_2) = 1$. These facts and the triangle inequality yield

$$\begin{aligned} 1 = |f(x_1) - f(x_2)| &= |f(x_1) - L + L - f(x_2)| \\ &\le |f(x_1) - L| + |L - f(x_2)| < \epsilon + \epsilon = 1. \end{aligned}$$

This time the contradiction is that $1 < 1$. So no limit at a can exist. ∎

Problems for Section 2.2

2.2.1. Using the ϵ, δ definition of limit, show that

a) $\lim_{x\to 2} 4x + 1 = 9$ **b)** $\lim_{x\to -1} 1 - 2x = 3$

c) $\lim_{x\to 5} \sqrt{x+4} = 3$ **d)** $\lim_{x\to 2} x^2 - x = 2$

e) $\lim_{x\to 3} \dfrac{1}{8 - 4x} = -1/4$ **f)** $\lim_{x\to 1} \dfrac{1}{\sqrt{x+3}} = 1/2$

g) $\lim_{x\to 0} x^2(\sin x + \cos x) = 0$ **h)** $\lim_{x\to 0} \dfrac{1}{x^2+1} = 1$

2.2.2. Assume that g is a bounded function, that is, $|g(x)| < B$ for all x. Prove that $\lim_{x\to 0} xg(x) = 0$.

2.2.3. Let $G(x)$ denote the greatest integer less than or equal to x.

a) Show that if $n \in \mathbf{Z}$, then $\lim_{x\to n} G(x)$ does not exist. (Hint: Draw the graph of $G(x)$ and use the type of argument given in Example 2.2.7.)

b) Show that if $a \notin \mathbf{Z}$, then $\lim_{x\to a} G(x) = G(a)$. (Hint: Begin by sufficiently restricting δ so that your attention is on a single "step" of $G(x)$.)

2.2.4. **a)** Show that $\lim_{x\to 0} \frac{1}{x}$ does not exist. (Hint: Review Example 2.2.8.)

b) Show that $\lim_{x\to 0} \frac{1}{x^2}$ does not exist.

c) Let $G(x)$ be the greatest integer function. Show that $\lim_{x\to 2} \frac{1}{G(x)}$ does not exist.

2.2.5. Consider the modified Dirichlet function $F(x) = \begin{cases} x, & \text{if } x \text{ is rational} \\ 0, & \text{if } x \text{ is irrational.} \end{cases}$

a) Show that $\lim_{x\to 0} F(x) = 0$.

b) Show that if $a \neq 0$, then $\lim_{x\to a} F(x)$ does not exist.

2.2.6. In showing that limits did not exist in Examples 2.2.7–9, we used the following idea: if $\lim_{x\to a} f(x) = L$ and if x_1 and x_2 are "sufficiently close" to a, both $f(x_1)$ and $f(x_2)$ must be "approximately" L. That is, $|f(x_1) - f(x_2)|$ must be "small." This can be made precise.

Theorem: *If $\lim_{x\to a} f(x) = L$, then for every $\epsilon > 0$ there is $\delta > 0$ such that, for every pair of numbers x_1 and x_2 that both satisfy $0 < |x - a| < \delta$, then $|f(x_1) - f(x_2)| < \epsilon$.*

Justify each step in the following proof.

a) Given $\epsilon > 0$, show that there is a $\delta > 0$ such that if $0 < |x - a| < \delta$, then $|f(x) - L| < \epsilon/2$.

b) Let x_1 and x_2 be any numbers that satisfy $0 < |x - a| < \delta$. Show that

$$|f(x_1) - f(x_2)| = |f(x_1) - L + L - f(x_2)| \leq |f(x_1) - L| + |f(x_2) - L|.$$

c) Show that $|f(x_1) - L| + |f(x_2) - L| < \epsilon$.

d) Show that $|f(x_1) - f(x_2)| < \epsilon$.

2.2.7. Can you prove the converse to the preceding theorem? What problem do you immediately encounter?

•2.2.8. **The Squeeze Theorem:** *Assume that $f(x) \leq g(x) \leq h(x)$ for all x in some open interval containing a, except perhaps at a itself. If $\lim_{x\to a} f(x) = \lim_{x\to a} h(x) = L$, then $\lim_{x\to a} g(x)$ exists and equals L, also.*

Justify each step in the following proof of the squeeze theorem.

a) Given $\epsilon > 0$ show that there exists a $\delta_1 > 0$ such that if $0 < |x - a| < \delta_1$, then $|f(x) - L| < \epsilon$.

b) Show that if $0 < |x - a| < \delta_1$, then $L - \epsilon < f(x)$.

c) Show that there exists a $\delta_2 > 0$ such that if $0 < |x - a| < \delta_2$, then $h(x) < L + \epsilon$.

d) Let $\delta = \min\{\delta_1, \delta_2\}$. Show that if $0 < |x - a| < \delta$, then $L - \epsilon < g(x) < L + \epsilon$.

e) Complete the proof by showing that if $0 < |x - a| < \delta$, then $|g(x) - L| < \epsilon$.

•2.2.9. **a)** Suppose that $f(x) \leq 0$ for all x (except perhaps at $x = a$). Show that if $\lim_{x\to a} f(x) = L$, then $L \leq 0$. (Hint: Assume instead that $L > 0$. Let $\epsilon = L/2$ and derive a contradiction.)

b) State and prove the analogue for $f(x) \geq 0$.

•2.2.10. Suppose that f is the constant function defined by $f(x) = c$ for all x. Show that for any real number a, $\lim_{x\to a} f(x) = c$.

•2.2.11. Show that for any real number a, $\lim_{x\to a} x = a$.

•2.2.12. Let $f(x) = \sin(\frac{1}{x})$. Use a proof by contradiction to show that $\lim_{x\to 0} f(x)$ does not exist. Hint: The key is to notice that $f(x)$ oscillates rapidly between -1 and 1 near the origin. In fact, if n is any integer,

$$f(\tfrac{1}{n\pi+\pi/2}) = \sin(n\pi + \pi/2) = \begin{cases} 1, & \text{if } n \text{ is even} \\ -1, & \text{if } n \text{ is odd.} \end{cases}$$

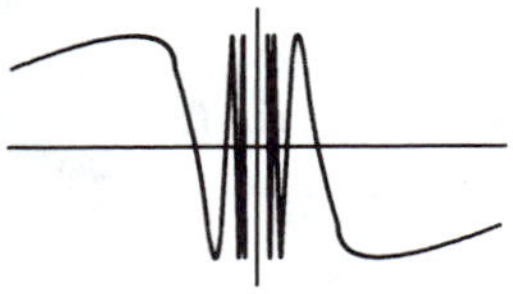

Figure 2.2.5

2.3 Limit Theory

In this section we prove a number of elementary results that make calculating certain limits quite easy. The first theorem says that a function can approach at most one limit at a. In other words, it is impossible to be "infinitesimally" close to two distinct numbers.

THEOREM 2.3.1 *If $\lim_{x\to a} f(x)$ exists, then this limit is unique. That is, if L and M are both limits of f at a, then $L = M$.*

PROOF The proof is by contradiction. Assume that $L \neq M$. Let $\epsilon = |L - M|/2$. L is a limit for f at a, so there is a $\delta_1 > 0$ such that

$$\text{if } \ 0 < |x-a| < \delta_1, \ \text{ then } \ |f(x) - L| < \epsilon.$$

M is also a limit for f at a, so there is a $\delta_2 > 0$ such that

$$\text{if } \ 0 < |x-a| < \delta_2, \ \text{ then } \ |f(x) - M| < \epsilon.$$

There is no need for δ_1 and δ_2 to be the same. (Why?) However, if we let $\delta = \min\{\delta_1, \delta_2\}$, both the inequalities are satisfied. That is,

$$\text{if } \ 0 < |x-a| < \delta, \ \text{ then } \ |f(x) - L| < \epsilon \ \text{ and } \ |f(x) - M| < \epsilon.$$

To finish the proof, use the method employed earlier to show that certain limits do not exist. If $0 < |x - a| < \delta$, then by the triangle inequality,

$$\begin{aligned} 2\epsilon = |L - M| &= |L - f(x) + f(x) - M| \\ &\leq |f(x) - L| + |f(x) - M| < \epsilon + \epsilon = 2\epsilon. \end{aligned}$$

We have reached the contradiction that $2\epsilon < 2\epsilon$. So our original assumption about L and M is incorrect and L must equal M. ∎

The next few theorems are concerned with what happens to functions and their limits under simple algebraic operations.

THEOREM 2.3.2 ***(The Sum Theorem)*** *If $\lim_{x \to a} f(x) = L$ and $\lim_{x \to a} g(x) = M$, then*

a) $\lim_{x \to a}(f + g)(x) = L + M$;

b) $\lim_{x \to a}(f - g)(x) = L - M$.

PROOF We prove **(a)** and leave **(b)** as an exercise. Let $\epsilon > 0$ be given. Since $(f + g)(x) = f(x) + g(x)$, we must find $\delta > 0$ such that

$$\text{if } \ 0 < |x - a| < \delta, \quad \text{then } \ |(f(x) + g(x)) - (L + M)| < \epsilon.$$

It's natural to re-associate terms and to use the triangle inequality here:

$$\begin{aligned} |(f(x) + g(x)) - (L + M)| &= |(f(x) - L) + (g(x) - M)| \\ &\leq |f(x) - L| + |g(x) - M|. \end{aligned}$$

If each of these last two terms can be made smaller than $\epsilon/2$, the proof can be completed. But this is easy! By the definition of limit, there is a $\delta_1 > 0$ such that

$$\text{if } \ 0 < |x - a| < \delta_1, \quad \text{then } \ |f(x) - L| < \epsilon/2,$$

and there is a $\delta_2 > 0$ such that

$$\text{if } \ 0 < |x - a| < \delta_2, \quad \text{then } \ |g(x) - M| < \epsilon/2.$$

Choosing $\delta = \min\{\delta_1, \delta_2\}$, both inequalities hold. If $0 < |x - a| < \delta$, then

$$|(f(x) + g(x)) - (L + M)| \leq |f(x) - L| + |g(x) - M| < \epsilon/2 + \epsilon/2 = \epsilon. \quad ∎$$

The first type of limit of a product that we consider is multiplying a function by a consant.

THEOREM 2.3.3 ***(The Constant Multiple Theorem)*** *Assume that* $\lim_{x\to a} f(x) = L$. *Then for any constant c,* $\lim_{x\to a} cf(x) = cL$.

The proof is an elementary exercise in working with the limit definition and is outlined in the problem section. We now consider the more general situation of the limit of a product of two functions.

THEOREM 2.3.4 ***(The Product Theorem)*** *Assume that* $\lim_{x\to a} f(x) = L$ *and that* $\lim_{x\to a} g(x) = M$. *Then* $\lim_{x\to a}(f \cdot g)(x) = LM$.

PROOF The triangle inequality provides the key to the proof. Notice that

$$\begin{aligned} |(f \cdot g)(x) - LM| &= |(f(x)g(x) - f(x)M) + (f(x)M - LM)| \\ &\leq |f(x)g(x) - f(x)M| + |f(x)M - LM| \\ &= |f(x)| \cdot |g(x) - M| + |M| \cdot |f(x) - L|. \end{aligned} \tag{1}$$

Now we show how to make each part of the last sum small.

Since $\lim_{x\to a} f(x) = L$, there exists $\delta_1 > 0$ such that

$$0 < |x - a| < \delta_1 \Longrightarrow |f(x) - L| < 1.$$

By problem 1.3.20, if $|f(x) - L| < 1$, then $|f(x)| < |L| + 1$. So

$$0 < |x - a| < \delta_1 \Longrightarrow |f(x)| < |L| + 1.$$

Now let $m = \max\{|L| + 1, |M|\}$. Then from (1) we obtain the inequality

$$|(f \cdot g)(x) - LM| \leq m|g(x) - M| + m|f(x) - L|.$$

Given any $\epsilon > 0$, there exists $\delta_2 > 0$ and $\delta_3 > 0$ such that

$$0 < |x - a| < \delta_2 \Longrightarrow |f(x) - L| < \epsilon/2m$$

and

$$0 < |x - a| < \delta_3 \Longrightarrow |g(x) - M| < \epsilon/2m.$$

Let $\delta = \min\{\delta_1, \delta_2, \delta_3\}$. Now we can complete the proof. Whenever $0 < |x - a| < \delta$, then

$$\begin{aligned} |f(x)g(x) - LM| &< m|g(x) - M| + m|f(x) - L| \\ &< m \cdot \frac{\epsilon}{2m} + m \cdot \frac{\epsilon}{2m} = \epsilon. \quad \blacksquare \end{aligned}$$

Using the results developed so far, we can show that limits involving polynomials are easy to calculate. From problem 2.2.10, $\lim_{x\to a} x = a$. So by the product theorem,

$$\lim_{x\to a} x^2 = (\lim_{x\to a} x)(\lim_{x\to a} x) = a \cdot a = a^2.$$

In fact a simple induction argument (see problem 1 at the end of this section) shows that for any positive integer n

$$\lim_{x \to a} x^n = a^n.$$

The constant multiple theorem implies that $\lim_{x \to a} cx^n = ca^n$. Now consider a general polynomial:

$$p(x) = c_n x^n + c_{n-1} x^{n-1} + \cdots + c_1 x + c_0.$$

Using the sum theorem and the results just established,

$$\lim_{x \to a} p(x) = c_n a^n + c_{n-1} a^{n-1} + \cdots + c_1 a + c_0 = p(a).$$

That is, we have just shown the following:

COROLLARY 2.3.5 *If $p(x)$ is a polynomial, then $\lim_{x \to a} p(x) = p(a)$. That is, a polynomial has a limit at every point a and this limit is just the value of the polynomial at that point.*

Recall how hard we worked to show that $\lim_{x \to 2} x^2 + 1 = 5$ in Example 2.2.5. This result is now a trivial application of Corollary 2.3.5. In contrast, to show that $\lim_{x \to 2} \frac{1}{2x+1} = \frac{1}{5}$ (Example 2.2.6) is not a consequence of the corollary or any of the preceding results, since $\frac{1}{2x+1}$ is a rational function and not a polynomial. We need a quotient theorem for limits to take care of this situation.

THEOREM 2.3.6 ***(The Quotient Theorem)*** *Assume that $\lim_{x \to a} f(x) = L$ and that $\lim_{x \to a} g(x) = M$ and $M \neq 0$. Then*

$$\lim_{x \to a} \left(\frac{f}{g} \right)(x) = \frac{L}{M}.$$

PROOF We proceed as in the proof of the product theorem for limits except that we must be careful to avoid division by 0. With this is mind, since $\lim_{x \to a} g(x) = M \neq 0$, there is a $\delta_1 > 0$ such that if $0 < |x - a| < \delta_1$, then $|g(x) - M| < |M|/2$. By problem 1.3.20 it follows that

$$|M|/2 < |g(x)|. \tag{2}$$

This shows that $g(x) \neq 0$. So if $0 < |x - a| < \delta_1$, by first using (2) and then the triangle inequality (after having added and subtracted the common term LM), we obtain

$$\begin{aligned}
\left|\frac{f(x)}{g(x)} - \frac{L}{M}\right| = \left|\frac{f(x)M - g(x)L}{g(x)M}\right| &< \frac{2|f(x)M - g(x)L|}{|M|^2} \\
&= \frac{2|f(x)M - LM + LM - g(x)L|}{|M|^2} \\
&\leq \frac{2|f(x)M - LM|}{|M^2|} + \frac{2|g(x)L - LM|}{M^2} \\
&= \frac{2|f(x) - L|}{|M|} + \frac{2|L||g(x) - M|}{M^2}. \qquad (3)
\end{aligned}$$

So the problem is reduced to making each of the last summands small.

Given any $\epsilon > 0$, there exists a $\delta_2 > 0$ such that if $0 < |x - a| < \delta_2$, then

$$|f(x) - L| < |M|\epsilon/4.$$

Similarly there exists a $\delta_3 > 0$ such that if $0 < |x - a| < \delta_3$ and $L \neq 0$, then

$$|g(x) - M| < M^2\epsilon/4|L|.$$

Let $\delta = \min\{\delta_1, \delta_2, \delta_3\}$. If $0 < |x-a| < \delta$, then by (3) and the foregoing inequalities,

$$\left|\frac{f(x)}{g(x)} - \frac{L}{M}\right| < \frac{2|f(x) - L|}{|M|} + \frac{2|L||g(x) - M|}{M^2} < \frac{2|M|\epsilon}{4|M|} + \frac{2|L|M^2\epsilon}{4|L|M^2} = \epsilon.$$

It is a simple exercise to modify the proof if L happens to be 0. ∎

Problems for Section 2.3

•2.3.1. Let **N** denote the set of natural numbers: $\mathbf{N} = \{1, 2, 3, \ldots\}$. A basic property of **N** is the **principle of mathematical induction**. Suppose that $P(n)$ means that property or formula P holds for the natural number n. The principle of mathematical induction says that $P(n)$ holds for every natural number n if the following two conditions are satisfied:

i) $P(1)$ holds;

ii) If $P(k)$ holds, then $P(k+1)$ holds.

It is easy to see why induction works. Using (i), $P(1)$ holds; now using (ii), since $P(1)$ is true, so is $P(2)$. Using (ii) again, since $P(2)$ is true, so is $P(3)$, and so on. In this way every natural number n will be reached in a finite number of steps; so $P(n)$ holds for every $n \in \mathbf{N}$.

a) Use induction to prove that for every natural number n, $\lim_{x\to a} x^n = a^n$. Here's an outline of how to proceed.

i) Let $P(n)$ be the statement $\lim_{x\to a} x^n = a^n$.
ii) Verify that $P(1)$ holds.
iii) Assume that $P(k)$ holds, that is, $\lim_{x\to a} x^k = a^k$. Show that this forces $P(k+1)$ to be true (i.e., $\lim_{x\to a} x^{k+1} = a^{k+1}$). (Hint: Factor x^{k+1} as $x \cdot x^k$ and use the product theorem for limits and the fact that $P(1)$ and $P(k)$ hold.)

b) Show that for any positive integer n, $\lim_{x\to a} x^{-n} = a^{-n}$ (for $a \neq 0$).

2.3.2. In Example 2.2.4 we proved that if $a > 0$, then $\lim_{x\to a} x^{1/2} = a^{1/2}$.

a) Use the product theorem to show that if $a > 0$, then $\lim_{x\to a} x^{3/2} = a^{3/2}$.

b) Use induction to show that for any $n \in \mathbf{N}$, if $a > 0$, then $\lim_{x\to a} x^{n+1/2} = a^{n+1/2}$.

c) Show that for any $n \in \mathbf{N}$, if $a > 0$, then $\lim_{x\to a} x^{-n+1/2} = a^{-n+1/2}$.

2.3.3. Let $r(x)$ be any rational function, that is, $r(x) = \frac{p(x)}{q(x)}$, where $p(x)$ and $q(x)$ are both polynomials. Show that if $q(a) \neq 0$, then $\lim_{x\to a} r(x) = r(a)$.

2.3.4. Suppose that $\lim_{x\to a} f(x) = L$ and that $\lim_{x\to a} f(x) + g(x) = M$. Show that $\lim_{x\to a} g(x) = M - L$.

2.3.5. For each set of conditions, find functions f and g that satisfy them.

a) $\lim_{x\to a}(f(x) + g(x)) = 0$, but $\lim_{x\to a} f(x)$ and $\lim_{x\to a} g(x)$ do not exist.

b) $\lim_{x\to a} f(x)g(x)$ exists, but $\lim_{x\to a} f(x)$ and $\lim_{x\to a} g(x)$ do not exist.

c) $\lim_{x\to a} \frac{f(x)}{g(x)}$ exists, but $\lim_{x\to a} \frac{1}{g(x)}$ does not exist.

•2.3.6. Suppose that $g(x) \leq h(x)$ for all x (except perhaps at a). If $\lim_{x\to a} g(x) = M$ and $\lim_{x\to a} h(x) = N$, prove that $M \leq N$. (Hint: Let $f(x) = h(x) - g(x)$ and then use problem 2.2.8.)

•2.3.7. Prove Theorem 2.3.3, the constant multiple theorem.

a) Show that if $c = 0$, then cf is a constant function and problem 2.2.9 applies.

b) Assume $c \neq 0$. Given $\epsilon > 0$, show that there is a $\delta > 0$ such that

$$\text{if } 0 < |x - a| < \delta, \text{ then } |f(x) - L| < \epsilon/|c|.$$

c) Complete the proof by showing that for this same δ

$$\text{if } 0 < |x - a| < \delta, \text{ then } |cf(x) - cL| < \epsilon.$$

d) Give another proof of the constant multiple theorem using the product theorem.

•2.3.8. **a)** Prove **(b)** of the Theorem 2.3.2 (the sum theorem) by mimicking the proof of **(a)** of the same theorem.

b) Give another proof of **(b)** of the sum theorem by using the constant multiple theorem, and the fact that $f(x) - g(x)$ can be written as $f(x) + (-1)g(x)$.

2.4 Other Types of Limits

2.4.1 One-Sided Limits

There will be several situations in which we will want to know about the "limit" of a function f at a point a even though there is no positive δ such that $f(x)$ is defined for all x satisfying $0 < |x - a| < \delta$. We have seen such an example already: $\lim_{x\to 0} \sqrt{x}$ does not exist, since $\sqrt{x}$ is not defined in any open interval that contains 0.

Still, we want to distinguish the controlled behavior of the square root function at 0 (see Figure 2.4.1) from the uncontrolled behavior of $\frac{1}{x}$, which also has no limit at 0 (Figure 2.4.2). A simple modification of the limit definition allows us to make this distinction.

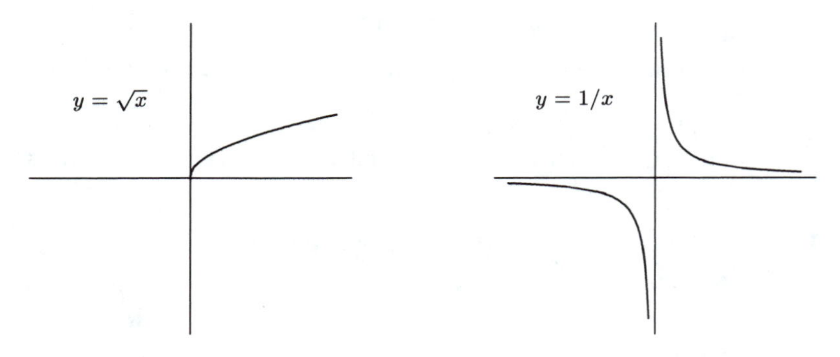

Figure 2.4.1 **Figure 2.4.2**

Definition 2.4.1 *Let f be defined on the open interval (a, c). The number L is the* ***limit of $f(x)$ as x approaches a from above*** *(or* ***from the right****) if for every $\epsilon > 0$, there is a $\delta > 0$ such that*

$$\text{if } \ 0 < x - a < \delta, \quad \text{then } \ |f(x) - L| < \epsilon.$$

When this limit exists, we will write $\lim_{x\to a^+} f(x) = L$.

Note that "$0 < x - a < \delta$" means the same as "$0 < |x - a| < \delta$ and

$x > a$." Similarly, when x approaches a via values smaller than a, we have the following:

DEFINITION 2.4.2 *Let f be defined on the open interval (b, a). The number L is the* **limit of $f(x)$ as x approaches a from below** *(or* **from the left***) if for every $\epsilon > 0$, there is a $\delta > 0$ such that*

$$\text{if} \quad -\delta < x - a < 0, \quad \text{then} \quad |f(x) - L| < \epsilon.$$

When this limit exists, we will write $\lim_{x \to a^-} f(x) = L$.

Of course, "$-\delta < x - a < 0$" means that "$0 < |x - a| < \delta$ and $x < a$."

Limits from above and below are also called **one-sided limits.**

EXAMPLE 2.4.1 Show that $\lim_{x \to 0^+} \sqrt{x} = 0$.

SOLUTION Given $\epsilon > 0$, find $\delta > 0$ such that

$$\text{if} \quad 0 < x - 0 < \delta, \quad \text{then} \quad |\sqrt{x} - 0| < \epsilon.$$

Working backward, since $0 < x$,

$$|\sqrt{x} - 0| < \epsilon \Longrightarrow \sqrt{x} < \epsilon \Longrightarrow x < \epsilon^2.$$

Choosing $\delta = \epsilon^2$ we then may say given $\epsilon > 0$ if $0 < x - 0 < \delta$, then

$$0 < x < \epsilon^2 \Longrightarrow 0 < \sqrt{x} < \epsilon \Longrightarrow |\sqrt{x} - 0| < \epsilon. \quad \square$$

Suppose now that

$$f(x) = \begin{cases} 1, & \text{if } x \geq 0 \\ -1, & \text{if } x < 0. \end{cases}$$

In Example 2.2.7 we saw that $\lim_{x \to 0} f(x)$ does not exist. However, in this case it is easy to see that both one-sided limits do exist at 0 but are different: $\lim_{x \to 0^+} f(x) = 1$, and $\lim_{x \to 0^-} f(x) = -1$. The next result says that this type of situation is not unusual. The *only* way both one-sided limits can exist *without* the limit existing is if the two one-sided limits are different. You are asked to prove this in the problem section.

THEOREM 2.4.3 *Both $\lim_{x \to a^+} f(x)$ and $\lim_{x \to a^-} f(x)$ exist and are equal if and only if $\lim_{x \to a} f(x)$ exists.*

2.4.2 Limits at Infinity

It will also be useful to describe the behavior of a function f as x increases without bound or as x decreases without bound. For example, we would like to make statements such as "when x gets large, x^{-1} approaches 0" mathematically precise. The limit definition is again easily modified to handle this case.

DEFINITION 2.4.4 *Let f be defined on the interval $(a, +\infty)$. The number L is the **limit of $f(x)$ at $+\infty$** if for every $\epsilon > 0$, there is a number N such that*

$$\text{if} \quad x > N, \quad \text{then} \quad |f(x) - L| < \epsilon.$$

When this limit exists, we will write $\lim_{x\to+\infty} f(x) = L$.

This time the condition "$0 < |x - a| < \delta$," which says that x is near a, has been replaced in Definition 2.4.4 by "$x > N$," which indicates that x is large. Similarly:

DEFINITION 2.4.5 *Let f be defined on the interval $(-\infty, b)$. The number L is the **limit of $f(x)$ at $-\infty$** if for every $\epsilon > 0$, there is a number N such that*

$$\text{if} \quad x < N, \quad \text{then} \quad |f(x) - L| < \epsilon.$$

When this limit exists, we will write $\lim_{x\to-\infty} f(x) = L$.

EXAMPLE 2.4.2 Show that $\lim_{x\to-\infty} \frac{1}{x} = 0$.

SOLUTION Given $\epsilon > 0$, we must find N such that

$$\text{if} \quad x < N, \quad \text{then} \quad \left|\frac{1}{|x|} - L\right| < \epsilon.$$

We may assume that $x < 0$. Therefore, working backward yields

$$\left|\frac{1}{x} - 0\right| < \epsilon \Longrightarrow -\frac{1}{x} < \epsilon \Longrightarrow \frac{1}{x} > -\epsilon \Longrightarrow x < -\frac{1}{\epsilon}.$$

So we may select $N = -\frac{1}{\epsilon}$ (a negative number). ◻

In a similar fashion, one can show that $\lim_{x\to+\infty} \frac{1}{x} = 0$. In fact, this observation allows us to transform questions about limits at infinity to problems about limits at 0. If we let $y = \frac{1}{x}$, then as $x \to +\infty$ we have $y \to 0+$ and vice versa. This leads to the following result:

THEOREM 2.4.6 *Assume that f is a function defined on an interval of the form $(a, +\infty)$. Let $y = 1/x$. Then*

$$\lim_{x \to +\infty} f(x) = L \iff \lim_{y \to 0+} f(1/y) = L.$$

The analogous result holds for limits as $x \to -\infty$.

PROOF We prove the necessity of the result. By definition $\lim_{x \to +\infty} f(x) = L$ if and only if given any $\epsilon > 0$ there exists N (which we may assume is greater than 0) such that if $x > N$, then $|f(x) - L| < \epsilon$. Let $\delta = 1/N$. If $0 < y < 1/N$, then $1/y > N$, and $|f(1/y) - L| < \epsilon$. Consequently $\lim_{y \to 0+} f(1/y) = L$. The converse follows in a similar fashion. ∎

EXAMPLE 2.4.3 Evaluate $\lim_{x \to +\infty} \frac{\sin x}{x}$.

SOLUTION Transform the problem to a limit at 0^+ as follows:

$$\lim_{x \to +\infty} \frac{\sin x}{x} = \lim_{y \to 0+} y \sin(\tfrac{1}{y}) = 0,$$

where the last equality follows from Example 2.2.3. □

There are still further variations on the limit concept, which you are asked to discuss in the problem section.

Problems for Section 2.4

2.4.1. Show that $\lim_{x \to 0+} f(x) = 1$ and that $\lim_{x \to 0-} f(x) = -1$ for the function f given by

$$f(x) = \begin{cases} 1, & \text{if } x \geq 0 \\ -1, & \text{if } x < 0 \, . \end{cases}$$

2.4.2. Show that $\lim_{x \to 2+} f(x) = 2$ and that $\lim_{x \to 2-} f(x) = 6$ for the function f given by

$$f(x) = \begin{cases} x, & \text{if } x \geq 2 \\ x^2 + x, & \text{if } x < 2 \, . \end{cases}$$

2.4.3. Show that $\lim_{x \to 3-} \sqrt{3 - x} = 0$.

2.4.4. Let $G(x)$ be the greatest integer function. Show that $\lim_{x \to -4-} G(x) = -5$ and that $\lim_{x \to -4+} G(x) = -4$.

2.4.5. Prove Theorem 2.4.3.

2.4.6. Show that $\lim_{x \to 3-} \frac{1}{x^2 - 1} = 1/8$. (Use Theorem 2.4.3.)

2.4.7. Let $p(x)$ be a polynomial. Use Theorem 2.4.3 to show that $\lim_{x\to a^+} p(x) = p(a)$ and that $\lim_{x\to a^-} p(x) = p(a)$.

2.4.8. Suppose that $\lim_{x\to a^+} f(x) = L$ and $\lim_{x\to a^+} g(x) = M$.

a) Prove that $\lim_{x\to a^+}(f(x)+g(x)) = L + M$.

b) If c is a constant, show that $\lim_{x\to a^+} cf(x) = cL$.

•2.4.9. **a)** Show that if $n \in \mathbf{N}$, then $\lim_{x\to+\infty} 1/x^n = 0$.

b) Show that if p is the polynomial $p(x) = a_n x^n + a_{n-1}x^{n-1} + \cdots + a_1 x + a_0$, then $\lim_{x\to+\infty} p(x)/x^n = a_n$.

c) Show that the result in **(b)** can be rephrased as follows: for any $\epsilon > 0$, there is an N such that if $x > N$, then $a_n - \epsilon < p(x)/x^n < a_n + \epsilon$.

d) Show that if $a_n > 0$, there is an N such that if $x > N$, then $a_n/2 < p(x)$. This result says that for sufficiently large x, $p(x)$ is bounded away from 0. (Hint: In **(c)** take $\epsilon = a_n/2$ and $x > 1$.)

•2.4.10. Show that if $\lim_{x\to a^+} f(x) = L$ and if c is a function such that $a < c(x) < x$ for all $x > a$, then $\lim_{x\to a^+} f(c(x)) = L$.

2.4.11. Notice that as $x \to 0$, $f(x) = x^{-2}$ increases without bound. Make this idea precise by defining the general expressions

a) $\lim_{x\to a} f(x) = +\infty$ **b)** $\lim_{x\to a} f(x) = -\infty$

2.4.12. Define the following one-sided limits and give examples of each.

a) $\lim_{x\to a^+} f(x) = +\infty$ **b)** $\lim_{x\to a^-} f(x) = +\infty$

c) $\lim_{x\to a^+} f(x) = -\infty$ **d)** $\lim_{x\to a^-} f(x) = -\infty$

2.4.13. Notice that as x gets large, $f(x) = x^2$ increases without bound. Carefully define the following four expressions:

a) $\lim_{x\to+\infty} f(x) = +\infty$ **b)** $\lim_{x\to-\infty} f(x) = +\infty$

c) $\lim_{x\to+\infty} f(x) = -\infty$ **d)** $\lim_{x\to-\infty} f(x) = -\infty$

2.5 Continuity

We have seen that for any polynomial $p(x)$, $\lim_{x\to a} p(x) = p(a)$. In problem 2.3.3 you showed that for any point a in the domain of a rational function $r(x)$, $\lim_{x\to a} r(x) = r(a)$. For such nicely behaved functions, evaluating limits is trivial. However, for a general function f it need not be the case that $\lim_{x\to a} f(x) = f(a)$. This can fail to happen in several ways:

1. $f(a)$ may not be defined;
2. $\lim_{x\to a} f(x)$ may not exist;
3. $\lim_{x\to a} f(x)$ and $f(a)$ may both exist but not be equal.

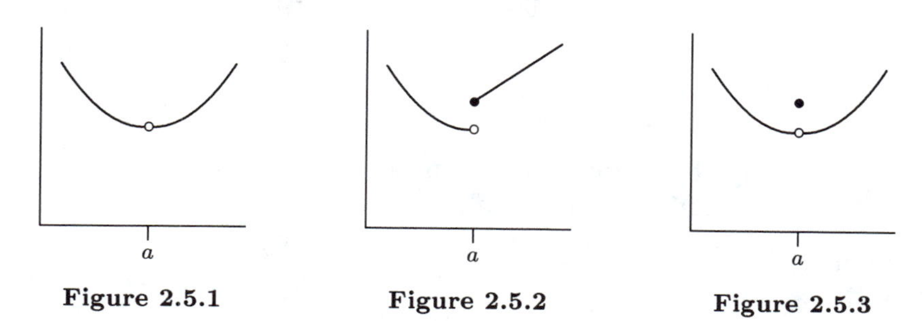

Figure 2.5.1 **Figure 2.5.2** **Figure 2.5.3**

We single out the well-behaved functions for special attention.

DEFINITION 2.5.1 *A function f is said to be* ***continuous at a point*** *a in its domain if $\lim_{x\to a} f(x) = f(a)$.*

EXAMPLE 2.5.1

a) Our foregoing comments indicate that polynomials and rational functions are continuous at any points in their domains.

b) Recall from Example 2.2.3 that $\lim_{x\to 0} x\sin(\frac{1}{x}) = 0$. However, the function $x\sin(\frac{1}{x})$ is *not* continuous at 0 because 0 is not in its domain.

c) Example 2.2.4 shows that $f(x) = \sqrt{x}$ is continuous at any $a > 0$.

d) Example 2.2.9 shows that that the Dirichlet function,

$$D(x) = \begin{cases} 1, & \text{if } x \text{ is rational} \\ 0, & \text{if } x \text{ is irrational} \end{cases}$$

is not continuous at any point! □

The definition of continuity can also be given directly in terms of an ϵ, δ description.

DEFINITION 2.5.1a *f is* ***continuous at*** *a if for every $\epsilon > 0$, there is a $\delta > 0$ such that*

$$\textit{if } \; |x - a| < \delta, \quad \textit{then } \; |f(x) - f(a)| < \epsilon.$$

Notice that since a must be in the domain of f for f to be continuous there, we no longer require that "$0 < |x - a|$." Secondly, we have $f(a)$ playing the role of L in the limit definition, since we want $\lim_{x\to a} f(x) = f(a)$. (Compare with Definition 2.2.1.)

If we let $h = x - a$ in the preceding discussion (so that $x = a + h$), we obtain the following pairs of equivalent statements:

$$0 < |x - a| < \delta \iff 0 < |h| < \delta$$

and

$$|f(x) - f(a)| < \epsilon \iff |f(a + h) - f(a)| < \epsilon.$$

Using these equivalences and the two previous definitions of continuity, we obtain a third formulation of the definition.

DEFINITION 2.5.1b *The function f is* ***continuous at*** *a if a is in the domain of f and*

$$\lim_{h \to 0} f(a + h) = f(a).$$

The eighteenth century view of continuity was quite different from the modern one. Continuity referred to the constancy of the analytic expression of the function (i.e., a continuous function was one given by a single "rule" throughout its entire domain). In this earlier view, a function could fail to be continuous in either of two ways. Louis Arbogast described this in a prize-winning essay he wrote for the St. Petersburg Academy in 1791 on the topic of integration and continuity.

> We shall call *discontinuous curves* as much those which are formed by the joining of several portions of curves as those which, traced by the free movement of the hand, are not submitted to any law for any part of their course; provided that all parts of the curves join together without interruption.... The law of continuity is again broken, when the different parts of a curve do not join together.... We call curves of this type, *discontiguous curves*, because all of their parts do not join up....[1]

For example, according to Arbogast's description the absolute value function would not be called continuous at 0 because its analytic expression changes at that point:

$$|x| = \begin{cases} x, & \text{if } x \geq 0 \\ -x, & \text{if } x < 0. \end{cases}$$

(In the problems you will show that $|x|$ is continuous at 0 in the modern sense.) Further any freely drawn curve would not be continuous because no "rule" or analytic expression could be given for it. What Arbogast

[1] Grattan-Guinness, "The Emergence of Mathematical Analysis," pp. 103–104. Also see Grabiner, *Origins of Cauchy's Calculus*, p. 91.

called "discontiguous functions" with their jumps or breaks more closely resemble the modern notion of discontinuous functions.

Early in the nineteenth century mathematicians began to focus in on the key elements of the current definition of continuity. Cauchy stated,

> The function $f(x)$ will remain continuous with respect to x between the given limits, if between these limits an infinitely small increase of the variable always produces an infinitely small increase of the function itself.[2]

Though the kernel of the modern notion of continuity is present notice that Cauchy uses the term "increase" instead of "change" in the variable and the corresponding function. Bolzano was more careful, saying that f would be continuous at x if "the difference $f(x+\omega)-f(x)$ can be made smaller than any given quantity, if one makes ω as small as one wishes."[3] Fifty years later Weierstrass finally made the notion of continuity precise.

Our next task is to show that continuous functions will remain "well behaved" under elementary algebraic operations.

THEOREM 2.5.2 *Assume that f and g are both continuous at a. Then*

a) *$f+g$ is continuous at a;*

b) *cf is continuous at a for any constant c;*

c) *$f \cdot g$ is continuous at a;*

d) *if $g(a) \neq 0$, then f/g is continuous at a.*

PROOF We give a proof of **(c)**; the rest are similar and you are asked to complete their proofs in the problem section. Since f and g are continuous at a, $\lim_{x \to a} f(x) = f(a)$ and $\lim_{x \to a} g(x) = g(a)$. By the product theorem for limits,

$$\lim_{x \to a}(f \cdot g)(x) = \lim_{x \to a} f(x)g(x) = f(a)g(a) = (f \cdot g)(a),$$

so $f \cdot g$ is continuous at a. ∎

EXAMPLE 2.5.2 Using Theorem 2.5.2 (how many times?) the function

$$h(x) = \frac{8x + \sqrt{x} + 1}{2x^2 + x + 9}$$

is continuous for all $x > 0$ because polynomials and the square root function are continuous on this set. Consider how much harder it would be to directly verify this with an ϵ, δ proof and you begin to appreciate the power of even the very simple Theorem 2.5.2. □

---

[2] Grattan-Guinness, "The Emergence of Mathematical Analysis," p. 110.

[3] Edwards, *The History of the Calculus*, p. 308.

However, Theorem 2.5.2 does not apply to the less complicated looking function $s(x) = \sqrt{2x^2+2}$. At this point we cannot say that $s(x)$ is continuous even though $2x^2 + 2$ is continuous (and positive) and the square root function is continuous (at positive values). What is required is a result relating the composition of functions to continuity.

THEOREM 2.5.3 *If g is continuous at a and f is continuous at $g(a)$, then $f \circ g$ is continuous at a.*

PROOF First, g is continuous at a, so $g(a)$ is defined. Second, f is continuous at $g(a)$, so $f(g(a))$ is defined. Thus a is in the domain of $f \circ g$. To show that $f \circ g$ is continuous at a, we need to show that $\lim_{x \to a} f \circ g(x) = f \circ g(a)$, which can be accomplished by a direct ϵ, δ proof.

Given $\epsilon > 0$, we must locate $\delta > 0$ such that

$$\text{if } |x - a| < \delta, \quad \text{then } |f(g(x)) - f(g(a))| < \epsilon.$$

Because f is continuous at $g(a)$, we only need to ensure that $g(x)$ is sufficiently close to $g(a)$ to obtain this last inequality. That is, f is continuous at $g(a)$, so there is a $\delta_1 > 0$ such that

$$\text{if } |t - g(a)| < \delta_1, \quad \text{then } |f(t) - f(g(a))| < \epsilon.$$

This means that whenever $g(x)$ is within δ_1 of $g(a)$, then $f(g(x))$ will be within ϵ of $f(g(a))$. That is,

$$|g(x) - g(a)| < \delta_1 \Longrightarrow |f(g(x)) - f(g(a))| < \epsilon. \tag{1}$$

Using the continuity of g we can satisfy the first inequality in (1). Because $\delta_1 > 0$, we can use it as the "epsilon" in the definition of the continuity of g at a. In particular, there is a $\delta > 0$ such that

$$\text{if } |x - a| < \delta, \quad \text{then } |g(x) - g(a)| < \delta_1. \tag{2}$$

Putting (1) and (2) together,

$$|x - a| < \delta \Longrightarrow |g(x) - g(a)| < \delta_1 \Longrightarrow |f(g(x)) - f(g(a))| < \epsilon. \quad \blacksquare$$

By combining several results, rather complicated functions can be shown to be continuous with relatively little effort.

EXAMPLE 2.5.3 The polynomial $g(x) = 6(x^3 - 1)^2 + 2$ is continuous and always positive, and $f(x) = \sqrt{x}$ is continuous when $x > 0$. Therefore the composite

$$f \circ g(x) = \sqrt{6(x^3 - 1)^2 + 2}$$

is continuous at any point. Furthermore, since $h(x) = (x + 1)^5$ is continuous at every point, so is the composite function:

$$h \circ (f \circ g) = h(f(g(x))) = \left(\sqrt{6(x^3 - 1)^2 + 2} + 1\right)^5. \quad \square$$

Much of what we do later requires that a function be continuous at every point in an entire interval. If f is continuous at each point in the open interval (a, b), then f is called **continuous on** (a, b). If f is continuous for every real number in its domain, f is called **continuous.** If f happens to be continuous for all real numbers, we sometimes will emphasize this by saying that f is **continuous everywhere**. Defining continuity on a closed interval $[a, b]$ is a bit trickier because of the endpoints.

DEFINITION 2.5.4 *f is **continuous on the closed interval** $[a, b]$ if the following two conditions are satisfied:*

i) f is continuous on the open interval (a, b);

ii) $\lim_{x \to a^+} f(x) = f(a)$ and $\lim_{x \to b^-} f(x) = f(b)$.

The second condition says that as either endpoint is approached from *within* the interval $[a, b]$, the appropriate one-sided limit exists and is the same as the function value at that endpoint.

EXAMPLE 2.5.4 $f(x) = \sqrt{x}$ is continuous on $[0, b]$ because we have seen that (i) f is continuous on $(0, b)$ (see Example 2.2.4); and (ii) $\lim_{x \to 0^+} \sqrt{x} = 0$ (see Example 2.4.1), and by Theorem 2.4.3 $\lim_{x \to b^-} \sqrt{x}$ exists because $\lim_{x \to b} \sqrt{x} = \sqrt{b}$ (again use Example 2.2.4). $\square$

EXAMPLE 2.5.5 $f(x) = x \sin(\frac{1}{x})$ is not continuous on $[0, 1]$ because $f(0)$ is not defined even though $\lim_{x \to 0^+} f(x) = 0$. $\square$

Problems for Section 2.5

2.5.1. Show that $|x|$ is continuous everywhere. (Hint: Problem 1.3.16d makes this easy.)

2.5.2. Suppose that $f(x)$ is continuous on the interval (a, b). Prove that $|f(x)|$ is continuous on (a, b).

2.5.3. Prove parts **(a)**, **(b)**, and **(d)** of Theorem 2.5.2.

2.5.4. Prove that $g(x) = (-2x^2 - \sqrt{x} - 1)^2$ is continuous on $[0, \infty)$.

2.5.5. Prove that $x^{3/2}$ is continuous on the closed interval $[0, 4]$ by completing the following steps:

a) Show that $x^{3/2}$ is continuous on the open interval $(0, 4)$.

b) Show by ϵ, δ calculation that $\lim_{x\to 0^+} x^{3/2} = 0$.

c) Show that $\lim_{x\to 4^-} x^{3/2} = 4^{3/2}$ either by direct calculation or by Theorem 2.4.3.

2.5.6. This problem uses Definition 2.5.1b of continuity.

a) Prove that if f is continuous at a, then $\lim_{h\to 0}(f(a+h) - f(a)) = 0$.

b) Prove that if $\lim_{h\to 0}(f(a+h) - f(a)) = 0$, then f is continuous at a.

•2.5.7. Show that if f is continuous at a and if $f(a) > 0$, then there is a positive number δ such that $f(x) > 0$ on the interval $(a - \delta, a + \delta)$. (Hint: Let $\epsilon = f(a)$.)

2.5.8. **a)** Notice that $x^{1/4} = \sqrt{\sqrt{x}}$. Show that if $a > 0$, then $x^{1/4}$ is continuous at a. (Avoid an ϵ, δ proof.)

b) Prove that $x^{1/4}$ is continuous on $[0, \infty)$.

c) Prove that if $a > 0$, then $\lim_{x\to a} x^{n/4} = a^{n/4}$ for all $n \in \mathbf{N}$.

2.5.9. Let $f(x) = \begin{cases} \sqrt{x+2}, & \text{if } x \geq 2 \\ 3x^2 - 5x, & \text{if } x < 2 \end{cases}$. Show that $f(x)$ is continuous at 2 by using Theorem 2.4.3.

⋆2.5.10. Let s be the function defined by

$$s(x) = \begin{cases} x\sin(\frac{1}{x}), & \text{if } x \neq 0 \\ 0, & \text{if } x = 0\,. \end{cases}$$

a) Show that $s(x)$ is continuous at 0. (Hint: Most of the hard work has already been done in Example 2.2.3. Compare with Example 2.5.5.)

b) Next, *assume* that $\sin x$ is continuous everywhere. Show that $s(x)$ is continuous for $x \neq 0$.

c) Show that $s(x)$ is continuous everywhere.

2.6 Continuity on Closed Intervals

This section is devoted to three major theorems about continuous functions whose proofs depend on the structural properties of the real numbers. In particular, the least upper bound axiom and the Heine-Borel theorem will be crucial here. Recall that the least upper bound axiom says that *every nonempty set S of real numbers that has an upper bound has a least upper bound β.* Problem 1.2.5 showed that this least upper bound β is characterized by two properties:

i) $\beta \geq x$ for all $x \in S$;
ii) for every $\delta > 0$, there is a number $x' \in S$ such that $x' > \beta - \delta$.

Note that (i) says that β is an upper bound for S, while (ii) says that $\beta - \delta$ is *not* an upper bound for S no matter how small δ is chosen. We will refer back to these two properties several times.

2.6.1 The Intermediate Value Theorem

On an intuitive level, the graph of a continuous function f is an "unbroken" curve. If f is continuous on the closed interval $[a, c]$ with $f(a) < 0$ and $f(c) > 0$, geometric intuition tells us that there is at least one point b between a and c such that $f(b) = 0$. When f is not continuous, it is evident that such a point b need not exist. (See Figures 2.6.1 and 2.6.2).

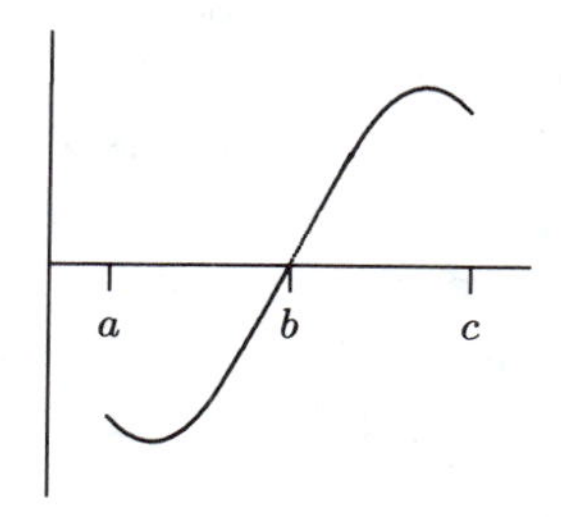

Figure 2.6.1

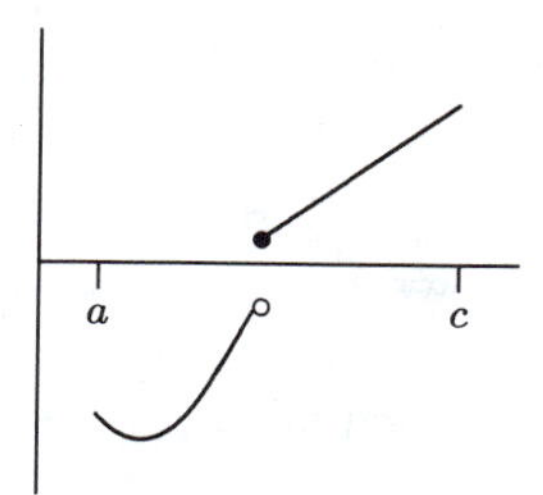

Figure 2.6.2

Early in the nineteenth century as mathematicians began to worry about the foundations of calculus, both Bolzano and Cauchy attempted to give purely analytic proofs of this "obvious" geometric result. But without a sufficiently clear conception of the real numbers, they were unable to give complete proofs.

THEOREM 2.6.1 ***(The Intermediate Value Theorem)*** *Let f be continuous on the closed interval $[a, c]$ and let y be any real number such that*

$$f(a) < y < f(c).$$

Then there exists at least one number b in the interval $[a, c]$ such that $f(b) = y$. (A similar theorem is obtained by replacing the hypothesis $f(a) < y < f(c)$ with $f(a) > y > f(c)$.)

PROOF Let S be the subset of all numbers x in $[a, c]$ that satisfy the inequality $f(x) < y$. S is nonempty because a is in S. S is bounded above since S is a subset of $[a, c]$. By the least upper bound axiom, S has a least upper bound b. Since S is bounded above by c, it follows that $b \leq c$. Because a is in S, $b \geq a$. So $a \leq b \leq c$. There are three possibilities:

$$\text{(i) } f(b) < y; \qquad \text{(ii) } f(b) > y; \qquad \text{(iii) } f(b) = y.$$

We will show that (i) and (ii) are impossible, so (iii) must hold and that will finish the proof.

Assume (i) holds, that is, $f(b) < y$. First, observe that we cannot have $b = c$ since $f(c) > y$. We already know that $b \leq c$; hence we may conclude that $b < c$. Now let $\epsilon = y - f(b)$. Then (i) implies that $\epsilon > 0$. Since f is continuous at b, there is a $\delta > 0$ such that if $|x - b| < \delta$, then $|f(x) - f(b)| < \epsilon$. Pick any x satisfying $b < x < b + \delta$. For any such x,

$$f(x) - f(b) \leq |f(x) - f(b)| < \epsilon = y - f(b).$$

Comparing the first and last terms of this inequality shows $f(x) < y$. This means that x belongs to S; but $b < x$. This contradicts the fact that b is an upper bound of S. So (i) must be false.

Suppose that (ii) holds instead; that is, $f(b) > y$. Because $f(a) < y$, we have $a \neq b$, and so we must have $a < b$. This time set $\epsilon = f(b) - y$, which is positive. Again the continuity of f at b allows us to select $\delta > 0$ such that if $|x - b| < \delta$, then $|f(x) - f(b)| < \epsilon$. Then for all x satisfying $b - \delta < x < b$,

$$f(b) - f(x) \leq |f(x) - f(b)| < \epsilon = f(b) - y.$$

This time we conclude that $-f(x) < -y$ or $f(x) > y$. This means that any x satisfying $b - \delta < x < b$ is not in S. But this means that b is not the *least* upper bound of S; so (ii) is false.

Hence (iii) must hold, $f(b) = y$. ∎

The intermediate value theorem is used when graphing a function f. If the derivative f' is continuous with $f' = 0$ at a and c and nowhere between, by the intermediate value theorem f' cannot switch sign between a and c. For if f' did switch from positive to negative (or vice versa), f' would be 0 at some point between a and c. So between the critical points a and c the function f is either always increasing or always decreasing.

COROLLARY 2.6.2 *If n is an odd positive integer, then the polynomial*

$$p(x) = x^n + a_{n-1}x^{n-1} + \cdots + a_1 x + a_0$$

has a root.

PROOF It is easy to check (as in problem 2.4.9) that

$$\lim_{x \to -\infty} \frac{p(x)}{x^n} = 1.$$

This means that for any $\epsilon > 0$, there is a negative number $-M$ such that

$$-\epsilon < \frac{p(x)}{x^n} - 1 < \epsilon \qquad (x < -M).$$

Taking $\epsilon = 1/2$ and using only the first inequality, it follows that

$$1/2 = 1 - \epsilon < \frac{p(x)}{x^n} \qquad (x < -M).$$

Since x is negative, we have

$$0 > x^n/2 > p(x) \qquad (x < -M).$$

In a similar fashion (see problem 2.4.9) there is a positive number N such that if $x > N$, then $0 < p(x)$. Now choose any $a < -M$ and any $b > N$. Since p is a polynomial, it is continuous on $[a, b]$. Since $p(a) < 0 < p(b)$, the intermediate value theorem says that there is a c in (a, b) such that $p(c) = 0$. That is, p has a root at c. ∎

2.6.2 Uniform Continuity

We turn next to examining a "stronger" type of continuity. Example 2.2.4 showed that for $a > 0$, $\lim_{x \to a} \sqrt{x} = \sqrt{a}$. In other words, $\sqrt{x}$ is continuous on $(0, +\infty)$. For a given ϵ, we found that if $\delta = \min\{\epsilon\sqrt{a}, a\}$, then $|\sqrt{x} - \sqrt{a}| < \epsilon$. Notice that this choice of δ depends on a. In such situations it would be more convenient to use a single value for δ no matter what a was, if possible. We now single out the functions where this is possible for special attention.

DEFINITION 2.6.3 *A function f is* ***uniformly continuous on an interval*** *I if for every $\epsilon > 0$, there is a $\delta > 0$ such that if x and y are in I and $|x - y| < \delta$, then $|f(x) - f(y)| < \epsilon$.*

Note that the interval I may be closed, open, or half-open. The definition says that for uniformly continuous functions we can find a single number δ such that *anytime* two points in the interval I are within δ of each other their function values are close, that is, $f(x)$ is within ϵ of $f(y)$. This number δ depends only on ϵ and is *independent* of x. In contrast, if f is merely continuous on I, δ will usually depend on both ϵ and the particular value a that is being approached.

It should be clear that a uniformly continuous function is continuous. The converse does not hold; there are continuous functions that are not uniformly continuous on a particular interval I, as the following example shows.

EXAMPLE 2.6.1 Let $f(x) = x^2$ on the interval $I = (-\infty, +\infty)$. Since f is a polynomial, it is continuous on I. We will show that f is not uniformly continuous on I. If x^2 were uniformly continuous, for $\epsilon = 1$ there would be a single value $\delta > 0$ such that if $|x - y| < \delta$, then $|x^2 - y^2| < 1$. In particular, if we choose $x = \frac{1}{\delta} > 0$ and $y = x + \frac{\delta}{2}$, then

$$|x - y| = |x - (x + \tfrac{\delta}{2})| = \delta/2 < \delta.$$

But

$$|x^2 - y^2| = |x - y| \cdot |x + y| = \tfrac{\delta}{2}(2x + \tfrac{\delta}{2}) > \delta x = 1,$$

which contradicts the definition of uniform continuity.

Simply put, when x is large small changes in x produce relatively large changes (bigger than $\epsilon = 1$) in $f(x)$. □

EXAMPLE 2.6.2 In contrast to the preceding example, we now show that $f(x) = x^2$ *is* uniformly continuous on the interval $[0, 1]$.

Given $\epsilon > 0$, we must find a $\delta > 0$ such that if x and y are in $[0, 1]$ and $|x - y| < \delta$, then $|x^2 - y^2| < \epsilon$. This is easy! Since $|x| < 1$ and $|y| < 1$,

$$|x^2 - y^2| = |x + y||x - y| \leq (|x| + |y|)|x - y| \leq 2|x - y|.$$

Select $\delta = \epsilon/3$. If x and y are in $[0, 1]$ and $|x - y| < \delta = \epsilon/3$, then from the preceding equation

$$|x^2 - y^2| \leq 2|x - y| < 2(\epsilon/3) < \epsilon.$$

So f is uniformly continuous on [0,1]. □

The preceding examples show that the same function may be uniformly continuous on one interval but not on another. The type of interval involved often plays a crucial role in determining whether a function is uniformly continuous. In the 1870s Eduard Heine first gave the definition of uniform continuity and then proved the following result:

THEOREM 2.6.4 *If f is continuous on the closed, bounded interval $[a, b]$, then f is uniformly continuous there.*

PROOF Given $\epsilon > 0$, we need to find a single value $\delta > 0$ such that if x and y are in $[a, b]$ and $|x - y| < \delta$, then $|f(x) - f(y)| < \epsilon$.

The general plan is as follows: (1) use continuity to generate an open cover of $[a, b]$; (2) use the Heine-Borel theorem to obtain a finite subcover of $[a, b]$; (3) use this finiteness to generate the required δ.

Since f is continuous on $[a, b]$, for each s in $[a, b]$ there is a number $\delta_s > 0$ such that if x is in $[a, b]$ and $|x - s| < \delta_s$, then $|f(x) - f(s)| < \epsilon/2$. Notice that at this stage each δ still depends on s.

For each s in $[a, b]$ form the open interval $(s - \frac{\delta_s}{2}, s + \frac{\delta_s}{2})$. The infinite collection of all such intervals forms an open cover of $[a, b]$. By the Heine-Borel theorem, there is a finite subcollection, C, of these intervals that covers $[a, b]$. Denote the intervals in C by

$$(s_1 - \tfrac{\delta_1}{2}, s_1 + \tfrac{\delta_1}{2}), \ldots, (s_n - \tfrac{\delta_n}{2}, s_n + \tfrac{\delta_n}{2}).$$

Next choose $\delta = \min\{\frac{\delta_1}{2}, \ldots, \frac{\delta_n}{2}\}$. Notice that δ is now independent of s. The final step is to show that δ satisfies the definition of uniform continuity.

If x is in $[a, b]$, it must lie in one of the intervals in the subcover C, say $(s_k - \frac{\delta_k}{2}, s_k + \frac{\delta_k}{2})$. This means that $|x - s_k| < \frac{\delta_k}{2}$. Next, if y is in $[a, b]$ and $|x - y| < \delta$, then

$$|y - s_k| = |y - x + x - s_k| \le |y - x| + |x - s_k| < \delta + \tfrac{\delta_k}{2} \le \tfrac{\delta_k}{2} + \tfrac{\delta_k}{2} = \delta_k.$$

But by the way δ_k was chosen, if $|y - s_k| < \delta_k$, then $|f(y) - f(s_k)| < \epsilon/2$, and if $|x - s_k| < \delta_k/2 < \delta_k$, then $|f(x) - f(s_k)| < \epsilon/2$. Combining these inequalities,

$$\begin{aligned} |f(x) - f(y)| &= |f(x) - f(s_k) + f(s_k) - f(y)| \\ &\le |f(x) - f(s_k)| + |f(s_k) - f(y)| < \epsilon/2 + \epsilon/2 = \epsilon. \end{aligned}$$

So f is uniformly continuous on $[a, b]$. ∎

Uniform continuity has a number of important consequences. One of the simplest is that a continuous function on a closed, bounded interval must attain a maximum and a minimum value. We will prove this result in two stages. First we show that such a function is bounded.

THEOREM 2.6.5 ***(The Boundedness Theorem)*** *If f is continuous on the closed interval $[a, b]$, then f is bounded above, that is, there is some number M such that for all x in $[a, b]$, $f(x) \leq M$.*

PROOF Choose any positive number ϵ. Since f is uniformly continuous on $[a, b]$, there is a $\delta > 0$ such that for any pair of numbers x and y in $[a, b]$ that satisfy $|x - y| < \delta$, we have $|f(x) - f(y)| < \epsilon$.

Since $[a, b]$ is of finite length, we can subdivide $[a, b]$ into a finite number of closed intervals, $I_1, I_2, \ldots, I_n$, each of length less than δ. For each interval I_k, choose any point x_k within it. Next, let

$$M' = \max\{f(x_1), f(x_2), \ldots, f(x_n)\}.$$

It is now easy to show that f is bounded above on $[a, b]$. If x is any point in $[a, b]$, it must lie in one of the closed subintervals just constructed, say $x \in I_k$. Since I_k has length less than δ and since $x_k \in I_k$, we have $|x - x_k| < \delta$. By uniform continuity, $|f(x) - f(x_k)| < \epsilon$, or equivalently,

$$f(x_k) - \epsilon < f(x) < f(x_k) + \epsilon.$$

Consequently f is bounded above because

$$f(x) < f(x_k) + \epsilon \leq M' + \epsilon.$$

That is, our upper bound M may be taken to be $M' + \epsilon$. ∎

There is an obvious companion theorem to the preceding result.

COROLLARY 2.6.6 *If f is continuous on the closed interval $[a, b]$, then f is bounded below; that is, there is some number m such that for all x in $[a, b]$, $m \leq f(x)$.*

To prove this one can mimic the preceding proof making the appropriate changes. For a quick way to prove the corollary see the problem section.

Uniform continuity is important in the proof of the boundedness theorem, so it is essential that the interval involved be closed for the proof to work. Consider the rational function $f(x) = \frac{x}{(x^2-1)}$. It is continuous on $(-1, 1)$ but is unbounded there since, as you should check, $\lim_{x\to 1^-} f(x) = +\infty$ and $\lim_{x\to -1^-} f(x) = -\infty$. In contrast, the next theorem shows that if f is continuous on a closed and bounded interval, not only must f be bounded above and below, but it is bounded above by one of its own values and it is bounded below by one of its own values. First we clarify these ideas in the following definition:

DEFINITION 2.6.7 *Let f be a function and let S be a subset of the domain of f. A point x in S is a* **maximum point** *for f on S if $f(x) \geq f(y)$ for every $y \in S$. The number $f(x)$ is called the* **maximum value** *of f on S. Similarly, a point x in S is a* **minimum point** *for f on S if $f(x) \leq f(y)$ for every $y \in S$, and $f(x)$ is called the* **minimum value** *of f on S. More generally, if x is either a maximum point or a minimum point, x is called an* **extreme point** *of f on S and $f(x)$ is called an* **extreme value** *of f on S.*

Notice that a function may have more than one point in S at which the maximum value occurs. That is, a function can have several maximum points but only a single maximum value. The same is true for minima.

THEOREM 2.6.8 ***(The Max-Min Theorem)*** *If f is continuous on the closed interval $[a, b]$, then there is a number x_0 in $[a, b]$ such that $f(x_0) \geq f(x)$ for all x in $[a, b]$. That is, f achieves a maximum value on $[a, b]$. Similarly, there is a number x_1 in $[a, b]$ such that $f(x_1) \leq f(x)$ for all x in $[a, b]$. That is, f achieves a minimum value on $[a, b]$.*

PROOF Let $S = \{f(x) \mid x \in [a, b]\}$. By Theorem 2.6.5, f is bounded above, so S is bounded above. By the least upper bound axiom, S has a least upper bound, M. That is, $M \geq f(x)$ for all x in $[a, b]$. There are two possibilities: (i) $M > f(x)$ for all x in $[a, b]$ and (ii) $M = f(x_0)$ for some number x_0 in $[a, b]$.

We show that (i) is impossible. If it were the case that $M > f(x)$ for all x in $[a, b]$, then $M - f(x)$ would be a continuous and positive function for all x in $[a, b]$. Hence the function

$$g(x) = \frac{1}{M - f(x)}$$

would be positive and, by Theorem 2.3.8, would be continuous on $[a, b]$. But then Theorem 2.6.5 would apply to $g(x)$, so $g(x)$ would be bounded above by some positive number L on $[a, b]$. That is, for all x in $[a, b]$,

$$g(x) = \frac{1}{M - f(x)} \leq L\,. \tag{1}$$

Solving for $f(x)$ in (4) shows that $f(x) \leq M - L^{-1}$ for all x in $[a, b]$. But this contradicts the fact that M is the least upper bound for S. So (i) is false. Hence (ii) holds, so there is some x_0 in $[a, b]$ for which $f(x_0) = M$. But since $M \geq f(x)$ for all x in $[a, b]$, we have $f(x_0) \geq f(x)$ for all x in $[a, b]$. In other words, f achieves its maximum at x_0.

A similar argument using greatest lower bounds shows that f also achieves its minimum. For a quicker proof see the problems. ∎

From a first course in calculus you should be familiar with results similar to the max-min theorem. In fact, you probably used a result somewhat like the following to solve max-min word problems.

> If f is differentiable on the closed interval $[a, b]$, then f achieves a maximum and a minimum on $[a, b]$. Further, these extreme values occur either at critical points of f or at one or both of the endpoints of the interval.

Compare this result with the max-min theorem. In the next chapter we will show that if f is differentiable, f is continuous (see Theorem 3.1.2). So the existence of maximum and minimum values for a differentiable function f is an immediate consequence of the max-min theorem. Yet the max-min theorem does not tell us how to locate these extreme values. If we strengthen the hypothesis by demanding that f be differentiable on $[a, b]$, it is possible to locate these extreme values; they occur either at critical points or at endpoints. The strengthened hypothesis produces a stronger result. Still the max-min theorem has its applications because there are continuous functions that are nowhere differentiable![1]

Problems for Section 2.6

2.6.1. Show that if f is uniformly continuous on (a, b), then f is continuous on (a, b). Do the same for $[a, b]$.

2.6.2. Show that a linear function $f(x) = mx + b$ is uniformly continuous on the interval $(-\infty, +\infty)$.

2.6.3. Is $|x|$ uniformly continuous on $(-\infty, +\infty)$? (Hint: See Problem 1.3.16d.)

2.6.4. Assume that f and g are continuous on $[a, b]$.

- **a)** Show that $f + g$ is uniformly continuous on $[a, b]$.
- **b)** Show that cf is uniformly continuous on $[a, b]$, where c is any constant.
- **c)** Show that $f \cdot g$ is uniformly continuous on $[a, b]$.
- **d)** If g is never 0 on $[a, b]$, show that f/g is uniformly continuous on $[a, b]$.

2.6.5. Assume now that I is any interval (and so not necessarily closed and bounded) and that f and g are both uniformly continuous on I.

- **a)** Show that $f + g$ is uniformly continuous on I.
- **b)** Show that cf is uniformly continuous on I, where c is any constant.
- **c)** Must $f \cdot g$ be uniformly continuous on I? (Hint: Consider the function x^2 on $I = (-\infty, +\infty)$ as a product of two functions.)

---

[1] Michael Spivak, *Calculus*, 2nd ed., (Berkeley, Calif.: Publish or Perish, Inc., 1980), pp. 144–145, 475–477.

2.6.6. **a)** Prove Corollary 2.6.6. (Hint: Show that $-f$ is bounded above on $[a, b]$. How is an upper bound for $-f$ related to a lower bound for f?)

b) Prove the second half of Theorem 2.6.8; that is, show that a continuous function f achieves a minimum on a closed interval $[a, b]$. (Hint: Again consider $-f$.)

2.6.7. **a)** Prove that $p(x) = 6x^4 + 4x^3 - 2x^2 - x - \pi$ has a root in the interval $[-2, 0]$.

b) Show that $p(x) = x^4 - x^3 + x^2 + x - 1$ has at least two roots in $[-1, 1]$.

2.6.8. Why doesn't the proof of Corollary 2.6.2 work when n is an even positive integer?

2.6.9. Use the intermediate value theorem to show that any positive number a has an nth root, that is, there is some real number x such that x^n=a (where $n \in \mathbf{N}$).

a) Let $p(x) = x^n$. Why is p continuous on $[0, a + 1]$?

b) Show that $p(0) < a < p(a + 1)$.

c) Prove that there is a number $x \in [0, a + 1]$ such that $p(x) = a$. Use this to finish the proof of the result.

2.6.10. **a)** Prove that if f is uniformly continuous on the bounded, open interval $I = (a, b)$, then f is bounded on (a, b).

b) In this same situation, does f necessarily achieve a maximum and minimum on I? Give examples.

c) Why must I be bounded in part **(a)**?

2.6.11. The following two trigonometric identities are well known: $|\sin \beta| \leq |\beta|$ and $\sin \alpha - \sin \beta = 2 \cos \frac{1}{2}(\alpha + \beta) \cdot \sin \frac{1}{2}(\alpha - \beta)$.

a) Use these two identities to show that $|\sin \alpha - \sin \beta| \leq |\alpha - \beta|$. (Hint: Use the second identity first and then bound the cosine term.)

b) Show that $\sin x$ is uniformly continuous on $(-\infty, +\infty)$.

2.6.12. **a)** We say that a function satisfies a **Lipschitz condition** if for all $x, y \in \mathbf{R}$ there is a real number M such that $|f(x) - f(y)| < M|x - y|$. Show that such a function f is uniformly continuous on $(-\infty, +\infty)$.

b) Show that the function $f(x) = (1 + |x|)^{-1}$ is uniformly continuous on $\mathbf{R}$.

c) A function satisfies a Lipschitz condition of order $\frac{1}{2}$ if for all $x, y \in \mathbf{R}$ there is a real number M such that $|f(x) - f(y)| < M|x - y|^{1/2}$. Show that such a function f is uniformly continuous on $(-\infty, +\infty)$.

⋆2.6.13. If f and g are uniformly continuous on I and both are bounded above and below, show that $f \cdot g$ is uniformly continuous on I.

3

Differentiation and Integration

Calculus was not created in a vacuum. Newton's work in particular was an attempt to describe and solve certain physical and mechanical problems. Morris Kline in *The History of Mathematical Thought from Ancient to Modern Times* observes that the major scientific problems of the seventeenth century provided the impetus for the development of calculus. In particular, he points to the following four questions:[1]

First, given a formula for the distance traveled by an object as a function of time, find the object's velocity and acceleration at any given instant. (Recall that this idea of instantaneous rate of change motivated the development of limits in Chapter 2.) Conversely, given the acceleration of a body, find the object's velocity and distance functions.

Second, given a curve, determine the "tangent line" to it at any of its various points. What at first seems to be a problem in pure geometry actually has a wide variety of applications. For example, by the "direction" of an object moving along a curved path at any instant one means the "direction" of the tangent vector (line) to the path at that point.

Third, given a function determine its maximum and minimum values.

Fourth, given a curve, find its length and the area it bounds. Given a surface, find its area and the volume it encloses.

---

[1] Morris Kline, *Mathematical Thought from Ancient to Modern Times* (New York: Oxford University Press, 1972), pp. 342–343.

The solutions to the third and fourth problems were of great importance to astronomers. For example, given an accurate description (observation) of the orbit of a planet about the sun, they could now calculate the length of the orbit as well as its aphelion (farthest point from the sun) and its perihelion (nearest point to the sun).

With a precise definition of limit now in hand it is a straightforward project to develop solutions to all of these problems.

3.1 The Derivative

Newton and Leibniz independently gave the first comprehensive treatments of calculus in the second half of the seventeenth century. But as was noted earlier, much of the rigor in analysis came about quite a bit later. Early in the eighteenth century Bolzano, and shortly thereafter Cauchy, gave the definition of the derivative that we still use.

3.1.1 The Definition of the Derivative

DEFINITION 3.1.1 *Let f be a function defined in an open interval containing a. The function f is* **differentiable at** *a if*

$$\lim_{h \to 0} \frac{f(a+h) - f(a)}{h}$$

exists. When the limit does exist, it is denoted by $f'(a)$ and is called the **derivative of** f **at** a. *If f is differentiable at every point in its domain, we will say that f is* **differentiable**.

The definition of the derivative can be rephrased as a limit as $x \to a$. Let $x = a + h$, then $h = x - a$. Now $h \to 0$ if and only if $x \to a$. Making the appropriate substitutions in the preceding definition gives

$$f'(a) = \lim_{x \to a} \frac{f(x) - f(a)}{x - a},$$

if this limit exists. We will use the two definitions interchangeably.

For any function f, we can form a new function called the **derivative of** f, which we will denote by f'. The domain of f' is the set of all real numbers a such that f is differentiable at a. The value of f' at any point a in its domain is

$$\lim_{h \to 0} \frac{f(a+h) - f(a)}{h}.$$

For $h \neq 0$, the average rate of change in f over the interval from a to $a+h$ is simply the difference quotient:

$$\frac{f(a+h)-f(a)}{(a+h)-a} = \frac{f(a+h)-f(a)}{h}.$$

For appropriately small values of h and "reasonable" functions, this average rate of change should not differ much from what we intuitively understand as the instantaneous rate of change in f at a. (Recall the lengthy discussion of this problem in Chapter 2, Section 2.) Hence we define the **instantaneous rate of change** in f at a to be $f'(a)$, that is,

$$\lim_{h \to 0} \frac{f(a+h)-f(a)}{h},$$

if this limit exists. In particular, if $s(t)$ represents the distance traveled by a particle along a straight line after an amount of time t has elapsed, $s'(a)$ represents the instantaneous velocity of the object at time $t = a$. Similarly, if $v(t)$ represents the velocity of a particle at time t, then $v'(a)$ represents the instantaneous acceleration of the object at $t = a$.

We can also interpret the derivative in a geometric way. For $h \neq 0$, the difference quotient

$$\frac{f(a+h)-f(a)}{h}$$

is the slope of a secant line to the graph of f that passes through the points $(a, f(a))$ and $(a+h, f(a+h))$, as in Figure 3.1.1. If f is differentiable at a, then as $h \to 0$, we get a sequence of secant lines whose slopes approach $f'(a)$, as in Figure 3.1.2. We define the **tangent line** to the graph of f at $(a, f(a))$ to be the line through $(a, f(a))$ that has slope $f'(a)$. Thus we have addressed the second of the four problems mentioned at the beginning of this chapter, at least for curves that are graphs of functions.

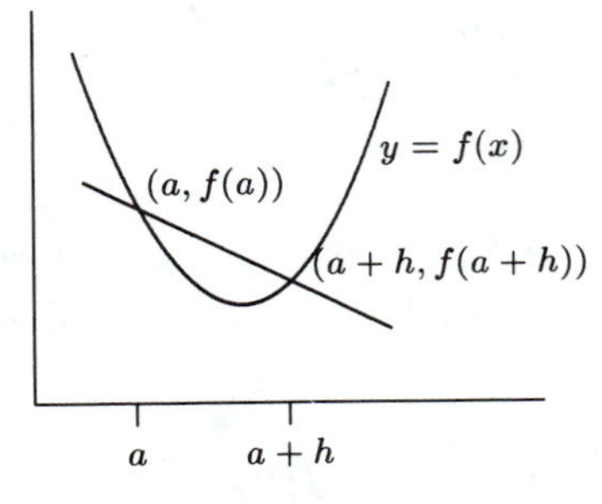

Figure 3.1.1

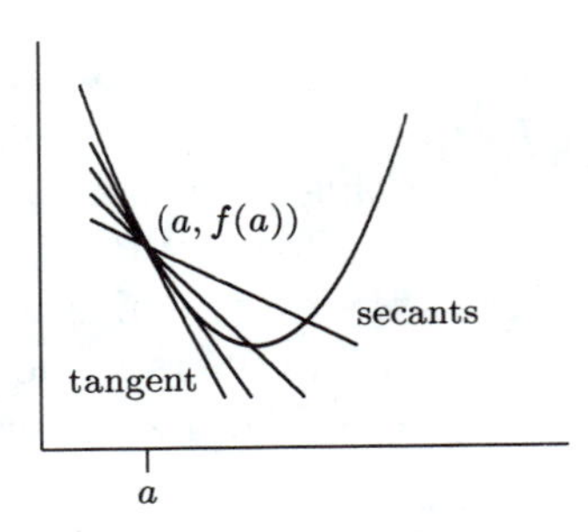

Figure 3.1.2

EXAMPLE 3.1.1 Show that function $f(x) = x^3$ is differentiable and that its derivative is the function $f'(x) = 3x^2$.

SOLUTION Using the definition of the derivative, we see that

$$\begin{aligned}\lim_{h\to 0}\frac{f(a+h)-f(a)}{h} &= \lim_{h\to 0}\frac{(a+h)^3-a^3}{h}\\ &= \lim_{h\to 0}\frac{a^3+3a^2h+3ah^2+h^3-a^3}{h}\\ &= \lim_{h\to 0}\frac{3a^2h+3ah^2+h^3}{h}\\ &= \lim_{h\to 0} 3a^2+3ah+h^2\\ &= 3a^2.\end{aligned}$$

Thus f is differentiable at any point a and $f'(a) = 3a^2$. □

Another common way of denoting the derivative f' is with the notation used by Leibniz:

$$\frac{d}{dx}(f(x)).$$

The advantage to this notation is that it can be used to concisely display both the original function and its derivative. For example, in problem 8 (at the end of this section) you are asked to show that for any $n \in \mathbf{N}$,

$$\frac{d}{dx}(x^n) = nx^{n-1}.$$

The next result relates differentiability and continuity.

THEOREM 3.1.2 *If f is differentiable at a, then f is continuous at a. (Equivalently, if f is not continuous at a, then f is not differentiable at a.)*

PROOF Using Definition 2.5.1b, it suffices to show that $\lim_{h\to 0} f(a+h) = f(a)$, or equivalently, that $\lim_{h\to 0} f(a+h) - f(a) = 0$. But

$$\begin{aligned}\lim_{h\to 0} f(a+h)-f(a) &= \lim_{h\to 0}\frac{f(a+h)-f(a)}{h}\cdot h\\ &= \lim_{h\to 0}\frac{f(a+h)-f(a)}{h}\cdot\lim_{h\to 0} h\\ &= f'(a)\cdot 0\\ &= 0. \blacksquare\end{aligned}$$

3.1.2 Extreme Values

We have just shown that a differentiable function f is necessarily continuous. So if a differentiable function f is restricted to a closed bounded interval $[a, b]$, by the max-min theorem f achieves a maximum and a minimum on $[a, b]$. Our goal is to use the differentiability of f to locate these extreme values.

To start, suppose that f is differentiable on the open interval (a, b) and has a maximum point at x in this interval. Then for any point c between a and x, since x is a maximum point, $f(x) \geq f(c)$, or equivalently, $f(x) - f(c) \geq 0$. Hence the secant line through the points $(c, f(c))$ and $(x, f(x))$ has nonnegative slope:

$$\frac{f(x) - f(c)}{x - c} \geq 0\,.$$

Similarly, if d is between x and b (so that $x - d < 0$), the slope of the secant through $(x, f(x))$ and $(d, f(d))$ is nonpositive:

$$\frac{f(x) - f(d)}{x - d} \leq 0\,.$$

This is illustrated in Figure 3.1.3.

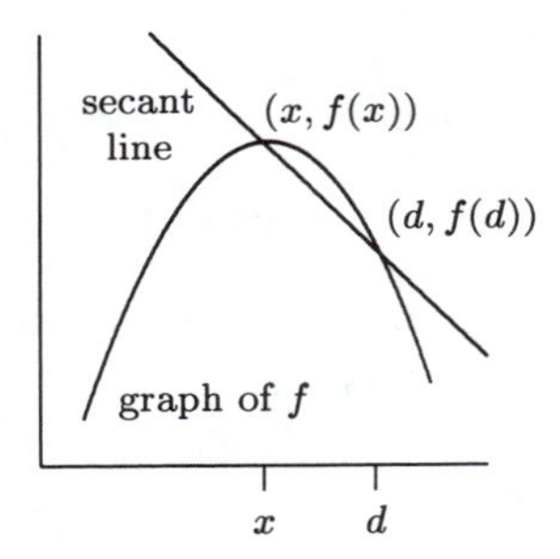

Figure 3.1.3

Since $f'(x)$ is the limit of such secant slopes, the only way it can be both nonnegative and nonpositive is to be zero. Analogous remarks hold for minimum points. This observation is the key element in the proof of the following result.

THEOREM 3.1.3 *Let f be defined on the open interval (a, b). If x is an extreme point of f on (a, b) and if f is differentiable at x, then $f'(x) = 0$. (Note: we do not assume that f is differentiable or even continuous at points of (a, b) other than x.)*

PROOF Let us formalize the foregoing argument. Assume that x is a maximum point. (A similar argument works for minimum points.) Then $f(x)$ is a maximum value for f on (a,b), so for any number h such that $x+h$ is in (a,b) we must have $f(x) \geq f(x+h)$, or equivalently, $f(x+h)-f(x) \leq 0$. When $h > 0$ this means that

$$\frac{f(x+h)-f(x)}{h} \leq 0\,.$$

So, by problem 2.3.6

$$\lim_{h\to 0+} \frac{f(x+h)-f(x)}{h} \leq 0\,,$$

if it exists. On the other hand, when $h < 0$,

$$\frac{f(x+h)-f(x)}{h} \geq 0\,.$$

So

$$\lim_{h\to 0^-} \frac{f(x+h)-f(x)}{h} \geq 0\,,$$

if it exists. But f is differentiable at x, so both one-sided limits not only exist but in fact must be equal to $f'(x)$ (Theorem 2.4.3). This means that $f'(x) \leq 0$ and $f'(x) \geq 0$. Consequently $f'(x) = 0$. ∎

It is important to note that the converse to Theorem 3.1.3 does not hold; $f'(x)$ may be zero without x being an extreme value of f. A simple example is provided by the function $f(x) = x^3$. In Example 3.1.1 we saw that $f'(x) = 3x^2$. Clearly $f'(0) = 0$, but 0 is neither a maximum nor a minimum point for f on any open interval (a,b) that contains 0 (see Figure 3.1.4). However, such points are still crucial in understanding the overall behavior of a function.

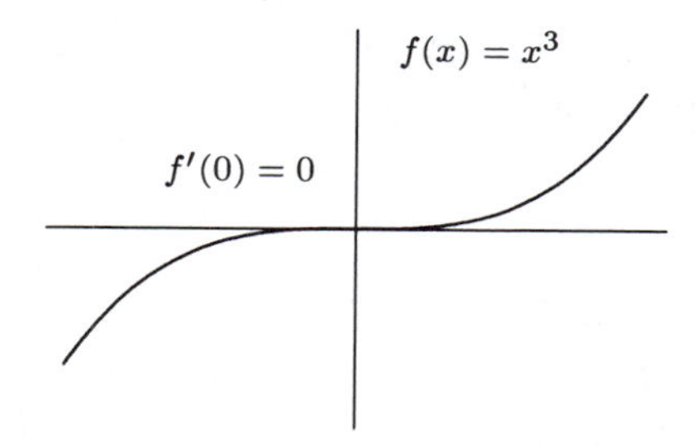

Figure 3.1.4

DEFINITION 3.1.4 *The number x is a* **critcal point** *of f if $f'(x) = 0$. In this case the number $f(x)$ is called a* **critical value** *of f.*

Theorem 3.1.3 can be used to refine the max-min theorem.

THEOREM 3.1.5 *If f is continuous on the closed, bounded interval $[a, b]$, then f has maximum and minimum points on $[a, b]$. Further, the only possible extreme points are:*

a) *critical points of f;*

b) *points where f' does not exist;*

c) *the endpoints of $[a, b]$.*

PROOF Since f is continuous, by the max-min theorem f has at least one maximum point and one minimum point on $[a, b]$. Let x be any such extreme point. If x is an endpoint, **(c)** holds. If x is not an endpoint, x is in the open interval (a, b). In this situation there are two possibilities. The first possibility is that f is differentiable at x. In this case, Theorem 3.1.3 says $f'(x) = 0$, so x is a critical point and **(a)** holds. The other possibility is that f is not differentiable at x, but then **(b)** automatically holds. Hence at any extreme point of f, one of the three conditions must hold. ∎

If we strengthen the hypothesis of Theorem 3.1.3 and assume additionally that f is differentiable on (a, b), conclusion **(b)** is impossible. That is:

COROLLARY 3.1.6 *If f is continuous on $[a, b]$ and differentiable on (a, b), then f has maximum and minimum points and these occur either at critical points of f or at endpoints of $[a, b]$.*

Theorem 3.1.5 and Corollary 3.1.6 show how to locate extreme points on closed, bounded intervals. First, find all critical points of f and points at which f is not differentiable. Next, evaluate f at all such points and at the endpoints. The largest and smallest of these values yield the maximum and minimum values of f. To extend these results to open or unbounded intervals see problem 6.

3.1.3 Examples

Theorem 3.1.1 says that if f is not continuous at a, then f cannot be differentiable at a. Even if f is continuous at a point, if f has a "corner" there, it will fail to be differentiable, as the next example shows.

EXAMPLE 3.1.2 Let $f(x) = |x|$. Show that f is not differentiable at 0 (even though it is continuous there).

SOLUTION If

$$f'(0) = \lim_{x \to 0} \frac{f(x) - f(0)}{x}$$

exists, both of the one-sided limits are equal. That is,

$$\lim_{x \to 0+} \frac{f(x) - f(0)}{x} = \lim_{x \to 0-} \frac{f(x) - f(0)}{x}.$$

But as $x \to 0^+$, $x > 0$, so $f(x) = |x| = x$ and

$$\lim_{x \to 0+} \frac{f(x) - f(0)}{x} = \lim_{x \to 0+} \frac{x}{x} = 1.$$

When $x \to 0^-$, $x < 0$, so $f(x) = |x| = -x$ and

$$\lim_{x \to 0-} \frac{f(x) - f(0)}{x} = \lim_{x \to 0-} \frac{-x}{x} = -1.$$

Since the one-sided limits are not equal,

$$\lim_{x \to 0} \frac{f(x) - f(0)}{x}$$

does not exist; therefore $f'(0)$ does not exist. ☐

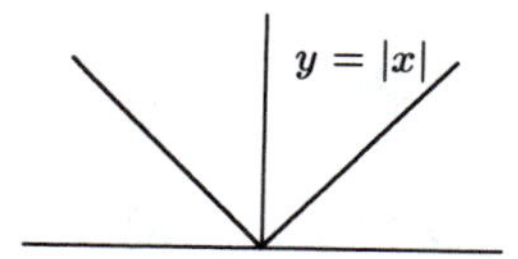

Figure 3.1.5

The "corner" in the graph of $f(x) = |x|$ at 0 is reflected in the two different one-sided derivatives at 0. If our intuitive understanding of continuity is that the graph of a function f is "unbroken," the intuitive meaning of f being differentiable is that the graph of f is "smooth" or without "corners." While the absolute value function is not differentiable at 0, it is differentiable at every other point (see problem 2). The next example displays precisely the opposite type of behavior.

EXAMPLE 3.1.3 Let

$$f(x) = \begin{cases} x^2, & \text{if } x \text{ is rational} \\ 0, & \text{if } x \text{ is irrational.} \end{cases}$$

Show that f is differentiable only at 0.

SOLUTION First we show that $f'(0) = 0$. Given $\epsilon > 0$, we must find a $\delta > 0$ such that

$$\text{if} \quad 0 < |x - 0| < \delta, \quad \text{then} \quad \left| \frac{f(x) - f(0)}{x} \right| < \epsilon.$$

Let $\delta = \epsilon$. Notice that $f(0) = 0$ and for any h we have $0 \leq f(x) \leq x^2$. Therefore if $0 < |x - 0| < \delta = \epsilon$, then

$$\left|\frac{f(x) - f(0)}{x}\right| = \left|\frac{f(x)}{x}\right| \leq \left|\frac{x^2}{x}\right| = |x| < \delta = \epsilon.$$

To show that f is not differentiable at any point a other than 0, we will use Theorem 3.1.2. We will show that f is not even continuous at a. If $x \neq 0$, then

$$\frac{1}{x^2} \cdot f(x) = \begin{cases} x^2/x^2 = 1, & \text{if } x \text{ is rational} \\ 0/x^2 = 0, & \text{if } x \text{ is irrational} \end{cases} = D(x).$$

But $1/x^2$ is continuous at every point except 0. If f were continuous at some nonzero point a, the product $f(x)/x^2 = D(x)$ would be continuous at a. But $D(x)$ is not continuous at any point, as we have already shown in Example 2.2.9. □

Problems for Section 3.1

•3.1.1. **a)** Show that a linear function $f(x) = mx + b$ is differentiable and that $f'(x) = m$.

b) Use **(a)** to prove Theorem 3.2.1 in the next section.

3.1.2. Show that $f(x) = |x|$ is differentiable at $a \neq 0$. Consider two cases: (i) $a > 0$ and (ii) $a < 0$.

3.1.3. **a)** Even though $f(x) = |x|$ is not differentiable at 0, show that the function $g(x) = x|x|$ is differentiable at 0.

b) More generally, show that if f is continuous at 0, then $g(x) = xf(x)$ is differentiable at 0. (Why doesn't the familiar product rule for derivatives apply here? Why is the continuity of f important?)

3.1.4. Where is the function

$$f(x) = \begin{cases} x, & \text{if } x \text{ is rational} \\ 0, & \text{if } x \text{ is irrational} \end{cases}$$

differentiable? (Hint: Review Example 3.1.3 and problem 2.2.5.)

3.1.5. Given a function f, we say that c is a **relative maximum** (or oppositely, **relative minimum**) point if there is some $\delta > 0$ such that c is a maximum (minimum) point for f on the interval $(c - \delta, c + \delta)$. In this situation, $f(c)$ is called a **relative maximum (relative minimum)** value of f.

a) Prove that if f is differentiable at a relative extreme point c, then c is a critical point of f.

b) Show that if c is a maximum (minimum) point for f on the open interval (a, b), then c is a relative maximum (minimum) point for f.

c) Now show that if f is differentiable on (a, b) and f has an extreme point c, then c is a critical point of f.

d) Consider the function $f(x) = x^3$ on the open, bounded interval $(-1, 1)$. Does f have any extreme points? Any critical points?

3.1.6. **a)** Define the terms **derivative from the left** and **derivative from the right** of f at a.

b) Show that if the derivative of f at a exists, both one-sided derivatives exist and are equal to $f'(a)$.

c) Show that if both one-sided derivatives exist and are equal, $f'(a)$ exists.

3.1.7. Show that $\sqrt{x}$ is differentiable at any $a > 0$.

•3.1.8. **a)** Recall that for any positive integer n and for any distinct reals x and a,

$$\frac{x^n - a^n}{x - a} = x^{n-1} + x^{n-2}a + \cdots + xa^{n-2} + a^{n-1}.$$

Use this fact to prove that if $f(x) = x^n$, then $f'(x) = nx^{n-1}$.

b) Show that if $g(x)$ is differentiable, then for any positive integer n, $(g(x))^n$ is differentiable and $\frac{d}{dx}(g(x))^n = n(g(x))^{n-1}g'(x)$. (Hint: Note that the numerator of $\dfrac{(g(x))^n - (g(a))^n}{x - a}$ can be factored to give

$$\frac{g(x) - g(a)}{x - a}((g(x))^{n-1} + (g(x))^{n-2}g(a) + \cdots + g(x)(g(a))^{n-2} + (g(a))^{n-1}).$$

Where does the continuity of g enter your argument?)

3.1.9. Show that if $g(x)$ is differentiable at a and $g(a) \neq 0$, then $\frac{1}{g(x)}$ is differentiable at a. How is continuity used in the proof?

3.2 Elementary Laws of Differentiation

This section is a brief overview of the relationship between the differentiation process and certain algebraic manipulations of differentiable functions. The goal throughout is to find formulas for the derivatives of these complicated functions in terms of their component functions. You have already proven the following result in the previous problem section.

THEOREM 3.2.1 *a) If $f(x) = c$ is a constant function, then f is differentiable and $f'(x) = 0$.*

b) If $f(x) = x$, then f is differentiable and $f'(x) = 1$.

Using the definition of derivative and the sum rule for limits, one easily verifies the following result.

THEOREM 3.2.2 ***(The Sum Rule)*** *Let f and g be functions differentiable at $x = a$. Then the sum $f + g$ is differentiable at $x = a$ and*

$$(f + g)'(a) = f'(a) + g'(a).$$

The following is only slightly more complicated to prove:

THEOREM 3.2.3 ***(The Product Rule)*** *If f and g are functions differentiable at $x = a$, then the product $f \cdot g$ is differentiable at $x = a$ and*

$$(f \cdot g)'(a) = f'(a) \cdot g(a) + f(a) \cdot g'(a).$$

PROOF First we rewrite the difference quotient for $f \cdot g$ by using the "trick" of adding 0 to it in the form

$$\frac{-f(a)g(x) + f(a)g(x)}{x - a}$$

in the second of the following steps:

$$\begin{aligned}
\frac{(f \cdot g)(x) - (f \cdot g)(a)}{x - a} &= \frac{f(x)g(x) - f(a)g(a)}{x - a} \\
&= \frac{f(x)g(x) - f(a)g(x) + f(a)g(x) - f(a)g(a)}{x - a} \\
&= \frac{f(x)g(x) - f(a)g(x)}{x - a} + \frac{f(a)g(x) - f(a)g(a)}{x - a} \\
&= \frac{f(x) - f(a)}{x - a} \cdot g(x) + f(a) \cdot \frac{g(x) - g(a)}{x - a}\,.
\end{aligned}$$

Taking limits in this last expression we find that

$$\begin{aligned}(f \cdot g)'(a) &= \lim_{x \to a} \frac{f(x) - f(a)}{x - a} \cdot g(x) + \lim_{x \to a} f(a) \cdot \frac{g(x) - g(a)}{x - a} \\ &= f'(a) \cdot \lim_{x \to a} g(x) + f(a)g'(a) \\ &= f'(a)g(a) + f(a)g'(a).\end{aligned}$$

The continuity of g is crucial at the last step. Since g is differentiable at a, then g is continuous there. So we are assured that $\lim_{x \to a} g(x) = g(a)$. ∎

COROLLARY 3.2.4 *If n is a positive integer and $f(x) = x^n$, then f is differentiable and $f'(x) = nx^{n-1}$.*

You are asked to prove this by using induction and the product rule in problem 8. (Also, see problem 3.1.8.)

A quotient $f(x)/g(x)$ can be expressed as a product: $f(x) \cdot 1/g(x)$; so to obtain a quotient rule for derivatives we only need to find an expression for the derivative of $1/g(x)$.

LEMMA 3.2.5 *If g is differentiable at a and $g(a) \neq 0$, then the function $r(x) = 1/g(x)$ is differentiable at a and*

$$r'(a) = \frac{-g'(a)}{(g(a))^2}.$$

PROOF Since g is differentiable at a, it is continuous at a. Since $g(a) \neq 0$, by problem 2.5.7 there is an interval $(a - \delta, a + \delta)$ containing a on which g is never 0. Therefore $r(x) = 1/g(x)$ is a well-defined function on the interval $(a - \delta, a + \delta)$. To calculate $r'(a)$ observe that

$$\begin{aligned}\lim_{x \to a} \frac{r(x) - r(a)}{x - a} &= \lim_{x \to a} \frac{\frac{1}{g(x)} - \frac{1}{g(a)}}{x - a} \\ &= \lim_{x \to a} \frac{g(a) - g(x)}{g(x)g(a) \cdot (x - a)} \\ &= \lim_{x \to a} \frac{-(g(x) - g(a))}{x - a} \cdot \frac{1}{g(x)g(a)} \\ &= \frac{-g'(a)}{(g(a))^2},\end{aligned}$$

as required. At the last step the continuity of g at a ensures that as $x \to a$, we have $g(x) \to g(a)$. ∎

The quotient rule for derivatives is now a simple consequence of the preceding result and the product rule.

THEOREM 3.2.6 ***(The Quotient Rule)*** *Suppose that f and g are both differentiable at a and that $g(a) \neq 0$. Then the quotient (f/g) is differentiable at $x = a$ and*

$$\left(\frac{f}{g}\right)'(a) = \frac{f'(a)g(a) - f(a)g'(a)}{\left(g(a)\right)^2}.$$

The final situation that we consider in this section is the differentiation of a composite function $f \circ g$. This will be the most "delicate" of the differentiation arguments tackled so far. Here's the rough idea of how to find $(f \circ g)'(a)$, assuming that g is differentiable at a and that f is differentiable at $g(a)$. If the derivative does exist, then

$$(f \circ g)'(a) = \lim_{x \to a} \frac{(f \circ g)(x) - (f \circ g)(a)}{x - a} = \lim_{x \to a} \frac{f(g(x)) - f(g(a))}{x - a}.$$

If we want this last expression to look like our usual difference quotients, the denominator should be $g(x) - g(a)$ rather than $x - a$. We can achieve this by a little bit of algebraic manipulation:

$$\lim_{x \to a} \frac{f(g(x)) - f(g(a))}{x - a} = \lim_{x \to a} \frac{f(g(x)) - f(g(a))}{g(x) - g(a)} \cdot \frac{g(x) - g(a)}{x - a}.$$

This works out nicely because now the second factor in the last expression is just $g'(a)$. Thus we expect

$$(f \circ g)'(a) = \lim_{x \to a} \frac{f(g(x)) - f(g(a))}{g(x) - g(a)} \cdot g'(a).$$

Because g is continuous at a, then $g(x) - g(a)$ goes to 0 as x goes to a. We expect the first term in the preceding limit to become $f'(g(a))$; thus $(f \circ g)'(a) = f'(g(a)) \cdot g'(a)$.

This argument is essentially correct unless we have divided by 0 in the form $g(x) - g(a)$ at the second stage. It might well be the case that $g(x) - g(a) = 0$, for example, if g were a constant function or if g were oscillating rapidly near a. With a bit of care we can even take care of this problem.

THEOREM 3.2.7 ***(The Chain Rule)*** *If g is differentiable at a and f is differentiable at $g(a)$, then $f \circ g$ is differentiable at a and*

$$(f \circ g)'(a) = f'(g(a)) \cdot g'(a).$$

PROOF The key is to define an auxiliary function F so that we can avoid division by 0 when $g(x) = g(a)$. Let

$$F(x) = \begin{cases} \frac{f(g(x))-f(g(a))}{g(x)-g(a)}, & \text{if } g(x) \neq g(a) \\ f'(g(a)), & \text{if } g(x) = g(a) . \end{cases}$$

Notice that the difference quotient for the derivative of $f \circ g$ can now be written as

$$\frac{f(g(x)) - f(g(a))}{x - a} = \begin{cases} \frac{f(g(x))-f(g(a))}{g(x)-g(a)} \cdot \frac{g(x)-g(a)}{x-a}, & \text{if } g(x) \neq g(a) \\ 0 = f'(g(a)) \cdot \frac{g(x)-g(a)}{x-a}, & \text{if } g(x) = g(a) . \end{cases}$$

$$= \begin{cases} F(x) \cdot \frac{g(x)-g(a)}{x-a}, & \text{if } g(x) \neq g(a) \\ F(x) \cdot \frac{g(x)-g(a)}{x-a}, & \text{if } g(x) = g(a) . \end{cases}$$

But this means that

$$\lim_{x \to a} \frac{f(g(x)) - f(g(a))}{x - a} = \lim_{x \to a} F(x) \cdot \frac{g(x) - g(a)}{x - a} .$$

We would like to use the product rule for limits to evaluate this last expression. In fact, in order for the chain rule to be correct, we need to show that $\lim_{x \to a} F(x) = f'(g(a))$. So choose any $\epsilon > 0$. Since f is differentiable at $g(a)$, by definition there is a $\delta_1 > 0$ such that if $0 < |y - g(a)| < \delta_1$, then

$$\left| \frac{f(y) - f(g(a))}{y - g(a)} - f'(g(a)) \right| < \epsilon .$$

Of course, instead of y in the last equation, we want $g(x)$. But since g is differentiable at a, g is continuous there. Using δ_1 as the "epsilon" in the definition of continuity of g, there is a $\delta > 0$ such that if $|x - a| < \delta$, then

$$|g(x) - g(a)| < \delta_1 .$$

Combining the last two inequlities, if $|x - a| < \delta$, then $|g(x) - g(a)| < \delta_1$. So

$$|F(x) - f'(g(a))| = \begin{cases} \left| \frac{f(g(x))-f(g(a))}{g(x)-g(a)} - f'(g(a)) \right| < \epsilon, & \text{if } g(x) \neq g(a), \\ |f'(g(a)) - f'(g(a))| = 0 < \epsilon, & \text{if } g(x) = g(a). \end{cases}$$

This proves that, $\lim_{x\to a} F(x) = f'(g(a))$. Now we can finish the proof.

$$\lim_{x\to a} \frac{f(g(x)) - f(g(a))}{x-a} = \lim_{x\to a} F(x) \cdot \frac{g(x)-g(a)}{x-a}$$
$$= \lim_{x\to a} F(x) \cdot \lim_{x\to a} \frac{g(x)-g(a)}{x-a} = f'(g(a)) \cdot g'(a),$$

where at the last step we have used the differentiability of g at a. Hence $(f \circ g)'(a) = f'(g(a)) \cdot g'(a)$. ▮

Problems for Section 3.2

3.2.1. If c is any constant and f is a function differentiable at a, the scalar product $cf(x)$ is differentiable at a and $(cf)'(a) = c \cdot f'(a)$.

a) Prove this result directly from the definition of derivative.

b) Prove it by using Theorems 3.2.1 and 3.2.3 instead.

3.2.2. Prove Theorem 3.2.2.

3.2.3. Prove the quotient rule by using the product rule and Lemma 3.2.5.

3.2.4. **a)** If $f+g$ is differentiable at a and f is differentiable at a, must g be differentiable at a?

b) If $f+g$ is differentiable at a, must f and g be differentiable at a?

3.2.5. If $f \cdot g$ is differentiable at a and f is differentiable at a, must g be differentiable at a? Do you need any further condition on f?

3.2.6. Assume only that g is continuous at 0 and that $g(0) \neq 0$. Let $f(x) = \dfrac{x^2}{g(x)}$.

a) Why doesn't the quotient rule for derivatives apply to $f(x)$?

b) Show by using the definition of the derivative that $f(x)$ is differentiable at 0 and calculate $f'(0)$.

3.2.7. **a)** Assume that g is a bounded function such that $|g(x)| < B$ for all x. Let $f(x) = x^2 g(x)$. Why doesn't the product rule for derivatives apply to $f(x)$?

b) Show that $f(x)$ is differentiable at 0 and find $f'(0)$.

3.2.8. Use induction to prove that for every natural number n, $\frac{d}{dx}x^n = nx^{n-1}$. Let $P(n)$ be the formula $\frac{d}{dx}x^n = nx^{n-1}$.

a) Verify that $P(1)$ holds, that is, show that $\frac{d}{dx}x = 1x^0 = 1$.

b) *Assume* that $P(k)$ holds, that is, $\frac{d}{dx}x^k = kx^{k-1}$. Show that this forces $P(k+1)$ to be true. That is, show that $\frac{d}{dx}x^{k+1} = (k+1)x^k$ by using the factorization $x^{k+1} = x \cdot x^k$, and the fact that $P(1)$ and $P(k)$ both hold and the product rule.

3.2.9. Use Lemma 3.2.5 and the preceding problem to prove that for any natural number n, if $f(x) = x^{-n}$, then f is differentiable and $f'(x) = -nx^{-n-1}$.

3.2.10. Prove that any polynomial $p(x) = a_n x^n + a_{n-1}x^{n-1} + \cdots + a_1 x + a_0$ is differentiable.

3.2.11. **a)** Verify that $(x^{1/3} - a^{1/3})(x^{2/3} + a^{1/3}x^{1/3} + a^{2/3}) = x - a$.

b) Use an ϵ, δ proof to show that for $a > 0$, $\lim_{x\to a} x^{1/3} = a^{1/3}$. (Hint: Use the factorization of $x - a$ in **(a)** in your proof.)

c) Use **(b)** to show that $x^{1/3}$ and $x^{2/3}$ are continuous for any $a > 0$.

d) Use the definition of derivative to show that if $f(x) = x^{1/3}$, then for any $a > 0$, $f'(a) = \frac{1}{3}a^{-2/3}$.

e) Can you modify the proof of the previous steps to show that the formula for $f'(a)$ is valid for $a < 0$? What happens at $a = 0$?

f) Can you modify the foregoing arguments to show that for all $n \in \mathbf{N}$, the function $x^{1/n}$ is differentiable for all $a > 0$?

3.2.12. Suppose that $f(x)$ is a differentiable function.

a) Prove that for any positive integer n, $(f(x))^n$ is differentiable and that its derivative is $n(f(x))^{n-1}f'(x)$. Give a different proof than in problem 3.1.8b.

b) Prove the analogous result for negative integers assuming $f(x) \neq 0$.

c) Let $g(x) = x^{1/n}$, where $n \in \mathbf{N}$. *Assume that $g(x)$ is differentiable for all $x > 0$.* Show that $g'(x) = \frac{1}{n}x^{(1-n)/n}$. (Hint: Calculate the derivative of $(g(x))^n$ directly by doing an algebraic simplification first. Then calculate the derivative of $(g(x))^n$ using **(a)**.)

d) Let $g(x) = x^{p/q}$, where p, $q \in \mathbf{N}$. *Assume that $g(x)$ is differentiable for all $x > 0$.* Show that $g'(x) = \frac{p}{q}x^{(p-q)/q}$.

3.2.13. Show that there exists a function $f(x)$ that is differentiable everywhere on $(-\infty, \infty)$, but that $f'(x)$ is *not* continuous. (Hint: Consider $f(x) = x^2 \sin\frac{1}{x}$.)

a) For $x \neq 0$, apply the product and chain rules to find an expression for $f'(x)$.

b) For $x = 0$, use either Definition 3.1.1 or Problem 3.1.3b to find $f'(0)$.

c) Show that $\lim_{x\to 0} f'(x)$ does not exist. (Hint: Use Example 2.2.3 to show that one term in the expression for $f'(x)$ from **(a)** is continuous and then use the same type of argument as in problem 2.2.12 to show that the other term of $f'(x)$ is not continuous.)

3.3 The Mean Value Theorem and Its Consequences

Theorem 3.2.1 says that if f is a constant function, its derivative is the zero function. Geometrically this is obvious: a constant function has zero slope or zero rate of change. We now turn this observation around and ask; if f is a differentiable function whose derivative is always 0, must f be a constant function? We can even give this question a physical interpretation. If an object has zero instantaneous velocity over a given interval, is the object stationary, that is, is the position of the object constant? Our intuition says yes, but how do we show it?

The problem we face in trying to prove such a result is that we must use information about f' to determine something about f. We can think of information about f' as infinitesimal or "local" information about how f is changing (or not changing, in our case) at some point. We need to use this local information to conclude something "global" about f (in this case that f is constant). Typically our previous results have used global information about f to say something local about f'. For example, "at an extreme point c (global information) of a differentiable function f we must have $f'(c) = 0$ (local information)."

3.3.1 The Mean Value Theorem

The mean value theorem is the critical result that relates local and global information. It has a wide variety of consequences that are crucial to the development of both differential and integral calculus. The mean value theorem says that if f is differentiable on $[a, b]$, then there is a point c between a and b such that

$$f'(c) = \frac{f(b) - f(a)}{b - a}.$$

Geometrically the mean value theorem is saying that there must be a point c between a and b whose tangent slope $f'(c)$ (local information) equals the slope of the secant through $(a, f(a))$ and $(b, f(b))$ (global information).

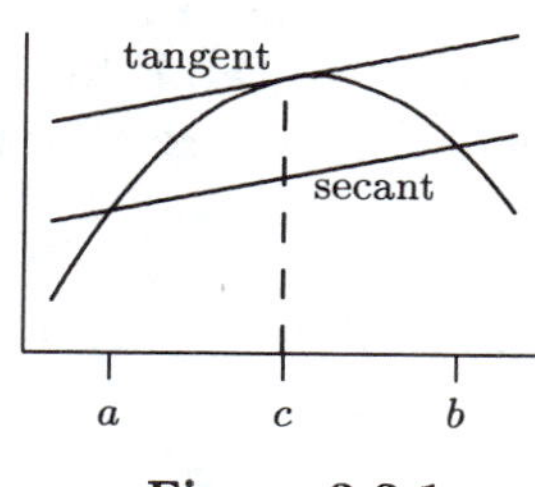

Figure 3.3.1

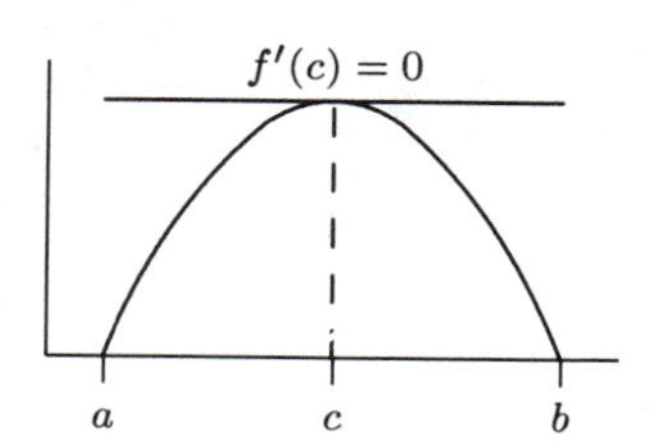

Figure 3.3.2

Despite being such a fundamental result, the proof of the mean value theorem is quite easy. We begin by proving a special case of the theorem.

THEOREM 3.3.1 ***(Rolle's Theorem)*** *Assume that f is continuous on the closed and bounded interval $[a, b]$, differentiable on (a, b), and that $f(a) = f(b) = 0$. Then there is a point c strictly between a and b such that $f'(c) = 0$.*

PROOF By Corollary 3.1.6, f must have maximum and minimum values on $[a, b]$. There are two possibilities: either (i) both of these extreme values occur at the endpoints of $[a, b]$ or (ii) at least one of the extreme values occurs at a critical point of f.

In case (i), $f(a) = f(b) = 0$ is both the minimum value and the maximum value of f on $[a, b]$. This can only happen if f is constant on the interval. But if f is constant, $f'(c) = 0$ for *any* point c between a and b. The situation in case (ii) is even more straightforward. If f has a critical point c between a and b, then $f'(c) = 0$. So either case leads to the desired result. ∎

The mean value theorem is a consequence of Rolle's theorem.

THEOREM 3.3.2 ***(The Mean Value Theorem)*** *Let f be continuous on $[a, b]$ and differentiable on (a, b). Then there is a point c strictly between a and b such that*

$$f'(c) = \frac{f(b) - f(a)}{b - a}.$$

PROOF We cannot apply Rolle's theorem directly to f because $f(a)$ or $f(b)$ may not be 0. So we use f to create an auxiliary function F that does satisfy the conditions of Rolle's theorem. To do this, let $L(x)$ be the linear function whose graph is the secant line that passes through the two points $(a, f(a))$ and $(b, f(b))$ on the graph of f (see Figure 3.3.1). Then $L(x)$ is continuous and differentiable, since it is linear. We consider the difference function

$$F(x) = f(x) - L(x).$$

$F(x) = f(x) - L(x)$ is continuous on $[a, b]$ and differentiable on (a, b), since f and L are. Since the line L passes through the point $(a, f(a))$, then $L(a) = f(a)$. Thus

$$F(a) = f(a) - L(a) = f(a) - f(a) = 0.$$

Similarly

$$F(b) = f(b) - L(b) = f(b) - f(b) = 0.$$

So F satisfies all three conditions of Rolle's theorem on $[a, b]$. Hence there is a point c between a and b such that $F'(c) = 0$. Now $F'(c) = f'(c) - L'(c)$. But L is linear, so its derivative is just the slope of the graph of L. Since the graph of L passses through the two points $(a, f(a))$ and $(b, f(b))$, the slope of L is

$$\frac{f(b) - f(a)}{b - a}.$$

Putting all of this together we have

$$0 = F'(c) = f'(c) - L'(c) = f'(c) - \frac{f(b) - f(a)}{b - a}.$$

Consequently

$$f'(c) = \frac{f(b) - f(a)}{b - a}. \quad \blacksquare$$

Even more general results are possible. For example, consider the following result due to Cauchy.

COROLLARY 3.3.3 ***(The Generalized Mean Value Theorem)*** *Suppose that f and g are continuous on the closed, bounded interval $[a, b]$ and are differentiable on (a, b). Then there is a point c strictly between a and b such that*

$$\big(g(b) - g(a)\big) f'(c) = \big(f(b) - f(a)\big) g'(c).$$

To see that Corollary 3.3.3 is a generalization of the mean value theorem, consider the special case when $g(x) = x$. Then $g(a) = a$, $g(b) = b$, and $g'(x) = 1$. Making these substitutions in the statement of the corollary yields the statement of the mean value theorem.

PROOF As in the last proof, the key here lies in finding an appropriate auxiliary function that, after applying the mean value theorem, will yield the desired equality. In this case our desired equation is equivalent to

$$(g(b) - g(a)) f'(c) - (f(b) - f(a)) g'(c) = 0. \tag{1}$$

Thus we are led naturally to consider the auxiliary function

$$h(x) = (g(b) - g(a))f(x) - (f(b) - f(a))g(x).$$

Since f and g are continuous on $[a, b]$ and differentiable on (a, b), h satisfies the conditions of the mean value theorem. The derivative of h is given by

$$h'(x) = (g(b) - g(a))f'(x) - (f(b) - f(a))g'(x). \tag{2}$$

The function h has the same values at each endpoint ($h(a) = h(b) = g(b)f(a) - f(b)g(a)$). Therefore, applying the mean value theorem to h we have for some $c \in (a, b)$

$$0 = \frac{h(b) - h(a)}{b - a} = h'(c) = (g(b) - g(a))f'(c) - (f(b) - f(a))g'(c),$$

which proves (1). ∎

We are now ready to prove that a function whose derivative is identically zero on an interval is constant. First, recall that x is an **interior point** of an interval I if x is not an endpoint of I.

THEOREM 3.3.4 *Suppose that f is continuous on an interval I and that $f'(x) = 0$ for all interior points x of I. Then f is constant on I.*

PROOF To show that f is constant on I, it suffices to show that if a and b are any two distinct points in I, then $f(a) = f(b)$. We may assume $a < b$. Applying the mean value theorem to f on $[a, b]$ we have

$$f'(c) = \frac{f(b) - f(a)}{b - a}$$

for some c between a and b. Since c is in I, we know that $f'(c) = 0$. It follows that $0 = f(b) - f(a)$, so $f(a) = f(b)$. ∎

COROLLARY 3.3.5 *Assume f and g are continuous on an interval I and that for each interior point x of I, $f'(x) = g'(x)$. Then there is a constant c such that $f = g + c$.*

PROOF Since $f'(x) = g'(x)$ at each interior point of I, $(f - g)'(x) = 0$ at each interior point. By Theorem 3.3.4 the function $f - g$ is a constant function c, that is, $f = g + c$. ∎

3.3.2 L'Hôpital's Rule

The generalized mean value theorem is of more than just theoretical interest. It is the key to proving a theorem that greatly simplifies evaluating limits of the form

$$\lim_{x \to a} \frac{f(x)}{g(x)},$$

where both $\lim_{x \to a} f(x) = 0$ and $\lim_{x \to a} g(x) = 0$. Notice that the quotient rule for limits is useless in such cases, since the denominator is going to 0. This is exactly the problem that arose with the difference quotient and led to the development of the derivative. Indeed, the limits of such quotients are calculated using derivatives, as we see in the next theorem.

THEOREM 3.3.6 ***(L'Hôpital's Rule)*** *Assume that f and g are differentiable on the open interval (a, b) and that*

$$\lim_{x \to a+} f(x) = 0 \qquad \text{and} \qquad \lim_{x \to a+} g(x) = 0.$$

Further, assume that $g'(x) \neq 0$ for all $x \in (a, b)$. If $\lim_{x \to a^+} f'(x)/g'(x)$ exists, then

$$\lim_{x \to a^+} \frac{f(x)}{g(x)} = \lim_{x \to a^+} \frac{f'(x)}{g'(x)}.$$

An analogous result holds for left-hand limits. Combining the two one-sided limit theorems, one obtains a similar result for two-sided limits for $x \to c$, where c is any point in the interval (a, b). In this latter case, f and g need not even be defined at c.

PROOF We would like to use the generalized mean value theorem , but f and g may not even be defined at a. Consequently we create two new functions that are defined at a and that at other points agree with f and g, respectively:

$$F(x) = \begin{cases} f(x), & \text{if } x \neq a \\ 0, & \text{if } x = a, \end{cases} \qquad \text{and} \qquad G(x) = \begin{cases} g(x), & \text{if } x \neq a \\ 0, & \text{if } x = a. \end{cases}$$

Since f and F are equal on (a, b) and f is differentiable, we may conclude that F is differentiable on (a, b). Further, since $\lim_{x \to a^+} F(x) = \lim_{x \to a^+} f(x) = 0 = F(0)$, we see that F is continuous on $[a, b)$. Consequently if $a < x < b$, then F is continuous on the closed interval $[a, x]$ and differentiable on (a, x). The same remarks apply to G, and thus F and G satisfy the conditions of the generalized mean value theorem on

$[a, x]$. Therefore there is a point c_x (note the dependence of this point on x) satisfying $a < c_x < x$ such that

$$(F(x) - F(a))G'(c_x) = (G(x) - G(a))F'(c_x).$$

But $F(a) = G(a) = 0$, and away from a we have $F = f$ and $G = g$. Therefore the last equation simplifies to

$$f(x)g'(c_x) = g(x)f'(c_x). \tag{1}$$

Now by hypothesis, $g'(c_x) \neq 0$. Further, $g(x) \neq 0$. (If $g(x) = 0$, we would have $G(x) = G(a) = 0$ and by Rolle's theorem there would be a point y between a and x such that $G'(y) = 0$. This would contradict the hypothesis that g' is never 0 on (a, b).) Therefore we may divide (1) by both $g'(c_x)$ and $g(x)$ and obtain

$$\frac{f(x)}{g(x)} = \frac{f'(c_x)}{g'(c_x)}.$$

As $x \to a$, the point $c_x \to a$ because $a < c_x < x$. Taking limits we find that

$$\lim_{x \to a+} \frac{f(x)}{g(x)} = \lim_{x \to a+} \frac{f'(c_x)}{g'(c_x)} = \lim_{x \to a+} \frac{f'(x)}{g'(x)},$$

where the last equality follows from problem 2.4.11. ∎

To apply L'Hôpital's rule, the functions involved must be differentiable in an open interval containing the point at which the limit is being taken (except perhaps at the point itself). You should always check that this is the case.

EXAMPLE 3.3.1 Evaluate $\lim_{x \to 0} \frac{x^2 + 2x}{\sin x}$ if it exists.

SOLUTION Notice that $\lim_{x \to 0} x^2 + 2x = \lim_{x \to 0} \sin x = 0$, and the functions are differentiable everywhere. Therefore we may apply L'Hôpital's rule.

$$\lim_{x \to 0} \frac{x^2 + 2x}{\sin x} = \lim_{x \to 0} \frac{2x + 2}{\cos x} = 2. \quad \square$$

EXAMPLE 3.3.2 Evaluate $\lim_{x \to 0} \frac{1 - \cos x}{x \sin x}$ if it exists.

SOLUTION This time, $\lim_{x \to 0} 1 - \cos x = \lim_{x \to 0} x \sin x = 0$, and the functions are differentiable everywhere. Therefore we may apply L'Hôpital's rule.

$$\lim_{x \to 0} \frac{1 - \cos x}{x \sin x} = \lim_{x \to 0} \frac{\sin x}{\sin x + x \cos x}.$$

This new limit is still of the "0/0" form. The functions involved are still differentiable, so we apply L'Hôpital's rule once again and find that

$$\lim_{x \to 0} \frac{1 - \cos x}{x \sin x} = \lim_{x \to 0} \frac{\sin x}{\sin x + x \cos x} = \lim_{x \to 0} \frac{\cos x}{\cos x + \cos x - x \sin x} = \frac{1}{2}. \quad \square$$

L'Hôpital's rule can be extended to limits at infinity.

COROLLARY 3.3.7 *Assume that f and g are differentiable on the interval $(a, +\infty)$ and assume that*

$$\lim_{x \to +\infty} f(x) = 0 \qquad \text{and} \qquad \lim_{x \to +\infty} g(x) = 0.$$

Further, assume that $g'(x) \neq 0$ for all $x \in (a, +\infty)$. If $\lim_{x \to +\infty} f'(x)/g'(x)$ exists, then

$$\lim_{x \to +\infty} \frac{f(x)}{g(x)} = \lim_{x \to +\infty} \frac{f'(x)}{g'(x)}.$$

An analogous result holds for limits as $x \to -\infty$.

PROOF By Theorem 2.4.6,

$$\lim_{y \to 0+} f\left(\frac{1}{y}\right) = \lim_{x \to +\infty} f(x) = 0 \quad \text{and} \quad \lim_{y \to 0+} g\left(\frac{1}{y}\right) = \lim_{x \to +\infty} g(x) = 0.$$

Since f and g are differentiable on the interval $(a, +\infty)$, it follows that the composite functions $f\left(\frac{1}{y}\right)$ and $g\left(\frac{1}{y}\right)$ are differentable on $(0, 1/a)$. Further,

$$\frac{d}{dy} g\left(\frac{1}{y}\right) = \frac{-g'\left(\frac{1}{y}\right)}{y^2},$$

which is never 0 on $(0, 1/a)$ since g' is never 0 on $(a, +\infty)$. So L'Hôpital's rule may be applied to

$$\lim_{y \to 0+} \frac{f\left(\frac{1}{y}\right)}{g\left(\frac{1}{y}\right)}.$$

Thus, using Theorem 2.4.6 and L'Hôpital's rule

$$\lim_{x \to +\infty} \frac{f(x)}{g(x)} = \lim_{y \to 0+} \frac{f\left(\frac{1}{y}\right)}{g\left(\frac{1}{y}\right)} = \lim_{y \to 0+} \frac{-y^{-2} f'\left(\frac{1}{y}\right)}{-y^{-2} g'\left(\frac{1}{y}\right)}.$$

Simplifying the last expression and using Theorem 2.4.6 again gives the desired result:

$$\lim_{x \to +\infty} \frac{f(x)}{g(x)} = \lim_{y \to 0+} \frac{f'\left(\frac{1}{y}\right)}{g'\left(\frac{1}{y}\right)} = \lim_{x \to +\infty} \frac{f'(x)}{g'(x)}. \quad \blacksquare$$

EXAMPLE 3.3.3 Evaluate $\lim_{x\to+\infty} \dfrac{1-\cos\frac{1}{x}}{\sin\frac{1}{x}}$ if it exists.

SOLUTION Both $1-\cos\frac{1}{x}$ and $\sin\frac{1}{x}$ are differentiable for $x > 0$ and both approach 0 as $x \to +\infty$. Further,

$$\frac{d}{dx}\sin\tfrac{1}{x} = -x^{-2}\cos\tfrac{1}{x},$$

which is never 0 for $x > 2/\pi$. Therefore the extension of L'Hôpital's rule applies here on the interval $(2/\pi, +\infty)$ and we obtain

$$\lim_{x\to+\infty}\frac{1-\cos\frac{1}{x}}{\sin\frac{1}{x}} = \lim_{x\to+\infty}\frac{x^{-2}\sin\frac{1}{x}}{-x^{-2}\cos\frac{1}{x}} = \lim_{x\to+\infty}\frac{\sin\frac{1}{x}}{-\cos\frac{1}{x}} = \frac{\sin 0}{\cos 0} = 0. \quad \square$$

Problems for Section 3.3

3.3.1. Let $f(x) = |x|$ on $[-3, 3]$. Show that $f(3) = f(-3)$, but that there is no number c in $(-3, 3)$ such that $f'(c) = 0$. Does this contradict the mean value theorem?

3.3.2. Suppose that $|f'(x)| \leq M$ on $[a, b]$, that is, that the derivative of f is bounded on $[a, b]$. Use the mean value theorem to prove that $|f(b) - f(a)| \leq M(b - a)$.

3.3.3. We say that a function f satisfies a **Lipschitz condition** at x if there exists a constant M and a $\delta > 0$ such that if $|x - y| < \delta$ then $|f(y) - f(x)| < M|y - x|$.

a) Assume that f is differentiable everywhere and that f' is continuous. Show that for any x and for any $\delta > 0$ there is an M such that $|f'(y)| \leq M$ for all points y in $[x - \delta, x + \delta]$.

b) Show that f satisfies a Lipschitz condition at x. Use the previous problem.

3.3.4. Use Rolle's theorem to show that there is a solution to $\cot x = x$ in the interval $(0, \pi/2)$. (Hint: Let $f(x) = x\cos x$ on $[0, \pi/2]$.)

3.3.5. Use Rolle's theorem to show that there is a solution to $\tan x = 1 - x$ in the interval $(0, 1)$. (Hint: Let $f(x) = (x - 1)\sin x$ on $[0, 1]$.)

3.3.6. Suppose that f is continuous on two disjoint intervals I and J and assume that $f'(x) = 0$ at each interior point of both I and J. Is f necessarily a constant function?

3.3.7. A function f is called **increasing on an open interval** I if whenever a and b are points in I such that $a < b$, then $f(a) < f(b)$. Prove that if $f'(x) > 0$ for all x in an open interval I, then f is increasing on I by completing the following steps.

a) Let a and b be any two points in I such that $a < b$. Use the mean value theorem to show that there is a point $c \in (a, b)$ such that $f(b) = f(a) + f'(c) \cdot (b - a)$.

b) Show that $f'(x) \cdot (b - a) > 0$ and finish the proof.

3.3.8. Define the expression **decreasing on an open interval** I and then state and prove the analogue to the preceding theorem for decreasing functions.

3.3.9. A function F is said to be an **antiderivative** of f on an open interval I if $F'(x) = f(x)$ for all $x \in I$. Prove that if F and G are both antiderivatives of f on the open interval I, then $F = G + c$ for some constant c. (In other words, two antiderivatives of the same function differ by at most a constant.)

3.3.10. A **fixed point** of a function is a point d such that $f(d) = d$. Suppose that f is differentiable everywhere and that $f'(x) < 1$ for all x. Show that there can be at most one fixed point for f. (Hint: Suppose that a and b were two fixed points, with $a < b$. Apply the mean value theorem to obtain a contradiction.)

3.3.11. Evaluate the following limits if they exist.

a) $\lim_{x\to 1} \dfrac{1-x^2}{1-\sqrt{x}}$ b) $\lim_{x\to 0} \dfrac{1-\cos x}{6x^2-x^4}$ c) $\lim_{x\to 0} \dfrac{4x^3+x^4}{1+x-e^x}$

d) $\lim_{x\to 0} \dfrac{\ln(1+x)}{x-x^2}$ e) $\lim_{x\to \pi} \dfrac{\sin x}{\sin 2x}$ f) $\lim_{x\to 0} \dfrac{\sqrt{1-x}-\sqrt{1+x}}{x}$

3.3.12. Let $f(x) = x^2$ and $g(x) = x^3$ on the interval $[-1, 1]$. Show that there is no point $c \in (-1, 1)$ such that

$$\frac{f'(c)}{g'(c)} = \frac{f(1)-f(-1)}{g(1)-g(-1)}.$$

Does this contradict the generalized mean value theorem?

3.3.13. This problem is adapted from Robert M. Genther, "A Simple Estimate of the Error in Linear Approximation" (*Amer. Math. Mo.*, 96 (1989), 522–523). The mean value theorem provides a way to estimate the value of a function f at x near a known value $f(a)$. Prove the following result: If f'' exists and is bounded on the interval I with endpoints a and x, then

$$f(x) = f(a) + f'(a)(x-a) + R,$$

where $|R| < M(x-a)^2$ and M is the least upper bound for $|f''|$ on I.

a) Show that there is a point c between a and x such that $f(x) = f(a) + f'(c)(x-a)$.

b) Show that there is a point c^* such that $f'(c) = f'(a) + f''(c^*)(c-a)$.

c) Show that $f(x) = f(a) + f'(a)(x-a) + f''(c^*)(c-a)(x-a)$.

d) Use the fact that $|c-a| < |x-a|$ to complete the proof.

3.3.14. **a)** Let $f(x) = x^2$ and $g(x) = x^3$ on the interval $[0, 1]$. Find the values of c guaranteed in the mean value theorem for each of these functions.

b) Find the value of c guaranteed by the generalized mean value theorem for the pair of functions f and g. (Note that all these values of c differ.)

3.4 Integration

In this section we will solve the problem of finding the area under certain curves. The ancients were greatly interested in such questions and devised various algorithms that allowed them to solve a range of such problems. The Babylonians and Egyptians knew the area formula for a triangle and thus were able to calculate the area of any region that could be completely subdivided into a finite number of nonoverlapping triangles. Such regions include rectangles and general polygons.

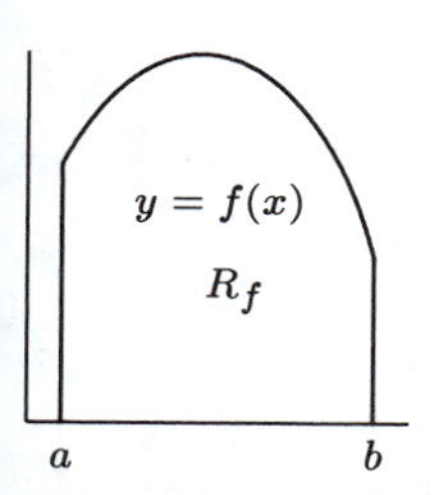

Figure 3.4.1

We will begin by considering the following general situation. Assume that a function f is continuous and nonnegative on the closed, bounded interval $[a, b]$. Let R_f denote the region that is bounded above by the graph of $y = f(x)$ and below by the x-axis and that lies between the vertical lines $x = a$ and $x = b$ (see Figure 3.4.1). The problem is to find the area of R_f. Among the general regions that can be described in this way are such familiar objects as rectangles, triangles, and semicircles. To see that this problem is not trivial, think about the familiar formula πr^2 for the area of a circle. How would you actually *prove* that this formula is correct?

3.4.1 Attempts to Solve the Area Problem

The Greeks made great advances in calculating areas and volumes of various regions. The most fruitful procedure they used was the *method of exhaustion*. It was devised by Eudoxus of Cnidos in the fourth century B.C. and used extensively by Archimedes to compute the areas of circles and ellipses and the volumes of spheres. Underlying any argument concerning area (volume), whether or not it used the method of exhaustion, were two basic assumptions:

1. if region A is contained in region B, then area $(A) \leq$ area (B);
2. if region C is the union of two nonoverlapping regions A and B, then area $(C) =$ area $(A) +$ area (B).

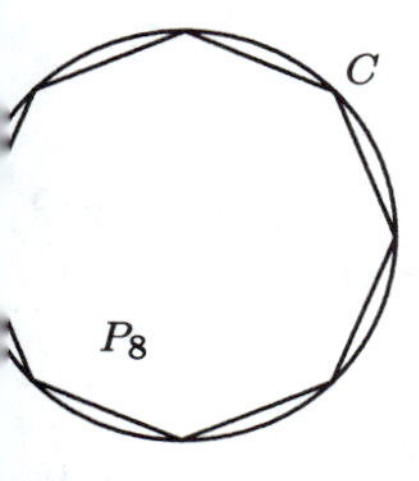

Figure 3.4.2

To find the area of a curvilinear region C (such as a circle) by the method of exhaustion, one had to construct a sequence of inscribed polygons $P_1, P_2, \ldots, P_n$ in C, where each successive polygon encompassed the previous one and came closer to completely filling ("exhausting") the area of C (see Figure 3.4.2).

The crucial step was showing that the area of the region inside C but not contained in P_n (denoted by $C - P_n$) could be made arbitrarily small by choosing n sufficiently large. Using limits, we might be tempted in hindsight to rewrite this as

$$\lim_{n\to\infty} \text{area}\,(P_n) = C.$$

Yet this would misrepresent the state of Greek mathematics at the time:

> The method of exhaustion, although equivalent to the type of argument now employed in proving the existence of a limit in the differential and integral calculus, does not represent the point of view involved in the passage to the limit.[1]

The Greeks had no concept corresponding to our completeness of the real number system, so passing to the limit as we know it was simply impossible. Instead, in an exhaustion proof the inscribed polygons would be made to "approach" the figure C but could never coincide with C. This was sufficient because by using a proof by contradiction "it could be shown that a ratio greater or less than that of equality was inconsistent with the principle that the difference [of the areas of C and of P_n] could be made as small as desired."[2]

Today, the concept of area seems simple, intuitive, and quite familiar. Yet reflection reveals that what is actually familiar is a catalog of area formulas for elementary regions including triangles, squares, rectangles, trapezoids, and circles. We know (have been told) the areas of certain classes of regions but we have not been told what area is itself. We have no definition of area. In fact we will see that such a definition is quite complicated, and in the end we too will only be able to define area for a restricted class of regions.

To define the area of a region such as R_f we could proceed as follows: We might begin by taking as an "axiom" that the area of a rectangle is given by the formula "base $\times$ height." Then a natural way to find the area of an arbitrary region R_f is to approximate R_f by a union of rectangles. A simple method of doing this is to divide the interval $[a, b]$ into smaller subintervals each of which will act as a base of a rectangle. Next, approximate f on each subinterval by an appropriate constant to form the height of the rectangle. The sum of the areas of these rectangles gives an approximate value for the area of R_f. (See Figure 3.4.3.) We now begin to make some of these ideas precise.

---

[1] Carl B. Boyer, *The History of the Calculus and Its Conceptual Development* (New York: Dover Publications, Inc., 1959), pp. 34–35.

[2] Boyer, p. 35.

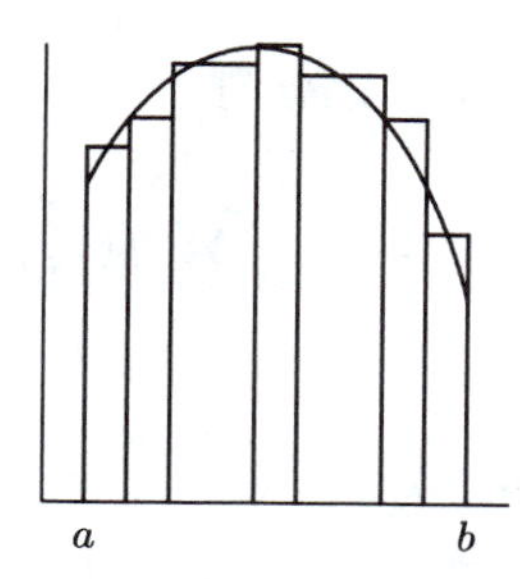

Figure 3.4.3

DEFINITION 3.4.1 *A **partition** P of the closed, bounded interval $[a, b]$ is a finite set of points $\{x_0, x_1, \ldots, x_n\}$ such that $a = x_0 < x_1 < \cdots < x_n = b$.*

The subintervals $[x_{i-1}, x_i]$ form the bases of the approximating rectangles. It is sometimes convenient to use Δx_i to denote the length of the base of the ith interval, that is,

$$\Delta x_i = x_i - x_{i-1}.$$

To approximate the varying height of the graph of f on the ith interval, we evaluate f at a single point within the interval. Pick any $x_i^* \in [x_{i-1}, x_i]$ and use $f(x_i^*)$ as the height of the ith rectangle. The area of the ith rectangle is then $f(x_i^*)\Delta x_i$. Taking the sum of the areas of all these rectangles gives

$$\text{area}\,(R_f) \approx \sum_{i=1}^{n} f(x_i^*)\Delta x_i.$$

Such a sum is called a **Riemann sum** for f over $[a, b]$.

If we include more points in our partition of $[a, b]$, it "appears" that the approximation to the area of R_f improves.[3]

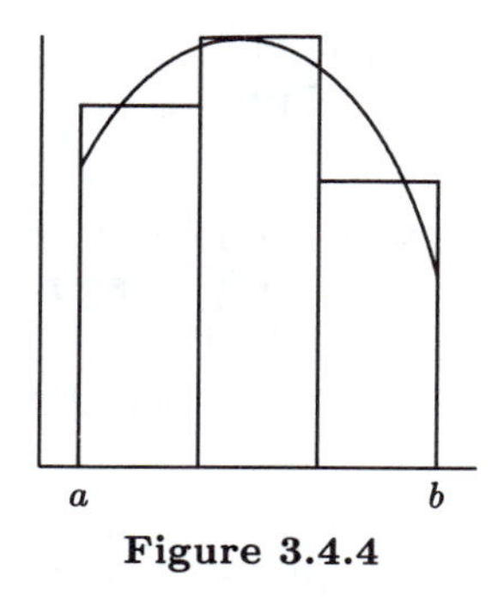

Figure 3.4.4

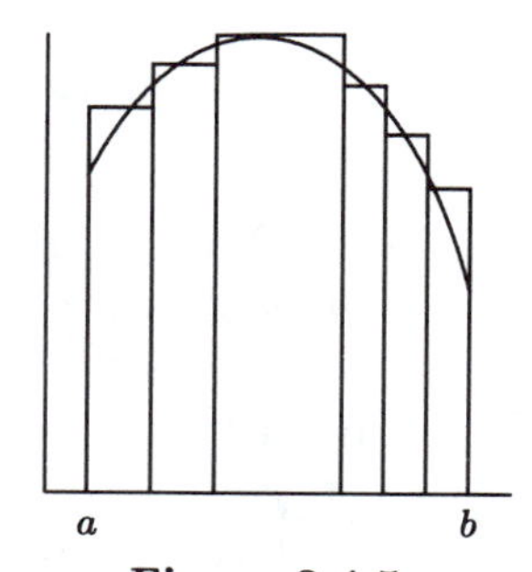

Figure 3.4.5

[3] This is not always the case. Sometimes additional partition points make the approximation worse, depending on the choice of points at which f is evaluated.

We expect that as the number of partition points becomes "infinite" we should approach the exact area of the region. That is,

$$\text{area}\,(R_f) = \lim_{n\to\infty} \sum_{i=1}^{n} f(x_i^*)\Delta x_i. \tag{1}$$

There are several difficulties with this limit expression. First, n is an integer and we have not yet defined limits for integer variables. Second, for the *same* integer n, $\sum_{i=1}^{n} f(x_i^*)\Delta x_i$ can have any of an infinite number of values depending on which points are chosen to be partition points and which points are then chosen to be the evaluation points x_i^*. Third, if the partition points are not reasonably distributed throughout the interval, the approximation in (1) will never be very good, no matter how many points are used (see Figure 3.4.6). What we really want is the widths of all the rectangles to go to 0. This will happen if we force the maximum width of the subintervals to go to 0. So we might try setting

$$\text{area}\,(R_f) = \lim_{\max\{\Delta x_i\}\to 0} \sum_{i=1}^{n} f(x_i^*)\Delta x_i. \tag{2}$$

But again we run into some of the same problems as with (1). For the same $\max\{\Delta x_i\}$ the approximation $\sum_{i=1}^{n} f(x_i^*)\Delta x_i$ can take on infinitely many values depending on partition and evaluation points.

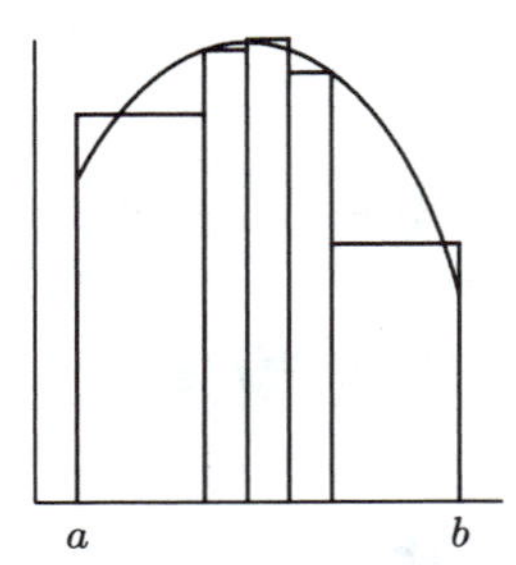

Figure 3.4.6

We are having trouble making our limit precise because *too many quantities are varying at once.* We can simplify this process by focusing on two particular types of Riemann sums.[4] The key is to choose the heights of the approximating rectangles in an especially convenient way.

[4] Several times in the text we will make use of Riemann sums of the type in (1) and (2) to *motivate* particular definitions. Despite their lack of precision, such expressions convey enormous amounts of geometric intuition. Historically the type of sums in (1) and (2) were used earlier to "define" the integral than the approach we will shortly take in the next section.

Since f is continuous on $[a, b]$ it is continuous on $[x_{i-1}, x_i]$. By the max-min theorem f will attain both a maximum and minimum on $[x_{i-1}, x_i]$ which we denote by

$$M_i = \max\{f(x) \mid x_{i-1} \le x \le x_i\}$$

and

$$m_i = \min\{f(x) \mid x_{i-1} \le x \le x_i\}.$$

The quantities $M_i(x_i - x_{i-1})$ and $m_i(x_i - x_{i-1})$ represent areas of rectangles each of whose base is the interval $[x_{i-1}, x_i]$ and whose heights are M_i and m_i, respectively (see Figure 3.4.7). For the partition P the sum

$$U(P) = \sum_{i=1}^{n} M_i(x_i - x_{i-1})$$

gives the total area of the "upper" rectangles whose union contains R_f and $U(P)$ is the largest possible value that a Riemann sum using partition P can have; it is the "worst" overestimate possible using P. The sum

$$L(P) = \sum_{i=1}^{n} m_i(x_i - x_{i-1})$$

gives the total area of the "lower" rectangles whose union is contained in R_f and $L(P)$ is the smallest value or "worst" underestimate that any Riemann sum for f can have using the partition P.

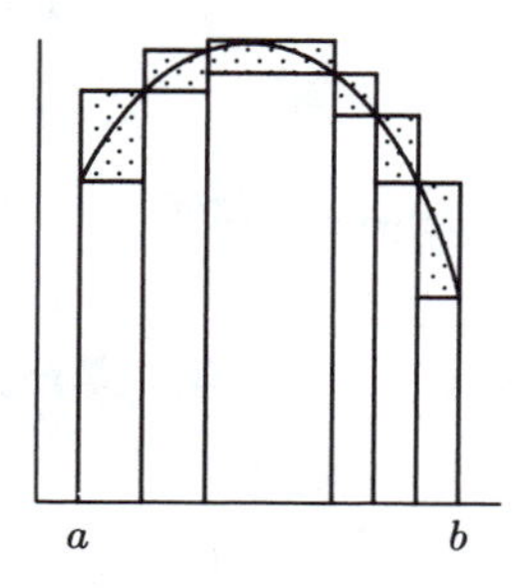

Figure 3.4.7

However we define the area of R_f, it seems reasonable that the area should lie between the extreme underestimate and overestimate,

$$L(P) \le \text{area of } R_f \le U(P), \tag{3}$$

no matter what partition P is used. Actually, we will turn this entire argument around and use the approximation process to define the meaning of the term "area" for regions such as R_f in terms of the sums $L(P)$ and $U(P)$.

3.4.2 Upper and Lower Sums

At this stage we have a rough idea of the problem. If we can find partitions that will make the difference $U(P) - L(P)$ arbitrarily small, we can "zero in" on the area of R_f in (3). In this way we can be led to a *definition* of area in terms of the sums $L(P)$ and $U(P)$. If possible, such a definition should preserve our intuitions about area, such as familiar formulas and properties. At the same time we should not implicitly assume in our arguments any facts about area that cannot be proven from our definition. For example, we should not assume without proof that if region B contains region A, then area$(A) \leq$ area(B). In this section we explore the properties of the sums $U(P)$ and $L(P)$ from which we will forge our definition of area.

For which sorts of functions do the sums $U(P)$ and $L(P)$ exist? Assume that f is defined on the closed, bounded interval $[a, b]$ and that $P = \{x_0, x_1, \ldots, x_n\}$ is a partition of $[a, b]$. If f is not continuous on $[a, b]$, then f need not achieve a maximum or minimum on each subinterval $[x_{k-1}, x_k]$. Still, if f is at least bounded on $[a, b]$, then f will have a least upper bound and a greatest lower bound on each subinterval as a consequence of the least upper bound axiom. In this context, different language is used. Let S be a bounded set of real numbers. The least upper bound of S is also called the **supremum** of S and is denoted by $\sup S$ (pronounced "soup S"). The greatest lower bound of S is also called the **infimum** of S and is denoted by $\inf S$.

DEFINITION 3.4.2 *Let f be a bounded function on $[a, b]$ and let $P = \{x_0, x_1, \ldots, x_n\}$ be a partition of $[a, b]$. Let*

$$m_i = \inf\{f(x) \mid x_{i-1} \leq x \leq x_i\} \text{ and } M_i = \sup\{f(x) \mid x_{i-1} \leq x \leq x_i\}.$$

*The **lower sum of f relative to the partition** P is*

$$L(P, f) = \sum_{i=1}^{n} m_i(x_i - x_{i-1}).$$

*The **upper sum of f relative to the partition** P is*

$$U(P, f) = \sum_{i=1}^{n} M_i(x_i - x_{i-1}).$$

When either the function f or the partition P is clear from the context, these sums will be denoted simply by $L(P)$ and $U(P)$ or $L(f)$ and $U(f)$.

These sums are clearly motivated by the earlier rectangle approach to area, but they do not depend on the idea of area for their meaning. In fact, f, $L(P)$, and $U(P)$ might very well be negative here, which would have been meaningless in the earlier discussion.

As one would expect, the lower sum for a partition P is less than or equal to the upper sum for P. That is, since $m_i \le M_i$, then $m_i(x_i - x_{i-1}) \le M_i(x_i - x_{i-1})$. Summing yields

$$L(P, f) = \sum_{i=1}^{n} m_i(x_i - x_{i-1}) \le \sum_{i=1}^{n} M_i(x_i - x_{i-1}) = U(P, f) \qquad (4)$$

But what happens when we try to compare $L(P, f)$ to $U(Q, f)$ when P and Q are different partitions? Guided by the area analogy, since $U(Q)$ overestimates area and $L(P)$ underestimates area, we expect that $L(P) \le U(Q)$. In fact this is the case, but we must provide an analytic proof that does not depend on the undefined notion of area. The easiest way to do this involves "refining" the partitions involved.

DEFINITION 3.4.3 *Let P and Q be partitions of $[a, b]$. Q is said to be **finer** than P if Q contains all the points of P. Q is then called a **refinement** of P.*

THEOREM 3.4.4 *If Q is finer than P, then*

$$L(P, f) \le L(Q, f) \le U(Q, f) \le U(P, f).$$

PROOF The middle inequality is just (4). So it remains to verify the other two inequalities. We will concentrate on showing $L(P) \le L(Q)$. A similar argument, whose details you should take the time to write out, works for showing $U(Q) \le U(P)$.

We treat the simplest case first. Suppose that Q has only one additional point, y, not in P. In particular, say $P = \{x_0, x_1, \ldots, x_n\}$ and $Q = \{x_0, \ldots, x_{k-1}, y, x_k, \ldots, x_n\}$. Let

$$m' = \inf\{f(x) \mid x_{k-1} \le x \le y\} \quad \text{and} \quad m'' = \inf\{f(x) \mid y \le x \le x_k\}.$$

As usual, let

$$m_k = \inf\{f(x) \mid x_{k-1} \le x \le x_k\}.$$

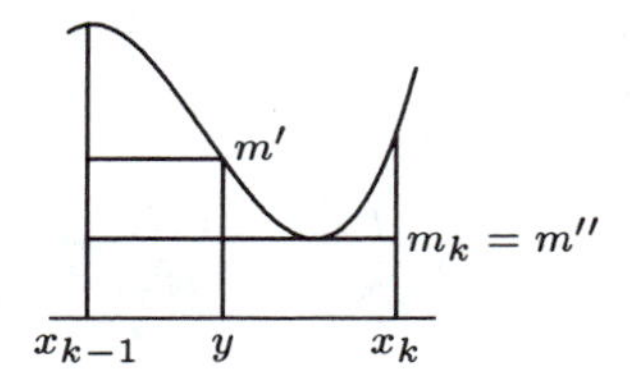

Figure 3.4.8

Since $[x_{k-1}, y]$ is contained in $[x_{k-1}, x_k]$, the set $\{f(x) \mid x_{k-1} \le x \le y\}$ is a subset of $\{f(x) \mid x_{k-1} \le x \le x_k\}$. Consequently $m_k \le m'$ and similarly $m_k \le m''$. Splitting the interval $[x_{k-1}, x_k]$ in two pieces yields

$$m_k(x_k - x_{k-1}) = m_k(y - x_{k-1}) + m_k(x_k - y) \le m'(y - x_{k-1}) + m''(x_k - y).$$

Therefore

$$\begin{aligned} L(P, f) &= \sum_{i=1}^{n} m_i(x_i - x_{i-1}) \\ &\le \sum_{i=1}^{k-1} m_i(x_i - x_{i-1}) + m'(y - x_{k-1}) \\ &\qquad\qquad + m''(x_k - y) + \sum_{i=k+1}^{n} m_i(x_i - x_{i-1}) \\ &= L(Q, f). \end{aligned}$$

So in the special case where Q is obtained from P by inserting a single point,

$$L(P, f) \le L(Q, f). \tag{5}$$

To prove the general case, if Q is obtained from P by inserting n points, simply employ an intermediate sequence of partitions that insert a single point at a time. That is, consider the partitions

$$P = P_0, P_1, P_2, \ldots, P_n = Q,$$

where P_k is obtained from P_{k-1} by inserting a single point. Applying equation (5) n times yields the desired result:

$$L(P, f) = L(P_0, f) \le L(P_1, f) \le \cdots \le L(P_n, f) = L(Q, f). \quad \blacksquare$$

The preceding theorem expresses a relationship between a lower and an upper sum when one partition is finer than another. But our initial goal was to relate any lower sum to any upper sum without any assumptions of a relationship between the partitions. This turns out to be a simple consequence of the last result.

Corollary 3.4.5 *Let f be a bounded function on $[a, b]$ and let P and R be any two partitions of $[a, b]$. Then*

$$L(P, f) \le U(R, f).$$

PROOF Let Q be the **common refinement** of both P and R, that is, let Q be the partition of $[a, b]$ whose points are exactly the union of all the points in P and all the points in Q. By equation (2) and by Theorem 3.4.4 applied to the partitions P and Q and then to R and Q, the corollary follows:

$$L(P, f) \leq L(Q, f) \leq U(Q, f) \leq U(R, f). \blacksquare$$

This last corollary is extremely important. It shows that the set of all upper sums is bounded below and that the set of all lower sums is bounded above. Specifically, for any partition P of $[a, b]$, Corollary 3.4.5 says that for every partition Q of $[a, b]$,

$$L(P, f) \leq U(Q, f).$$

Since Q was an arbitrary partition, $L(P, f)$ is a lower bound for all the upper sums, that is,

$$L(P, f) \leq \inf\{U(Q, f) \mid Q \text{ partitions } [a, b]\}.$$

But P itself was arbitrary, so in turn $\inf\{U(Q, f)\}$ is an upper bound for all the lower sums, that is,

$$\sup\{L(P, f) \mid P \text{ partitions } [a, b]\} \leq \inf\{U(P, f) \mid P \text{ partitions } [a, b]\}.$$

The definition of infimum and supremum leads to the following:

THEOREM 3.4.6 *If Q_1 and Q_2 are any partitions of $[a, b]$, then*

$$L(Q_1, f) \leq \sup_P\{L(P, f)\} \leq \inf_P\{U(P, f)\} \leq U(Q_2, f).$$

A word about notation is in order here. To indicate that we are taking supremums and infimums of sets where the *partition* P is varying we have used the subscript P. That is,

$$\sup_P\{L(P, f)\} \equiv \sup\{L(P, f) \mid P \text{ partitions } [a, b]\}$$

and

$$\inf_P\{U(P, f)\} \equiv \inf\{U(P, f) \mid P \text{ partitions } [a, b]\}.$$

3.4.3 The Riemann Integral

Returning to the area question, if we have a strict inequality in Theorem 3.4.6,

$$\sup_P\{L(P, f)\} < \inf_P\{U(P, f)\},$$

any number that lay between these two values will lie between every pair of upper and lower sums. Geometrically this would mean that the upper and lower sums do not "zero in" on the area under the graph of f; thus we could not define an area under such a function in terms of upper and lower sums. Therefore we are interested in functions such that

$$\sup_P\{L(P, f)\} = \inf_P\{U(P, f)\}.$$

DEFINITION 3.4.7 *A bounded function f on the interval $[a, b]$ is called* ***Riemann integrable*** *on $[a, b]$ if*

$$\sup_P\{L(P, f)\} = \inf_P\{U(P, f)\}.$$

If this is the case, the common value is called the ***integral of*** *f on $[a, b]$ and is denoted by*

$$\int_a^b f \qquad \text{or by} \qquad \int_a^b f(x)\,dx.$$

During the eighteenth century integration was viewed as the inverse process of differentiation. To integrate a function f on $[a, b]$ one simply(!) located a function F such that $F' = f$ on $[a, b]$. Then

$$\int_a^b f(x)\,dx = F(b) - F(a)$$

(see the first fundamental theorem of calculus). This point of view had originated with Newton. In contrast, Leibniz had taken a summation approach to integration (though in the end antidifferentiation was used to evaluate these sums). Leibniz's view was out of favor for over a century. However, as mathematicians began to use more general sorts of functions, they found more and more problems with the definition of integration as antidifferentiation and were forced to reconsider the integral as a sum.[5]

[5] I. Grattan-Guinness, *From the Calculus to Set Theory, 1630–1910* (London: Gerald Duckworth & Co. Ltd., 1980), p. 5.

It was Cauchy in particular who revived the summation point of view during the 1820s. He gave the following definition of the integral, which we recognize as the type of sum we first used to approximate the area under a curve (see (1)).[6] Given a partition $P = \{x_0, x_1, \ldots, x_n\}$ of $[a, b]$,

$$\int_a^b f = \lim_{n\to\infty} \sum_{i=1}^{n} f(x_{i-1})(x_i - x_{i-1}).$$

Instead of using upper and lower sums, Cauchy used the value of f at x_{i-1}, the left-hand endpoint of the ith subinterval, as the estimate of f over the entire interval. In the 1850s Bernhard Riemann generalized Cauchy's idea by defining

$$\int_a^b f = \lim_{\delta\to 0} \sum_{i=1}^{n} f(\xi_i)(x_i - x_{i-1}),$$

where ξ_i is now any point in the ith subinterval and δ is the maximum length of any subinterval in the partition P.[7] (This is essentially formula (2).) During the 1870s Riemann's view was reformulated by several mathematicians into the upper and lower sum approach we have taken here.

When f is nonnegative we return to the area problem that motivated our entire discussion. In view of Definition 3.4.7, we are now able to define area for certain regions.

DEFINITION 3.4.8 *Assume that f is nonnegative and Riemann integrable on $[a, b]$. Then the* **area** *of the region bounded above by the graph $y = f(x)$ and below by the x-axis and which lies between the vertical lines $x = a$ and $x = b$ is defined to be $\int_a^b f$.*

In the 1820s Cauchy proved that any continuous function is integrable (see Theorem 3.5.1). For us, this means that the area question posed at the outset (area under f where f is continuous and nonnegative on $[a, b]$) can always be answered by

$$\text{area} = \int_a^b f.$$

By the middle of the nineteenth century, mathematicians were dealing with much less well-behaved functions than continuous ones (Dirichlet's function, for example). At this time, Bernhard Riemann formulated the definition of integrability mentioned earlier, which generalized

[6] C. H. Edwards, Jr., *The Historical Development of the Calculus* (New York: Springer-Verlag, Inc., 1979), p. 318.

[7] Edwards, p. 323.

Cauchy's ideas and which was readily applicable to some (highly) discontinuous functions. Some of these functions were found to be integrable in Riemann's sense of the term (see Example 3.5.1 for an integrable function with an infinite number of discontinuities) and some were not (Dirichlet's function in Example 3.4.2). There have been further generalizations of the integral concept since Riemann's work. In particular, Henri Lebesgue (1875–1941) discovered a reformulation of the integral under which Dirichlet's function is integrable. Lebesgue's more abstract definition of the integral is based on an area of analysis called measure theory, which is studied in advanced courses in real analysis.

EXAMPLE 3.4.1 The simplest integrable function is a constant function $f(x) = c$ on $[a, b]$. Let $P = \{x_0, x_1, \ldots, x_n\}$ be any partition of $[a, b]$. Because f is constant, on each subinterval we have $m_i = M_i = c$. Therefore

$$\begin{aligned} L(P, f) &= \sum_{i=1}^{n} c(x_i - x_{i-1}) \\ &= c\big((x_1 - x_0) + (x_2 - x_1) + \cdots + (x_n - x_{n-1})\big) \\ &= c(x_n - x_0) \\ &= c(b - a). \end{aligned}$$

We will see many such *telescoping* sums, that is, sums that reduce to the difference of the last and first terms; be on the lookout for them. Similarly

$$U(P, f) = \sum_{i=1}^{n} c(x_i - x_{i-1}) = c(b - a).$$

Since all upper and lower sums are equal, we have

$$\sup_P\{L(P)\} = \inf_P\{U(P)\} = c(b - a) = \int_a^b c\,dx.$$

This yields the familiar formula for the area of a rectangle when c is positive. ☐

EXAMPLE 3.4.2 Let D be the Dirichlet function on the interval $[0, 1]$:

$$D(x) = \begin{cases} 1 & \text{if } x \text{ is rational} \\ 0 & \text{if } x \text{ is irrational.} \end{cases}$$

Show that D is not Riemann integrable.

SOLUTION Let $P = \{x_0, x_1, \ldots, x_n\}$ be any partition of $[0, 1]$. By the density of the rationals and irrationals we know that between any two points x_{i-1} and x_i there are both rational and irrational numbers. Hence $m_i = 0$ and $M_i = 1$. Therefore

$$U(P, D) = \sum_{i=1}^{n} 1(x_i - x_{i-1}) = 1(1 - 0) = 1,$$

and

$$L(P, D) = \sum_{i=1}^{n} 0(x_i - x_{i-1}) = 0.$$

But this result holds for every partition P, so $\sup_P\{L(P)\} = 0$ and $\inf_P\{U(P)\} = 1$. Consequently the Dirichlet function is not integrable. □

This last example shows that not all bounded functions are integrable. One of our goals is to find conditions on f that ensure its integrability. The next theorem expresses the meaning of integrability in a form that is useful in this process.

THEOREM 3.4.9 *Assume that f is bounded on $[a, b]$. Then f is integrable on $[a, b]$ if and only if for every $\epsilon > 0$ there is a partition Q such that $U(Q, f) - L(Q, f) < \epsilon$.*

PROOF We will prove the necessity ($\Leftarrow$). A proof of the sufficiency is outlined in the problem section.

Assume that f is bounded on $[a, b]$ and that for any $\epsilon > 0$ there is a partition Q_ϵ of $[a, b]$ such that $U(Q_\epsilon) - L(Q_\epsilon) < \epsilon$. To show that f is integrable on $[a, b]$ we will use a proof by contradiction. If f is *not* integrable on $[a, b]$, by definition $\sup_P\{L(P)\} \neq \inf_P\{U(P)\}$. Therefore $\sup_P\{L(P)\} < \inf_P\{U(P)\}$. Then let $\epsilon = \inf_P\{U(P)\} - \sup_P\{L(P)\} > 0$. By our hypothesis there is a partition Q_ϵ such that $U(Q_\epsilon) - L(Q_\epsilon) < \epsilon$. Theorem 3.4.6 says that

$$L(Q_\epsilon) \leq \sup_P\{L(P)\} \leq \inf_P\{U(P)\} \leq U(Q_\epsilon).$$

It follows that

$$\inf_P\{U(P)\} - \sup_P\{L(P)\} \leq U(Q_\epsilon) - L(Q_\epsilon) < \epsilon.$$

But this contradicts our assumption that $\epsilon = \inf_P\{U(P)\} - \sup_P\{L(P)\}$, so f must be integrable. ∎

EXAMPLE 3.4.3 Show that $f(x) = x$ is integrable on $[a, b]$.

SOLUTION By Theorem 3.4.9 we can do this by finding a partition P such that $U(P) - L(P) < \epsilon$, where $\epsilon > 0$ is arbitrary.

Let $P_n = \{x_0, x_1, \ldots, x_n\}$ be the partition of $[a, b]$ into n equal subintervals of length $x_i - x_{i-1} = \frac{(b-a)}{n}$. Since f is continuous and increasing, on the ith subinterval the minimum occurs at the left endpoint x_{i-1} and the maximum occurs at the right endpoint x_i. That is,

$$M_i = f(x_i) = x_i \qquad \text{and} \qquad m_i = f(x_{i-1}) = x_{i-1}.$$

Therefore

$$\begin{aligned}
U(P_n) - L(P_n) &= \sum_{i=1}^{n} x_i(x_i - x_{i-1}) - \sum_{i=1}^{n} x_{i-1}(x_i - x_{i-1}) \\
&= \sum_{i=1}^{n} (x_i - x_{i-1})(x_i - x_{i-1}) \\
&= \sum_{i=1}^{n} \frac{(b-a)}{n} \cdot \frac{(b-a)}{n} \\
&= n\left(\frac{(b-a)}{n}\right)^2 = \frac{(b-a)^2}{n}.
\end{aligned}$$

Since n is arbitrary, by taking n sufficiently large we can make $\frac{(b-a)^2}{n}$ smaller than ϵ. (How large must n be?) Hence $U(P_n) - L(P_n) < \epsilon$, which makes f integrable.

Even though we have shown that $f(x) = x$ is integrable, we have not solved the harder problem of actually evaluating $\int_a^b x\, dx$. In this case we can evaluate $\int_a^b x\, dx$ using the following argument:

Let $P = \{x_0, x_1, \ldots, x_n\}$ be any partition of $[a, b]$. The average of $L(P)$ and $U(P)$ lies between the two, that is,

$$L(P) \leq \frac{L(P) + U(P)}{2} \leq U(P).$$

But

$$\begin{aligned}
U(P) + L(P) &= \sum_{i=1}^{n} x_i(x_i - x_{i-1}) + \sum_{i=1}^{n} x_{i-1}(x_i - x_{i-1}) \\
&= \sum_{i=1}^{n} (x_i + x_{i-1})(x_i - x_{i-1}) \\
&= \sum_{i=1}^{n} x_i^2 - x_{i-1}^2 \\
&= x_n^2 - x_0^2 = b^2 - a^2.
\end{aligned}$$

We see that the average is a constant value $\frac{L(P)+U(P)}{2} = \frac{b^2-a^2}{2}$ for all partitions P. So for *any* partition P we have $L(P,f) \leq \frac{b^2-a^2}{2} \leq U(P,f)$ and consequently

$$\sup_P\{L(P)\} \leq \frac{b^2-a^2}{2} \leq \inf_P\{U(P)\}. \tag{6}$$

Since f is integrable we actually have equalities in (6),

$$\int_a^b x\,dx = \frac{b^2-a^2}{2}.$$

This averaging process worked only because the average of $L(P,f)$ and $U(P,f)$ was, in fact, independent of the particular partition P. This is generally not the case. ◻

Problems for Section 3.4

3.4.1. Complete the proof of Theorem 3.4.4 by showing that $U(Q,f) \leq U(P,f)$ when Q is finer than P.

•3.4.2. Show that if Q is a refinement of a partition P, then for any function f, $0 \leq U(Q) - L(Q) \leq U(P) - L(P)$. (Hint: Use Theorem 3.4.4.)

•3.4.3. Finish the proof of Theorem 3.4.9 by showing that if f is integrable on $[a,b]$ with $\int_a^b f = I$, then given $\epsilon > 0$ there is a partition P such that $U(P,f) - L(P,f) < \epsilon$.

- **a)** Why is $I = \sup_P\{L(P,f)\} = \inf_P\{U(P,f)\}$?
- **b)** Show that there is a partition Q such that $I - (\epsilon/2) < L(Q,f) \leq I$. (Use a property of l.u.b.'s.)
- **c)** Show that there is a partition R such that $I \leq U(R,f) < I + (\epsilon/2)$.
- **d)** Let P be the common refinement of Q and R. What result says $L(Q) \leq L(P) \leq U(P) \leq U(R)$?
- **e)** Now show that $I - (\epsilon/2) < L(P) \leq U(P) < I + (\epsilon/2)$.
- **f)** Show that $U(P) - L(P) < \epsilon$ to complete the proof.

•3.4.4.

- **a)** Suppose that f is integrable on $[a,b]$. Then by definition f must be bounded on $[a,b]$. In particular, suppose that $m \leq f(x) \leq M$ for all $x \in [a,b]$. Show that $m(b-a) \leq \int_a^b f \leq M(b-a)$. (Hint: Compute $U(P)$ and $L(P)$ where $P = \{a,b\}$ is the trivial partition.)
- **b)** Show that if f is nonnegative and integrable on $[a,b]$, then $0 \leq \int_a^b f$.

•3.4.5. Show that the "skyscraper" function (graph it),

$$s(x) = \begin{cases} 1, & \text{if } x = 0 \\ 0, & \text{if } 0 < x \le 1, \end{cases}$$

is integrable on $[0,1]$ and that $\int_0^1 s(x)\,dx = 0$.

a) For $n \in \mathbf{N}$ let P_n be the partition $\{0, \frac{1}{n}, 1\}$. Compute $L(P_n, s)$ and $U(P_n, s)$.

b) Show that s is integrable by showing that for any $\epsilon > 0$, there is a sufficiently large $n \in \mathbf{N}$ such that $U(P_n, s) - L(P_n, s) < \epsilon$.

c) Show that $\int_0^1 s = 0$.

3.4.6. Show that if f is continuous and increasing on $[a,b]$, then f is integrable on $[a,b]$. (Hint: Mimic the work in Example 3.4.3.)

3.4.7. Prove that if f is integrable on $[a,b]$, then for every $\epsilon > 0$ there is a partition P such that $0 \le U(P,f) - \int_a^b f < \epsilon$ and $0 \le \int_a^b f - L(P,f) < \epsilon$.

•3.4.8. Show that if f and g are integrable on $[a,b]$ and $g(x) \le f(x)$ for all $x \in [a,b]$, then $\int_a^b g \le \int_a^b f$.

3.4.9. Suppose that f is an integrable function on $[a,b]$. Suppose that g is a bounded function on $[a,b]$ such that for every partition $P = \{x_0, x_1, \ldots, x_n\}$ of $[a,b]$ we have

$$M_i^g - m_i^g \le M_i^f - m_i^f,$$

where $M_i g$, M_i^f, m_i^g, and m_i^f represent the supremums and infimums of g and f, respectively, on the interval $[x_{i-1}, x_i]$. Show that g is integrable on $[a,b]$.

3.4.10. Let $f(x) = 3x - x^2$. Draw an accurate graph of f over the interval $[0,4]$. Let P be the partition of $[0,4]$ consisting of $\{0, 1/2, 1, 2, 3, 7/2, 4\}$. Evaluate $U(P,f)$ and $L(P,f)$.

•3.4.11. Assume that f is integrable on $[a,b]$. Suppose that J is a real number such that $L(P,f) \le J \le U(P,f)$ for every partition P of $[a,b]$. Show that $J = \int_a^b f$. (Hint: Use properties of supremums and infimums and the definition of integrable.)

3.4.12. Suppose that f is differentiable on $[a,b]$ with $|f'(x)| < k$. Let P be any partition of $[a,b]$. Prove that $U(P,f) - L(P,f) \le k(b-a)^2$. (Hint: Let M_i and m_i denote the maximum and minimum values of f on the subinterval $[x_{i-1}, x_i]$. Use the mean value theorem to show that $M_i - m_i \le k\Delta x_i \le k(b-a)$.)

3.5 Properties of the Integral

This section presents a careful treatment of some of the basic properties of the integral. We begin by exhibiting two different classes of functions that are integrable.

3.5.1 Integrability Conditions

THEOREM 3.5.1 *If f is continuous on $[a, b]$, then f is integrable on $[a, b]$.*

PROOF Fix $\epsilon > 0$. By Theorem 3.4.9, to show that f is integrable we must produce a partition P such that $U(P, f) - L(P, f) < \epsilon$. Let $\epsilon' = \frac{\epsilon/}{b-a}$. Since f is continuous on $[a, b]$, by Theorem 2.6.4 f is uniformly continuous on $[a, b]$. So there is a $\delta > 0$ such that for any x and y in $[a, b]$, if $|x - y| < \delta$, then $|f(x) - f(y)| < \epsilon'$.

Let $P = \{x_0, x_1, \ldots, x_n\}$ be any partition of $[a, b]$ such that each subinterval $[x_{i-1}, x_i]$ has length less than δ. Then using the uniform continuity of f, for any x and y in $[x_{i-1}, x_i]$ we have $|f(x) - f(y)| < \epsilon'$.

In particular, if $M_i = f(x)$ and $m_i = f(y)$ are the maximum and minimum values of f on $[x_{i-1}, x_i]$, it follows that $M_i - m_i < \epsilon'$. Therefore

$$
\begin{aligned}
U(P, f) - L(P, f) &= \sum_{i=1}^{n} (M_i - m_i)(x_i - x_{i-1}) \\
&< \epsilon' \sum_{i=1}^{n} (x_i - x_{i-1}) = \epsilon'(b - a) = \epsilon \,. \quad \blacksquare
\end{aligned}
$$

Notice that uniform continuity is essential to the proof; it allows us to bound the values $(M_i - m_i)$ for all i with a single uniform bound. From this we are able to force $U(P) - L(P)$ to be arbitrarily small. When Cauchy first stated and "proved" this theorem in the 1820s, the concept of uniform continuity was unknown and his demonstration was less than rigorous. In fact, even after redefining the integral in the 1850s Riemann did not make the general statement that all continuous functions were integrable. The result in Theorem 3.5.1 became clear in the 1870s only after Heine had defined uniform continuity and showed that all continuous functions on closed, bounded intervals were uniformly continuous. We have shown that the familiar continuous functions are integrable on closed, bounded intervals. However, a function need not be continuous to be integrable, as problem 3.4.5 shows. Further, the next theorem states that any monotone function, continuous or not, is integrable.

THEOREM 3.5.2 *If f is nondecreasing on $[a,b]$, then f is integrable on $[a,b]$. (The same result holds for non-increasing functions.)*

PROOF We must first show that f is bounded. But since f is nondecreasing, for any $x \in [a,b]$ we have $f(a) \leq f(x) \leq f(b)$. Hence f is bounded below by $f(a)$ and above by $f(b)$ on $[a,b]$.

Using Theorem 3.4.9, for any $\epsilon > 0$ we must locate a partition such that $U(P) - L(P) < \epsilon$. Consider the partition $P_n = \{x_0, x_1, \ldots, x_n\}$ of $[a,b]$ into n equal subintervals each of length $x_i - x_{i-1} = \frac{b-a}{n}$. Since f is nondecreasing, on each subinterval $[x_{i-1}, x_i]$, f achieves its minimum at x_{i-1} and its maximum at x_i. Therefore $m_i = f(x_{i-1})$ and $M_i = f(x_i)$.

Thus the difference between the upper and lower sums for P is

$$\begin{aligned} U(P_n, f) - L(P_n, f) &= \sum_{i=1}^{n} \big(f(x_i) - f(x_{i-1})\big)(x_i - x_{i-1}) \\ &= \frac{b-a}{n} \sum_{i=1}^{n} (f(x_i) - f(x_{i-1})) \\ &= \frac{b-a}{n} \Big(f(b) - f(a)\Big). \end{aligned}$$

Since n was arbitrary, it can be chosen sufficiently large so that

$$U(P_n, f) - L(P_n, f) = \frac{b-a}{n} \Big(f(b) - f(a)\Big) < \epsilon.$$

By Theorem 3.4.9, f is integrable. ∎

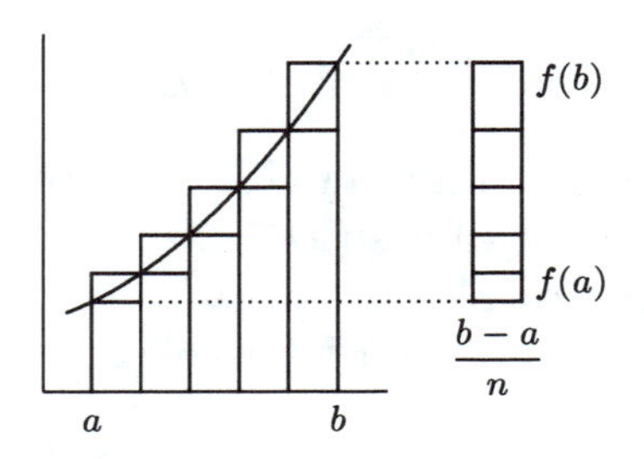

Figure 3.5.1

EXAMPLE 3.5.1 We will now show that a function may be highly discontinuous and yet still be integrable.

Define the step function f on $[0,1]$ as follows. For any natural number n, let

$$f(x) = \begin{cases} 0, & \text{if } x = 0 \\ 1/n, & \text{if } 1/(n+1) < x \leq 1/n. \end{cases}$$

Observe that f has an infinite number of discontinuities that occur at the points $1/n$ (for $n > 1$). Clearly Theorem 3.5.1 does not apply, yet f is obviously nondecreasing, so by Theorem 3.5.2 f is integrable on $[0, 1]$. (In fact, it can be shown that $\int_0^1 f(x)\,dx = \frac{\pi^2}{6} - 1$.) ☐

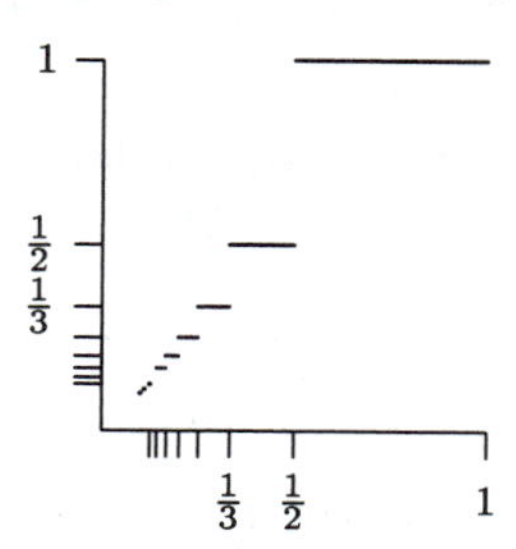

Figure 3.5.2

3.5.2 The Additivity and Linearity of the Integral

EXAMPLE 3.5.2 We would like to combine the preceding results to allow us to integrate even wider classes of functions. Suppose that f is defined on $[0, 2]$ by

$$f(x) = \begin{cases} x^2, & \text{if } 0 \le x \le 1 \\ 1.5 - x, & \text{if } 1 < x \le 2 \,. \end{cases}$$

Neither Theorem 3.5.1 nor Theorem 3.5.2 applies to show that f is integrable. Yet f is continuous on $[0, 1]$ and f is nonincreasing on $[1, 2]$, so f is integrable on $[0, 1]$ and $[1, 2]$ separately. Somehow we should be able to combine these facts to show that f is integrable on the entire interval $[0, 2]$.

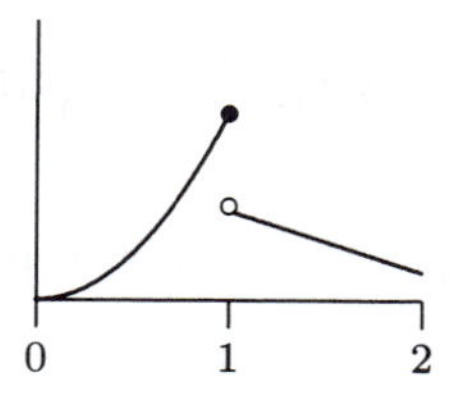

Figure 3.5.3

THEOREM 3.5.3 ***(Additivity)*** *If f is integrable over $[a, b]$ and $[b, c]$, then f is integrable over $[a, c]$. Conversely, if f is integrable over $[a, c]$ and if $a < b < c$, then f is integrable over both $[a, b]$ and $[b, c]$. In addition,*

$$\int_a^c f = \int_a^b f + \int_b^c f.$$

PROOF We will prove the first part of the theorem and leave the converse as an exercise (which is outlined in the problem section).

Assume that f is integrable on $[a,b]$ and $[b,c]$. Let $P_1 = \{a = x_0, \ldots, x_k = b\}$ be any partition of $[a,b]$ and $P_2 = \{b = x_k, \ldots, x_n = c\}$ be any partition of $[b,c]$. Note the labeling: $x_k = b$. Let $P = \{x_0, x_1, \ldots, x_n\}$ be the partition of $[a,c]$ obtained by taking the union of the points in P_1 and P_2. As usual, let $m_i = \inf\{f(x) \mid x_{i-1} \le x \le x_i\}$ and let $\Delta x_i = x_i - x_{i-1}$. Then

$$\begin{aligned} L(P,[a,c]) = \sum_{i=1}^{n} m_i \Delta x_i &= \sum_{i=1}^{k} m_i \Delta x_i + \sum_{i=k+1}^{n} m_i \Delta x_i \\ &= L(P_1,[a,b]) + L(P_2,[b,c]). \end{aligned}$$

(Note the change in the lower sum notation to distinguish between the intervals.) But for any partition P of $[a,c]$

$$L(P,[a,c]) \le \sup_Q\{L(Q,[a,c])\}.$$

Combined with the preceding equation, this means that

$$L(P_1,[a,b]) + L(P_2,[b,c]) \le \sup_Q\{L(Q,[a,c])\}$$

where P_1 and P_2 are any partitions of $[a,b]$ and $[b,c]$, respectively. That is, $\sup_Q\{L(Q,[a,c])\}$ is an upper bound for the set of *all* sums of the form $L(P_1,[a,b]) + L(P_2,[b,c])$. By problem 1.2.6,[1]

$$\sup_{P_1}\{L(P_1,[a,b])\} + \sup_{P_2}\{L(P_2,[b,c])\} \le \sup_Q\{L(Q,[a,c])\}.$$

Since f is integrable on $[a,b]$ and $[b,c]$, this last equation can be rewritten as

$$\int_a^b f + \int_b^c f \le \sup_Q\{L(Q,[a,c])\}.$$

A similar argument for upper sums shows that

$$\inf_Q\{U(Q,[a,c])\} \le \int_a^b f + \int_b^c f.$$

[1] Since a supremum is the same as a least upper bound, problem 1.2.6 showed that $\sup\{a+b \mid a \in A,\ b \in B\} = \sup A + \sup B$, where A and B are any (nonempty) bounded sets.

These last two inequalities combine to give

$$\int_a^b f + \int_b^c f \leq \sup_Q\{L(Q,[a,c])\} \leq \inf_Q\{U(Q,[a,c])\} \leq \int_a^b f + \int_b^c f.$$

It follows that all four terms in the inequality are equal, which means that f is integrable on $[a,c]$ and further that

$$\int_a^c f = \sup_Q\{L(Q,[a,c])\} = \inf_Q\{U(Q,[a,c])\} = \int_a^b f + \int_b^c f. \quad \blacksquare$$

So far $\int_a^b f$ is defined only when $a < b$. It will prove convenient to extend this definition.

DEFINITION 3.5.4 *If f is integrable over $[a,b]$ (where $a < b$), then we define*

$$\int_b^a f = -\int_a^b f.$$

For any number a in the domain of f, we define

$$\int_a^a f = 0.$$

Using these conventions, Theorem 3.5.3 can be generalized.

THEOREM 3.5.5 *If f is integrable on an interval I containing a, b, and c, then*

$$\int_a^c f = \int_a^b f + \int_b^c f$$

(even if b is not between a and c).

The proof is similar to Theorem 3.5.3 but requires a case-by-case check for each possible ordering of the points a, b, and c.

Theorem 3.5.5 expresses the additivity of the integral with respect to the interval under consideration, but the integral possesses another type of additivity.

THEOREM 3.5.6 ***(Linearity I)*** *If f and g are integrable on $[a,b]$, then $f+g$ is integrable on $[a,b]$ and*

$$\int_a^b (f+g) = \int_a^b f + \int_a^b g.$$

PROOF If we knew that f and g were continuous, $f+g$ would be continuous and therefore integrable by Theorem 3.5.1. To prove the general result is not difficult but requires quite a bit more care.

Choose $\epsilon > 0$. Since f and g are integrable, by Theorem 3.4.9 there are partitions Q and R of $[a,b]$ such that $U(Q,f) - L(Q,f) < \epsilon/2$ and $U(R,g) - L(R,g) < \epsilon/2$. Let $P = \{x_0, x_1, \ldots, x_n\}$ be the common refinement of Q and R. By Theorem 3.4.4,

$$U(P,f) - L(P,f) \leq U(Q,f) - L(Q,f) < \epsilon/2$$

and

$$U(P,g) - L(P,g) \leq U(R,g) - L(R,g) < \epsilon/2.$$

Summing these two inequalities yields

$$U(P,f) + U(P,g) - \big(L(P,f) + L(P,g)\big) < \epsilon. \tag{1}$$

Next we compare $U(P, f+g)$ to $U(P,f) + U(P,g)$. To accomplish this let

$$\begin{aligned} M_i &= \sup\{(f+g)(x) \mid x_{i-1} \leq x \leq x_i\}, \\ M_i^f &= \sup\{f(x) \mid x_{i-1} \leq x \leq x_i\}, \\ M_i^g &= \sup\{g(x) \mid x_{i-1} \leq x \leq x_i\}. \end{aligned}$$

Then for any x in $[x_{i-1}, x_i]$,

$$(f+g)(x) = f(x) + g(x) \leq M_i^f + M_i^g. \tag{2}$$

This means that $M_i^f + M_i^g$ is an upper bound for the function $f+g$ on $[x_{i-1}, x_i]$, therefore

$$M_i \leq M_i^f + M_i^g. \tag{3}$$

(Why can we fail to have equality here?) Multiplying each term in (3) by $\Delta x_i = x_i - x_{i-1}$ and then summing gives

$$\sum_{i=1}^{n} M_i \Delta x_i \leq \sum_{i=1}^{n} (M_i^f + M_i^g) \Delta x_i = \sum_{i=1}^{n} M_i^f \Delta x_i + \sum_{i=1}^{n} M_i^g \Delta x_i.$$

That is, for any partition P of $[a,b]$

$$U(P, f+g) \leq U(P,f) + U(P,g). \tag{4}$$

In an entirely similar fashion it follows that

$$L(P,f) + L(P,g) \leq L(P, f+g). \tag{5}$$

Subtracting (5) from (4) and then using (2) yields

$$U(P,f+g) - L(P,f+g) \leq (U(P,f) + U(P,g)) - (L(P,f) + L(P,g)) < \epsilon.$$

By Theorem 3.4.9, since ϵ was arbitrary, $f+g$ is integrable on $[a,b]$.

Showing that $\int_a^b f+g = \int_a^b f + \int_a^b g$ is now a straightforward exercise, which is outlined in the problem section. ∎

THEOREM 3.5.7 ***(Linearity II)*** *If c is any constant and f is integrable over $[a,b]$, then cf is integrable over $[a,b]$ and*

$$\int_a^b cf = c\int_a^b f.$$

PROOF We will assume that $c > 0$ and leave the cases $c < 0$ and $c = 0$ as exercises. Let $P = \{x_0, x_1, \ldots, x_n\}$ be any partition of $[a,b]$. Let

$$M_i = \sup\{f(x) \mid x_{i-1} \le x \le x_i\},$$
$$\overline{M}_i = \sup\{cf(x) \mid x_{i-1} \le x \le x_i\}.$$

By problem 1.2.3, $\overline{M}_i = cM_i$, so

$$U(P, cf) = \sum_{i=1}^{n} cM_i \Delta x_i = cU(P, f).$$

Using problem 1.2.3 again and the integrability of f it follows that

$$\inf_P\{U(P, cf)\} = \inf_P\{cU(P, f)\} = c \cdot \inf_P\{U(P, f)\} = c\int_a^b f.$$

Similarly $\sup_P\{L(P, cf)\} = c\int_a^b f$. Consequently cf is integrable on $[a,b]$ and $\int_a^b cf = c\int_a^b f$. ∎

From work in linear algebra you might recognize the two linearity properties of the integral as the defining characteristics of a linear transformation. Let $\mathcal{C}[a,b]$ denote the set of continuous functions on $[a,b]$. It is a straightforward exercise to show that $\mathcal{C}[a,b]$ is a real vector space. In fact, Theorem 2.5.2 shows that $\mathcal{C}[a,b]$ is closed under addition and scalar multiplication. From Theorems 3.5.6 and 3.5.7 it follows that Riemann integration is a linear transformation from $\mathcal{C}[a,b] \to \mathbf{R}$ by

$$f \to \int_a^b f(x)\,dx.$$

This "linear algebra" approach to analysis leads to many interesting and powerful results and has become a dominant theme in many areas of higher level analysis.

Problems for Section 3.5

3.5.1. For each of the following functions state a theorem that shows it is integrable on the given interval.

a) $f(x) = |x|$ on $[-1, 2]$.

b) $g(x) = x^{1/4}$ on $[2, 9]$.

c) $h(x) = \begin{cases} x^2, & \text{if } 0 \le x \le 1 \\ 2x + 3, & \text{if } 1 < x \le 2 \end{cases}$.

3.5.2. Justify in one sentence your answers to each of the following questions.

a) If f is a function that has a derivative at each point of $[a, b]$, is f integrable on $[a, b]$?

b) If $f(x) = x^{1/4}$ and $g = x^2 + 1$, is $f \circ g$ integrable on $[a, b]$?

c) If f and g are both continuous on $[a, b]$, is $f \cdot g$ integrable on $[a, b]$?

3.5.3. **a)** Which theorems guarantee that if f and g are integrable on $[a, b]$, then $f - g$ is also integrable on $[a, b]$?

b) Show that if f and g are integrable on $[a, b]$ and $g(x) \le f(x)$ for all $x \in [a, b]$, then $\int_a^b g \le \int_a^b f$. (Hint: Consider the function $f - g$.)

3.5.4. Prove Theorem 3.5.2 for nonincreasing functions.

•3.5.5. Prove Theorem 3.5.7 in the cases $c = 0$ and $c < 0$. The case $c < 0$ requires care in relating $\sup f$ or $\inf f$ to $\sup cf$ or $\inf cf$ on each subinterval of a partition of $[a, b]$.

3.5.6. Verify that (5) holds in the proof of Theorem 3.5.6. That is, show that $L(P, f) + L(P, g) \le L(P, f + g)$. (Hint: Let $m_i = \inf\{(f + g)(x) \mid x_{i-1} \le x \le x_i\}$, $m_i^f = \inf\{f(x) \mid x_{i-1} \le x \le x_i\}$, $m_i^g = \inf\{g(x) \mid x_{i-1} \le x \le x_i\}$. Show that $m_i^f + m_i^g \le m_i$ and then mimic the proof of (4).)

3.5.7. Are there bounded functions f and g on $[a, b]$ such that $\int_a^b (f + g)$ exists but the integrals $\int_a^b f$ and $\int_a^b g$ do not? How does this relate to Theorem 3.5.6?

3.5.8. Is there a function f and a constant c such that $\int_a^b cf$ exists but $c \int_a^b f$ does not? How is Theorem 3.5.7 related to this question?

•3.5.9. The second part of the **additivity theorem** states that *if $a < b < c$ and f is integrable over $[a, c]$, then f is integrable over $[a, b]$ and $[b, c]$ and $\int_a^c f = \int_a^b f + \int_b^c f$.* Prove this result. Start by taking $\epsilon > 0$. Since f is integrable on $[a, c]$, by Theorem 3.4.9 there is a partition P such that $U(P) - L(P) < \epsilon$. Let Q be the refinement of P obtained by adjoining the point b to P. Then $Q = \{x_0, \ldots, x_k = b, \ldots, x_n\}$. (If P should happen to already contain b, simply take $Q = P$.)

a) Show that $0 \le U(Q) - L(Q) \le U(P) - L(P) < \epsilon$.

b) Next, split up the partition Q into two pieces $Q_1 = \{x_0, \ldots, x_k\}$ and $Q_2 = \{x_k, \ldots, x_n\}$ to obtain partitions of $[a, b]$ and $[b, c]$, respectively. Verify that $L(Q_1, [a, b]) + L(Q_2, [b, c]) = L(Q, [a, c])$. (Similarly $U(Q_1, [a, b]) + U(Q_2, [b, c]) = U(Q, [a, c])$.)

c) Now show $(U(Q_1, [a, b]) - L(Q_1, [a, b])) + (U(Q_2, [b, c]) - L(Q_2, [b, c])) = U(Q, [a, c]) - L(Q, [a, c])$.

d) Show that f is integrable over $[a, b]$ and over $[b, c]$ separately by showing that $U(Q_1, [a, b]) - L(Q_1, [a, b]) < \epsilon$ and $U(Q_2, [b, c]) - L(Q_2, [b, c]) < \epsilon$.

e) Notice that the hypothesis of the first half of the additivity theorem is now satisfied. Use this to show that $\int_a^c f = \int_a^b f + \int_b^c f$.

•3.5.10. Complete the proof of Theorem 3.5.6 by showing that $\int_a^b f + g = \int_a^b f + \int_a^b g$, where f and g are both integrable on $[a, b]$. (Hint: Given $\epsilon > 0$, let P be the partition chosen as in the proof of the first part of Theorem 3.5.6 in the text.)

a) Show that $L(P, f) + L(P, g) \le \int_a^b f + g \le U(P, f) + U(P, g)$.

b) Show that $L(P, f) + L(P, g) \le \int_a^b f + \int_a^b g \le U(P, f) + U(P, g)$.

c) To complete the proof, subtract the inequality in **(b)** from the inequality in **(a)** and use inequality (1) in the text to show that

$$\left| \int_a^b f + g - \left(\int_a^b f + \int_a^b g \right) \right| \le U(P, f + g) - L(P, f + g) < \epsilon.$$

•3.5.11. We know that continuous functions are integrable. Yet we also know that continuity is not a requirement for integrability. Example 3.5.1 shows that even a function with an infinite number of discontinuities can be integrable. However, other functions discontinuous at an infinite number of points, such as the Dirichlet function, are not integrable (see Example 3.4.2). In this problem we will show that if a bounded function is continuous except at a *finite* number of points in a closed, bounded interval, then the function is integrable. The heart of the matter is settled in the following result: **Lemma:** *Assume f is bounded on $[a, b]$ and continuous on $[a, b]$ except at a. Then f is integrable on $[a, b]$.*

a) To prove the lemma begin by enclosing the discontinuity in a small subinterval $[a, a + \delta]$ as follows. Note first that since f is bounded, it has a supremum M and an infimum m on $[a, b]$. Now given any $\epsilon > 0$ choose a $\delta > 0$ such that $(M - m)\delta < \epsilon/2$ and $a + \delta < b$. Show that on the first subinterval $[a, a + \delta]$ the upper and lower sums differ by less than $\epsilon/2$. (Illustrate with a picture.)

b) Next consider the upper and lower sums on the remaining portion of the interval $[a + \delta, b]$. Show that there is a partition Q of $[a + \delta, b]$ such that $U(Q, [a + \delta, b]) - L(Q, [a + \delta, b]) < \epsilon/2$.

c) Let $P = \{a, a + \delta, \dots, b\}$ be the partition of $[a, b]$ obtained from the union of the single point a with the points of the partition Q of $[a + \delta, b]$. Complete the proof of the lemma by showing that $U(P, [a, b]) - L(P, [a, b]) < \epsilon$.

d) Clearly the lemma would hold if f were discontinuous at b instead of a. Prove the following. **Theorem.** *If f is bounded on $[a,b]$ and continuous on $[a,b]$ except at a finite number of points, then f is integrable on $[a,b]$.* (Hint: Subdivide $[a,b]$ at each of the points of discontinuity and then again at the midpoints of these subintervals. Now use the lemma and the additivity of integrals [Theorem 3.5.3].)

3.5.12. Show that if f is integrable on $[a,b]$ and g is bounded on $[a,b]$ and $f = g$ except at a finite number of points in $[a,b]$, then g is integrable on $[a,b]$ and $\int_a^b g = \int_a^b f$. (Hint: First use the previous problem to show that $g - f$ is integrable. Combine this with the fact that f is integrable to show that g must also be integrable.)

3.5.13. Assume that g is a continuous function on the interval $[0,1]$. If $g(0) = 2$ and $|g(x) - g(y)| \leq |x - y|$ for all x and y in $[0,1]$, determine upper and lower bounds for $\int_0^1 g(x)$.

3.5.14. Recall (problem 2.1.7) that if f and g are functions defined on the interval $[a,b]$, the function $\max(f,g)$ is defined as follows: for each $x \in [a,b]$,

$$\max(f,g)(x) = \begin{cases} f(x), & \text{if } f(x) \geq g(x) \\ g(x), & \text{if } g(x) > f(x). \end{cases}$$

Show that if f and g are integrable on $[a,b]$, then $\max(f,g)$ is integrable on $[a,b]$.

a) Let $P = \{x_0, x_1, \ldots, x_n\}$ be any partition of $[a,b]$. Let M_i^f, M_i^g, M_i, m_i^f, m_i^g, and m_i denote the suprema and infima of f, g, and $\max(f,g)$, respectively, on $[x_{i-1}, x_i]$. Show that $M_i = \max\left\{M_i^f, M_i^g\right\}$.

b) Show that if $M_i = M_i^f$, then $M_i - m_i \leq M_i^f - m_i^f$. (Hint: Start by using the fact that $\max(f,g)(x) \geq f(x)$ to show that $m_i \geq m_i^f$.) Similarly show that if $M_i = M_i^g$, then $M_i - m_i \leq M_i^g - m_i^g$.

c) Use **(b)** to show that in either situation $M_i - m_i \leq (M_i^f - m_i^f) + (M_i^g - m_i^g)$.

d) Show that $U(P, \max(f,g)) - L(P, \max(f,g)) \leq U(P,f) - L(P, f+g) + U(P,g) - L(P,g)$.

e) Use the integrability of f and g and Theorem 3.4.9 to finish the proof.

3.5.15. Show that if f is integrable on $[a,b]$, then $|f|$ is integrable on $[a,b]$ and $\int_a^b f \leq \int_a^b |f|$. (Hint: Recall [problem 2.1.7] that $|f| = \max(f, -f)$. Use the previous problem to show that $|f|$ is integrable and use problem 3.4.8 to obtain the inequality.)

3.6 The Fundamental Theorems of Calculus

The two fundamental theorems of calculus express the basic relationship that exists between integration and differentiation. These theorems depend on the following simple method for constructing new functions: Suppose that f is integrable on $[a, b]$. By Theorem 3.5.3 if $a \leq x \leq b$, then f is integrable on the interval $[a, x]$. So we may define a new function F whose domain is $[a, b]$ by

$$F(x) = \int_a^x f\,.$$

It is the relationship between the two functions f and F that ties differentiation and integration together.

3.6.1 The First Fundamental Theorem of Calculus

First we show that the function F we have constructed is continuous.

THEOREM 3.6.1 *If f is integrable on $[a, b]$ and F is defined on $[a, b]$ by*

$$F(x) = \int_a^x f(t)\,dt,$$

then F is continuous on $[a, b]$.

PROOF Fix $\epsilon > 0$. Since f is integrable on $[a, b]$, it is bounded on $[a, b]$. Therefore there is a positive number M such that for all t in $[a, b]$

$$-M \leq f(t) \leq M.$$

To show that F is continuous on $[a, b]$ we will prove a stronger result, namely, that F is *uniformly* continuous on $[a, b]$. Let $\delta = \epsilon/M$. Assume that x and y are in $[a, b]$ and that $|x - y| < \delta$. We may assume that $x \geq y$ (else rename the values). Then by the definition of F and Theorem 3.5.3,

$$F(x) - F(y) = \int_a^x f - \int_a^y f = \int_y^x f\,.$$

For all $t \in [a, b]$, $-M \leq f(t) \leq M$, and it follows from problem 3.4.4 applied to f on the interval $[y, x]$ that

$$-M(x - y) \leq \int_y^x f(t)\,dt \leq M(x - y),$$

or equivalently,

$$\left| \int_y^x f(t)\, dt \right| \leq M|x - y|. \tag{1}$$

Thus if $|x - y| < \delta$, then by (1)

$$|F(x) - F(y)| = \left| \int_y^x f(t)\, dt \right| \leq M|x - y| < M\delta = M\epsilon/M = \epsilon,$$

so f is uniformly continuous and therefore continuous. ∎

Even though f might not be continuous (and in fact might be highly discontinuous, as in Example 3.5.1), if f is integrable, as we have just seen F must be continuous. In other words, the integration process results in a function that is "smoother" than the initial one. What happens if we start with a continuous function f? Then, as we will show in Theorem 3.6.5, the resulting F is not only continuous but is also differentiable. Again there is an "improvement." The next theorem employs the mean value theorem to extend this connection between integration and differentiation further.

THEOREM 3.6.2 ***(The First Fundamental Theorem of Calculus)*** *If f is integrable on $[a, b]$ and if there is a function g such that $g' = f$ on $[a, b]$, then*

$$\int_a^b f = g(b) - g(a).$$

PROOF Let $P = \{x_0, x_1, \ldots, x_n\}$ be any partition of $[a, b]$. As usual, set

$$m_i = \inf\{f(x) \mid x_{i-1} \leq x \leq x_i\},$$

and

$$M_i = \sup\{f(x) \mid x_{i-1} \leq x \leq x_i\}.$$

The key idea is to write $g(b) - g(a)$ as a telescoping sum:

$$g(b) - g(a) = \sum_{i=1}^{n} g(x_i) - g(x_{i-1}). \tag{2}$$

The function g is differentiable on $[a, b]$, so by the mean value theorem there is a point $x_i^* \in [x_{i-1}, x_i]$ such that

$$g(x_i) - g(x_{i-1}) = g'(x_i^*)(x_i - x_{i-1}) = f(x_i^*)(x_i - x_{i-1}). \tag{3}$$

Combining (2) and (3) we now have

$$g(b) - g(a) = \sum_{i=1}^{n} g(x_i) - g(x_{i-1}) = \sum_{i=1}^{n} f(x_i^*)(x_i - x_{i-1}). \tag{4}$$

But $m_i \leq f(x_i^*) \leq M_i$, so it follows from (4) that

$$\sum_{i=1}^{n} m_i(x_i - x_{i-1}) \leq \sum_{i=1}^{n} f(x_i^*)(x_i - x_{i-1}) \leq \sum_{i=1}^{n} M_i(x_i - x_{i-1}).$$

This means that for any partition P

$$L(P, f) \leq g(b) - g(a) \leq U(P, f).$$

Since f is integrable by hypothesis, we must have

$$\int_a^b f = g(b) - g(a). \quad \blacksquare$$

To use the first fundamental theorem we must know two things about f: that it is integrable and that there is a function g that is an antiderivative of f. One class of functions that we know to be integrable are the continuous functions (Theorem 3.5.1).

Corollary 3.6.3 *If f is continuous on $[a, b]$ and $f = g'$ on $[a, b]$, then*

$$\int_a^b f = g(b) - g(a).$$

This result makes calculating certain integrals almost trivial. For example, let $f(x) = x^n$ on $[a, b]$, where n is a positive integer. Since f is continuous and since $g(x) = \frac{x^{n+1}}{n+1}$ is an antiderivative of f,

$$\int_a^b x^n \, dx = \frac{b^{n+1}}{n+1} - \frac{a^{n+1}}{n+1}.$$

Compare this to the work we did to evaluate $\int_a^b x \, dx$ in Example 3.4.3 and you will begin to appreciate the power of the first fundamental theorem. You should be able to manufacture many other such examples yourself.

Corollary 3.6.3 does not say which continuous functions have antiderivatives. It turns out that all do. In fact, as indicated earlier, integrating a continuous function yields a differentiable function. To prove this result it is helpful to look back at the proof of Theorem 3.6.2. Two crucial steps and their corresponding hypotheses are apparent:

1. we were able to apply the mean value theorem to g (and then use the hypothesis that $f = g'$);
2. f was integrable, so the upper and lower sums for f converged to $g(b) - g(a)$.

We now show that there is a mean value theorem for integrals of continuous functions.

THEOREM 3.6.4 ***(The Integral Mean Value Theorem)*** *Suppose that f is continuous on $[a, b]$. Then there is a point c in $[a, b]$ such that*

$$\int_a^b f = f(c)(b-a).$$

PROOF Let m and M be the minimum and maximum values of f on $[a, b]$. By Problem 3.4.4,

$$m(b-a) \le \int_a^b f \le M(b-a).$$

Dividing this inequality by the quantity $(b - a)$ gives

$$m \le \frac{1}{b-a} \int_a^b f \le M.$$

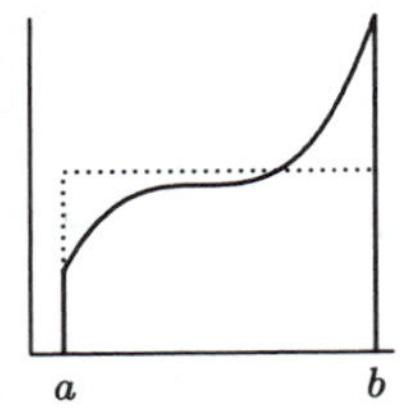

Figure 3.6.1

But f is continuous on $[a, b]$, so by the intermediate value theorem f assumes every value between m and M. Therefore there is a point $c \in [a, b]$ such that

$$f(c) = \frac{1}{b-a} \int_a^b f.$$

The result follows from multiplication by $(b - a)$. ∎

The value $\frac{1}{b-a} \int_a^b f(t)\, dt$ is called the **average** or **mean value** of f on $[a, b]$. It follows from the integral mean value theorem that if f is continuous and nonnegative on $[a, b]$, the area under the graph of f is the same as the area of the rectangle with height $f(c)$. A physical example of this is given by a cross section of a wave in a tank. Once we allow the wave to settle, the final water level is the average value of the height of the original wave. (See Figure 3.6.1.)

3.6.2 The Second Fundamental Theorem of Calculus

If we set $F(x) = \int_a^x f$ and observe that $F(a) = 0$, the conclusion of the integral mean value theorem can be rewritten as

$$f(c)(b-a) = \int_a^b f = F(b) - F(a),$$

or equivalently,

$$\frac{F(b) - F(a)}{b-a} = f(c).$$

This looks like the original mean value theorem, and in fact it would be the mean value theorem if we knew that $F' = f$. Using the integral mean value theorem, we can now show that this is the case when f is continuous.

THEOREM 3.6.5 ***(The Second Fundamental Theorem of Calculus)*** *Assume that f is continuous on $[a, b]$ and define F on $[a, b]$ by*

$$F(x) = \int_a^x f.$$

Then for all x in $[a, b]$, $F'(x) = f(x)$. (If $x = a$ or $x = b$, then $F'(x)$ should be interpreted as the appropriate one-sided derivative.)

PROOF We will assume that x is in (a, b) and leave the minor modifications of the proof in the cases $x = a$ and $x = b$ to the reader. We must show that

$$\lim_{h \to 0} \frac{F(x+h) - F(x)}{h} = f(x).$$

That is, given $\epsilon > 0$, we must produce a $\delta > 0$ such that

$$\text{if} \quad 0 < |h| < \delta, \quad \text{then} \quad \left| \frac{F(x+h) - F(x)}{h} - f(x) \right| < \epsilon \qquad (5).$$

Let's look more closely at this expression that we wish to make less than ϵ. If $h > 0$, by Theorem 3.5.3

$$F(x+h) - F(x) = \int_a^{x+h} f - \int_a^x f = \int_x^{x+h} f.$$

Applying the integral mean value theorem to f on $[x, x+h]$, there is a point c_h between x and $x+h$ such that

$$F(x+h) - F(x) = \int_x^{x+h} f = f(c_h)\big((x+h) - x\big) = f(c_h) \cdot h.$$

Similarly if $h < 0$, then $x + h < x$ and we may write

$$F(x+h) - F(x) = \int_x^{x+h} f = -\int_{x+h}^{x} f.$$

Applying the integral mean value theorem to f over $[x+h, x]$, gives

$$F(x+h) - F(x) = -\int_{x+h}^{x} f = -f(c_h)\big(x - (x+h)\big) = f(c_h)\cdot h$$

for some c_h between $x+h$ and x. Thus in either case ($h < 0$ or $h > 0$) proving (5) reduces to finding a δ such that

$$\left|\frac{F(x+h) - F(x)}{h} - f(x)\right| = \left|\frac{f(c_h)\cdot h}{h} - f(x)\right| = |f(c_h) - f(x)| < \epsilon, \quad (6)$$

where $|c_h - x| < |(x+h) - x| = |h|$. Because f is continuous at x, we can find a $\delta > 0$ such that when $0 < |t - x| < \delta$, then $|f(t) - f(x)| < \epsilon$. If in (5) we take $0 < |h| < \delta$, then $|c_h - x| < h < \delta$; therefore $|f(c_h) - f(x)| < \epsilon$. ∎

Although we now know that any continuous function f has an antiderivative (namely $F(x) = \int_a^x f$), if we use this antiderivative in Corollary 3.6.3 to compute a definite integral we simply obtain the identity $\int_a^b f = \int_a^b f$. Thus this theorem is not helpful in actually *evaluating* definite integrals, and it remains to locate a computable antiderivative. However, as is illustrated by problem 3.6.5, some very important functions are defined in terms of integrals, and the second fundamental theorem of calculus plays a central role in studying these functions.

As a further application of the ideas of this section, consider the following reinterpretation of Corollary 3.6.3.

LEMMA 3.6.6 *Suppose that f has a continuous derivative on $[a, b]$. Then for any x in $[a, b]$*

$$f(x) = f(a) + \int_a^x f'(t)\,dt.$$

PROOF By Corollary 3.6.3, since f' is continuous,

$$f(a) + \int_a^x f'(t)\,dt = f(a) + \big(f(x) - f(a)\big) = f(x). \quad ∎$$

The importance of this observation is that it expresses f in terms of its derivative. (Consider again our discussion preceding the mean value theorem about global versus local behavior of a function.) Using the lemma we may be able to deduce global properties of f from the local information provided by f'. For example:

THEOREM 3.6.7 *Suppose that f' is continuous and nonnegative (nonpositive) on an interval I. Then f is a nondecreasing (nonincreasing) function on I.*

PROOF Let a and x be in I, with $x > a$. Then since f' is nonnegative, by problem 3.4.4

$$\int_a^x f'(t)\,dt \geq 0\,.$$

Then f must be nondecreasing because by Lemma 3.6.6

$$f(x) = f(a) + \int_a^x f'(t)\,dt \geq f(a).$$

A similar argument works when f' is nonpositive. ∎

Problems for Section 3.6

3.6.1. Show that for all real x, $\int_0^x |t|\,dt = \frac{1}{2}x|x|$. (Hint: Split the problem into two cases: $x > 0$ and $x < 0$.)

3.6.2.

a) What is $f(2)$ if f is continuous everywhere and $\int_0^x f(t)\,dt = x^2(1+x)$?

b) If $F(x) = \int_0^x t\sqrt{t+1}\,dt$, what is $F'(3)$?

3.6.3. Suppose that f and g are differentiable functions on $[a, b]$ and that f' and g' are continuous on $[a, b]$.

a) Prove that the derivative (with respect to x) of $\int_a^x f(t)g'(t)\,dt$ is the same as the derivative of $f(x)g(x) - f(a)g(a) - \int_a^x f'(t)g(t)\,dt$.

b) Show that $\int_a^b f(t)g'(t)\,dt = f(b)g(b) - f(a)g(a) - \int_a^b f'(t)g(t)\,dt$.

•3.6.4. Alter Theorem 3.6.7 as follows: Suppose that f' is continuous and *positive* on the interval I. Prove that f is an increasing function on I. (This will require you to alter problem 3.4.4 appropriately.)

•3.6.5. In this problem we define the natural logarithm function and deduce some of its basic properties. Let $f(t) = 1/t$ on the interval $(0, \infty)$. We define the **natural logarithm** of x (denoted $\ln x$) by the equation

$$\ln x = \int_1^x \frac{1}{t}\,dt, \qquad \text{for any } x > 0.$$

a) Show that $\ln x$ is well defined, that is, that $\int_1^x 1/t\,dt$ exists for all $x > 0$.

b) Show that $\ln x$ is continuous.

c) Show that $(\ln x)' = 1/x$.

d) Show that $\ln x$ is an increasing function.

e) We say that f is a **one-to-one** function if whenever $x \neq y$ are in the domain of f, we have $f(x) \neq f(y)$. Show that $\ln x$ is one-to-one.

f) Let $Q = \{1, 3/2, 2\}$ be a partition of $[1, 2]$. Show that $U(Q, 1/t) < 1$ and conclude from this that $\ln 2 = \int_1^2 t^{-1}\, dt < 1$.

g) Now let P be the partition of $[1, 3]$ given by $\{1, 7/6, 8/6, \ldots, 17/6, 3\}$. Show that $L(P, 1/t) > 1$ and conclude from this that $\ln 3 > 1$.

h) Show that there is a unique real number e such that $\ln e = 1$ and that $2 < e < 3$.

i) Prove the familiar rule for logarithms: $\ln(ab) = \ln a + \ln b$ for all $a,\ b > 0$. (Hint: Define a new function $L(x) = \ln(ax)$. Show that $L'(x) = (\ln x)'$, so by Corollary 3.3.5 $\ln(ax) = L(x) = \ln x + c$, and then show that $c = \ln a$.)

j) Show that $\ln(x^n) = n \ln x$ for all $x > 0$ and $n \in \mathbf{N}$.

k) Prove that $\ln(a/b) = \ln a - \ln b$ for all $a,\ b > 0$.

l) Show that $\ln(x^{-n}) = -n \ln x$ for all $x > 0$ and $n \in \mathbf{N}$.

3.6.6. The following are properties of functions.

1. f is continuous on $[-2, 4]$.
2. f uniformly continuous on $[-2, 4]$.
3. f is differentiable on $[-2, 4]$.
4. f is integrable on $[-2, 4]$.
5. f is decreasing on $[-2, 4]$.
6. f has no maximum on $[-2, 4]$.
7. f is bounded on $[-2, 4]$.
8. f is a polynomial.

For each of the following find a function f that satisfies the required restrictions or else briefly state why no such function exists.

a) f satisfies 2 but not 4.
b) f satisfies 3 but not 4.
c) f satisfies 4 but not 3.
d) f satisfies 5 but not 4.
e) f satisfies 5 but not 4.
f) f satisfies 6 and 1.
g) f satisfies 7 but not 4.
h) f satisfies 2 but not 3.

3.6.7. Suppose that $f(x)$ is defined as the sum of two integrable functions $g(x)$ and $h(x)$ on the interval $[a, b]$. Show that the average value for f on $[a, b]$ is the sum of the average values for g and h.

3.6.8. **a)** Show that if f is continuous, then $\frac{d}{dx} \int_x^b f(t)\, dt = -f(x)$.

b) Using the chain rule, show that if f is continuous and g is differentiable, then $\frac{d}{dx} \int_a^{g(x)} f(t)\, dt = f(g(x))g'(x)$.

c) If f is continuous and g and h are differentiable, find $\frac{d}{dx} \int_{h(x)}^{g(x)} f(t)\, dt$.

3.6.9. Suppose that f is integrable on $[a, b]$. Define $F(x) = \int_a^x f(t)\, dt$ for all x in $[a, b]$. Now define $G(x) = \int_a^x F(t)\, dt$ for all x in $[a, b]$. Prove that G is differentiable on $[a, b]$.

3.6.10. Prove the **generalized integral mean value theorem:** *If f and g are continuous on $[a,b]$ and $g(x) > 0$ on $[a,b]$, then there is a $c \in [a,b]$ such that $\int_a^b fg = f(c)\int_a^b g$.*

a) Let m and M be the minimum and maximum values of f on $[a,b]$. Show that

$$m\int_a^b g \leq \int_a^b fg \leq M\int_a^b g$$

(Hint: Determine how the functions $mg(x)$, $f(x)g(x)$ and $Mg(x)$ are related and use problem 3.4.8.)

b) Finish the proof by mimicking the proof of the integral mean value theorem with $\int_a^b g$ playing the role of $b-a$.

c) Where in the proof do you need the hypothesis that $g > 0$? How would the theorem change if $g < 0$ on $[a,b]$?

d) Show that the integral mean value theorem is a special case of this theorem by choosing an appropriate function g.

3.6.11. Prove the following **change of variable** or **substitution** formula: *Let u be a differentiable function on $[a,b]$ and let u' be integrable on $[a,b]$. Assume that f is continuous on the range of u. If $u(a) = c$ and $u(b) = d$, then*

$$\int_a^b (f \circ u)(x)u'(x)\,dx = \int_c^d f(x)\,dx.$$

(Hint: As in Theorem 3.6.1, for x in the range of u, define $F(x) = \int_c^x f(t)\,dt$. Let $g(x) = (F \circ u)(x)$ for $x \in [a,b]$. Now write $\int_a^b g'(x)\,dx$ in two ways.)

3.7 Taylor Polynomials

Polynomials are very simple functions with which to work. Their values are easily computed and they are continuous and differentiable everywhere. We presume the reader is familiar with the calculus of the so-called *elementary functions*, such as the trigonometric functions and the natural logarithm and natural exponential functions. Despite the name "elementary," these functions are often difficult to evaluate except at a few isolated points. This section is devoted to proving a few basic results that will allow us to approximate such functions by evaluating an appropriate polynomial.

Recall that the tangent line to a differentiable function f is already a linear or first-degree polynomial approximation to f. As an example, suppose that $f(x) = \sqrt{x+1}$. Since f is differentiable at 0, the equation of the tangent line there is given by

$$L(x) = f'(0)\cdot x + f(0) = \frac{1}{2}x + 1.$$

Figure 3.7.1 shows that $L(x)$ provides a very rough approximation to $f(x)$ near 0. By definition, f and L have the same value, 1, when $x = 0$, and they have the same slope there, namely $\frac{1}{2}$. This agreement in value and slope between a function and its tangent line at a given point provides a starting place for constructing an infinite sequence of approximations to f.

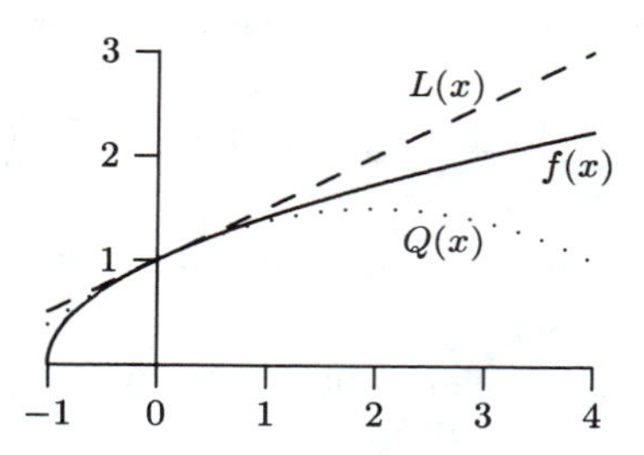

Figure 3.7.1

As a second approximation to f, consider the quadratic polynomial

$$Q(x) = -\frac{1}{8}x^2 + \frac{1}{2}x + 1.$$

Q agrees with f and its first *two* derivatives at 0. That is, $Q(0) = f(0) = 2$, $Q'(0) = f'(0) = \frac{1}{2}$, and $Q''(0) = f''(0) = -\frac{1}{4}$ as you can easily check. Further, $Q(x)$ seems to provide a better aproximation to $f(x)$, at least for x near 0 (see Figure 3.7.1). We might try to improve the approximation by using polynomials that agree with f up through its third or some higher derivative. However, none of these polynomials are truly useful in approximating f until there is some way of estimating the error in the approximation. Rather than going further with this specific example, we will start to construct the framework of a general theory.

Suppose that the function f has derivatives up to order n (where $n \geq 1$) at $x = 0$. As before, we try to construct a polynomial P of degree not greater than n that agrees with f and all of its derivatives up through order n at $x = 0$. In other words, we want

$$P(0) = f(0), \quad P'(0) = f'(0), \quad P''(0) = f''(0), \ \ldots, \ P^{(n)}(0) = f^{(n)}(0). \tag{1}$$

In general, such a polynomial has the form

$$P(x) = a_0 + a_1x^1 + a_2x^2 + \cdots + a_nx^n. \tag{2}$$

We need to determine, if possible, the $n + 1$ coefficients in $P(x)$ by using the constraints imposed on P and its derivatives at $x = 0$ that are listed in (1). First, evaluating P at $x = 0$, we see that $P(0) = a_0$. But it is also required that $P(0) = f(0)$ to satisfy (1). Consequently $a_0 = f(0)$. Next, using (2), we compute $P'(x)$ and evaluate it at $x = 0$ and find that $P'(0) = a_1$. Thus we must have $a_1 = f'(0)$ to satisfy (1). Similarly, computing $P''(x)$ and evaluating it at $x = 0$ yields $P''(0) = 2a_2$. But we need $P''(0) = f''(0)$, so $a_2 = f''(0)/2$. If we differentiate $P(x)$ a total of k times and evaluate $P^{(k)}$ at 0, we find that $P^{(k)}(0) = k!a_k$. (This formula is even valid for $k = 0$ if we interpret $P^{(0)}(0)$ as $P(0)$ and set $0! = 1$.) We also need $P^{(k)}(0) = f^{(k)}(0)$ to satisfy (1). So in general we must have

$$a_k = \frac{f^{(k)}(0)}{k!}, \qquad (k = 1, 2, \ldots, n). \tag{3}$$

Thus if any polynomial of degree $\leq n$ is to satisfy the conditions in (1), its coefficients are given by (3). Note that $P(x)$ will be a degree n polynomial if and only if $f^{(n)}(0) \neq 0$. Conversely it is easy to check that the polynomial $P(x)$ whose coefficients are determined by (3) satisfies the conditions in (1) and so agrees with f and its derivatives up through order n at $x = 0$.

This discussion is easily generalized to allow for approximations of f at a point a other than 0. We simply alter (1) and (2) to reflect this. Assuming that f is differentiable n times at a, we now ask that

$$P(a) = f(a), \quad P'(a) = f'(a), \quad P''(a) = f''(a), \ldots, \quad P^{(n)}(a) = f^{(n)}(a).$$

It now makes sense to write the polynomial P in powers of $(x - a)$, so

$$P(x) = a_0 + a_1(x-a)^1 + a_2(x-a)^2 + \cdots + a_n(x-a)^n.$$

As before, we solve for the $n + 1$ coefficients of P by successively differentiating $P(x)$, evaluating at $x = a$, and then comparing the result with the corresponding derviative of f. This yields

$$a_k = \frac{f^{(k)}(a)}{k!}, \qquad (k = 1, 2, \ldots, n).$$

DEFINITION 3.7.1 *Assume that f is differentiable n times at $x = a$. Then the polynomial*

$$\begin{aligned} p_{n,a}(x) &= \sum_{k=0}^{n} \frac{f^{(k)}(a)}{k!}(x-a)^k \\ &= f(a) + f'(a)(x-a) + \frac{f''(a)}{2}(x-a)^2 + \cdots + \frac{f^{(n)}(a)}{n!}(x-a)^n \end{aligned}$$

is called the nth ***Taylor polynomial for f at*** *a.*

EXAMPLE 3.7.1 Consider, again, the function $f(x) = \sqrt{x+1}$. Find $p_{4,0}(x)$ for f and find a general expression for the nth Taylor polynomial at 0.

SOLUTION The values of f and its first four derivatives at 0 are

$$
\begin{aligned}
f(x) &= (x+1)^{1/2}, & f(0) &= 1; \\
f'(x) &= \frac{1}{2}(x+1)^{-1/2}, & f'(0) &= \frac{1}{2}; \\
f''(x) &= (-1)\frac{1}{2^2}(x+1)^{-3/2}, & f''(0) &= -\frac{1}{4}; \\
f^{(3)}(x) &= (-1)^2\frac{1\cdot 3}{2^3}(x+1)^{-5/2}, & f^{(3)}(0) &= \frac{3}{8}; \\
f^{(4)}(x) &= (-1)^3\frac{1\cdot 3\cdot 5}{2^4}(x+1)^{-7/2}, & f^{(4)}(0) &= -\frac{15}{16}.
\end{aligned}
$$

Consequently

$$
\begin{aligned}
p_{4,0}(x) &= \sum_{k=0}^{4} \frac{f^{(k)}(0)}{k!}(x-0)^k \\
&= 1 + \frac{1}{2}x - \frac{1}{2^2\cdot 2!}x^2 + \frac{3}{2^3\cdot 3!}x^3 - \frac{15}{2^4\cdot 4!}x^4 \\
&= 1 + \frac{1}{2}x - \frac{1}{8}x^2 + \frac{1}{16}x^3 - \frac{5}{128}x^4.
\end{aligned}
$$

To find the general Taylor polynomial at 0, observe that there is a pattern to the derivatives of f. In particular, we see that if $k \geq 2$, then

$$f^{(k)}(x) = (-1)^{k-1}\frac{1\cdot 3\cdots(2k-3)}{2^k}(x+1)^{-(2k-1)/2},$$

and

$$f^{(k)}(0) = (-1)^{k-1}\frac{1\cdot 3\cdots(2k-3)}{2^k}.$$

Accounting for the terms when $k = 0$ and 1 separately, we find that

$$p_{n,0}(x) = \sum_{k=0}^{n} \frac{f^{(k)}(0)}{k!}(x-0)^k = 1 + \frac{x}{2} + \sum_{k=2}^{n}(-1)^{k-1}\frac{1\cdot 3\cdots(2k-3)}{2^k k!}x^k. \; \square$$

If the nth Taylor polynomial at a exists for the function f, the error in the approximation of f is given by the difference $f(x) - p_{n,a}(x)$. This difference is called the **remainder term** and will be denoted by $r_{n,a}(x)$. Thus

$$
\begin{aligned}
f(x) &= p_{n,a}(x) + r_{n,a}(x) \\
&= f(a) + f'(a)(x-a) + \cdots + \frac{f^{(n)}(a)}{n!}(x-a)^n + r_{n,a}(x).
\end{aligned}
$$

The next theorem gives two descriptions of this remainder.

THEOREM 3.7.2 ***(Taylor's Theorem with Remainder)*** *Assume that f and its first $n+1$ derivatives are defined and continuous on the closed interval $[a,b]$. For any $x \in [a,b]$, the remainder term $r_{n,a}(x) = f(x) - p_{n,a}(x)$ may be written as*

$$r_{n,a}(x) = \int_a^x f^{(n+1)}(t)\frac{(x-t)^n}{n!}\,dt.$$

PROOF Since f is continuously differentiable $n+1$ times on $[a,b]$, it follows from Lemma 3.6.6 that for any fixed x in $[a,b]$

$$f(x) = f(a) + \int_a^x f'(t)\,dt.$$

Now apply integration by parts (problem 3.6.3) to $\int_a^x f'(t)\,dt$ using $u = f'(t)$ and $v = -(x-t)$. This gives

$$\begin{aligned}\int_a^x f'(t)\,dt &= -f'(t)\cdot(x-t)\Big|_a^x - \int_a^x -f''(t)\cdot(x-t)\,dt \\ &= f'(a)\cdot(x-a) + \int_a^x f''(t)\cdot(x-t)\,dt.\end{aligned}$$

It now follows that

$$f(x) = f(a) + f'(a)(x-a) + \int_a^x f''(t)(x-t)\,dt,$$

where the integral in this expression is exactly the remainder $r_{1,a}(x)$. If we apply integration by parts again to this integral using the funtions $u = f''(t)$ and $v = \frac{-(x-t)^2}{2}$, we obtain

$$f(x) = f(a) + f'(a)(x-a) + f''(a)\frac{(x-a)^2}{2} + \int_a^x f'''(t)\frac{(x-t)^2}{2}\,dt.$$

We now have the second Taylor polynomial and its remainder $r_{2,a}(x)$.

Thus, repeated integration by parts simply cranks out the Taylor polynomial for f with an expression for the remainder. It is now clear how to supply the induction step to prove the result. Assume that the remainder formula holds for $n = k$. That is,

$$f(x) = f(a) + f'(a)(x-a) + \cdots + f^{(k)}(a)\frac{(x-a)^k}{k!} + \int_a^x f^{(k+1)}(t)\frac{(x-t)^k}{k!}\,dt.$$

Apply integration by parts to the integral on the right using $u = f^{(k+1)}(t)$ and $v = \frac{-(x-t)^k}{k!}$. This yields

$$\begin{aligned} f(x) &= f(a) + f'(a)(x-a) + f''(a)\frac{(x-a)^2}{2} + \cdots + f^{(k)}(a)\frac{(x-a)^k}{k!} \\ &\quad + f^{(k+1)}(t)\frac{(x-t)^{k+1}}{(k+1)!}\Big|_a^x - \int_a^x -f^{(k+2)}(t)\frac{(x-t)^{k+1}}{(k+1)!}\,dt \\ &= f(a) + f'(a)(x-a) + \cdots + f^{(k+1)}(a)\frac{(x-a)^{k+1}}{(k+1)!} \\ &\quad + \int_a^x f^{(k+2)}(t)\frac{(x-t)^{k+1}}{(k+1)!}\,dt. \quad \blacksquare \end{aligned}$$

The expression for the remainder in the preceding theorem is called the **integral form** of the remainder. Other forms of the remainder are possible.

COROLLARY 3.7.3 *Assume that f and its first $n+1$ derivatives are all defined and continuous on the closed interval $[a, b]$. Then the remainder term for any $x \in (a, b)$ may be written as*

$$r_{n,a}(x) = \frac{f^{(n+1)}(t)}{n!}(x-t)^n(x-a) \quad \textit{for some } t \textit{ in } (a, x);$$

or as

$$r_{n,a}(x) = \frac{f^{(n+1)}(t)}{(n+1)!}(x-a)^{n+1} \quad \textit{for some } t \textit{ in } (a, x).$$

(These are known as the Cauchy and Lagrange forms of the remainder, respectively.)

It is remarkably easy to derive the other two forms of the remainder from the integral form using the integral mean value theorem and its generalization. The proofs are outlined in the problem section.

REMARK We note that the hypothesis of the continuity of $f^{(n+1)}$ is necessary in the proof of the integral form of Taylor's theorem. It guarantees that the integrand is continuous, which ensures the integrability of the expression for the remainder. The proofs of the Lagrange and Cauchy forms of the remainders outlined in the problem section also make use of this continuity condition. However, it is possible to drop the restriction on the continuity of $f^{(n+1)}$ if one makes use of the mean value theorem instead of the integral mean value theorem (see problem 6).[1]

---

[1] Also see Michael Spivak, *Calculus*, 2nd ed., (Berkeley, Calif.: Publish or Perish, Inc., 1980), p. 392–394.

Taylor's theorem can also be viewed as a generalization of the mean value theorem. For when we take $n = 0$, the Lagrange and Cauchy forms of the remainder reduce to the same thing, namely, $r_{n,a}(x) = f'(t)$ for some t in (a, x). But then Taylor's theorem becomes

$$f(x) = f(a) + f'(t)(x - a).$$

This is just another way of stating the mean value theorem.

When we use the remainder theorem we are generally interested only in an upper bound on the remainder and not in the actual point t that gives the precise value of the remainder.

EXAMPLE 3.7.2 As an illustration, we will approximate $\sqrt{2}$ using Taylor polynomials (a considerable problem before the advent of hand-held calculators). In Example 3.7.1 we found the general expression for the nth Taylor polynomial at 0 for $f(x) = \sqrt{x+1}$ to be

$$p_{n,0}(x) = 1 + \frac{x}{2} + \sum_{k=2}^{n}(-1)^{k-1}\frac{1 \cdot 3 \cdots (2k-3)}{2^k k!}x^k.$$

Since $f(1) = \sqrt{2}$, we obtain approximations for $\sqrt{2}$ by evaluating $p_{n,0}(1)$ for various n. For example, the fourth Taylor polynomial yields a fairly good approximation:

$$\sqrt{2} \approx 1 + \frac{1}{2} - \frac{1}{8} + \frac{5}{128} = 1.4140625, \tag{3}$$

The remainder theorem provides a way to get a bound for the error in such approximations. To apply the theorem in this case, the polynomials are centered at $a = 0$ and the approximation is taking place at $x = 1$, so the interval under consideration is $[0, 1]$. Using the Lagrange form of the error term, we know that there is some t in $(0, 1)$, so the approximation in (3) is off by

$$r_{4,0}(1) = \frac{f^{(5)}(t)}{5!}(1-0)^5 = \frac{85(t+1)^{-9/2}}{2^5 \cdot 5!}.$$

We can bound the absolute value of the error term by noting that $(t + 1)^{-9/2} < 1$ for all t in $(0, 1)$. Consequently

$$|r_{4,0}(1)| = \left|\frac{85(t+1)^{-9/2}}{2^5 \cdot 5!}\right| < \frac{85}{2^5 \cdot 5!} = \frac{17}{768} < 0.0222\,.$$

From this observation it follows that approximating $\sqrt{2}$ by 1.4140625 has an error of *at most* 0.0222 (though here again you should recognize that the approximation is actually much better than that). ◻

EXAMPLE 3.7.3 Generally, when doing such approximations one has a desired degree of accuracy in mind. The problem then is to determine the Taylor polynomial of smallest degree that will supply the desired accuracy. For example, determine the general form of the Taylor polynomial for $f(x) = \cos x$ at 0 and show that for any x, the remainder can be made smaller than any positive ϵ.

SOLUTION Recall from previous course work that the derivatives of $\cos x$ repeat themselves in cycles of length 4.

$$\begin{aligned} f(x) &= \cos x, & f(0) &= 1; \\ f'(x) &= -\sin x, & f'(0) &= 0; \\ f''(x) &= -\cos x, & f''(0) &= -1; \\ f'''(x) &= \sin x, & f'''(0) &= 0; \\ f''''(x) &= \cos x, & f''''(0) &= 1. \end{aligned}$$

If k is odd, $f^{(k)}(0) = 0$, so that all the odd degree terms drop out of the Taylor polynomials. When k is even, $f^{(k)}(0)$ alternates in value; first it is $+1$ and then it is -1. Putting all of this together, the $(2n)$th Taylor polynomial at 0 is

$$p_{2n,0}(x) = \sum_{k=0}^{2n} \frac{f^{(k)}(0)}{k!}(x-0)^k = \sum_{k=0}^{n} \frac{(-1)^k}{(2k)!}x^{2k}.$$

If we use the Lagrange form of the remainder, we see that for some t in the interval $(0, x)$ (or $(x, 0)$ if $x < 0$)

$$r_{2n,0}(x) = \frac{f^{(2n+1)}(t)}{(2n+1)!}(x-0)^{2n+1} = \frac{(-1)^{n+1}\sin t}{(2n+1)!}x^{2n+1}.$$

Let us show that this remainder can be made arbitrarily small by choosing n to be sufficiently large. First, note that $|\sin t| \leq 1$ for any t. Next, choose an integer N such that $N > \max\{|x|,\ x^2,\ 1/\epsilon\}$. Then

$$\begin{aligned} r_{2N,0}(x) = \left|\frac{(-1)^{N+1}\sin t}{(2N+1)!}x^{2N+1}\right| &\leq \frac{|x| \cdot (x^2)^N}{1 \cdot 2 \cdots N \cdot (N+1) \cdots 2N \cdot (2N+1)} \\ &\leq \frac{N \cdot N^N}{N \cdot (N+1) \cdots 2N \cdot (2N+1)} \\ &\leq \frac{N^N}{N^N \cdot (2N+1)} \\ &\leq \frac{1}{2N+1} < \frac{1}{N} < \frac{1}{(1/\epsilon)} = \epsilon. \end{aligned}$$

When a particular value of x and a degree of tolerance ϵ are specified, calculations such as the foregoing are greatly simplified. For example, if we wanted to determine the value of $\cos(1)$ to within a tolerance of 10^{-6}, the preceding discussion would have us choose $N > \max\{1, 1, 10^6\}$. Thus choosing $N = 1,000,001$, we would be forced to evaluate $p_{2000002,0}(1)$. (Good luck!) In fact, the problem is much simpler than that. In the particular case of $x = 1$ the $(2n)$th Taylor polynomial simplifies to

$$p_{2n,0}(1) = \sum_{k=0}^{n} \frac{(-1)^k}{(2k)!},$$

and the corresponding remainder term becomes

$$r_{2n,0}(1) = \frac{(-1)^{n+1} \sin t}{(2n+1)!}.$$

This time the remainder is easy to bound. We have $|\sin t| \leq 1$ for all t, so

$$|r_{2n,0}(1)| = \left| \frac{(-1)^{n+1} \sin t}{(2n+1)!} \right| \leq \frac{1}{(2n+1)!}.$$

Now choose $2n + 1$ sufficiently large so that $1/(2n+1)! < 10^{-6}$. Observe that $11! = 39,916,800$ suffices, so the tenth Taylor polynomial (which has only 6 nonzero terms) will give an acceptable approximation.

$$p_{10,0}(1) = 1 - \frac{1}{2!} + \frac{1}{4!} - \frac{1}{6!} + \frac{1}{8!} - \frac{1}{10!} \approx 0.5403023\,. \quad \square$$

We have just shown that $\cos x$ can be approximated to any degree of accuracy by using a Taylor polynomial centered at the origin of sufficiently high degree. In other words, we can write

$$\cos x \approx 1 - \frac{x^2}{2!} + \frac{x^4}{4!} - \frac{x^6}{6!} + \cdots + \frac{(-1)^n x^{2n}}{(2n)!}.$$

To improve the accuracy of this approximation, one simply adds more terms to this Taylor polynomial. If we add an infinite number of terms together does our approximation become an equality, so that the remainder term vanishes? In other words, are these elementary functions expressible as some sort of "infinite sums"? For example, can we write

$$\cos x = 1 - \frac{x^2}{2!} + \frac{x^4}{4!} - \frac{x^6}{6!} + \frac{x^8}{8!} - \cdots ?$$

What do we mean by such an infinite sum? In Chapter 4 we will develop a theory for infinite sums and infinite polynomials and we will find that many lovely formulas such as the preceding expression for $\cos x$ do indeed hold. In formulating appropriate definitions for infinite sums we will again see the crucial role played by the concept of limits and the completeness of the real numbers.

Problems for Section 3.7

3.7.1. Find $p_{4,0}(x)$ and $p_{n,0}(x)$ for the following functions:

a) $\sin x$ **b)** e^x **c)** $\ln(1+x)$ **d)** e^{-2x} **e)** $\cos(3x)$

3.7.2. Write out the nth Taylor polynomials for the following functions.

a) $g(x) = \cos x$ at $a = 0, n = 8$, on $[-2, 2]$.

b) $h(x) = e^{-2x}$ at $a = 0, n = 5$, on $[-3, 3]$.

c) $f(x) = \sqrt{x}$ at $a = 1, n = 4$, on $[.5, 1.5]$.

d) Use the Lagrange form of the remainder to find an upper bound on the difference between these same functions and their Taylor polynomials.

e) Suppose that you wished to estimate $\cos 2$ to within an error of .001. How many terms of the Taylor series would you need? Compare your Taylor series estimate with a calculator value for $\cos 2$.

3.7.3. **a)** Use L'Hôpital's rule twice to show that $\lim_{x\to a} \frac{f(x) - p_{2,a}(x)}{(x-a)^2} = 0$, where $p_{2,a}(x)$ is the second Taylor polynomial of f.

b) Prove again that $\lim_{x\to a} \frac{f(x) - p_{n,a}(x)}{(x-a)^n} = 0$. This time use the fact that $f(x) - p_{n,a}(x) = r_{n,a}(x)$ with the Lagrange form of the remainder. (What must you assume about $f^{(n+1)}$?)

3.7.4. In problem 3.3.13 we saw that if f'' exists and is bounded on $[a, x]$, then $f(x) = f(a) + f'(a)(x-a) + R$, where $R < |M|(x-a)^2$ and M is the least upper bound of $|f''|$ on I. Compare this estimate of R with the estimate of the remainder from Taylor's theorem for $n = 1$.

3.7.5. Find the nth Taylor polynomial for the function $(x+1)^n$ about the point $a = 0$. What is the remainder term? Do you recognize the formula?

3.7.6. It is remarkably easy to derive the other two forms of the remainder from the integral remainder using the integral mean value theorem and its generalization.

a) Use Theorem 3.6.4 to derive the Cauchy form of the remainder,

$$\int_a^x f^{(n+1)}(t)\frac{(x-t)^n}{n!}\,dt = f^{(n+1)}(t^*)\frac{(x-t^*)^n}{n!}(x-a)$$

b) Use the generalized integral mean value theorem (problem 3.6.10) to derive the Lagrange form of the remainder,

$$\int_a^x f^{(n+1)}(t)\frac{(x-t)^n}{n!}\,dt = f^{(n+1)}(t^*)\frac{(x-a)^{n+1}}{(n+1)!}.$$

3.7.7. Assume that f'' exists on $[a, x]$ and let $R(x) = f(x) - f(a) - f'(a)(x - a)$ be the difference between f and its first Taylor polynomial at x. In this problem we will use Rolle's theorem to derive the Lagrange form of the remainder for the first Taylor polynomial. (We do not require that f'' be continuous on $[a, b]$. Compare this with Theorem 3.7.2 and Corollary 3.7.3.) Think of x as fixed and let

$$h(t) = f(x) - f(t) - f'(t)(x - t) - R(x)\frac{(x - t)^2}{(x - a)^2}$$

a) First show that $h(x) = 0$ and $h(a) = 0$. (For the second part you will need to use the expression for $R(x)$).

b) Verify that you can apply Rolle's theorem to h on the interval $[a, x]$.

c) Now compute $h'(t)$. (Remember that x is a constant.)

d) Rolle's theorem says that h' must be 0 for some $t^* \in [a, x]$. Set $h'(t)$ equal to 0 and show that $R(x) = \frac{f''(t^*)}{2}(x - a)^2$. This is the Lagrange form of the remainder.

e) Can you generalize the problem to get the remainder for the second Taylor polynomial? What would $R(x)$ be? What form would $h(t)$ have?

4

Sequences and Series

At the end of Section 3.7 we saw that if the function $f(x)$ were differentiable n times at a, its nth Taylor polynomial at a was

$$p_{n,a}(x) = \sum_{k=0}^{n} \frac{f^{(k)}(a)}{k!}(x-a)^k.$$

For example, let $f(x) = e^x$ and let $a = 0$. Since $f^{(k)}(x) = e^x$ for all k,

$$\begin{aligned} p_{n,0}(x) &= \sum_{k=0}^{n} \frac{f^{(k)}(0)}{k!}(x-0)^k = \sum_{k=0}^{n} \frac{e^0}{k!}(x)^k \\ &= \sum_{k=0}^{n} \frac{x^k}{k!} = 1 + \frac{x}{1!} + \frac{x^2}{2!} + \cdots + \frac{x^n}{n!}. \end{aligned}$$

As n gets larger, more terms are added to the polynomial. As $n \to \infty$, we get

$$p_{\infty,0}(x) = 1 + x + \frac{x^2}{2!} + \frac{x^3}{3!} + \frac{x^4}{4!} + \cdots.$$

How do we make sense of such infinite sums? Can we add, subtract, multiply, and divide such expressions? For example, at the end of Section 3.7 we saw that the "infinite" Taylor polynomial for $\cos x$ at $a = 0$ was

$$q_{\infty,0}(x) = 1 - \frac{x^2}{2!} + \frac{x^4}{4!} - \frac{x^6}{6!} + \frac{x^8}{8!} - \cdots.$$

To write

$$p_{\infty,0}(x) + q_{\infty,0}(x) = 2 + x + \frac{x^3}{3!} + \frac{2x^4}{4!} + \frac{x^5}{5!} + \frac{x^7}{7!} + \frac{2x^8}{8!} + \frac{x^9}{9!} + \cdots,$$

we must use the associative property an infinite number of times. However, the axioms for the real numbers outlined in Chapter 1 apply only to a finite number of operations.

Historically such questions were troublesome. It took the mathematical community until the first part of the nineteenth century, with the work of Bolzano and Cauchy, to develop an adequate theory to deal with infinite sums.

Here's a simple example of the sort of thing that was problematic. Suppose that

$$S = 1 + x + x^2 + x^3 + x^4 + \cdots, \tag{1}$$

then

$$xS = x + x^2 + x^3 + x^4 + \cdots. \tag{2}$$

Subtracting (2) from (1) gives $S - xS = 1$ or $(1-x)S = 1$. Consequently

$$S = \frac{1}{1-x}. \tag{3}$$

Taking $x = 2$, by comparing (1) and (3) we obtain

$$S = -1 = \frac{1}{1-2} = 1 + 2 + 2^2 + 2^3 + 2^4 + \cdots.$$

Can a negative number possibly be the sum of positive numbers? The material in this chapter outlines a theory of infinite sums that will permit us to make sense of such questions.

4.1 Infinite Sequences

The idea of a sequence is a simple and familiar one. Informally, a sequence is an ordered succession of objects, events, or numbers. Sequences of numbers often provide a first encounter with the infinite. For example, in the sequence

$$1/2,\ 3/4,\ 7/8,\ 15/16,\ \ldots$$

we are given enough terms to determine its pattern so that we could continue writing out terms indefinitely. General sequences of numbers are frequently written using subscripted letters such as

$$a_1,\ a_2,\ a_3,\ \ldots$$

This notation provides a clue as to the formal definition of a sequence. The subscripts indicate that a sequence associates to each positive integer n a single number a_n. This is the language of the definition of a function.

DEFINITION 4.1.1 *A* ***sequence*** *is a function whose domain is the set* **N** *of natural numbers.*

If a is a sequence, its value $a(n)$ at n is usually denoted by a_n and is called the nth **term** of the sequence. The entire sequence a is often denoted by $\{a_n\}_{n=1}^{\infty}$. For example, the sequence

$$\{a_n\}_{n=1}^{\infty} = 1,\ 4,\ 9,\ 16,\ \ldots$$

is defined by $a_n = n^2$. Thus we might also write $\{a_n\}_{n=1}^{\infty} = \{n^2\}_{n=1}^{\infty}$. Other simple sequences are

$$\{1/n\}_{n=1}^{\infty} = 1,\ 1/2,\ 1/3,\ \ldots \qquad \{(-1)^n\}_{n=1}^{\infty} = -1,\ 1,\ -1,\ 1,\ \ldots$$

and

$$\{n!\}_{n=1}^{\infty} = 1,\ 2,\ 6,\ 24,\ \ldots$$

These particular sequences have readily discernible and predictable behavior. For example, it is clear that the terms of the sequence $\{\frac{1}{n}\}_{n=1}^{\infty}$ approach 0 as n gets large. In other words, the terms of this sequence become arbitrarily close to 0 if we go out far enough in the sequence. In contrast, the terms of the alternating sequence $\{(-1)^n\}_{n=1}^{\infty}$ bounce back and forth between -1 and 1 and so do not approach a single value. On the other hand, the terms of $\{n!\}_{n=1}^{\infty}$ and $\{n^2\}_{n=1}^{\infty}$ continue to get larger without bound. The notion of a limit is lurking close by.

Though sequences are functions, their domains are not intervals, so the definition of $\lim_{x\to a} f(x) = L$ does not make sense here. Instead of trying to express the notion of x being taken close enough to a to make $f(x)$ within ϵ of L, we need to express the notion of n *being sufficiently large* to make a_n within ϵ of some limit L. We have seen this idea before when working with limits at infinity where "being sufficiently large" was expressed formally by saying that "there is a number N such that if $x > N$, then $f(x)$ is close to L." This language is easily carried over to sequences.

DEFINITION 4.1.2 *A sequence $\{a_n\}_{n=1}^{\infty}$ is said to* ***converge to the real number*** *A if for each $\epsilon > 0$ there exists a positive integer N such that for all $n > N$, $|a_n - A| < \epsilon$. This is denoted by writing $\lim_{n\to\infty} a_n = A$. If $\{a_n\}_{n=1}^{\infty}$ does not converge, it is said to* ***diverge.***

EXAMPLE 4.1.1 Show that if $p > 0$, then $\{1/n^p\}_{n=1}^{\infty}$ converges to 0.

SOLUTION Given $\epsilon > 0$, we must find N such that if $n > N$, then $|(1/n^p) - 0| < \epsilon$. But

$$\left|\frac{1}{n^p} - 0\right| < \epsilon \iff n^p > \frac{1}{\epsilon} \iff n > \left(\frac{1}{\epsilon}\right)^{\frac{1}{p}}.$$

By the Archimedean principle, we can choose $N > \left(\frac{1}{\epsilon}\right)^{\frac{1}{p}}$. ☐

EXAMPLE 4.1.2 Show that $\left\{\dfrac{3n}{n+2\sqrt{n}}\right\}_{n=1}^{\infty}$ converges to 3.

SOLUTION Given $\epsilon > 0$, we must find N such that

$$n > N \Longrightarrow \left|\frac{3n}{n+2\sqrt{n}} - 3\right| < \epsilon.$$

Notice that

$$\left|\frac{3n}{n+2\sqrt{n}} - 3\right| = \left|\frac{3n - 3n - 6\sqrt{n}}{n+2\sqrt{n}}\right| = \left|\frac{-6\sqrt{n}}{n+2\sqrt{n}}\right| < \left|\frac{-6\sqrt{n}}{n}\right| = \frac{6}{\sqrt{n}}.$$

But $6/\sqrt{n} < \epsilon$ whenever $n > 36/\epsilon^2$. Therefore if we choose $N > 36/\epsilon^2$, for any $n > N$ we have

$$\left|\frac{3n}{n+2\sqrt{n}} - 3\right| < \frac{6}{\sqrt{n}} < \frac{6}{\sqrt{N}} < \frac{6}{6/\epsilon} = \epsilon. \quad \square$$

Many of the basic limit theorems for functions discussed in Chapter 2 carry over to this context. The proof of the next result is similar to the proof of Theorem 2.3.1 and is left as an exercise.

THEOREM 4.1.3 *If a sequence converges, its limit is unique.*

EXAMPLE 4.1.3 Show that $\{a_n\}_{n=1}^{\infty} = \{(-1)^n\}_{n=1}^{\infty}$ diverges.

SOLUTION Observe that the values of this sequence alternate: when n is even, $a_n = 1$; and when is n odd, $a_n = -1$. The idea is to show that since the sequence is not approaching a single value as n gets large, it cannot have a limit. We employ a proof by contradiction. Assume that A is the limit of $\{a_n\}_{n=1}^{\infty}$. Since the distance between -1 and 1 is 2, A cannot be within 1 of both a_n and a_{n+1} for any n. We use this to obtain a contradiction. Take $\epsilon = 1$. Since A is the limit of $\{(-1)^n\}_{n=1}^{\infty}$, there must be an N such that if $n > N$, then $|a_n - A| < \epsilon = 1$. But for any $n > N$ we have $|a_{n+1} - a_n| = 2$, so by the triangle inequality

$$2 = |a_{n+1} - a_n| = |(a_{n+1} - A) + (A - a_n)| \le |a_{n+1} - A| + |A - a_n| < \epsilon + \epsilon = 2,$$

which is a contradiction. $\square$

In many of the proofs of limit theorems and in working out explicit limit examples in Chapter 2, it was crucial to find bounds for certain factors that arose during the calculations. The same is true for sequences. We say that a sequence $\{a_n\}_{n=1}^{\infty}$ is **bounded** if there exists some number M such that $|a_n| \le M$ for all n. For example, $\{1/n\}_{n=1}^{\infty}$ is bounded, but $\{n!\}_{n=1}^{\infty}$ is not. The following simple result is often quite useful.

Lemma 4.1.4 *Every convergent sequence is bounded.*

PROOF Suppose that $\lim_{n\to\infty} a_n = A$. Taking $\epsilon = 1$, there exists N such that if $n > N$, then $|a_n - A| < 1$. Using the triangle inequality, if $n > N$, then

$$|a_n| = |a_n - A + A| \leq |a_n - A| + |A| < 1 + |A|.$$

Thus the infinite tail end of the sequence is bounded by $1 + |A|$. The first N terms of the sequence are easily bounded by inspection. If we let

$$M = \max\{|a_1|, |a_2|, \ldots, |a_N|, 1 + |A|\},$$

we have $|a_n| \leq M$ for all n. ∎

The next result translates the basic limit theorems for functions of a real variable into the language of sequences.

Theorem 4.1.5 *Assume that $\{a_n\}_{n=1}^{\infty}$ and $\{b_n\}_{n=1}^{\infty}$ are convergent sequences with limits A and B, respectively. Then*

a) $\lim_{n\to\infty} a_n + b_n = A + B$;

b) $\lim_{n\to\infty} ca_n = cA$, *for any constant c in* $\mathbf{R}$;

c) $\lim_{n\to\infty} a_n b_n = AB$;

d) $\lim_{n\to\infty}(a_n/b_n) = A/B$, *provided that $b_n \neq 0$ for all n and $B \neq 0$.*

PROOF The proofs of **(a)** to **(c)** are similar to the proofs of Theorems 2.3.2 to 2.3.4 and are left as exercises. The proof of **(d)** is similar to the proof of the quotient rule for limits of functions, Theorem 2.3.6. We begin by bounding b_n away from 0. First, $B \neq 0$, so $|B|/2 > 0$. Since $\{b_n\}_{n=1}^{\infty}$ converges to B, there exists N_1 such that if $n > N_1$, then $|B - b_n| < |B|/2$. By the triangle inequality, $|B| - |b_n| \leq |B - b_n|$. It follows that

$$n > N_1 \Rightarrow |B|/2 < |b_n|. \tag{1}$$

We need to show that $\left|\frac{a_n}{b_n} - \frac{A}{B}\right|$ can be made arbitrarily small by taking n sufficiently large. By using (1) and then the triangle inequality (having added and subtracted the common term AB), we see that

$$\begin{aligned}
\left|\frac{a_n}{b_n} - \frac{A}{B}\right| = \left|\frac{a_n B - A b_n}{b_n B}\right| &< \frac{2|a_n B - A b_n|}{|B|^2} \\
&\leq \frac{2|a_n B - AB|}{|B|^2} + \frac{2|AB - A b_n|}{|B|^2} \\
&= \frac{2}{|B|}|a_n - A| + \frac{2|A|}{B^2}|B - b_n|.
\end{aligned} \tag{2}$$

To finish the proof, we now make each of the last two summands in (2) small. Since $\{a_n\}_{n=1}^{\infty}$ converges to A, we can choose N_2 such that

$$\text{if } n > N_2, \text{then } |a_n - A| < \frac{|B|\epsilon}{4}.$$

Similarly since $\{b_n\}_{n=1}^{\infty}$ converges to B (and assuming $A \neq 0$), we can choose N_3 such that

$$\text{if } n > N_3, \text{then } |b_n - B| < \frac{B^2\epsilon}{4|A|}.$$

Let $N = \max\{N_1, N_2, N_3\}$. If $n > N$, by (2) and the foregoing inequalities

$$\left|\frac{a_n}{b_n} - \frac{A}{B}\right| < \frac{\epsilon}{2} + \frac{\epsilon}{2} = \epsilon.$$

It is a simple exercise to modify the proof if $A = 0$. ∎

EXAMPLE 4.1.4 Show that $\{t_n\}_{n=1}^{\infty} = \left\{\frac{3n+1}{2n^2 - n}\right\}_{n=1}^{\infty}$ converges to 0.

SOLUTION We would like to apply **(d)** of the preceding theorem to evaluate the limit. However, both the numerator and denominator of t_n increase without bound, so they do not converge and the theorem is not immediately applicable. However, if we divide both the numerator and denominator by the highest power of n in the denominator, that is, by n^2, we find that

$$t_n = \frac{\frac{3}{n} + \frac{1}{n^2}}{2 - \frac{1}{n}}.$$

Using Example 4.1.1 and Theorem 4.1.5, we see that

$$\lim_{n\to\infty} t_n = \lim_{n\to\infty} \frac{\frac{3}{n} + \frac{1}{n^2}}{2 - \frac{1}{n}} = \frac{0}{2} = 0. \quad \square$$

Problems for Section 4.1

4.1.1. Prove Theorem 4.1.3.

4.1.2. Prove the first three parts of Theorem 4.1.5.

4.1.3. For $n \geq 1$, define $a_n = 1 + \frac{1}{2} + \frac{1}{2^2} + \cdots + \frac{1}{2^n}$.

a) Show that $a_n = 2 - (1/2^{n+1})$.

b) Show that $\{a_n\}_{n=1}^{\infty}$ converges to 2.

4.1.4. Find the limits of the following sequences.

a) $\left\{\dfrac{n^2+n-3}{n^2+\sqrt{n}}\right\}_{n=1}^{\infty}$ b) $\left\{\dfrac{n^2-2n+1}{3n^3-n^2-6}\right\}_{n=1}^{\infty}$

c) $\left\{\dfrac{2n+\sqrt{n}+4}{3n+2}\right\}_{n=1}^{\infty}$ d) $\left\{\dfrac{5n^2+\sqrt[3]{n}+\cos n}{2n^2-2n+1}\right\}_{n=1}^{\infty}$

•4.1.5. Assume that $\lim_{n\to\infty} a_n = a$. Suppose that the function f is continuous at a and that its domain includes all a_n. Form the sequence $\{f(a_n)\}_{n=1}^{\infty}$. Show that $\lim_{n\to\infty} f(a_n) = f(a)$. That is, $\lim_{n\to\infty} f(a_n) = f(\lim_{n\to\infty} a_n)$, so the limit can be evaluated before or after we apply the function f. The proof is similar to showing that the composition of continuous functions is continuous (Theorem 2.5.3).

a) Let $\epsilon > 0$ be given. Show that there exists $\delta > 0$ such that if $|x-a| < \delta$, then $|f(x)-f(a)| < \epsilon$. Show that there is an N such that if $n > N$, then $|a_n - a| < \delta$.

b) To complete that proof, show that if $n > N$, then $|f(a_n)-f(a)| < \epsilon$.

c) Continuity is crucial. Let $\{a_n\}_{n=1}^{\infty} = \{1/n\}_{n=1}^{\infty}$ and let

$$f(x) = \begin{cases} 0, & \text{if } x = 0 \\ 1, & \text{if } x \neq 0 \end{cases}.$$

Show that $\lim_{n\to\infty} f(a_n) \neq f(\lim_{n\to\infty} a_n)$. Why doesn't this contradict what you just proved?

4.1.6. There is an obvious similarity in the definition of $\lim_{x\to+\infty} f(x)$ and the definition of $\lim_{n\to\infty}\{a_n\}_{n=1}^{\infty}$. Prove the following result, which makes this connection explicit.

Theorem: *Let $\{a_n\}_{n=1}^{\infty}$ be a sequence and let $f(x)$ be a function defined on an interval of the form $[b,\infty)$ such that $f(n) = a_n$ for all positive integers $n \geq b$. If $\lim_{x\to+\infty} f(x) = L$, then $\{a_n\}_{n=1}^{\infty}$ converges and $\lim_{n\to\infty} a_n = L$.*

•4.1.7. Show that $\{a_n\}_{n=1}^{\infty} = \left\{\ln(1+\frac{1}{n})^n\right\}_{n=1}^{\infty}$ converges to 1 by making use of the preceding problem as follows.

a) Define the function $f(x) = \ln\left(1+\dfrac{1}{x}\right)^x$ on the interval $[1,\infty)$ and prove that $\lim_{x\to+\infty} f(x) = 1$. (Hint: Rewrite f as $\frac{\ln(1+\frac{1}{x})}{\frac{1}{x}}$ and use L'Hôpital's rule.)

b) Verify that the hypotheses of the theorem in the preceding problem are satisfied and conclude that $\lim_{n\to\infty} a_n = 1$. Why can't L'Hôpital's rule be applied directly to the sequence $\{a_n\}_{n=1}^{\infty}$?

4.1.8. Show that the sequence $\left\{\frac{1+2+\cdots+n}{n^2}\right\}_{n=1}^{\infty}$ converges to 1/2.

4.1.9. **Theorem:** *If $\{a_n\}_{n=1}^{\infty}$ is a sequence of nonpositive numbers that converges to a limit A, then $A \leq 0$.*

a) Prove this result by modifying the argument in problem 2.2.9.

b) Assume that $\{b_n\}_{n=1}^{\infty}$ is a sequence of real numbers that converges to B. Assume further that $b_n \leq M$ for all n. Show that $B \leq M$.

4.1.10. Show that $\{a_n\}_{n=1}^{\infty} = \left\{\frac{2^n}{n!}\right\}_{n=1}^{\infty}$ converges to 0. (Hint: Show by induction that $a_n \leq 2 \cdot 2/n$.)

4.1.11. **a)** If $\{a_n\}_{n=1}^{\infty}$ converges and $\{b_n\}_{n=1}^{\infty}$ diverges, prove that $\{a_n + b_n\}_{n=1}^{\infty}$ diverges.

b) If $\{a_n\}_{n=1}^{\infty}$ and $\{b_n\}_{n=1}^{\infty}$ both diverge, must $\{a_n + b_n\}_{n=1}^{\infty}$ diverge?

4.1.12. **a)** Show that the sequence $\{a_n\}_{n=1}^{\infty} = \left\{\sqrt{n+1} - \sqrt{n}\right\}_{n=1}^{\infty}$ converges.

b) Show that $\left\{\sqrt{n}a_n\right\}_{n=1}^{\infty}$ also converges.

4.1.13. Let $\{s_n\}_{n=1}^{\infty}$ be defined by $s_n = \frac{1}{n^2} + \frac{2}{n^2} + \frac{3}{n^2} + \cdots + \frac{n}{n^2}$. Draw a graph that shows that s_n can be thought of as a Riemann sum for $\int_0^1 x\,dx$. To what value does $\{s_n\}_{n=1}^{\infty}$ converge?

4.1.14. Let $\{s_n\}_{n=1}^{\infty}$ be defined by $s_n = \frac{1}{n^3} + \frac{2^2}{n^3} + \frac{3^2}{n^3} + \cdots + \frac{n^2}{n^3}$. Draw a graph that shows that s_n can be thought of as a Riemann sum for $\int_0^1 x^2\,dx$. To what value does $\{s_n\}_{n=1}^{\infty}$ converge?

4.1.15. **a)** Define $\{a_n\}_{n=1}^{\infty}$ by $a_n = \int_1^n x^{-2}\,dx$. Does $\{a_n\}_{n=1}^{\infty}$ converge? If so, to what?

b) Define $\{a_n\}_{n=1}^{\infty}$ by $a_n = \int_{1/n}^1 \frac{1}{\sqrt{x}}\,dx$. Does $\{a_n\}_{n=1}^{\infty}$ converge? If so, to what?

c) Define $\{a_n\}_{n=1}^{\infty}$ by $a_n = \int_0^1 x^n\,dx$. Does $\{a_n\}_{n=1}^{\infty}$ converge? If so, to what?

4.2 Monotone and Cauchy Sequences

The convergence or divergence of the sequences we have considered so far has been a relatively straightforward matter. In those cases where the sequence did converge, its limit was either easily guessed or provided for you. However, sequences can be quite complicated. For example, consider the sequence

$$\left\{\left(1 + \frac{1}{n}\right)^n\right\}_{n=1}^{\infty} = 2, \frac{9}{4}, \frac{64}{27}, \frac{625}{256}, \ldots$$

Whether this sequence converges and what its limit might be is not immediately clear from what we have done so far. The nth power tends to make the terms in this sequence larger, while the base $1 + \frac{1}{n}$ is itself getting smaller. The material in this section provides a way to handle such problems. In particular, we will prove a result that will tell us precisely when a sequence converges, even if we do not know its limit!

4.2.1 Monotone Sequences

We begin by examining certain simple types of sequences.

DEFINITION 4.2.1 *A sequence $\{a_n\}_{n=1}^{\infty}$ of real numbers is **increasing** if $a_n \leq a_{n+1}$ for all $n \in \mathbf{N}$ and is **decreasing** if $a_n \geq a_{n+1}$ for all $n \in \mathbf{N}$. A sequence is **monotone** if it is either increasing or decreasing.*

Examples of monotone sequences are easy to manufacture. Since the natural log function is increasing, the sequence

$$\{a_n\}_{n=1}^{\infty} = \left\{\frac{1}{\ln(n+1)}\right\}_{n=1}^{\infty}$$

is decreasing. The sequence $\{2 - 2^{-n}\}_{n=1}^{\infty}$ is an increasing sequence, whereas $\{1/n^2\}_{n=1}^{\infty}$ is decreasing. A constant sequence $c, c, c, \ldots$ is *both* increasing and decreasing. There are, of course, more complicated sequences for which it may be difficult to determine whether the sequence is monotone.

EXAMPLE 4.2.1 As was just shown, the first few terms of the sequence $\{a_n\}_{n=1}^{\infty} = \left\{\left(1 + \frac{1}{n}\right)^n\right\}_{n=1}^{\infty}$ are increasing, as shown above. But it is not at all clear that the entire sequence is increasing. Let us supply a proof of this.

The binomial theorem[1] gives an expansion of a_n as a sum of $n + 1$ terms:

$$\begin{aligned} a_n &= 1 + \tfrac{n}{1} \cdot \tfrac{1}{n} + \tfrac{n(n-1)}{2!} \cdot \tfrac{1}{n^2} + \cdots + \tfrac{n(n-1)(n-2)\cdots 1}{n!} \cdot \tfrac{1}{n^n} \\ &= 1 + \tfrac{1}{1} \cdot \tfrac{n}{n} + \tfrac{1}{2!} \cdot \tfrac{n(n-1)}{n^2} + \cdots + \tfrac{1}{n!} \cdot \tfrac{n(n-1)(n-2)\cdots 1}{n^n} \\ &= 1 + 1 + \tfrac{1}{2!}\left(1 - \tfrac{1}{n}\right) + \cdots + \tfrac{1}{n!}\left(1 - \tfrac{1}{n}\right)\left(1 - \tfrac{2}{n}\right)\cdots\left(1 - \tfrac{n-1}{n}\right). \end{aligned}$$

[1] Recall that the binomial theorem states that for any $n \in \mathbf{N}$,

$$\begin{aligned} (1+a)^n = \sum_{k=0}^{n} \frac{n!}{k!(n-k)!} a^k &= 1 + \frac{n!}{1(n-1)!} a + \frac{n!}{2!(n-2)!} + \cdots + \frac{n!}{n!0!} a^n \\ &= 1 + \frac{n}{1} a + \frac{n(n-1)}{2!} + \cdots + \frac{n!}{n!} a^n. \end{aligned}$$

Analogously, for a_{n+1} we obtain an expansion as a sum of $n+2$ terms:

$$a_{n+1} = 1 + 1 + \frac{1}{2!}\left(1 - \frac{1}{n+1}\right) + \cdots$$
$$+ \tfrac{1}{(n+1)!}\left(1 - \tfrac{1}{n+1}\right)\left(1 - \tfrac{2}{n+1}\right)\cdots\left(1 - \tfrac{n}{n+1}\right).$$

Each of the first $n+1$ terms in the expansion of a_{n+1} is greater than or equal to the corresponding term in the expansion of a_n. Further, the final term of the expansion for a_{n+1} is positive, so $a_n < a_{n+1}$. Although this proves that $\{a_n\}_{n=1}^{\infty}$ is increasing, it is still not clear whether $\{a_n\}_{n=1}^{\infty}$ converges. □

The next theorem tells us when such a monotone sequence converges.

THEOREM 4.2.2 *A monotone sequence converges if and only if it is bounded.*

PROOF We have already shown (Lemma 4.1.4) that *any* convergent sequence, monotone or not, is bounded. Therefore we need only show that a sequence that is both monotone and bounded must necessarily converge.

First assume that $\{a_n\}_{n=1}^{\infty}$ is a bounded, increasing sequence. Let S denote the set of all terms in the sequence: $\{a_n \mid n \in \mathbf{N}\}$. Since $\{a_n\}_{n=1}^{\infty}$ is bounded, so is S. Hence S has a least upper bound, which we denote by A. We will now show that $\{a_n\}_{n=1}^{\infty}$ converges to A.

Consider any $\epsilon > 0$. Since A is the least upper bound for S and $A - \epsilon < A$, there must exist some element a_N of S such that $A - \epsilon < a_N$ (problem 1.2.5), or equivalently, $A - a_N < \epsilon$. Because the sequence is increasing, we know that if $n > N$, then $a_N \leq a_n \leq A$. Consequently for $n > N$ we have

$$|A - a_n| = A - a_n \leq A - a_N < \epsilon.$$

Therefore $\lim_{n\to\infty} a_n = A$.

In the case where $\{a_n\}_{n=1}^{\infty}$ is decreasing, let A be the greatest lower bound of S and proceed in a similar fashion. ∎

Sometimes a sequence $\{a\}_{n=1}^{\infty}$ fails to be monotone because of the first few terms. That is, a sequence may be increasing or decreasing for all $n \geq M$ where $M > 1$. In this case the infinite tail of the sequence is monotone and Theorem 4.2.2 still applies: such a sequence converges if and only if it is bounded.

The preceding proof shows not only that a monotone, bounded sequence converges, but that if the sequence is increasing, it converges to the least upper bound of the set of values of the sequence; and if it is decreasing, it converges to the greatest lower bound of the set of values of the sequence.

EXAMPLE 4.2.2 Reconsider the last example in the introduction to this chapter. Let r be a real number such that $0 \leq r < 1$. Let $s_n = 1 + r + r^2 + \cdots + r^n$. Show that the sequence $\{s_n\}_{n=1}^{\infty}$ converges.

SOLUTION The sequence is clearly increasing, so to show that it converges we only need to show that $\{s_n\}_{n=1}^{\infty}$ is bounded above. But

$$(1-r)s_n = (1-r)(1 + r + r^2 + \cdots + r^n) = 1 - r^{n+1}.$$

Consequently $\{s_n\}_{n=1}^{\infty}$ is bounded above and converges, since

$$s_n = \frac{1 - r^{n+1}}{1 - r} < \frac{1}{1 - r}.$$

In problem 4.2.4 you are asked to show that $\lim_{n\to\infty} r^n = 0$. With this result, it follows that

$$\begin{aligned}\lim_{n\to\infty} 1 + r + r^2 + \cdots + r^n &= \lim_{n\to\infty} s_n \\ &= \lim_{n\to\infty} \frac{1 - r^{n+1}}{1 - r} = \frac{1}{1 - r} \quad (0 \leq r < 1). \ \square\end{aligned}$$

4.2.2 Cauchy Sequences

Sequences converge because their terms eventually get close to some (possibly unknown) limit value. As the terms begin to approach a common limit, these terms themselves cannot differ very much from each other. This idea is made precise in the following definition.

DEFINITION 4.2.3 *A sequence $\{a_n\}_{n=1}^{\infty}$ of real numbers is a* ***Cauchy sequence*** *if for each $\epsilon > 0$ there exists a number N such that whenever $m > N$ and $n > N$, then $|a_n - a_m| < \epsilon$.*

LEMMA 4.2.4 *Every convergent sequence is a Cauchy sequence.*

PROOF The proof is another application of the triangle inequality. Given $\epsilon > 0$, suppose that $\{a_n\}_{n=1}^{\infty}$ converges to A. Then there is an integer N such that if $n > N$, then $|a_n - A| < \epsilon/2$. Now for $m,\ n > N$, the triangle inequality implies that

$$|a_n - a_m| = |a_n - A + A - a_m| \leq |a_n - A| + |a_m - A| < \epsilon/2 + \epsilon/2 = \epsilon. \ \blacksquare$$

LEMMA 4.2.5 *Every Cauchy sequence is bounded.*

PROOF The proof is similar to the proof of Lemma 4.1.4. Suppose that $\{a_n\}_{n=1}^{\infty}$ is a Cauchy sequence. Letting $\epsilon = 1$, there is an integer N such that $|a_n - a_m| < 1$ for any n, $m > N$. In particular, if we let $m = N + 1$, using the triangle inequality we find that for any $n > N$,

$$|a_n| = |a_n - a_{N+1} + a_{N+1}| \leq |a_n - a_{N+1}| + |a_{N+1}| < 1 + |a_{N+1}|.$$

That is, for $n > N$, we see that $1 + |a_{N+1}|$ acts as a bound for $|a_n|$. As in the proof of Lemma 4.1.4, we can bound the first N terms of the sequence by inspection. If we let

$$B = \max\{|a_1|, |a_2|, \ldots, |a_N|, 1 + |a_{N+1}|\},$$

then B is a bound for $\{a_n\}_{n=1}^{\infty}$. ∎

We now are ready to prove the main result of this section.

THEOREM 4.2.6 *Every Cauchy sequence converges.*

This theorem in conjunction with Lemma 4.2.4 shows that a sequence converges if and only if it is a Cauchy sequence.

PROOF Assume that $\{a_n\}_{n=1}^{\infty}$ is a Cauchy sequence. Let S be the set whose elements are the terms of the sequence:

$$S = \{a_n \mid n \in \mathbf{N}\}.$$

From the preceeding lemma we know that $\{a_n\}_{n=1}^{\infty}$ is bounded, so S is a bounded set. There are two cases to consider: either S is a finite set (because terms in the sequence are repeating) or it is an infinite set.

Assume first that S is a finite set with k distinct elements,

$$S = \{s_1, s_2, \ldots, s_k\}.$$

Let ϵ denote the minimum distance between two elements of S, that is,

$$\epsilon = \min\{|s_i - s_j| \mid 1 \leq i < j \leq k\}. \tag{1}$$

S is a finite set, so $\epsilon > 0$. Since $\{a_n\}_{n=1}^{\infty}$ is a Cauchy sequence, there exists N such that if m, $n > N$, then $|a_n - a_m| < \epsilon$. In particular, for any $n > N$, we have

$$|a_n - a_{N+1}| < \epsilon. \tag{2}$$

But both a_n and a_{N+1} are elements of S. If a_n and a_{N+1} are distinct, then by (3)

$$|a_n - a_{N+1}| \geq \epsilon.$$

This contradicts (4), so we must have $a_n = a_{N+1}$ for all $n > N$. Thus the sequence converges to a_{N+1}.

Now assume that S is an infinite set. Since $\{a_n\}_{n=1}^{\infty}$ is a Cauchy sequence, by Lemma 4.2.4 we know that S is bounded. By the Bolzano-Weierstrass theorem, S has an accumulation point A. We will show that $\{a_n\}_{n=1}^{\infty}$ converges to A. Because $\{a_n\}_{n=1}^{\infty}$ is a Cauchy sequence, given $\epsilon > 0$ there exists an integer N such that if $m, n > N$ then $|a_n - a_m| < \epsilon/2$. Since A is an accumulation point of S, by Lemma 1.4.5 there are infinitely many points of S within $\epsilon/2$ of A. Since there are infinitely many such points, there must be an a_M with $M > N$ such that $|a_M - A| < \epsilon/2$. Now by the triangle inequality, for any $n > N$,

$$|a_n - A| = |a_n - a_M + a_M - A| \leq |a_n - a_M| + |a_M - A| < \epsilon/2 + \epsilon/2 = \epsilon.$$

Consequently $\{a_n\}_{n=1}^{\infty}$ converges to A. ∎

It is often hard to tell whether a sequence is Cauchy, so the theorem can be difficult to apply to determine whether a particular sequence converges. The fact that the convergent sequences are the Cauchy sequences is used primarily when trying to prove general results about sequences.

In a Cauchy sequence, for any $\epsilon > 0$ we must have $|a_n - a_m| < \epsilon$ for all m and n greater than some N. It is not sufficient that the difference in successive terms be small. That is, even if $|a_{n+1} - a_n| < \epsilon$ for all $n > N$, the sequence might still diverge. For example, let $\{a_n\}_{n=1}^{\infty} = \{\sqrt{n}\}_{n=1}^{\infty}$. Then $\sqrt{n}$ increases without bound and so does not converge. However,

$$|a_{n+1} - a_n| = \sqrt{n+1} - \sqrt{n} = \frac{(\sqrt{n+1} - \sqrt{n})(\sqrt{n+1} + \sqrt{n})}{\sqrt{n+1} + \sqrt{n}} < \frac{1}{2\sqrt{n}}.$$

Thus $|a_{n+1} - a_n|$ can be made as small as we like, yet $\{a_n\}_{n=1}^{\infty}$ diverges.

However, the next result does give a test for convergence stated in terms of the size of successive differences of successive terms of the sequence. This test is easier to apply to certain sequences than the more general Cauchy criterion.

COROLLARY 4.2.7 ***(The Contraction Principle)*** *Let $\{a_n\}_{n=1}^{\infty}$ be a sequence of real numbers. If r is any real constant such that $0 < r < 1$ and $|a_{n+2} - a_{n+1}| \leq r|a_{n+1} - a_n|$, then $\{a_n\}_{n=1}^{\infty}$ converges.*

PROOF We will show that $\{a_n\}_{n=1}^{\infty}$ is a Cauchy sequence. Notice that if $n > m$, the difference $a_n - a_m$ can be written as a telescoping sum of successive differences

$$a_n - a_m = \sum_{k=m}^{n-1} (a_{k+1} - a_k).$$

By applying the hypothesis $k-1$ times, we can approximate $|a_{k+1} - a_k|$ in terms of r and the initial difference $|a_2 - a_1|$.

$$|a_{k+1} - a_k| \le r|a_k - a_{k-1}| \le r^2|a_{k-1} - a_{k_2}| \le \cdots \le r^{k-1}|a_2 - a_1|.$$

Therefore, by the triangle inequality

$$|a_n - a_m| = \left|\sum_{k=m}^{n-1} (a_{k+1} - a_k)\right| \le \sum_{k=m}^{n-1} |a_{k+1} - a_k| = \sum_{k=m}^{n-1} r^{k-1}|a_2 - a_1|.$$

Now factor $r^{m-1}|a_2 - a_1|$ out of the sum and adjust the index to obtain

$$\begin{aligned}\sum_{k=m}^{n-1} r^{k-1}|a_2 - a_1| &= r^{m-1}|a_2 - a_1| \sum_{k=0}^{n-m-1} r^{k-1} \\ &= |a_2 - a_1| \cdot r^{m-1} \cdot \frac{1 - r^{n-m}}{1 - r} \\ &< |a_2 - a_1| \cdot r^{m-1} \cdot \frac{1}{1 - r},\end{aligned}$$

where the inequality follows from Example 4.2.2. But $\lim_{n\to\infty} r^n = 0$ (see problem 4.2.4). Therefore, given any $\epsilon > 0$ there is an N such that if $m - 1 > N$, then $r^{m-1} < \epsilon(1 - r)/|a_2 - a_1|$. Using this bound for r^{m-1}, it now follows that if $m - 1 > N$ and $n > m$, then

$$|a_n - a_m| < |a_2 - a_1| \cdot r^{m-1} \cdot \frac{1}{1 - r} < \epsilon.$$

Therefore $\{a_n\}_{n=1}^{\infty}$ is a Cauchy sequence and converges. ∎

The corollary can often be applied to show that certain recursively defined sequences converge.

EXAMPLE 4.2.3 Show that the following recursively defined sequence converges and find its limit.

$$a_1 = 1, \qquad a_n = 1 + \frac{1}{a_{n-1}} \qquad (n > 1).$$

SOLUTION This sequence is not monotone, since $\{a_n\}_{n=1}^{\infty} = 1, 2, 3/2, 5/3, 8/5, \cdots$. However, we will show that the contraction principle applies.

First, by induction $a_n \geq 1$. (The induction starts since $a_1 = 1$, and if $a_{n-1} \geq 1$, then $a_n = 1 + \frac{1}{a_{n-1}} > 1$.) Next, in recursively defined sequences it is especially easy to write the difference $|a_{n+2} - a_{n+1}|$ in terms of $|a_{n+1} - a_n|$. In this case,

$$|a_{n+2} - a_{n+1}| = \left|1 + \frac{1}{a_{n+1}} - 1 - \frac{1}{a_n}\right| = \left|\frac{1}{a_{n+1}} - \frac{1}{a_n}\right| = \left|\frac{a_n - a_{n+1}}{a_{n+1}a_n}\right|.$$

Using the recursive definition of a_{n+1} in the denominator of the last term,

$$|a_{n+2} - a_{n+1}| = \left|\frac{a_{n+1} - a_n}{\left(1 + \frac{1}{a_n}\right) a_n}\right| = \left|\frac{a_{n+1} - a_n}{a_n + 1}\right| < \frac{|a_{n+1} - a_n|}{2},$$

where the inequality at the last step is a consequence of $a_n > 1$. Thus by the contraction principle $\{a_n\}_{n=1}^{\infty}$ converges to some number ϕ.

To evaluate ϕ we use the recursive definition of the sequence and the fact that $a_n \to \phi$ and $a_{n+1} \to \phi$.

$$\phi = \lim_{n\to\infty} a_n = \lim_{n\to\infty} a_{n+1} = \lim_{n\to\infty} 1 + \frac{1}{a_n} = 1 + \frac{1}{\phi}.$$

Therefore $\phi^2 - \phi - 1 = 0$, so $\phi = \frac{1\pm\sqrt{5}}{2}$. Since $a_n > 1$ for all n, the negative value is impossible, so $\phi = \frac{1+\sqrt{5}}{2}$. The number ϕ was called the **golden ratio** by the Greeks. The interested reader may wish to consult Chapter 11 of H. M. S. Coxeter, *Introduction to Geometry* (New York: Wiley, 1969) for further information. ∎

4.2.3 Further Examples

We now have a number of techniques to show that sequences converge.

EXAMPLE 4.2.4 Show that the sequence $\{s_n\}_{n=1}^{\infty}$ is convergent, where

$$s_n = 1 + \frac{1}{1!} + \frac{1}{2!} + \cdots + \frac{1}{n!}.$$

SOLUTION $\{s_n\}_{n=1}^{\infty}$ is clearly increasing, so we could show that it converges by showing that it is bounded. However, it is even easier to show that it satisfies the contraction principle. Notice that $|s_{n+2} - s_{n+1}| = \frac{1}{(n+2)!} = \frac{1}{(n+2)(n+1)!}$ and $|s_{n+1} - s_n| = \frac{1}{(n+1)!}$. Therefore

$$|s_{n+2} - s_{n+1}| = \frac{1}{(n+2)(n+1)!} = \frac{|s_{n+1} - s_n|}{n+2} < \frac{|s_{n+1} - s_n|}{2}.$$

In Section 4.6 we will show that the limit of this sequence is e, the base of the natural logarithm function. ∎

EXAMPLE 4.2.5 Show that $\{a_n\}_{n=1}^{\infty} = \left\{\left(1+\frac{1}{n}\right)^n\right\}_{n=1}^{\infty}$ converges.

SOLUTION Example 4.2.1 showed that this sequence is increasing. Therefore it suffices to show that it is bounded above. Using the expansion for a_n in Example 4.2.1,

$$\begin{aligned} a_n &= 1 + 1 + \tfrac{1}{2!}\left(1-\tfrac{1}{n}\right) + \cdots + \tfrac{1}{n!}\left(1-\tfrac{1}{n}\right)\left(1-\tfrac{2}{n}\right)\cdots\left(1-\tfrac{n-1}{n}\right) \\ &\leq 1 + 1 + \tfrac{1}{2!} + \cdots + \tfrac{1}{n!} \\ &= s_n, \end{aligned}$$

where $\{s_n\}_{n=1}^{\infty}$ is the sequence in Example 4.2.4. Since $\{s_n\}_{n=1}^{\infty}$ converges, it is bounded above by some number s. Consequently we have $a_n < s_n \leq s$; therefore $\{a_n\}_{n=1}^{\infty}$ is also bounded by s and must converge to some limit a.

In fact, the limit of this sequence is again e. To show that $a = e$ we must show that $\ln a = 1$. Since $\ln x$ is a continuous function on $(0, \infty)$, problems 4.1.5 and 4.1.7 show that

$$\ln a = \lim_{n\to\infty} \ln(1+\tfrac{1}{n})^n = 1. \quad \square$$

EXAMPLE 4.2.6 Show that $\{\sqrt[n]{n}\}_{n=1}^{\infty}$ converges to 1.

SOLUTION First we show that the sequence is decreasing for $n \geq 3$. By taking the $n(n+1)$ power of both $\sqrt[n+1]{n+1}$ and $\sqrt[n]{n}$, we see that showing $\sqrt[n+1]{n+1} < \sqrt[n]{n}$ is equivalent to showing that $(n+1)^n < n^{n+1}$. Example 4.2.5 showed that $\left(1+\frac{1}{n}\right)^n$ is an increasing sequence that converges to e. Therefore, for $n \geq 3$

$$(n+1)^n = n^n \left(\frac{n+1}{n}\right)^n = n^n \left(1+\tfrac{1}{n}\right)^n < n^n e < n^n n = n^{n+1},$$

so $\{\sqrt[n]{n}\}_{n=1}^{\infty}$ is decreasing. Since $\{\sqrt[n]{n}\}_{n=1}^{\infty}$ is also bounded below by 1 (why?), it must converge to some number $A \geq 1$. Thus we may write $A = 1 + \epsilon$, where $\epsilon \geq 0$. For $n \geq 3$, since $\{\sqrt[n]{n}\}_{n=1}^{\infty}$ decreases to A, we must have $\sqrt[n]{n} > A = 1 + \epsilon$. Taking nth powers and using the binomial theorem, for $n \geq 3$ we have

$$n > (1+\epsilon)^n = 1 + n\epsilon + \frac{n(n-1)}{2}\epsilon^2 + \cdots + \epsilon^n > \frac{n(n-1)}{2}\epsilon^2.$$

Solving for ϵ^2 by using the first and last terms of the inequality yields

$$\frac{2}{n-1} > \epsilon^2.$$

By taking n large, we see that ϵ^2 must be arbitrarily small. Hence $\epsilon = 0$, so $A = 1$. $\square$

Problems for Section 4.2

4.2.1. Prove Theorem 4.2.2 when $\{a\}_{n=1}^{\infty}$ is decreasing.

4.2.2. For each of the following cases find a sequence that satisfies the given restrictions. If such a combination is impossible briefly state why.

a) A sequence that is monotone but not convergent.

b) A sequence that is not bounded but is convergent.

c) A sequence that is monotone but not Cauchy.

d) A sequence that is monotone and bounded but not Cauchy.

4.2.3. Prove that the sequence $\{p_n\}_{n=1}^{\infty}$ converges where $p_n = \dfrac{1 \cdot 3 \cdot 5 \cdots (2n-1)}{2 \cdot 4 \cdot 6 \cdots (2n)}$.

•4.2.4. Consider the sequence $\{a_n\}_{n=1}^{\infty} = \{r^n\}_{n=1}^{\infty}$, where $0 \le r < 1$. Show that this sequence converges to 0 by completing the following steps.

a) Using Theorem 4.2.2, show that $\{a_n\}_{n=1}^{\infty}$ converges and call its limit A.

b) Next, show that $\{ra_n\}_{n=1}^{\infty}$ converges both to rA and to A. (For the latter, use $ra_n = r^{n+1}$.)

c) Use the uniqueness of limits to show that $A = 0$.

d) Can you now show that if $0 \le |r| < 1$, then $\{r^n\}_{n=1}^{\infty}$ converges to 0?

4.2.5. Let $\{a_n\}_{n=1}^{\infty}$ be defined inductively as follows:

$$a_1 = 1; \qquad a_n = 1 + \frac{a_{n-1}}{4} \qquad (n > 1).$$

a) Show that if the sequence has a limit, it must be 4/3. (Hint: Use $\lim_{n\to\infty} a_n = \lim_{n\to\infty} a_{n+1}$ and the recursive definition of the sequence.)

b) Show by induction that $\{a_n\}_{n=1}^{\infty}$ is bounded above by 4/3.

c) Show that the sequence is increasing, and then use Theorem 4.2.2 to show that $\{a_n\}_{n=1}^{\infty}$ actually does converge to 4/3.

d) Prove this same result using the contraction principle.

4.2.6. Let $\{a_n\}_{n=1}^{\infty}$ be defined inductively as follows: let a_1 and a_2 be any real numbers and let

$$a_n = \frac{a_{n-1} + a_{n-2}}{2} \qquad (n > 2).$$

a) Pick values for a_1 and a_2 and examine the terms of the resulting sequences. Do these sequences appear to be monotone or bounded? What is happenning to the distance between successive terms?

b) Use the contraction principle to show that $\{a_n\}_{n=1}^{\infty}$ converges. (Hint: Only use the recursion formula on a_{n+2}.)

4.2.7. Let $\{a_n\}_{n=1}^{\infty}$ be defined inductively as follows:

$$a_1 = 1; \qquad a_n = 1 + \frac{1}{1 + a_{n-1}} \qquad (n > 1).$$

a) Show that the sequence is not monotone by writing out the first few terms.

b) Show that $\{a_n\}_{n=1}^{\infty}$ converges by using the contraction principle. (Hint: First show that $a_n \geq 1$ for all n. Next show that

$$|a_{n+2} - a_{n+1}| = \left| \frac{a_{n+1} - a_n}{(1 + a_{n+1})(1 + a_n)} \right| \leq \frac{|a_{n+1} - a_n|}{4}.)$$

c) Use $\lim_{n\to\infty} a_n = \lim_{n\to\infty} a_n + 1$ to show that $\{a_n\}_{n=1}^{\infty}$ converges to $\sqrt{2}$.

d) In 1572 Raphael Bombelli discovered this result though he wrote it in the form of a continued fraction:

$$\sqrt{2} = 1 + \cfrac{1}{2 + \cfrac{1}{2 + \cfrac{1}{2 + \cfrac{1}{2 + \cdots}}}}.$$

Show how this fraction can be derived from the definition of $\{a_n\}_{n=1}^{\infty}$.

4.2.8. Let $a_1 = \sqrt{2}$ and $a_{n+1} = \sqrt{2a_n}$ for $n \geq 1$. Show that $\{a\}_{n=1}^{\infty}$ converges to 2. (Hint: Begin by using induction to show that $a_n \geq \sqrt{2}$ for all n. Then show that the sequence converges by using the contraction principle. Finally, evaluate the limit by using recursion.)

4.2.9. Sometimes it is useful for sequences to begin at some integer other than 1 (often 0 is used). The Fibonacci sequence $\{F\}_{n=0}^{\infty}$ is defined inductively as follows:

$$F_0 = F_1 = 1; \qquad F_{n+2} = F_{n+1} + F_n \qquad (n \geq 0).$$

The first few terms of the Fibonacci sequence are 1, 1, 2, 3, 5, 8, ... Show that $\{a_n\}_{n=1}^{\infty} = \{F_n/F_{n-1}\}_{n=1}^{\infty}$ converges to the golden ratio $\phi = \frac{1+\sqrt{5}}{2}$. (Hint: For $n > 1$ show that that $a_n = 1 + \frac{1}{a_{n-1}}$ and use Example 4.2.3.)

4.2.10. Define the sequence $\{a_n\}_{n=1}^{\infty}$ by setting $a_n = \dfrac{1}{1^2} + \dfrac{1}{2^2} + \cdots + \dfrac{1}{n^2}$.

a) Show that $\{a_n\}_{n=1}^{\infty}$ converges. (Hint: Show that the sequence is bounded above by 2 by using induction to show that $a_n \leq 2 - \frac{1}{n}$.)

b) Does this sequence converge by the contraction principle?

•4.2.11. Let k be a positive integer. Show that $\lim_{n\to\infty} \sqrt[n]{n^k} = 1$. (Hint: Use induction on k and Example 4.2.6.)

4.2.12. We say that the sequence $\{a_n\}_{n=1}^{\infty}$ **diverges to** $+\infty$ if for any real number M there exists an integer N such that $a_n > M$ whenever $n > M$.

a) Show that $\left\{\dfrac{n^2+1}{2n-1}\right\}_{n=1}^{\infty}$ diverges to $+\infty$.

b) Assume that $\{a_n\}_{n=1}^{\infty}$ is an increasing sequence that is not bounded above. Prove that $\{a_n\}_{n=1}^{\infty}$ diverges to $+\infty$.

4.2.13. Let b be a positive number and define the sequence $\{a_n\}_{n=1}^{\infty}$ by

$$a_1 = b, \qquad a_2 = b^b, \qquad a_3 = b^{b^b}, \ldots, \qquad a_n = b^{a_{n-1}}, \ldots$$

a) Show that if $b \geq 1$, then $\{a_n\}_{n=1}^{\infty}$ is increasing. Further, show that if $b \geq 2$, then $\{a_n\}_{n=1}^{\infty}$ diverges.

b) $\{a_n\}_{n=1}^{\infty}$ is constant (and convergent) if $b = 1$. What happens if $1 < b < 2$? Use a calculator to investigate $\{a_n\}_{n=1}^{\infty}$ when $b = 1.25, 1.5, \ldots$

c) Show that if $1 < b \leq e^{1/e}$, then $\{a_n\}_{n=1}^{\infty}$ converges to some number a. (Hint: Show by induction that $a_n \leq e$ for all n.) Further, show that $a = b^a$. (Hint: Use $\lim_{n\to\infty} a_n = \lim_{n\to\infty} a_{n+1}$.)

d) Evaluate $\sqrt{2}^{\sqrt{2}^{\sqrt{2}^{\cdot^{\cdot^{\cdot}}}}}$.

4.2.14. Show that e is irrational by using a proof by contradiction. Suppose that it were rational: $e = \frac{P}{Q}$, for integers P and Q.

a) Write out the nth Taylor polynomial $p_{n,0}$ for the function $f(x) = e^x$.

b) Apply Taylor's theorem to show that for some $t \in [0, 1]$

$$\frac{P}{Q} - \left(1 + \frac{1}{1!} + \cdots + \frac{1}{n!}\right) = \frac{e^t}{(n+1)!}. \qquad (*)$$

c) Multiply both sides of $(*)$ by $n!$ and show that the left side of your new equation is an integer when $n > Q$.

d) We know that $0 < e < 3$ (see problem 3.6.5). Use this to show that the right side of your new equation from part **(b)** is not an integer when $n > 3$. Conclude that e is irrational.

4.3 Infinite Series and Convergence Tests

We turn our attention now to the problem of making sense of "infinite sums" such as

$$1 + \frac{1}{1!} + \frac{1}{2!} + \frac{1}{3!} + \cdots.$$

As we noted earlier, such expressions are problematic because addition is defined only for a finite number of terms. However, this expression should look familiar, since in Example 4.2.4 we considered the sequence of *finite* sums of the form

$$1+\frac{1}{1!}+\frac{1}{2!}+\frac{1}{3!}+\cdots+\frac{1}{n!}$$

and showed that it converged (in fact, it converges to e). It is exactly this process of taking limits of finite sums that allows us to define the sum of infinitely many terms.

4.3.1 Definitions and Examples

Any sequence $\{a_n\}_{n=1}^{\infty}$ of real numbers can be used to generate a **sequence $\{s_n\}_{n=1}^{\infty}$ of partial sums**, where

$$s_1 = a_1, \quad s_2 = a_1 + a_2, \quad s_3 = a_1 + a_2 + a_3, \ldots$$

so that in general the **nth partial sum** is $s_n = \sum_{k=1}^{n} a_k$.

DEFINITION 4.3.1 *The sequence $\{s_n\}_{n=1}^{\infty}$ of partial sums formed from a sequence $\{a_n\}_{n=1}^{\infty}$ is called an **infinite series**, or more simply a **series**, and is denoted by either $a_1 + a_2 + a_3 + \cdots$ or*

$$\sum_{k=1}^{\infty} a_k .$$

For example, the series generated by taking the partial sums of the sequence $\{a_n\}_{n=1}^{\infty} = \{1/n\}_{n=1}^{\infty}$ is the sequence $\{s_n\}_{n=1}^{\infty}$, for which

$$s_n = \sum_{k=1}^{n} \frac{1}{k} = 1 + \frac{1}{2} + \cdots + \frac{1}{n}.$$

This particular series is known as the **harmonic series**.

The letter k used in the symbol $\sum_{k=1}^{\infty} a_k$ is a dummy index and we will often use other letters such as j, m, or n. Frequently it will be convenient to start a series at an index value other than 1 (0 is often a natural starting point). This presents no real problem. For example, the series

$$1+\frac{1}{1!}+\frac{1}{2!}+\frac{1}{3!}+\cdots$$

has a particularly simple form if we start the summation at index 0:

$$\sum_{k=0}^{\infty} \frac{1}{k!}.$$

Notice that a series that begins with $k = q$, such as $\sum_{k=q}^{\infty} a_k$, can be rewritten as a series beginning with $k = 1$ by using $\sum_{k=1}^{\infty} b_k$, where $b_k = a_{q+k-1}$. Usually we will not bother to do this.

Having defined an infinite series as a sequence of partial sums, it is natural to regard the limit of such a sequence, if it exists, as the "infinite sum" of the series.

DEFINITION 4.3.2 *The series $\sum_{k=1}^{\infty} a_k$ is* ***convergent*** *if there is a real number S such that*

$$\lim_{n\to\infty} s_n = S.$$

If $\{s_n\}_{n=1}^{\infty}$ does not converge, we say that the series $\sum_{k=1}^{\infty} a_k$ is ***divergent***.

A word about terminology and notation is in order here. When an infinite series converges to S, we will usually denote this by writing

$$\sum_{k=1}^{\infty} a_k = S,$$

and we will say that the series has **sum** S. The word "sum" is used here in a very specific sense. The sum of a convergent series is obtained by taking the limit of the sequence of partial sums and not by ordinary term by term addition. Also notice that we have used the symbol $\sum_{k=1}^{\infty} a_k$ to denote both the sequence of partial sums and the sum of the series itself. This sum is a number and cannot "diverge" or "converge." It will be clear from the context whether we are referring to the series itself or to its sum.

EXAMPLE 4.3.1 Show that

$$\frac{1}{1\cdot 2} + \frac{1}{2\cdot 3} + \frac{1}{3\cdot 4} + \cdots = \sum_{k=1}^{\infty} \frac{1}{k(k+1)} = 1.$$

SOLUTION Notice that when $k \geq 1$

$$\frac{1}{k(k+1)} = \frac{1}{k} - \frac{1}{k+1}.$$

Using this fact, the partial sums telescope:

$$\begin{aligned} s_n &= \left(1 - \frac{1}{2}\right) + \left(\frac{1}{2} - \frac{1}{3}\right) + \cdots + \left(\frac{1}{n-1} - \frac{1}{n}\right) + \left(\frac{1}{n} - \frac{1}{n+1}\right) \\ &= 1 - \frac{1}{n+1}. \end{aligned}$$

Thus the series converges to 1 because

$$\lim_{n\to\infty} s_n = \lim_{n\to\infty} 1 - \frac{1}{n+1} = 1. \quad \square$$

Since series are special types of sequences, the theorems that we have worked out for sequences can be carried over to this context. It is a straightforward matter to demonstrate the linearity of convergent series by making use of Theorem 4.1.5.

THEOREM 4.3.3 *Assume that both of the series $\sum_{k=1}^{\infty} a_k$ and $\sum_{k=1}^{\infty} b_k$ are convergent. Then*

$$\sum_{k=1}^{\infty}(a_k + b_k) = \sum_{k=1}^{\infty} a_k + \sum_{k=1}^{\infty} b_k,$$

and for any constant c

$$\sum_{k=1}^{\infty} c \cdot a_k = c \cdot \sum_{k=1}^{\infty} a_k\,.$$

Using Cauchy sequences provides another way to characterize convergent series. $\sum_{k=1}^{\infty} a_k$ converges if and only if the sequence of partial sums $\{s_n\}_{n=1}^{\infty}$ converges. But by Lemma 4.2.5 and Theorem 4.2.7, $\{s_n\}_{n=1}^{\infty}$ converges if and only if $\{s_n\}_{n=1}^{\infty}$ is a Cauchy sequence. This means that for any $\epsilon > 0$, there is an integer N such that

$$m,\, n > N \Longrightarrow |s_n - s_m| < \epsilon.$$

Assuming that $n > m$, then

$$|s_n - s_m| = \left|\sum_{k=1}^{n} a_k - \sum_{k=1}^{m} a_k\right| = |a_{m+1} + a_{m+2} + \cdots + a_n|.$$

Therefore, the **Cauchy criterion for the convergence of a series** is written as follows:

$\sum_{k=1}^{\infty} a_k$ converges if and only if for every $\epsilon > 0$ there is an N such that

$$\text{if } n > m > N, \text{ then } |a_{m+1} + a_{m+2} + \cdots + a_n| < \epsilon. \qquad (1)$$

EXAMPLE 4.3.2 Show that the harmonic series, $\displaystyle\sum_{k=1}^{\infty} \frac{1}{k}$, diverges.

SOLUTION Notice that for any n

$$s_{2n} - s_n = \frac{1}{n+1} + \frac{1}{n+2} + \cdots + \frac{1}{2n} \geq n \cdot \frac{1}{2n} = \frac{1}{2}.$$

Taking $\epsilon = 1/2$ we have $|s_{2n} - s_n| \geq 1/2 = \epsilon$ for all values of n. Hence the sequence of partial sums is not Cauchy, so $\sum_{k=1}^{\infty} 1/k$ diverges. $\square$

Since a convergent series must satisfy (1) for all $m, n > N$ it must hold when $m = n - 1$. In this case (1) becomes $n - 1 > N \Longrightarrow |a_n| < \epsilon$. In other words, $\lim_{n\to\infty} a_n = 0$, which proves the following result.

THEOREM 4.3.4 ***(The nth-Term Test)*** *If $\sum_{k=1}^{\infty} a_k$ converges, then $\lim_{n\to\infty} a_n = 0$. Equivalently, if $\lim_{n\to\infty} a_n \neq 0$, then $\sum_{k=1}^{\infty} a_k$ is divergent.*

This theorem gives us a *necessary condition* for a series to converge: the terms must go to 0. However, this is not a *sufficient condition* for convergence. Notice that $\lim_{n\to\infty} \frac{1}{n} = 0$, but we have shown that the harmonic series, $\sum_{k=1}^{\infty} 1/k$, diverges.

A particularly simple type of series are the **geometric series**. Such series have the form

$$\sum_{k=0}^{\infty} r^n = 1 + r + r^2 + r^3 + \cdots,$$

where r is a real number.

EXAMPLE 4.3.3 If $|r| < 1$, show that $\sum_{k=0}^{\infty} r^n$ converges to $\frac{1}{1-r}$. If $|r| \geq 1$, show that the series diverges.

SOLUTION Notice that if $|r| \geq 1$, then $\lim_{n\to\infty} r^n \neq 0$, so the series diverges by the nth-term test. For $|r| < 1$, as in Example 4.2.2,

$$s_n = (1 + r + r^2 + r^3 \cdots + r^n) = \frac{1 - r^{n+1}}{1 - r}.$$

But $\lim_{n\to\infty} r^n = 0$, since $|r| < 1$. Consequently

$$\lim_{n\to\infty} s_n = \lim_{n\to\infty} \frac{1 - r^{n+1}}{1 - r} = \frac{1}{1 - r}. \quad \square$$

These simple geometric series form the foundation of many important tests for the convergence of series.

4.3.2 The Comparison Test

For the time being, we will restrict our attention to **nonnegative series**, that is, series of the form $\sum_{k=1}^{\infty} a_k$, where $a_k \geq 0$ for all k. (Occasionally it will be useful to consider **positive series**, where $a_k > 0$ for all k.) Notice that the sequence of partial sums for a nonnegative series is an increasing sequence, since

$$s_{n+1} = s_n + a_{n+1} \geq s_n.$$

We know from Theorem 4.2.2 that such a monotone sequence converges if and only if it is bounded. Therefore we have the following result.

LEMMA 4.3.5 *A nonnegative series converges if and only if the sequence of partial sums is bounded.*

This simple boundedness criterion, in conjunction with the convergence tests that we are about to develop, will turn out to be extremely useful. It may be hard to determine whether a particular series, by itself, is bounded. But often we are able to compare such series, term by term, to series that we know converge (certain geometric series, for example). This gives us a new way to determine the convergence of series.

THEOREM 4.3.6 ***(The Comparison Test)*** *Assume that $\sum_{k=1}^{\infty} a_k$ and $\sum_{k=1}^{\infty} b_k$ are nonnegative series and that $a_k \le b_k$ for all k.*

a) *If $\sum_{k=1}^{\infty} b_k$ converges, then so does $\sum_{k=1}^{\infty} a_k$.*

b) *If $\sum_{k=1}^{\infty} a_k$ diverges, then so does $\sum_{k=1}^{\infty} b_k$.*

PROOF The result hinges on the comparison of the partial sums for the two series. Let

$$s_n = a_1 + a_2 + \cdots + a_n \quad \text{and} \quad t_n = b_1 + b_2 + \cdots + b_n.$$

Because $a_k \le b_k$ for all k, it follows that $0 \le s_n \le t_n$. For **(a)**, $\{t_n\}_{n=1}^{\infty}$ is bounded above, since $\sum_{k=1}^{\infty} b_k$ converges. Therefore $\{s_n\}_{n=1}^{\infty}$ is bounded above and it converges by Lemma 4.3.5.

For **(b)**, if $\sum_{k=1}^{\infty} b_k$ were convergent, $\sum_{k=1}^{\infty} a_k$ would converge by **(a)**. ∎

EXAMPLE 4.3.4 Determine whether the following series is convergent or divergent:

$$\sum_{k=1}^{\infty} \frac{\sqrt{2} + \cos(k^2+1)}{k + \pi^k}.$$

SOLUTION Observe that

$$0 \le \frac{\sqrt{2} + \cos(k^2+1)}{k + \pi^k} < \frac{3}{k + \pi^k} < \frac{3}{\pi^k}.$$

But

$$\sum_{k=1}^{\infty} \frac{3}{\pi^k} = 3 \sum_{k=1}^{\infty} \frac{1}{\pi^k},$$

which converges by the geometric series test. Consequently the original series converges by comparison. □

EXAMPLE 4.3.5 Determine whether $\sum_{k=1}^{\infty} \frac{4k+\sin^2 k}{3k^2-k-1}$ converges or diverges.

SOLUTION Observe that

$$\frac{4k+\sin^2 k}{3k^2-k-1} > \frac{4k}{3k^2-k-1} > \frac{4k}{3k^2} > \frac{1}{k} > 0.$$

But we saw in Example 4.3.2 that the harmonic series diverges, so the series in question diverges by comparison. □

The fact that a nonnegative series converges if and only if the sequence of partial sums is bounded (Lemma 4.3.5) can sometimes be applied directly to show that certain series converge or diverge. If p is a real number, we say that the nonnegative series

$$\sum_{k=1}^{\infty} 1/k^p = 1 + \frac{1}{2^p} + \frac{1}{3^p} + \cdots$$

is a ***p*-series**. For example, when $p = 1$, we obtain the harmonic series.

COROLLARY 4.3.7 *The p-series $\sum_{k=1}^{\infty} \frac{1}{k^p}$ converges if and only if $p > 1$.*

PROOF If $p < 1$, comparison with the harmonic series ($p = 1$) shows that $\sum_{k=1}^{\infty} 1/k^p$ diverges. However, if $p > 1$,

$$\begin{aligned}
\sum_{k=1}^{\infty} \frac{1}{k^p} &= \frac{1}{1^p} + \left(\frac{1}{2^p} + \frac{1}{3^p}\right) + \left(\frac{1}{4^p} + \frac{1}{5^p} + \frac{1}{6^p} + \frac{1}{7^p}\right) + \cdots \\
&\le \frac{1}{1^p} + 2\left(\frac{1}{2^p}\right) + 4\left(\frac{1}{4^p}\right) + 8\left(\frac{1}{8^p}\right) + \cdots \\
&= 1 + \frac{1}{2^{p-1}} + \frac{1}{4^{p-1}} + \frac{1}{8^{p-1}} + \cdots \\
&= \sum_{k=1}^{\infty} \left(\frac{1}{2^{p-1}}\right)^k .
\end{aligned}$$

Since $p > 1$, we have that $\left|\frac{1}{2^{p-1}}\right| < 1$; consequently the geometric series $\sum_{k=1}^{\infty} \left(\frac{1}{2^{p-1}}\right)^k$ converges. It follows that the sequence of the partial sums of $\sum_{k=1}^{\infty} 1/k^p$ is bounded above. By Lemma 4.3.5, the p-series converges. ∎

4.3.3 The Ratio and Root Tests

There are two very general tests for convergence that are based on using geometric series in conjunction with the comparison test. Both tests were developed by Cauchy.

THEOREM 4.3.8 ***(The Ratio Test)*** *Suppose that $\sum_{k=1}^{\infty} a_k$ is a positive series such that*

$$\lim_{n\to\infty} \frac{a_{n+1}}{a_n} = r.$$

If $r < 1$, then the series converges, whereas if $r > 1$, then the series diverges. If $r = 1$, then the test is inconclusive.

PROOF Assume first that $r < 1$. Choose any number R such that $r < R < 1$. Since

$$\lim_{n\to\infty} \frac{a_{n+1}}{a_n} = r,$$

there is an integer N such that for all $n > N$ we have $\frac{a_{n+1}}{a_n} < R$, or equivalently,

$$a_{n+1} < Ra_n.$$

As usual, let s_k be the kth partial sum of the series. If $n > N$,

$$|s_{n+2} - s_{n+1}| = a_{n+2} < Ra_{n+1} = R|s_{n+1} - s_n|.$$

By the contraction principle (Corollary 4.2.7), the sequence of partial sums converges.

On the other hand, if $r > 1$, choose a number R such that $1 < R < r$. This time there is some integer N such that

$$\frac{a_{n+1}}{a_n} > R \quad \text{for all } n \geq N.$$

Hence

$$a_{n+1} > Ra_n > a_n \quad \text{for all } n \geq N,$$

so, for all $k > 0$

$$a_{N+k} > a_{N+k-1} > a_{N+k-2} > \cdots > a_N.$$

Thus a_n does not approach 0 and the series diverges by the nth term test.

To see that the test is inconclusive when $r = 1$, consider the two series $\sum_{k=1}^{\infty} 1/k$ and $\sum_{k=1}^{\infty} 1/k^2$. In both cases $\lim_{n\to\infty} a_{n+1}/a_n = 1$. Yet the harmonic series diverges, while $\sum_{k=1}^{\infty} 1/k^2$ converges by Corollary 4.3.7. ∎

EXAMPLE 4.3.6 Determine the positive values of r for which the series $\sum_{k=1}^{\infty} kr^k$ converges.

SOLUTION Here we have

$$\lim_{n\to\infty} \frac{a_{n+1}}{a_n} = \frac{(n+1)r^{n+1}}{nr^n} = \lim_{n\to\infty} \frac{n+1}{n} \cdot r = r.$$

So by the ratio test, $\sum_{k=1}^{\infty} kr^k$ converges for $0 < r < 1$ and diverges when $r > 1$. If $r = 1$, the series is simply $\sum_{k=1}^{\infty} k$, which diverges. ∎

A second test that is similar in spirit to the ratio test is the root test. Its proof is outlined in the problem section.

THEOREM 4.3.9 ***(The Root Test)*** *Suppose that $\sum_{k=1}^{\infty} a_k$ is a nonnegative series such that*

$$\lim_{n\to\infty} \sqrt[n]{a_n} = r.$$

If $r < 1$, then the series converges, whereas if $r > 1$, then the series diverges. If $r = 1$, then the test is inconclusive.

EXAMPLE 4.3.7 Determine whether $\displaystyle\sum_{k=1}^{\infty} \frac{k^2}{2^k}$ converges or diverges.

SOLUTION Observe that

$$\lim_{n\to\infty} \sqrt[n]{\frac{n^2}{2^n}} = \lim_{n\to\infty} \frac{(\sqrt[n]{n})^2}{2} = \frac{1}{2},$$

so by the root test the series converges. ∎

The ratio and root tests have an advantage over the other tests discussed in that these tests are phrased entirely in terms of the series in question. To apply them one does not need to know anything about the convergence or divergence of other series, as in the comparison test. Their disadvantage is that $\lim a_{n+1}/a_n$ or $\lim_{n\to\infty} \sqrt[n]{a_n}$ may fail to exist, or as often happens, these limits equal 1. In either situation, the test is inconclusive. Despite these problems, the ratio and root tests will play important roles in our examination of power series in Section 4.6.

Problems for Section 4.3

4.3.1.
a) If $\{a_n\}_{n=1}^{\infty}$ converges, show that $\sum_{k=1}^{\infty}(a_{k+1} - a_k)$ converges.
b) Now prove the converse: if $\sum_{k=1}^{\infty}(a_{k+1} - a_k)$ converges, $\{a_n\}_{n=1}^{\infty}$ converges.
c) Does $\sum_{k=1}^{\infty} \sqrt{k+1} - \sqrt{k}$ converge or diverge?
d) Does $\sum_{k=1}^{\infty} \sqrt{k+2} - 2\sqrt{k+1} + \sqrt{k}$ converge or diverge? (Hint: Try letting $\{a_n\}_{n=1}^{\infty} = \left\{\sqrt{n+1} - \sqrt{n}\right\}_{n=1}^{\infty}$.)

4.3.2. Using the following outline prove the **limit comparison test:** *Let $\sum_{k=1}^{\infty} a_k$ and $\sum_{k=1}^{\infty} b_k$ be nonnegative series. Assume that*

$$\lim_{n\to\infty} \frac{a_n}{b_n} = L > 0.$$

Then $\sum_{k=1}^{\infty} a_k$ converges if and only if $\sum_{k=1}^{\infty} b_k$ converges.

a) Since $\lim_{n\to\infty} \frac{a_n}{b_n} = L$, there is an integer N such that if $n > N$, then $\left|\frac{a_n}{b_n} - L\right| < \frac{L}{2}$. Use this to show that if $n > N$, then $a_n < \frac{3}{2}Lb_n$.

b) Now assume that the series $\sum_{k=1}^{\infty} b_k$ converges. Show that $\sum_{k=N+1}^{\infty} \frac{3}{2}Lb_k$ converges. (Note the starting index.)

c) Show that $\sum_{k=N+1}^{\infty} a_k$ converges, and that the entire series $\sum_{k=1}^{\infty} a_k$ therefore converges.

d) Show that the converse follows immediately. (Hint: $\lim_{n\to\infty} \frac{b_n}{a_n} = L^{-1}$.)

4.3.3. Determine whether the following series converge or diverge. Be explicit about which tests apply.

a) $\displaystyle\sum_{k=1}^{\infty} \frac{1}{k + |\sin k|}$ **b)** $\displaystyle\sum_{k=1}^{\infty} \frac{1}{k! - \sin^2 k}$ **c)** $\displaystyle\sum_{k=1}^{\infty} \frac{k!}{5^k}$

d) $\displaystyle\sum_{k=1}^{\infty} \frac{k+1}{2k+3}$ **e)** $\displaystyle\sum_{k=1}^{\infty} \frac{2k+1}{10^6(k+2)}$ **f)** $\displaystyle\sum_{k=1}^{\infty} \frac{k+1}{10^6 k^2}$

g) $\displaystyle\sum_{k=1}^{\infty} \frac{2^k}{k^5}$ **h)** $\displaystyle\sum_{k=1}^{\infty} \frac{k^2 - 3}{2 + k^5}$ **i)** $\displaystyle\sum_{k=1}^{\infty} \pi^{-k}$

j) $\displaystyle\sum_{k=1}^{\infty} \frac{1}{\sqrt{k^5+1}}$ **k)** $\displaystyle\sum_{k=1}^{\infty} \frac{k}{2k+5}$ **l)** $\displaystyle\sum_{k=1}^{\infty} \frac{(2k)!}{(k!)^2}$

4.3.4. Prove the root test (Theorem 4.3.9) by completing the following steps:

a) First, assume that $r < 1$. Choose a number R such that $r < R < 1$. Show that we can choose N such that $\sqrt[n]{a_n} < R$ for all $n > N$.

b) Show that $a_n < R^n$ for all $n > N$.

c) Show that $\sum_{K=N+1}^{\infty} a_k$ converges, and that the entire series $\sum_{k=1}^{\infty} a_k$ therefore converges.

d) Now assume that $r > 1$. Choose R such that $1 < R < r$. Show that there is an integer N such that if $n > N$, then $a_n > R^n$.

e) Show that $\sum_{k=1}^{\infty} a_k$ diverges.

f) To show that the ratio test is inconclusive when $r = 1$, examine the same two series used for this purpose in the proof of the ratio test.

4.3.5. If $\sum_{k=1}^{\infty} a_k$ converges and $\sum_{k=1}^{\infty} b_k$ diverges, prove that $\sum_{k=1}^{\infty} a_k + b_k$ diverges.

4.3.6. **a)** Assume that $\sum_{k=1}^{\infty} a_k$ is a convergent series with positive terms. Show that $\sum_{k=1}^{\infty} {a_k}^2$ also converges. (Hint: What can you say about the size of a_n when n is large?)

b) Show that the converse is false. Find an example of a series $\sum_{k=1}^{\infty} {a_k}^2$ that converges while $\sum_{k=1}^{\infty} a_k$ diverges.

4.3.7. Assume that $\sum_{k=1}^{\infty} a_k$ and $\sum_{k=1}^{\infty} b_k$ are convergent with positive terms.

a) Show that $\sum_{k=1}^{\infty} \sqrt{{a_k}^2 + {b_k}^2}$ converges.

b) Does the converse hold?

•4.3.8. Show that if $\sum_{k=1}^{\infty} a_k$ is a convergent series, then there exists an integer N such that if $n > N$, then $\left|\sum_{k=n+1}^{\infty} a_k\right| < \epsilon$. That is, the infinite tail of the series can be made arbitrarily small.

4.3.9. Let f be a twice differentiable, nonnegative function. Consider the series $\sum_{k=1}^{\infty} f(\frac{1}{k})$.

a) Show that if $\sum_{k=1}^{\infty} f(\frac{1}{k})$ converges, then $f(0) = 0$. (Hint: Use the n-th term test and problem 4.1.5.)

b) Show that if $f(0) = 0$ and $\sum_{k=1}^{\infty} f(\frac{1}{k})$ converges, then $f'(0) = 0$. (Hint: Apply the limit comparison test (problem 2) using the series $\sum_{k=1}^{\infty} \frac{1}{k}$.)

c) Show that if $f(0) = 0$ and $f'(0) = 0$, then $\sum_{k=1}^{\infty} f(\frac{1}{k})$ converges. (Hint: Apply the limit comparison test using the series $\sum_{k=1}^{\infty} \frac{1}{k^2}$.)

4.4 Absolute and Conditional Convergence

So far we have concentrated on nonnegative series. However, the tests we have developed may be applied indirectly to series with only nonpositive terms. If $\sum_{k=1}^{\infty} a_k$ is a nonpositive series, $\sum_{k=1}^{\infty}(-a_k)$ is a nonnegative series and

$$\sum_{k=1}^{\infty} a_k = -\sum_{k=1}^{\infty}(-a_k)\,.$$

In this way, a question about the convergence of a nonpositive series is reformulated into a question about the convergence of a nonnegative series.

4.4.1 Alternating Series

Many series have some positive and some negative terms. Sometimes this mixture of terms occurs in an orderly fashion, as is the case when the terms alternate in sign. Such series are called **alternating series** and they have the form

$$\sum_{k=1}^{\infty}(-1)^{k+1}a_k = a_1 - a_2 + a_3 - a_4 + \cdots$$

or

$$\sum_{k=1}^{\infty}(-1)^{k}a_k = -a_1 + a_2 - a_3 + a_4 - \cdots,$$

where $a_k > 0$.

Alternating series arise in a number of contexts. A particularly simple example is the **alternating harmonic series**,

$$\sum_{k=1}^{\infty}(-1)^{k+1}\frac{1}{k} = 1 - \frac{1}{2} + \frac{1}{3} - \frac{1}{4} + \cdots.$$

Shortly we will show that this series converges, which may be something of a surprise since the harmonic series itself diverges! Later we will show that the alternating harmonic series converges to $\ln 2$. An alternating series of similar form was used by Leibniz in 1673 when calculating the area of a unit circle. He found that

$$\sum_{k=1}^{\infty}(-1)^{k+1}\frac{1}{2k-1} = 1 - \frac{1}{3} + \frac{1}{5} - \frac{1}{7} + \cdots$$

converges to $\frac{\pi}{4}$. Another example of an alternating series is

$$\sum_{k=0}^{\infty}(-1)^{k}\frac{1}{k!} = 1 - 1 + \frac{1}{2!} - \frac{1}{3!} + \cdots,$$

which we will show converges to e^{-1}.

Although we will not deal extensively with alternating series, the following theorem gives a simple test that suffices to prove the convergence of the three alternating series just mentioned.

THEOREM 4.4.1 ***(Leibniz's Theorem)*** *Assume that $\{a_n\}_{n=1}^{\infty}$ is a decreasing sequence that converges to 0. Then the alternating series $\sum_{k=1}^{\infty}(-1)^{k+1}a_k$ converges.*

PROOF $\{a_n\}_{n=1}^{\infty}$ is decreasing. Consequently the partial sums s_{2n} (which are the sum of an even number of terms) form an increasing sequence, since

$$s_{2n+2} = s_{2n} + a_{2n+1} - a_{2n+2} \geq s_{2n}.$$

Further, s_{2n} is bounded above by a_1, since

$$\begin{aligned} s_{2n} &= a_1 - a_2 + a_3 + \cdots + a_{2n-1} - a_{2n} \\ &= a_1 - (a_2 - a_3) - \cdots - (a_{2n-1} - a_{2n}) < a_1. \end{aligned}$$

Therefore $\{s_{2n}\}_{n=1}^{\infty}$ converges to some limit S.

Now consider the sequence of partial sums s_{2n+1} consisting of an odd number of terms. Notice that

$$\lim_{n\to\infty} s_{2n+1} = \lim_{n\to\infty} s_{2n} + a_{2n+1} = S + 0 = S.$$

Since both the even and odd partial sums approach S, it is easy to show that the series converges to S. Given $\epsilon > 0$, there exists N_1 such that if $n > N_1$, then $|s_{2n+1} - S| < \epsilon$ and there exists N_2 such that if $n > N_2$, then $|s_{2n} - S| < \epsilon$. Let $N = \max\{2N_1 + 1, 2N_2\}$. If $n > N$, then $|s_n - S| < \epsilon$, so the series converges to S. ∎

EXAMPLE 4.4.1 The alternating harmonic series $\displaystyle\sum_{k=1}^{\infty}(-1)^{k+1}\frac{1}{k}$ converges, since $\{1/n\}_{n=1}^{\infty}$ decreases and converges to 0. □

EXAMPLE 4.4.2 Show that the alternating series $\displaystyle\sum_{k=1}^{\infty}(-1)^{k}\frac{\ln k}{k}$ converges.

SOLUTION Notice that by L'Hôpital's rule,

$$\lim_{n\to\infty}\frac{\ln n}{n} = \lim_{x\to\infty}\frac{\ln x}{x} = \lim_{x\to\infty}\frac{1/x}{1} = 0.$$

Although the first few terms of $\left\{\frac{\ln n}{n}\right\}_{n=1}^{\infty}$ are not decreasing, we will show that for $n \geq 3$, the sequence is decreasing (and clearly this is sufficient to apply Leibniz's theorem). To see that the sequence eventually becomes decreasing, let

$$f(x) = \frac{\ln x}{x}.$$

Then

$$f'(x) = \frac{1 - \ln x}{x^2}.$$

Since $1 - \ln x < 0$ for $x > e$, then $f'(x) < 0$ for $x > e$; thus f is a decreasing function for $x > e$. Consequently

$$\frac{\ln(n+1)}{n+1} = f(n+1) < f(n) = \frac{\ln n}{n} \qquad \text{for } n \geq 3.$$

By Leibniz's Theorem, the series converges. □

4.4.2 Conditional versus Absolute Convergence

Not all series with both positive and negative terms display the orderly changes of sign that alternating series exhibit. In this section we will develop another test for convergence that will allow us to determine whether some of these more general series converge.

One approach is to (blissfully) ignore the signs of the various terms in the series and treat all the terms as if they were positive and apply the many tests we have for such series. If we start with a general series $\sum_{k=1}^{\infty} a_k$, this approach amounts to examining $\sum_{k=1}^{\infty} |a_k|$ instead. Now we know that this idea has some limitations; we have seen that the alternating harmonic series $\sum_{k=1}^{\infty}(-1)^{k-1}1/k$ converges, while $\sum_{k=1}^{\infty} 1/k$ diverges. So the divergence of $\sum_{k=1}^{\infty} |a_k|$ does not necessarily imply the divergence of $\sum_{k=1}^{\infty} a_k$. However, we can say the following.

THEOREM 4.4.2 *If $\sum_{k=1}^{\infty} |a_k|$ converges, then $\sum_{k=1}^{\infty} a_k$ converges.*

PROOF The idea is quite simple. We let

$$b_k = a_k + |a_k| = \begin{cases} 2a_k, & \text{if } a_k \geq 0 \\ 0, & \text{if } a_k < 0 \end{cases}.$$

Notice that $0 \leq b_k \leq 2|a_k|$. By the hypothesis, $\sum_{k=1}^{\infty} 2|a_k|$ converges so by the comparison test $\sum_{k=1}^{\infty} b_k$ converges. Since $a_k = b_k - |a_k|$,

$$\sum_{k=1}^{\infty} a_k = \sum_{k=1}^{\infty} b_k - \sum_{k=1}^{\infty} |a_k|$$

also converges by Theorem 4.3.3. ∎

DEFINITION 4.4.3 *The series $\sum_{k=1}^{\infty} a_k$ is said to be* ***absolutely convergent*** *if $\sum_{k=1}^{\infty} |a_k|$ is convergent. If $\sum_{k=1}^{\infty} a_k$ converges but $\sum_{k=1}^{\infty} |a_k|$ diverges, then $\sum_{k=1}^{\infty} a_k$ is said to be* ***conditionally convergent****.*

Theorem 4.4.2 says that any absolutely convergent series is convergent. The alternating harmonic series provides an example of a series that is conditionally convergent, since the series converges, but not absolutely.

EXAMPLE 4.4.3 Show that the series $\displaystyle\sum_{k=1}^{\infty} \frac{\cos k}{k^2}$ converges.

SOLUTION The series has both positive and negative terms, but it is not alternating. However, since $|\cos(k)/k^2| \leq 1/k^2$, the series $\sum_{k=1}^{\infty} |\cos(k)/k^2|$ converges by the comparison test. Since $\sum_{k=1}^{\infty} \cos(k)/k^2$ is absolutely convergent, it converges. □

The ratio and root tests combined with Theorem 4.4.2 can be used to find a range of values for which a series of some general form will converge. For example, instead of asking if a particular series such as $\sum_{k=1}^{\infty} \frac{2^k}{k!}$ converges, we can ask for which values of x the series $\sum_{k=1}^{\infty} \frac{x^k}{k!}$ converges. Since x may be negative, we apply the ratio test to the series of absolute values $\sum_{k=1}^{\infty} \left|\frac{x^k}{k!}\right|$. We find that for any arbitrary but fixed value of x,

$$\lim_{n\to\infty} \frac{a_{n+1}}{a_n} = \lim_{n\to\infty} \frac{|x^{n+1}/(n+1)!|}{|x^n/n!|} = \lim_{n\to\infty} \frac{|x|}{n+1} = 0.$$

By the ratio test, $\sum_{k=1}^{\infty} \left|\frac{x^k}{k!}\right|$ converges for all values of x, so $\sum_{k=1}^{\infty} \frac{x^k}{k!}$ converges for all values of x. We will pursue these ideas in much greater detail in the final section in this chapter.

4.4.3 So You Think You're Confused

This brief introduction to sequences and series can seem a bit overwhelming, especially when tests for convergence are presented one after another. It is tempting to merely memorize results rather than to stop to understand what the various results mean. For centuries mathematicians struggled with the problems and paradoxes arising from infinite sequences and series. Your own struggle to become comfortable with this material should be viewed as part of this heritage. It is worth taking a moment to reflect on some of the questions that led mathematicians to develop sequences and series and also to examine some of the missteps that were made along the way.

Greek philosophers and mathematicians as early as the fifth century B.C. were struggling with problems of the infinite. During this period there were two competing views of the nature of space and time. One view held that space and time were infinitely divisible, so that motion was a smooth (continuous) process. The opposing view held that space and time were made up of a large but finite number of indivisible particles or intervals (like the frames in a movie) and that motion proceeded by a rapid succession of small jumps. Certainly with our modern notion of limits (and the associated ϵ, δ arguments), the former point of view may seem more natural to us. However, the Greeks encountered logical difficulties with both views. These difficulties were clearly expressed in the paradoxes of the philosohper Zeno (ca. 450 B.C.). These paradoxes were intended to show that motion is impossible whether we assume that a magnitude is infinitely divisible or that it is made up of a large finite number of parts. What follows is a description of these paradoxes.[1]

The first paradox, Achilles and the Tortoise, assumes that magnitudes are infinitely divisible. The paradox purports to show that a slow-moving object with a head start (the tortoise) can never be overtaken by a faster moving object (Achilles). Zeno argues that before Achilles can overtake the tortoise, Achilles must first reach the point where the tortoise started. During that time, the tortoise has made additional progress. While Achilles makes up this new distance, the tortoise has again made progress, and so on. In short, Achilles always remains behind.

The second paradox concerns the flight of an arrow. It assumes that time is composed of a large number of indivisible instants and it purports to show that motion is impossible in this situation. For if time is composed of indivisible instants, an arrow "in flight" is actually at rest because at any particular moment the arrow is in a fixed position. Since this is true of every moment, the arrow is never in motion.

Aristotle gave refutations of these paradoxes (he would not even grant the supposition of indivisible units of time). Though infinite series were not part of Aristotle's arsenal, we can rephrase the first paradox to show its relation to an infinite series. Suppose that we give the tortoise a 100 meter lead and we assume that Achilles runs five times as fast as the tortoise. While Achilles makes up the first 100 meters, the tortoise covers 20 meters. While Achilles makes up the 20 meters, the tortoise travels another 4 meters, and so on. Let t denote the number of seconds it takes Achilles to cover the first 100 meters. Then Achilles covers the next

[1] For a more complete discussion of these and other paradoxes, see Morris Kline, *Mathematical Thought from Ancient to Modern Times* (New York: Oxford University Press, 1972), pp. 34–37

20 meters in $t/5$ seconds; he covers the next 4 meters in $t/5^2$ seconds, and so on. If we sum all of these time intervals together, we obtain a convergent geometric series:

$$t + \frac{t}{5} + \frac{t}{5^2} + \cdots = \sum_{k=1}^{\infty} \frac{t}{5^k} = t \sum_{k=1}^{\infty} \frac{1}{5^k} = \frac{t}{1 - \frac{1}{5}} = \frac{5t}{4}.$$

Since Achilles travels 100 meters in t seconds, in $5t/4$ seconds Achilles travels 125 meters. The tortoise traverses one-fifth this distance, or 25 meters. When combined with its 100 meter headstart, we see that the tortoise has been overtaken by Achilles exactly in $5t/4$ seconds. An infinite number of intervals of time (and distance) have been covered in a finite amount of time because the series of time intervals is convergent. (It was not until the middle of the seventeenth century that Gregory of Saint Vincent showed that the Achilles and the Tortoise paradox could be resolved in this way by using geometric series.)

Even after the idea of a series had been introduced, it took a long time before the distinction between convergent and divergent series was fully appreciated. Late in the seventeenth century, James Bernoulli showed that the harmonic series diverges. He also pointed out that this meant a series could still diverge even though its terms were vanishing in the limit. In other words, as we have noted, the converse of the nth-term test fails. Until this time many mathematicians believed that this was not possible.[2] The simple alternating geometric series

$$\sum_{k=0}^{\infty} (-1)^k = 1 - 1 + 1 - 1 + \cdots$$

was also problematic for mathematicians of this period. There were two "obvious" ways to group terms so that the sum might be readily computed. First, one might write

$$(1-1) + (1-1) + (1-1) + \cdots,$$

which apparently sums to 0. Or one might try

$$1 - (1-1) - (1-1) - \cdots,$$

[2] For more detailed as well as fascinating discussions of these matters, the reader may consult Kline, *Mathematical Thought*, pp. 436–450; Judith V. Grabiner, *The Origins of Cauchy's Rigorous Calculus* (Cambridge, Mass.: MIT Press, 1981), pp. 97–113; or I. Grattan-Guinness, *The Development of the Foundations of Mathematical Analysis from Euler to Riemann* (Cambridge, Mass: MIT Press, 1970), pp. 68–85.

for which the sum appears to be 1. James Bernoulli combined these two approaches. Letting s denote the sum resulting from the first grouping of the terms, he observed that the second grouping gives $1 - s$. Since the two groupings ought to represent the same series, we should have $s = 1 - s$ and therefore that $\sum_{k=1}^{\infty}(-1)^k = 1/2$. Even Euler (1707–1783), who did much to develop the theory of sequences and series, had difficulty here. Euler argued that

$$\frac{1}{1-x} = 1 + x + x^2 + x^3 + \cdots. \tag{1}$$

This is just the geometric series formula from Example 4.3.3 for $-1 < x < 1$. However, Euler applied it without restrictions on x. Mathematicians of this era generally treated such equations in a "formal" way like polynomials of infinite degree. A proof of the validity of (1) would require multiplication by $1-x$, which would require an infinite set of operations on the right side of the equation. (Compare this to our partial sum approach.) With (1) in hand, Euler concluded that when $x = -1$,

$$\frac{1}{2} = \frac{1}{1-(-1)} = 1 - 1 + 1 - 1 + \cdots.$$

Leibniz also considered the alternating geometric series and agreed that its sum should be 1/2. His argument was based on examining the sequence of partial sums for the series, which is 1, 0, 1, 0,... The partial sums 1 and 0 occur equally often, so one should take their mean as the sum of the series. This solution was accepted by many mathematicians even though as Kline points out, "Leibniz conceded that his argument was more metaphysical than mathematical but went on to say that there was more metaphysical truth in mathematics than was generally recognized."[3]

To make matters worse, Nicholas Bernoulli (1687–1759) in a letter to Euler noted that if $r = 2$ in (1), one obtains

$$-1 = 1 + 2 + 4 + 8 + 16 + \cdots,$$

even though one would expect this latter sum to be infinite (and positive). Euler had previously obtained this same result in another way and had concluded that infinity must in some way be like 0 in that it acts as a sort of limit between the positive and negative numbers.

During the eighteenth century a formal and manipulative view of series predominated. While convergence and divergence were not totally ignored, they were not the central issue. It was not until the nineteenth century with Bolzano (whose work was not well known) and Cauchy that the modern theory of series was developed.

---

[3] Kline, *Mathematical Thought*, p. 446.

Problems for Section 4.4

4.4.1. Can a nonnegative series converge conditionally? How about a nonpositive series? Can a geometric series converge conditionally?

4.4.2. If $|r| < 1$, prove that $\sum_{k=1}^{\infty}(k+1)r^k$ converges. (Can you show that the sum is $(1-r)^{-2}$? This will be easy after reading Section 4.6. See problem 4.6.3d.)

4.4.3. Classify each of these series as conditionally or absolutely convergent or as divergent.

a) $\displaystyle\sum_{k=1}^{\infty}(-1)^{k-1}\frac{1}{2k-1}$ **b)** $\displaystyle\sum_{k=0}^{\infty}(-1)^k\frac{1}{k!}$ **c)** $\displaystyle\sum_{k=0}^{\infty}(-2)^{-k}$

d) $\displaystyle\sum_{k=1}^{\infty}\frac{(-k)^k}{k!}$ **e)** $\displaystyle\sum_{k=1}^{\infty}\frac{k!}{(-k)^k}$ **f)** $\displaystyle\sum_{k=1}^{\infty}\pi^{-2k}\sin k$

4.4.4. Intuitively, an absolutely convergent series converges because its terms are getting small fast whereas a conditionally convergent series converges only because there is enough cancellation taking place between the positive and negative terms. This can be made precise. Given a general series $\sum_{k=1}^{\infty} a_k$, let us separate it into its positive and negative terms. Define

$$b_n = \begin{cases} a_n & \text{if } a_n > 0 \\ 0 & \text{if } a_n \leq 0. \end{cases} \quad \text{and} \quad c_n = \begin{cases} 0 & \text{if } a_n > 0, \\ a_n & \text{if } a_n \leq 0. \end{cases}$$

a) Show that $2b_n = a_n + |a_n|$ and that $2c_n = a_n - |a_n|$.

b) Prove the following: If $\sum_{k=1}^{\infty} a_k$ converges absolutely, then both $\sum_{k=1}^{\infty} b_k$ and $\sum_{k=1}^{\infty} c_k$ converge. However, if $\sum_{k=1}^{\infty} a_k$ converges conditionally, then both $\sum_{k=1}^{\infty} b_k$ and $\sum_{k=1}^{\infty} c_k$ diverge. (Hint: The comparison test is useful for the first part; try a proof by contradiction for the second part [or use problem 4.3.5].)

•4.4.5. Assume that $\{a_n\}_{n=1}^{\infty}$ satisfies the hypotheses of Leibniz's theorem 4.4.1. Assume that $\sum_{k=1}^{\infty}(-1)^{k+1}a_k$ converges to s and, as usual, denote the nth partial sum of the series by s_n. Prove that $|s - s_n| \leq a_{n+1}$ as follows.

a) Recall that we showed in the proof of Leibniz's theorem that the even partial sums, s_{2n}, are increasing and that $s_{2n} \leq s_{2n+2} \leq s$. Show that the odd partial sums, s_{2n-1}, are decreasing and that $s < s_{2n+1} \leq s_{2n-1}$.

b) Use the two inequalities in **(a)** to show that $0 \leq s - s_{2n} \leq s_{2n+1} - s_{2n} = a_{2n+1}$ and that $0 \leq s_{2n-1} - s \leq s_{2n-1} - s_{2n} = a_{2n}$.

c) Use **(b)** to show that $|s - s_n| \leq a_{n+1}$.

d) Use this result to approximate $s = \sum_{k=1}^{\infty}(-1)^{k+1}/k^3$ to within .001.

4.4.6. A **rearrangement** of a series $\sum_{k=1}^{\infty} a_k$ is a series $\sum_{k=1}^{\infty} b_k$ whose terms are the same as those of $\sum_{k=1}^{\infty} a_k$ but which occur in a different order. For finite sums, the order of addition is unimportant. It can be shown that rearranging an absolutely convergent series has no effect on its sum. However, rearranging a conditionally convergent series can have a dramatic effect. For example, consider a rearrangement of the conditionally convergent alternating harmonic series, $\sum_{k=1}^{\infty}(-1)^{k+1}/k$. Let

$$s = 1 - \frac{1}{2} + \frac{1}{3} - \frac{1}{4} + \frac{1}{5} - \frac{1}{6} + \frac{1}{7} - \frac{1}{8} + \cdots .$$

Now rearrange the terms so that each odd term is followed by a pair of even ones. If we assume that rearrangement has no effect on the sum, then

$$\begin{aligned} s &= 1 - \frac{1}{2} - \frac{1}{4} + \frac{1}{3} - \frac{1}{6} - \frac{1}{8} + \frac{1}{5} - \frac{1}{10} - \frac{1}{12} + \frac{1}{7} - \frac{1}{14} - \frac{1}{16} + \cdots \\ &= \left(1 - \frac{1}{2}\right) - \frac{1}{4} + \left(\frac{1}{3} - \frac{1}{6}\right) - \frac{1}{8} \\ &\qquad + \left(\frac{1}{5} - \frac{1}{10}\right) - \frac{1}{12} + \left(\frac{1}{7} - \frac{1}{14}\right) - \frac{1}{16} + \cdots \\ &= \frac{1}{2} - \frac{1}{4} + \frac{1}{6} - \frac{1}{8} + \frac{1}{10} - \frac{1}{12} + \frac{1}{14} - \frac{1}{16} + \cdots \\ &= \frac{1}{2}\left(1 - \frac{1}{2} + \frac{1}{3} - \frac{1}{4} + \frac{1}{5} - \frac{1}{6} + \frac{1}{7} - \frac{1}{8} + \cdots\right) \\ &= \frac{1}{2}s. \end{aligned}$$

a) Show that $s = 0$.

b) Now obtain a contradiction by considering the sequence of partial sums and showing that $\sum_{k=1}^{\infty}(-1)^{k+1}/k = s > \frac{1}{2}$. Thus, conditionally convergent sequences may not be rearranged.

In fact, it can be shown that if $\sum_{k=1}^{\infty} a_k$ converges conditionally, then for any number β there is a rearrangement $\sum_{k=1}^{\infty} b_k$ that converges to β. The interested reader may wish to consult Michael Spivak, *Calculus*, 2nd ed., (Berkeley, Calif.: Publish or Perish, Inc., 1980), pp. 450–52.

4.4.7. Define the sequence $\{a_n\}_{n=1}^{\infty}$ by $a_1 = 1$, $a_2 = \int_1^2 \frac{1}{x}\,dx$, $a_3 = \frac{1}{2}$, $a_4 = \int_2^3 \frac{1}{x}\,dx$, ..., where in general

$$a_{2n-1} = \frac{1}{n}, \qquad a_{2n} = \int_n^{n+1} \frac{1}{x}\,dx.$$

a) Verify that $\{a_n\}_{n=1}^{\infty}$ converges to 0 by showing that the sequence of odd terms $\{a_{2n-1}\}_{n=1}^{\infty}$ and the sequence of even terms $\{a_{2n}\}_{n=1}^{\infty}$ both converge to 0.

b) Verify that $\{a_n\}_{n=1}^{\infty}$ is decreasing.

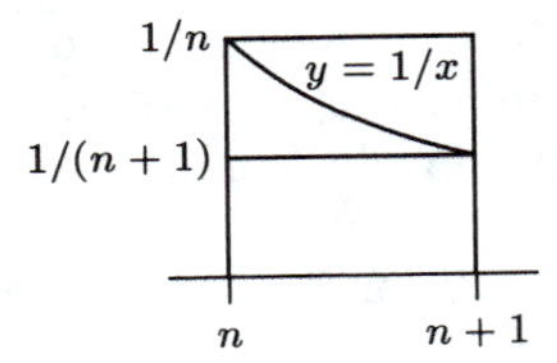

Use this figure to show that $a_{2n-1} < a_{2n} < a_{2n+1}$.

Figure 4.4.1

c) Show that $\sum_{k=1}^{\infty}(-1)^{k+1}a_k$ converges to some number γ.

d) Let s_{2n-1} be the $(2n-1)$ partial sum of the series. Then

$$s_{2n-1} = 1 - \int_1^2 \tfrac{1}{x}\,dx + \tfrac{1}{2} - \int_2^3 \tfrac{1}{x}\,dx + \cdots + \tfrac{1}{n-1} - \int_1^2 \tfrac{1}{x}\,dx + \tfrac{1}{n}.$$

Show that $s_{2n-1} = 1 + \frac{1}{2} + \frac{1}{n-1} + \frac{1}{n} - \ln n$.

e) Show that $\gamma = \lim_{n\to\infty}(1 + \frac{1}{2} + \frac{1}{n-1} + \frac{1}{n} - \ln n)$. The constant γ is called Euler's contstant and its approximate value is .57722. It is unknown whether γ is rational or irrational.

4.5 Sequences of Functions

In this section we will generalize the notion of sequences of numbers to sequences of functions. Our ultimate goal is to examine infinite sums of functions of the form

$$f(x) = f_1(x) + f_2(x) + f_3(x) + \cdots.$$

Just as series of real numbers were viewed as sequences of partial sums of real numbers, series of functions will be thought of as sequences of partial sums of functions.

We are interested in sequences $\{f_n\}_{n=1}^{\infty}$ whose terms are real-valued functions having a common domain in $\mathbf{R}$. A natural way to describe convergence of such a sequence of functions is to reduce the problem to the convergence of sequences of real numbers. Specifically, for each x in the common domain of the sequence $\{f_n\}_{n=1}^{\infty}$, we can form the sequence of real numbers $\{f_n(x)\}_{n=1}^{\infty}$ whose terms are just the values of the functions at the point x. Let D denote the set of all points x for which this sequence converges. Then we may define a new function f by setting

$$f(x) = \lim_{n\to\infty} f_n(x), \qquad \text{if } x \in D.$$

The function f is called the **limit function** of $\{f_n\}_{n=1}^{\infty}$, and we say that the sequence $\{f_n\}_{n=1}^{\infty}$ **converges pointwise** to f on the set D.

4.5.1 Some Problems with Pointwise Convergence

The fundamental problem is to determine which basic properties carry over from the functions in the sequence to the limit function f. Unfortunately, pointwise convergence turns out not to be sufficiently restrictive to preserve the properties of greatest interest. For example, suppose that f is the limit function for some sequence $\{f_n\}_{n=1}^{\infty}$ on some interval I. If each f_n in the sequence is continuous, we might hope that the limit function f would be continuous. This turns out to be false, as the following example illustrates.

EXAMPLE 4.5.1 Consider the sequence $\{f_n\}_{n=1}^{\infty}$, where

$$f_n(x) = \begin{cases} nx, & \text{if } 0 \le x \le 1/n \\ 1, & \text{if } 1/n \le x \le 1. \end{cases}$$

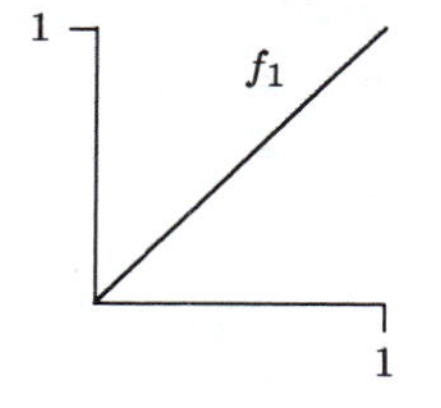

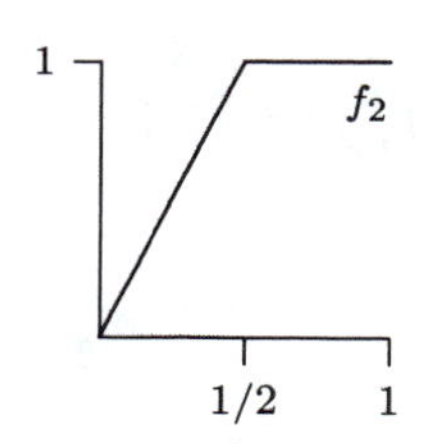

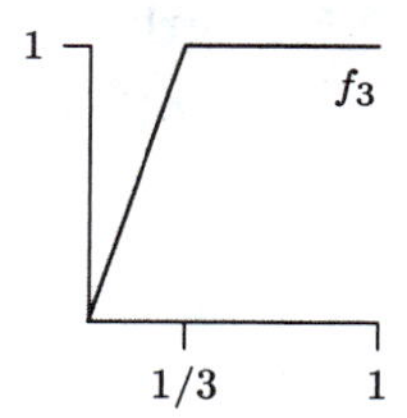

Figure 4.5.1

Each function f_n is continuous on the interval $[0,1]$, and $\{f_n\}_{n=1}^{\infty}$ converges pointwise to

$$f(x) = \begin{cases} 0, & \text{if } x = 0 \\ 1, & \text{if } 0 < x \le 1. \end{cases}$$

Consequently we have a sequence of continuous functions that converges pointwise to a discontinuous limit function. □

Integrability fares no better under pointwise convergence. Assume that $\{f_n\}_{n=1}^{\infty}$ converges pointwise to f on $[a,b]$ and that each function f_n is integrable there. If $\lim_{n\to\infty} \int_a^b f_n$ exists, we might hope that $\lim_{n\to\infty} \int_a^b f_n = \int_a^b \lim_{n\to\infty} f_n = \int_a^b f$. But counterexamples are easy to manufacture.

EXAMPLE 4.5.2 Consider the sequence $\{f_n\}_{n=1}^{\infty}$ defined by

$$f_n(x) = \begin{cases} n^2 x, & \text{if } 0 \le x \le 1/n \\ 2n - n^2 x, & \text{if } 1/n \le x \le 2/n \\ 0, & \text{if } 2/n \le x \le 2. \end{cases}$$

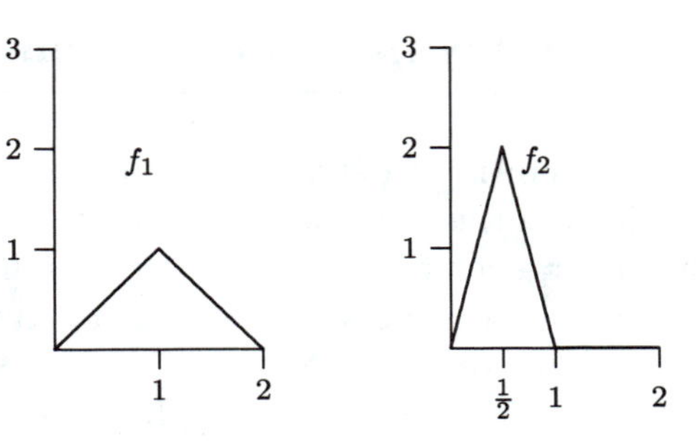

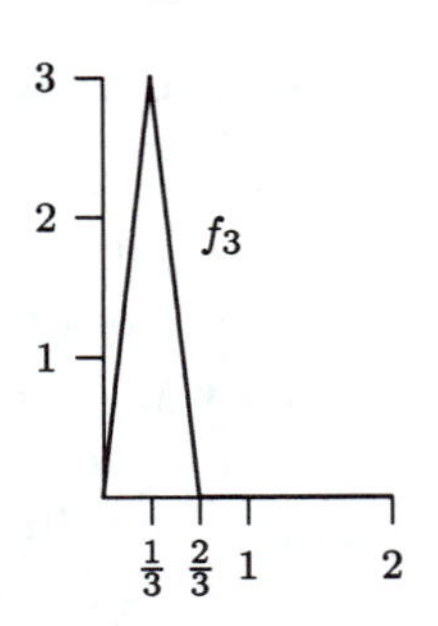

Figure 4.5.2

Each f_n is continuous and therefore integrable on the interval $[0,2]$. As n increases the spike in the graph of f_n moves back toward the origin. As a result, $\lim_{n\to\infty} f_n(x) = 0$ for all x in $[0,2]$. That is, $\{f_n\}_{n=1}^{\infty}$ converges pointwise to the constant function, $f(x) = 0$ for all x in $[0,2]$. Observe, however, that $\int_0^2 f_n$ is just the area of the triangle with base of length $2/n$ and height $f(\frac{1}{n}) = n$. Consequently $\int_0^2 f_n = 1$ for all n. On the other hand, $\int_0^2 f = \int_0^2 0 = 0$. Thus we see that

$$1 = \lim_{n\to\infty} \int_0^2 f_n \neq \int_0^2 \lim_{n\to\infty} f_n = 0.$$

Thus the order in which the limit and the integration processes occur matters. In general, they cannot be interchanged. ◻

Now consider differentiability. Suppose that f is the pointwise limit of a sequence of differentiable functions on an interval I. The following example shows that f need not be differentiable at each point of I.

EXAMPLE 4.5.3 On the interval $(0,\infty)$, let $\{f_n\}_{n=1}^{\infty}$ be the sequence of differentiable functions

$$f_n(x) = \frac{2x^n}{1+x^n}.$$

The limit function for this sequence is

$$f(x) = \begin{cases} 0, & \text{if } 0 < x < 1 \\ 1, & \text{if } x = 1 \\ 2, & \text{if } x > 1. \end{cases}$$

The limit function is not even continuous at 1 and so is not differentiable there. We see that differentiability is not preserved by pointwise convergence. ◻

4.5.2 Uniform Convergence

Examples 4.5.1 to 4.5.3 share a common feature. For each of the sequences, $\lim_{n\to\infty} f_n(x)$ exists at each point. But for different values of x, the values of $f_n(x)$ are moving toward their limits at very different rates. For arbitrarily large values of n, there are points on the graph of f_n that are relatively far from their limit points on the graph of f. The graphs of f_n are not moving *uniformly* toward the graph of the limit function f. This is at least part of the reason why continuity, integrability, and differentiability are not preserved. To prevent such behavior, we refine the notion of convergence for sequences of functions.

DEFINITION 4.5.1 *Let $\{f_n\}_{n=1}^{\infty}$ be a sequence of functions defined on a domain D. Then $\{f_n\}_{n=1}^{\infty}$* **converges uniformly** *on D to a function f defined on D if for every $\epsilon > 0$ there is an N such that for all x in D*

$$\text{if } n > N, \text{then } |f(x) - f_n(x)| < \epsilon.$$

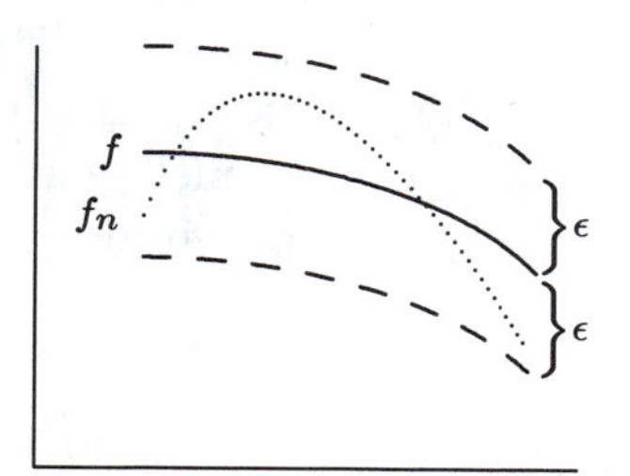

Figure 4.5.3

Clearly, uniform convergence implies pointwise convergence, but the converse is not true. The difference between the two types of convergence is similar to the difference between continuity and uniform continuity. To say that $\{f_n\}_{n=1}^{\infty}$ converges pointwise to f on D means that $\lim_{n\to\infty} f_n(x) = f(x)$ for each x in D. That is, *for each x in D* given any $\epsilon > 0$ there is an N (which in general will depend on both ϵ and x) such that if $n > N$, then $|f(x) - f_n(x)| < \epsilon$. But if $\{f_n\}_{n=1}^{\infty}$ converges uniformly to f on D, given any $\epsilon > 0$, there is an N (which depends only on ϵ) such that if $n > N$, then $|f(x) - f_n(x)| < \epsilon$ *for every x in D*. In the definition of pointwise convergence the phrase "for each x in D" occurs before the existence of N is asserted, so N may well depend on the value of x as well as on the value of ϵ. In the definition of uniform convergence, the reference to each x in D occurs after the existence of N is asserted, so that N depends only on ϵ and works for all x at the same time. This is illustrated geometrically in Figure 4.5.3. For sufficiently large n, the graph of f_n (dotted) will lie within a narrow ϵ band of the graph of the limit function f.

We now examine continuity, integrability, and differentiability under uniform convergence.

THEOREM 4.5.2 *Let $\{f_n\}_{n=1}^{\infty}$ be a sequence of continuous functions defined on an interval I that converges uniformly to a function f on I. Then f is continuous on I.*

PROOF Let $a \in I$. By the triangle inequality, for any $x \in I$,

$$|f(x) - f(a)| \leq |f(x) - f_n(x)| + |f_n(x) - f_n(a)| + |f_n(a) - f(a)|.$$

Uniform convergence will be used to make $|f(x) - f_n(x)|$ and $|f_n(a) - f(a)|$ small by choosing n to be sufficiently large. After n has been selected, the continuity of f_n will be used to make $|f_n(x) - f_n(a)|$ small by taking x sufficiently close to a.

Fix $\epsilon > 0$. Uniform convergence implies that there exists an integer N such that if $n > N$, then $|f(x) - f_n(x)| < \epsilon/3$ and $|f(a) - f_n(a)| < \epsilon/3$. In particular, both of these inequalities hold when $n = N+1$. But f_{N+1} is continuous at a, so there is a $\delta > 0$ such that if x in I and $|x - a| < \delta$, then $|f_{N+1}(x) - f_{N+1}(a)| < \epsilon/3$. Combining these results, if x in I and $|x - a| < \delta$, then

$$\begin{aligned}|f(x) - f(a)| &\leq |f(x) - f_{N+1}(x)| + |f_{N+1}(x) - f_{N+1}(a)| \\ &\quad + |f_{N+1}(a) - f(a)| \\ &< \epsilon/3 + \epsilon/3 + \epsilon/3 = \epsilon.\end{aligned}$$

Therefore f is continuous at a. ∎

The theorem shows that the convergence in Example 4.5.1 cannot be uniform, since each f_n is continuous but the limit is not.

EXAMPLE 4.5.4 Consider the sequence $\{f_n\}_{n=1}^{\infty}$ of continuous functions defined on $\mathbf{R}$ by

$$f_n(x) = \sqrt{x^2 + n^{-2}}.$$

The sequence converges (pointwise) to the function $f(x) = |x|$. To see that the convergence is uniform, let $\epsilon > 0$ be given. Choose N such that

$1/N < \epsilon$. Then for any $n > N$ and any x, observe that

$$\begin{aligned}
\left| |x| - \sqrt{x^2 + n^{-2}} \right| &= \sqrt{x^2 + n^{-2}} - |x| \\
&= \frac{(\sqrt{x^2 + n^{-2}} - |x|)(\sqrt{x^2 + n^{-2}} + |x|)}{\sqrt{x^2 + n^{-2}} + |x|} \\
&= \frac{n^{-2}}{\sqrt{x^2 + n^{-2}} + |x|} \\
&\le \frac{n^{-2}}{\sqrt{n^{-2}}} \\
&= \frac{1}{n} < \frac{1}{N} < \epsilon. \quad \square
\end{aligned}$$

THEOREM 4.5.3 *Assume that $\{f_n\}_{n=1}^{\infty}$ is a sequence of functions that are integrable on $[a, b]$, and that $\{f_n\}_{n=1}^{\infty}$ converges uniformly to an integrable function f on $[a, b]$. Then*

$$\lim_{n\to\infty} \int_a^b f_n = \int_a^b f.$$

PROOF Fix $\epsilon > 0$. By problem 3.5.15,

$$\begin{aligned}
\left| \int_a^b f(x)\,dx - \int_a^b f_n(x)\,dx \right| &= \left| \int_a^b f(x) - f_n(x)\,dx \right| \\
&\le \int_a^b |f(x) - f_n(x)|\,dx.
\end{aligned}$$

Uniform convergence implies that there exists an N such that

$$n > N \Longrightarrow |f(x) - f_n(x)| < \frac{\epsilon}{b-a}, \quad \text{for all } x \text{ in } [a, b].$$

So if $n > N$,

$$\left| \int_a^b f(x)\,dx - \int_a^b f_n(x)\,dx \right| \le \int_a^b |f(x) - f_n(x)|\,dx \le \int_a^b \frac{\epsilon}{b-a}\,dx = \epsilon.$$

Since ϵ was arbitrary, $\lim_{n\to\infty} \int_a^b f_n = \int_a^b f$. ∎

The hypotheses in Theorem 4.5.3 can be relaxed. It can be shown that if $\{f_n\}_{n=1}^{\infty}$ is a uniformly convergent sequence of functions integrable on $[a, b]$, the limit function f will still be integrable. (See problem 4.5.9.)

From Theorem 4.5.3 we know that the convergence in Example 4.5.2 cannot be uniform, since each f_n and the limit function is integrable, but $\lim_{n\to\infty} \int_a^b f_n \neq \int_a^b f$.

Obtaining differentiablity "in the limit" is a more delicate matter. In Example 4.5.4 each function in the sequence is differentiable and the convergence to the limit function is uniform on all of $\mathbf{R}$. However, the limit function is $f(x) = |x|$, which is not differentiable at the origin. Additional assumptions beyond uniform convergence are required to preserve differentiability.

THEOREM 4.5.4 *Suppose that $\{f_n\}_{n=1}^{\infty}$ is a sequence of differentiable functions that converges (pointwise) to f on $[a, b]$. Assume also that each f_n' is continuous on $[a, b]$ and that $\{f_n'\}_{n=1}^{\infty}$ converges uniformly on $[a, b]$. Then f is differentiable and*

$$\lim_{n\to\infty} f_n'(x) = f'(x).$$

PROOF Let g be the function to which the sequence $\{f_n'\}_{n=1}^{\infty}$ converges uniformly. Since the functions f_n' are continuous (and therefore integrable), Theorem 4.5.2 implies that g is continuous. By Theorem 4.5.3, g is integrable and for any x in $[a, b]$,

$$\int_a^x g(t)\,dt = \lim_{n\to\infty} \int_a^x f_n'(t)\,dt.$$

By the first fundamental theorem of calculus,

$$\int_a^x f_n'(t)\,dt = f_n(x) - f_n(a).$$

Therefore

$$\int_a^x g(t)\,dt = \lim_{n\to\infty} \int_a^x f_n'(t)\,dt = \lim_{n\to\infty} \big(f_n(x) - f_n(a)\big) = f(x) - f(a).$$

Solving for $f(x)$ in this equation, we find that

$$f(x) = f(a) + \int_a^x g(t)\,dt \quad \text{for all } x \text{ in } [a, b].$$

Since g is continuous on $[a, b]$, by the second fundamental theorem of calculus f is differentiable on $[a, b]$ and $f'(x) = g(x) = \lim_{n\to\infty} f'(x)$. ∎

Problems for Section 4.5

4.5.1. **a)** Find the pointwise limit of the sequence $\{x^n\}_{n=1}^{\infty}$ on the interval $[0, 1]$.

b) Does this sequence converge uniformly? (Use Theorem 4.5.2.)

4.5.2. **a)** Find the pointwise limit of the sequence $\{f_n\}_{n=1}^{\infty}$, where $f_n(x) = \frac{1}{x^n}$ on $[1, 2]$.

b) Show that this sequence does not converge uniformly on $[1, 2]$.

c) Show that the sequence does converge uniformly on $[2, 3]$.

d) What can you say about convergence on the interval $[2, \infty)$?

4.5.3. **a)** Find the pointwise limit of the sequence $\left\{\frac{1}{1+e^{-nx}}\right\}_{n=1}^{\infty}$ on the entire real line.

b) Does this sequence converge uniformly?

4.5.4. For each of the following sequences of functions determine the pointwise limit.

a) $\{f_n\}_{n=1}^{\infty}$, where $f_n(x) = \sqrt[n]{x}$ on $[0, 2]$.

b) $\{g_n\}_{n=1}^{\infty}$, where $g_n(x) = (1 + x^{2n})^{-1}$ on $(-\infty, \infty)$.

c) $\{h_n\}_{n=1}^{\infty}$, where $h_n(x) = (1 + (x - n)^2)^{-1}$ on $(-\infty, \infty)$.

d) $\{j_n\}_{n=1}^{\infty}$, where $j_n(x) = \left(\dfrac{1}{n} + x\right)^{-1}$ on $(0, 1)$.

4.5.5. Show that none of the sequences in problem 4 converges uniformly on the intervals given. (Hint: Use Theorem 4.5.2.)

4.5.6. Consider the sequence $\{f_n\}_{n=1}^{\infty}$, where $f_n(x) = \dfrac{x}{1 + n^2x^2}$.

a) Show that this sequence converges uniformly to $f(x) = 0$ on $(-\infty, \infty)$. (Hint: Find the maximum and minimum values of f_n on $(-\infty, \infty)$.)

b) Next, examine the derivatives of f_n. Show that for each x the sequence $\{f_n'(x)\}_{n=1}^{\infty}$ converges.

c) Show that $\lim_{n\to\infty} f_n'(0) \neq f'(0)$. Does the sequence of derivatives converge uniformly?

4.5.7. **a)** Let $\{f_n\}_{n=1}^{\infty}$ be a sequence of functions defined on $[a, b]$. Assume that each f_n is bounded, that is, $|f_n(x)| < M_n$ for all $x \in [a, b]$. Show that if $\{f_n\}_{n=1}^{\infty}$ converges uniformly to f, then f must also be bounded.

b) Show that the hypothesis of uniform convergence is necessary by devising a sequence of functions in which each one is bounded, but which converges pointwise to a function that is not bounded.

4.5.8. Let $\{f_n\}_{n=1}^{\infty}$ be a sequence of increasing functions that converges pointwise to f. Show that f must be increasing by arguing as follows.

a) Suppose that f were not increasing, that is, for some $x_1 > x_0$ we have $f(x_1) < f(x_0)$. Choose $\epsilon < \frac{1}{2}(f(x_0) - f(x_1))$. Show that there is an N such that if $n > N$, then both $|f_n(x_0) - f(x_0)| < \epsilon$ and $|f_n(x_1) - f(x_1)| < \epsilon$.

b) Show that this would imply that f_n was not increasing.

4.5.9. Prove the following generalization of Theorem 4.5.3. Let $\{f_n\}_{n=1}^{\infty}$ be a sequence of integrable functions that converges uniformly to $f(x)$ on $[a,b]$. Then f is integrable on $[a,b]$.

a) Given $\epsilon > 0$, show that there is an n such that

$$f_n(x) - \frac{\epsilon}{3(b-a)} < f(x) < f_n(x) + \frac{\epsilon}{3(b-a)}.$$

b) Show that if P is any partition, then $L(P, f_n) - \frac{\epsilon}{3} < L(P, f)$. (Hint: Show that $L(P, f_n - \frac{\epsilon}{3(b-a)}) = L(P, f_n) - \frac{\epsilon}{3}$, using the inequality in **(a)**.)

c) Similarly show that $U(P, f) < U(P, f_n) + \frac{\epsilon}{3}$.

d) Now show that if P is any partition, then

$$U(P, f) - L(P, f) < U(P, f_n) - L(P, f_n) + \frac{2\epsilon}{3}.$$

e) Show that there is a partition P^* with $U(P^*, f_n) - L(P^*, f_n) < \frac{\epsilon}{3}$. Show that $U(P^*, f) - L(P^*, f) < \epsilon$ and finish the proof.

4.5.10. In Example 4.5.4 we showed that $\left\{\sqrt{x^2 + n^{-2}}\right\}_{n=1}^{\infty}$ converged uniformly on $\mathbf{R}$ to $|x|$. Does $\{f_n'\}_{n=1}^{\infty}$ also converge uniformly on $\mathbf{R}$?

4.5.11. Consider the sequence $\{h_n\}_{n=1}^{\infty}$ on $[0, 2\pi]$, where $h_n(x) = \frac{1}{n}\sin(n^2 x)$.

a) Show that each h_n is differentiable.

b) Show that $\{h_n\}_{n=1}^{\infty}$ converges uniformly to 0 on $[0, 2\pi]$.

c) Now show that the sequence of derivatives $\{h_n'\}_{n=1}^{\infty}$ does not converge on $[0, 2\pi]$.

4.5.12. Consider the sequence $g_n(x) = (n^2 + n)x^{n-1}(1 - x)$.

a) Verify that each g_n is integrable and compute its integral $\int_0^1 g_n(x)\, dx$.

b) Show that for each $x \in [0,1]$ that $\lim_{n\to\infty} g_n(x) = 0$. (Hint: For $x \in (0,1)$, show that $\sum_{k=1}^{\infty} g_k$ converges, so the nth terms must go to 0.)

c) Show that $\lim_{n\to\infty} \int_0^1 g_n(x)\, dx \neq \int_0^1 0\, dx$. What went wrong?

4.5.13. For students familiar with improper integrals: Let

$$f_n(x) = \begin{cases} \frac{1}{n} & \text{for } 0 \le x \le n \\ 0 & \text{for } x > n. \end{cases}$$

a) Show that $f_n(x)$ converges uniformly to $f(x) = 0$ on $[0, \infty)$

b) Show that $\lim_{n\to\infty} \int_0^{\infty} f_n(t)\, dt \neq \int_0^{\infty} f(t)\, dt$. Why doesn't this contradict Theorem 4.5.3?

4.5.14. Prove **Dini's theorem:** *Let $\{f_n\}_{n=1}^{\infty}$ be a sequence of continuous functions defined on the closed, bounded interval $[a,b]$ such that*

$$f_1(x) \leq f_2(x) \leq \cdots \leq f_n(x) \leq \cdots \qquad (x \in [a,b]). \qquad (*)$$

If $\{f_n\}_{n=1}^{\infty}$ converges pointwise to the continuous function f on $[a,b]$, then $\{f_n\}_{n=1}^{\infty}$ converges uniformly to f on $[a,b]$.

a) Begin by defining $g_n = f - f_n$ for each n. Show that

$$g_1(x) \geq g_2(x) \geq \cdots \geq g_n(x) \geq \cdots \qquad (x \in [a,b])$$

and that $\{g_n\}_{n=1}^{\infty}$ converges pointwise to 0 on $[a,b]$.

b) The next few steps will show that $\{g_n\}_{n=1}^{\infty}$ converges uniformly to 0 on $[a,b]$. Fix $\epsilon > 0$. If $x \in [a,b]$, show that there exists an integer $N(x) > 0$ such that $0 \leq g_{N(x)}(x) < \epsilon/2$.

c) Show that there is a $\delta_x > 0$ such that if $y \in [a,b]$ and $|y-x| < \delta_x$, then $|g_{N(x)}(y) - g_{N(x)}(x)| < \epsilon/2$.

d) Let I_x be the open δ_x-interval $(x - \delta_x, x + \delta_x)$. Show that if $y \in [a,b]$ and $y \in I_x$, then $g_{N(x)}(y) < \epsilon$.

e) Clearly the collection of all open intervals I_x cover $[a,b]$. Show that a finite subcover, say $\{I_{x_1}, I_{x_2}, \ldots, I_{x_n}\}$, of $[a,b]$ exists.

f) Let $N = \max\{N(x_1), N(x_2), \ldots, N(x_n)\}$. If $y \in [a,b]$, then $y \in I_{x_j}$ for some j. Therefore $g_{N(x_j)}(y) < \epsilon$. Show that $g_N(y) \leq g_{N(x_j)}(y)$.

g) Show that $0 \leq g_N(y) < \epsilon$ for all $y \in [a,b]$.

h) Show that $\{g_n\}_{n=1}^{\infty}$ converges uniformly to 0 on $[a,b]$ and consequently that $\{f_n\}_{n=1}^{\infty}$ converges uniformly to f on $[a,b]$.

i) Prove that Dini's theorem remains true if all of the inequality signs in $(*)$ are reversed.

4.6 Series of Functions

In this section we complete our work with sequences and series by generalizing the notion of series to functions. When we define a function f by

$$f(x) = \sum_{k=1}^{\infty} f_k(x) = f_1(x) + f_2(x) + f_3(x) + \cdots,$$

we mean that

$$f(x) = \lim_{n\to\infty} f_1(x) + f_2(x) + \cdots + f_n(x)$$

exists. That is, the sequence of partial sums of the series

$$f_1(x),\ f_1(x)+f_2(x),\ f_1(x)+f_2(x)+f_3(x),\ldots,\sum_{k=1}^{n} f_k(x),,\ldots$$

has a pointwise limit f. Of course, the major results of the last section require uniform convergence. In fact, the three fundamental results (Theorems 4.5.2 to 4.5.4) concerning uniform convergence of sequences of functions are easily translated into results about series of functions.

4.6.1 Basic Results

DEFINITION 4.6.1 *The series $\sum_{k=1}^{\infty} f$ **converges uniformly** to f on D if the sequence of partial sums $\{\sum_{k=1}^{n} f_k\}_{n=1}^{\infty}$ converges uniformly on D.*

As before we will allow the summation in a series of functions to begin at an index value other than 1.

EXAMPLE 4.6.1 From Example 4.3.3, the geometric series $\sum_{k=0}^{\infty} x^k$ converges for $-1 < x < 1$. We now show that the convergence is uniform on any interval of the form $[-r, r]$ where $0 < r < 1$.

Fix $\epsilon > 0$. Since $0 < r < 1$, $\lim_{n\to\infty} r^{n+1} = 0$. Therefore we may choose N such that if $n > N$, then $r^{n+1} < \epsilon(1-r)$. If $n > N$, then from Example 4.3.3, for any $x \in [-r, r]$,

$$\left|\sum_{k=0}^{\infty} x^k - \sum_{k=0}^{n} x^k\right| = \left|\frac{1}{1-x} - \frac{1-x^{n+1}}{1-x}\right| = \frac{|x^{n+1}|}{1-x} \le \frac{r^{n+1}}{1-r} < \epsilon.$$

This shows that the convergence is uniform on $[-r, r]$. □

THEOREM 4.6.2 *Assume that $\sum_{k=1}^{\infty} f_k$ converges uniformly to f on $[a, b]$. If each f_k is continuous on $[a, b]$, then f is continuous on $[a, b]$.*

PROOF Since each f_k is continuous on $[a, b]$, so is each partial sum $\sum_{k=1}^{n} f_k$. The sequence of partial sums, $\{\sum_{k=1}^{n} f_k\}_{n=1}^{\infty}$, converges uniformly to f on $[a, b]$, so by Theorem 4.5.2 f is continuous. ∎

EXAMPLE 4.6.2 Consider the series $\sum_{k=1}^{\infty} x(1+x)^{-k}$ on $[0, 1]$. Show that it converges pointwise to a function f but not uniformly.

SOLUTION First, $f(0) = \sum_{k=1}^{\infty} 0(1+0)^{-k} = 0$. The limit function f is easy to determine for the other points of the interval. Notice that if $0 < x \le 1$, then $\frac{1}{2} \leq (1+x)^{-1} < 1$. Thus we may apply the geometric series formula to evaluate $f(x)$. For $x \in (0,1]$,

$$f(x) = \sum_{k=1}^{\infty} \frac{x}{(1+x)^k} = x \sum_{k=1}^{\infty} \left(\frac{1}{(1+x)}\right)^k$$
$$= x \cdot \frac{1}{1 - \frac{1}{1+x}} = x \cdot \frac{1+x}{1+x-1} = 1 + x.$$

Thus the limit fuction is

$$f(x) = \begin{cases} 0, & \text{if } x = 0 \\ 1 + x, & \text{if } 0 < x \leq 1. \end{cases}$$

Since f is not continuous on $[0,1]$, even though each f_n is, by Theorem 4.6.2 the convergence cannot be uniform. □

THEOREM 4.6.3 *Assume that $\sum_{k=1}^{\infty} f_k$ converges uniformly to f on $[a,b]$. If f and each f_k is integrable on $[a,b]$, then $\int_a^b f = \sum_{k=1}^{\infty} \int_a^b f_k$.*

PROOF Since each f_k is integrable on $[a,b]$, any partial sum $\sum_{k=1}^{n} f_k$ is integrable on $[a,b]$. By the linearity of the Riemann integral,

$$\int_a^b \sum_{k=1}^{n} f_k = \sum_{k=1}^{n} \int_a^b f_k \,.$$

Since the convergence to f is uniform, Theorem 4.5.3 applies and

$$\int_a^b f = \lim_{n\to\infty} \int_a^b \sum_{k=1}^{n} f_k = \lim_{n\to\infty} \sum_{k=1}^{n} \int_a^b f_k = \sum_{k=1}^{\infty} \int_a^b f_k \,. \quad ∎$$

THEOREM 4.6.4 *Assume that $\sum_{k=1}^{\infty} f_k$ converges (pointwise) to f on $[a,b]$ and that each f_k is differentiable on $[a,b]$. Assume further that each f_k' is continuous on $[a,b]$ and that $\sum_{k=1}^{\infty} f_k'$ converges uniformly on $[a,b]$. Then f is differentiable for all x in $[a,b]$ and*

$$f'(x) = \sum_{k=1}^{\infty} f_k'(x) \,.$$

PROOF Each partial sum $\sum_{k=1}^{n} f_k$ is differentiable with derivative $\sum_{k=1}^{n} f'_k$. Since each f'_k is continuous on $[a, b]$, so are the partial sums $\sum_{k=1}^{n} f'_k$. Since these partial sums of the derivatives converge uniformly, Theorem 4.5.4 applies: f is differentiable and

$$f'(x) = \lim_{n\to\infty} \sum_{k=1}^{n} f'_k = \sum_{k=1}^{\infty} f'_k . \quad \blacksquare$$

There is a particularly simple and useful test for uniform convergence of series of functions.

THEOREM 4.6.5 ***(The Weierstrass M-Test)*** *Assume that $\{f_n\}_{n=1}^{\infty}$ is a sequence of functions defined on D and suppose that $\{M_n\}_{n=1}^{\infty}$ is a sequence of positive real constants such that*

$$|f_n(x)| \leq M_n$$

for all x in D. Assume further that $\sum_{k=1}^{\infty} M_k$ converges. Then $\sum_{k=1}^{\infty} f_k$ converges uniformly on D.

PROOF Choose $\epsilon > 0$. Since the positive series $\sum_{k=1}^{\infty} M_k$ converges, by problem 4.3.8 there exists $N > 0$ such that for $n > N$, we have $0 \leq \sum_{k=n+1}^{\infty} M_k < \epsilon$. The comparison test shows that the series $\sum_{k=1}^{\infty} f_k(x)$ converges absolutely (pointwise) at each x in D. We call the limit $f(x)$. Therefore for any $n > N$ and any x in D

$$\left| f(x) - \sum_{k=1}^{n} f_k(x) \right| = \left| \sum_{k=n+1}^{\infty} f_k(x) \right| \leq \sum_{k=n+1}^{\infty} |f_k(x)| \leq \sum_{k=n+1}^{\infty} M_k < \epsilon,$$

which shows that the convergence is uniform. $\blacksquare$

EXAMPLE 4.6.3 Show that if $0 < r < 1$, then $\sum_{k=1}^{\infty} kx^k$ converges uniformly on $[-r, r]$.

SOLUTION Let x be any number in $[-r, r]$. Then $|f_k(x)| = |kx^k| \leq kr^k$. Since $0 < r < 1$, by Example 4.3.6, $\sum_{k=1}^{\infty} kr^k$ converges. By the Weierstrass M-test, $\sum_{k=1}^{\infty} kx^k$ converges uniformly on $[-r, r]$. $\square$

4.6.2 Power Series

A particularly simple but important type of series may be formed as follows: Let $f_n(x)$ be the polynomial $a_n(x - a)^n$ and form the series

$$f(x) = \sum_{n=0}^{\infty} f_n(x) = \sum_{n=0}^{\infty} a_n(x - a)^n.$$

Such a series is called a **power series centered at a.** The series in Examples 4.6.1 and 4.6.3 are both power series centered at 0.

Among the most important and interesting power series are those derived from certain functions. Specifically, assume that f has derivatives of all orders at a. Then the **Taylor series for f at a** is

$$\sum_{n=0}^{\infty} \frac{f^{(n)}(a)}{n!}(x-a)^n.$$

Recall from Section 3.7 that an infinitely differentiable function can be expressed as a degree n Taylor polynomial plus a remainder or error term:

$$f(x) = \sum_{k=0}^{n} \frac{f^{(k)}(a)}{k!}(x-a)^k + r_{n,a}(x).$$

The Taylor series for f converges to $f(x)$ if and only if the remainder term goes to 0.

More generally, notice that any power series $\sum_{k=0}^{\infty} a_k(x-a)^k$ centered at a converges to a_0 when $x = a$. The problem is to determine at what other values of x (if any) the series converges and where it does so uniformly, much as we did in Examples 4.6.1 and 4.6.3. The main result that we will prove is that each power series $\sum_{n=0}^{\infty} a_n(x-a)^n$ has an associated **interval of convergence** that is centered at a and extends a distance r in both directions. The number r is called the **radius of convergence** for the power series. (In Examples 4.6.1 and 4.6.3, $r = 1$.) Within the interval $(a-r, a+r)$ the series converges absolutely, and outside the interval in the regions $(-\infty, a-r)$ and $(a+r, \infty)$ it diverges. At each of the endpoints $a-r$ and $a+r$ the series may or may not coverge. Two extreme cases are possible: Sometimes $r = 0$ and the interval of convergence is reduced to the single point a. The other extreme is when the power series converges for all x, in which case we say the radius of convergence is $r = +\infty$.

For the sake of simplicity, we will state and prove results about power series centered at 0,

$$f(x) = \sum_{k=0}^{\infty} a_k x^k.$$

Such results are easily translated into statements about power series centered at a. The existence of the interval of convergence for a power series is based on the following result:

LEMMA 4.6.6 *Given the power series $\sum_{k=0}^{\infty} a_k x^k$.*

a) *If the series converges at some number $x_1 \neq 0$, then it converges absolutely for all $|x| < |x_1|$. Furthermore, the convergence is uniform on any interval of the form $[-s, s]$ where $0 < s < |x_1|$.*

b) *If the series diverges at some number x_2, then it diverges for all $x > |x_2|$.*

PROOF Since $\sum_{n=0}^{\infty} a_n {x_1}^n$ converges, we must have $\lim_{n\to\infty} a_n {x_1}^n = 0$. In particular, there is an integer N such that for all $n > N$, we have $a_n {x_1}^n < 1$.

Choose any positive number s such that $0 < s < |x_1|$. If $x \in [-s, s]$ and $n > N$, then

$$|a_n x^n| = |a_n {x_1}^n| \left|\frac{x}{x_1}\right|^n \leq \left|\frac{x}{x_1}\right|^n \leq \left|\frac{s}{x_1}\right|^n.$$

If we let $r = |s/x_1|$, then $0 < r < 1$. The preceding inequalities show that $|a_n x^n| < r^n$ whenever $n > N$. But the geometric series $\sum_{n=0}^{\infty} r^n$ is convergent, so by comparison $\sum_{k=0}^{\infty} a_k x^k$ is absolutely convergent on $[-s, s]$. By the Weierstrass M-test, the convergence is uniform on $[-s, s]$. Further, since any x with $|x| < |x_1|$ lies in some interval of the form $[-s, s]$ with $s < |x_1|$, this also shows that the power series converges absolutely for all x with $|x| < |x_1|$.

To prove **(b)**, take any y with $|y| > |x_2|$. If the power series converged at y, by **(a)** the series would converge absolutely for all x with $|x| < |y|$, including $x = x_2$, which contradicts the hypothesis. ∎

THEOREM 4.6.7 *Assume that the power series $\sum_{k=0}^{\infty} a_k x^k$ converges for some number $x_1 \neq 0$ and diverges for some number x_2. Then there exists a positive real number r such that the series converges absolutely for $|x| < r$ and diverges if $|x| > r$.*

PROOF We will use the least upper bound axiom to determine the radius of convergence r. Let S denote the set of all x for which the power series $\sum_{k=0}^{\infty} a_k x^k$ converges. S is not the empty set, since it contains x_1. By Lemma 4.6.6, S is bounded above by $|x_2|$. Consequently S has a least upper bound r. Using the lemma again, the power series converges for all x with $|x| < |x_1|$, so $r \geq |x_1| > 0$.

Now we show that the power series converges for any x such that $|x| < r$. If $|x| < r$, since r is the least upper bound of S there is some number y in S such that $|x| < y < r$. By Lemma 4.6.6, if the series converges at y, it converges at x.

Finally, observe that if the series converged at some x with $|x| > r$, we could choose z such that $r < z < |x|$. By the lemma the series would now converge at z, which contradicts the fact that r is the least upper bound of S. Consequently $\sum_{k=0}^{\infty} a_k x^k$ diverges for all x such that $|x| > r$. ∎

The preceeding theorem assumes that the power series in question converges at some finite number $x_1 \neq 0$ and diverges at some finite number x_2. We know that any power series $\sum_{k=0}^{\infty} a_k x^k$ must converge at 0. Consequently there are two cases not covered by the theorem. The power series may converge only at $x = 0$ and diverge everywhere else. In this case we say that the radius of convergence r is 0. Or the power series may not diverge at any x; that is, it converges everywhere and we say that its radius of convergence is infinite. In cases where the radius of convergence is finite, the series may or may not converge at the endpoints of the interval as the following example indicates.

EXAMPLE 4.6.4 Determine the interval of convergence for the power series

$$\sum_{k=0}^{\infty} \frac{x^k}{2^k k}.$$

SOLUTION We use the root test to determine where the series converges absolutely.

$$\lim_{n\to\infty} \sqrt[n]{\frac{|x^n|}{2^n n}} = \frac{|x|}{2} < 1 \iff |x| < 2.$$

Thus the radius of convergence for the series is 2 and the series converges on the interval $(-2, 2)$. We still need to check whether the series converges at the endpoints of this interval. When $x = 2$, the series becomes the familiar harmonic series, $\sum_{k=1}^{\infty} 1/k$, which diverges. When $x = -2$, the series becomes $\sum_{k=1}^{\infty} -1^k/k$. But the alternating harmonic series converges. Therefore the interval of convergence for the series is $[-2, 2)$. Finally, by Lemma 4.6.6, the series converges uniformly on any interval of the form $[-s, s]$ where $0 < s < 2$. □

Within its interval of convergence, each power series defines a function f whose value at x is given by

$$f(x) = \sum_{k=0}^{\infty} a_k x^k.$$

We say that the power series **represents the function** f within the interval of convergence, or we say that we have a **power series expansion** for f.

Given a power series expansion, what are the basic properties of the function f it represents? If a power series $\sum_{k=0}^{\infty} a_k x^k$ has radius of convergence r, by Lemma 4.6.6 it converges uniformly on any interval of the form $[-s, s]$ where $0 < s < r$. By Theorem 4.6.3 this series can be integrated term by term on the interval $[-s, s]$ and therefore can be integrated term by term over any closed subinterval of its interval of convergence. Further, since each term of a power series is continuous everywhere, it follows from Theorem 4.6.2 that the series is continuous on $[-s, s]$ and therefore over its interval of convergence. Consequently we have the following result:

THEOREM 4.6.8 *Assume that a function f is represented by the power series*

$$f(x) = \sum_{k=0}^{\infty} a_k x^k$$

in the interval $(-r, r)$. Then f is continuous on this interval and can be integrated on any closed subinterval by integrating the power series term by term.

Next we show that a power series may be differentiated term by term within its interval of convergence.

THEOREM 4.6.9 *Suppose that the power series $\sum_{k=0}^{\infty} a_k x^k$ represents the function f on the interval $(-r, r)$, where $r > 0$. Then the differentiated series $\sum_{k=1}^{\infty} k a_k x^{k-1}$ converges on $(-r, r)$ to $f'(x)$. Further, the differentiated series has exactly the same radius of convergence as $\sum_{k=0}^{\infty} a_k x^k$.*

PROOF To use Theorem 4.6.4 in the proof, we must first show that the differentiated series converges uniformly. We will show that $\sum_{k=1}^{\infty} k a_k x^{k-1}$ converges uniformly on every closed subinterval of the form $[-s, s]$ where $0 < s < r$. For any such s, there is a positive number x_0 such that $s < x_0 < r$. Since the series $\sum_{k=0}^{\infty} a_k {x_0}^k$ converges, its terms are bounded. That is, there is some number M such that for all k

$$|a_k| {x_0}^k \leq M.$$

If x is in the interval $[-s, s]$, for each term in the differentiated series $\sum_{k=1}^{\infty} k a_k x^{k-1}$ we have

$$|k a_k x^{k-1}| \leq k |a_k| s^{k-1} = k \frac{|a_k| {x_0}^k s^{k-1}}{{x_0}^k} \leq k M \frac{s^{k-1}}{{x_0}^k} = k \frac{M}{s} \left(\frac{s}{x_0} \right)^k.$$

But $0 < s/x_0 < 1$, so by Example 4.3.6 the series

$$\sum_{k=1}^{\infty} k \frac{M}{s} \left(\frac{s}{x_0} \right)^k = \frac{M}{s} \sum_{k=1}^{\infty} k \left(\frac{s}{x_0} \right)^k$$

converges. By the Weierstrass M-test, the series $\sum_{k=1}^{\infty} ka_k x^{k-1}$ converges uniformly on $[-s, s]$. Consequently, by Theorem 4.6.4,

$$f'(x) = \sum_{k=1}^{\infty} ka_k x^{k-1}$$

on $[-s, s]$. Since this is true for any s such that $0 < s < r$, this proves that the radius of convergence of the differentiated series is at least r.

To show that the radius of convergence does not exceed r, observe that $|a_k x^k| < |ka_k x^{k-1}|$ as soon as $k > |x|$. That is, the differentiated series eventually dominates the original series, so the radius of convergence r of $\sum_{k=0}^{\infty} a_k x^k$ is at least as large as the radius of convergence of $\sum_{k=1}^{\infty} ka_k x^{k-1}$. Thus the two series have the same radius of convergence. ∎

Notice that the differentiated series $f'(x) = \sum_{k=1}^{\infty} ka_k x^{k-1}$ itself continues to satisfy the hypotheses of Theorem 4.6.9. Therefore the second derivative of f may be computed term by term as

$$f''(x) = \sum_{k=2}^{\infty} k(k-1)a_k x^{k-2},$$

where the interval of convergence is again r. We can continue in this fashion to compute derivatives of all orders of the power series representing f, each derived series having radius of convergence r. Differentiating $f(x) = \sum_{k=0}^{\infty} a_k x^k$ a total of k times and then setting $x = 0$ in the resulting series, we find that $f^{(k)}(0) = k! a_k$. Consequently

$$a_k = f^{(k)}(0)/k!.$$

If we interpret $f^{(0)}(0)$ to mean $f(0)$, we may write

$$f(x) = \sum_{k=0}^{\infty} a_k x^k = \sum_{k=0}^{\infty} \frac{f^{(k)}(0)}{k!} x^k,$$

where this last expression is just the **Taylor series** for f at 0. In other words, we have shown that a convergent power series centered at 0 is the Taylor series centered at 0 for the function that it represents.

THEOREM 4.6.10 *If two power series $\sum_{k=0}^{\infty} a_k x^k$ and $\sum_{k=0}^{\infty} b_k x^k$ represent the same function f in the interval $(-r, r)$, then the two series are equal term by term and $a_k = b_k = f^{(k)}(0)/k!$. That is, both series are just the Taylor series for f centered at 0.*

We can now determine which power series represent some of the more familiar functions. As mentioned earlier, when f is infinitely differentiable at a, by the definition of the remainder term (see Theorem 3.7.2)

$$f(x) = \sum_{k=0}^{n} \frac{f^{(k)}(a)}{k!}(x-a)^k + r_{n,a}(x).$$

Letting $n \to \infty$, the Taylor series for f converges to $f(x)$ if and only if the remainder term $r_{n,a}(x) \to 0$.

EXAMPLE 4.6.5 Show that for all x, the Taylor series centered at 0 for $f(x) = e^x$ converges to e^x.

SOLUTION Notice that $f^{(k)}(x) = e^x$ for all $k \geq 0$. Therefore $f^{(k)}(0) = 1$. Consequently the Taylor series centered at 0 for e^x is

$$\sum_{k=0}^{\infty} \frac{f^{(k)}(0)}{k!}x^k = \sum_{k=0}^{\infty} \frac{x^k}{k!}.$$

From Corollary 3.7.3, for any fixed x,

$$r_{n,0}(x) = \frac{f^{(n+1)}(t)}{(n+1)!}(x-0)^{n+1} = \frac{e^t}{(n+1)!}x^{n+1}$$

for some t between 0 and x. Since $0 < |t| < |x|$,

$$|r_{n,0}(x)| < \frac{e^{|x|}}{(n+1)!}|x|^{n+1}.$$

Using the ratio test and the nth-term test we see that for each fixed x, $\lim_{n\to\infty} |x|^{n+1}/(n+1)! = 0$, and it follows that $\lim_{n\to\infty} r_{n,0}(x) = 0$. Therefore

$$e^x = \sum_{k=0}^{\infty} \frac{x^k}{k!}.$$

In particular,

$$e = e^1 = 1 + 1 + \frac{1}{2!} + \frac{1}{3!} + \frac{1}{4!} + \cdots,$$

and

$$\frac{1}{e} = e^{-1} = 1 - 1 + \frac{1}{2!} - \frac{1}{3!} + \frac{1}{4!} - \cdots \qquad \square$$

EXAMPLE 4.6.6 Show that for all x we have $\cos x = 1 - \frac{x^2}{2!} + \frac{x^4}{4!} - \frac{x^6}{6!} + \cdots$.

SOLUTION First we determine the Taylor series for $f(x) = \cos x$ centered at 0. In Example 3.7.3 we saw that $f^{(2k)}(0) = (-1)^k$ and $f^{(2k+1)}(0) = 0$. Thus the Taylor series for $\cos x$ centered at 0 is

$$\sum_{k=0}^{\infty} \frac{f^{(k)}(0)}{k!} x^k = \sum_{k=0}^{\infty} \frac{(-1)^k}{(2k)!} x^{2k} = 1 - \frac{x^2}{2!} + \frac{x^4}{4!} - \frac{x^6}{6!} + \cdots.$$

For any fixed x, the nth remainder term at 0 is

$$r_{n,0}(x) = \frac{f^{(n+1)}(t)}{(n+1)!}(x-0)^{n+1}$$

for some t between 0 and x. But $f^{(n+1)}(t)$ is either $\pm \cos t$ or $\pm \sin t$, so $|f^{(n+1)}(t)| \leq 1$. Therefore

$$|r_{n,0}(x)| \leq \frac{|x|^{n+1}}{(n+1)!}.$$

As in the preceding example, $\lim_{n\to\infty} |x|^{n+1}/(n+1)! = 0$, and it follows that $\lim_{n\to\infty} r_{n,0}(x) = 0$. Therefore for any x

$$\cos x = \sum_{k=0}^{\infty} \frac{(-1)^k}{(2k)!} x^{2k}. \quad \square$$

Problems for Section 4.6

4.6.1. Show that $\ln 2$ is the sum of the alternating harmonic series by examining the Taylor series for $\ln(1+x)$ centered at 0.

4.6.2. Show that any power series $\sum_{k=0}^{\infty} a_k x^k$ and the power series obtained by integrating term by term $\sum_{k=0}^{\infty} a_k x^{k+1}/(k+1)$ have the same interval of convergence. (Hint: Use Theorem 4.6.9.)

4.6.3. Find the Taylor series centered at 0 for each of the following functions.

a) $f(x) = \dfrac{1}{1-x}$, where $-1 < x < 1$.

b) $f(x) = \dfrac{1}{1+x}$, where $-1 < x < 1$.

c) $f(x) = \dfrac{1}{1-x^2}$, where $-1 < x < 1$.

d) $f(x) = \dfrac{1}{(1-x)^2}$, where $-1 < x < 1$. This is simple if you differentiate the function in **(a)**. Also see problem 4.4.2.

4.6.4. Find the Taylor series centered at 0 for each of the following functions.

a) $f(x) = \sin x$.

b) $f(x) = e^{2x}$.

c) $f(x) = (x-a)^{-1}$, where $a \neq 0$.

d) $f(x) = \sinh x = \frac{1}{2}(e^x - e^{-x})$.

4.6.5. Which familiar functions do the following power series represent?

a) $1 + x^3 + x^6 + x^9 + x^{12} + \cdots$. (Hint: Look back at problem 3a and c.)

b) $x + \frac{x^2}{2} + \frac{x^3}{3} + \frac{x^4}{4} + \cdots$. (Hint: Differentiate.)

c) $1 + 2x + 3x^2 + 4x^3 + 5x^4 + \cdots$. (Hint: Integrate.)

d) $1 - x + \dfrac{x^2}{2!} - \dfrac{x^3}{3!} + \dfrac{x^4}{4!} - \cdots$.

4.6.6. Find the pointwise limit of the series $\sum_{k=1}^{\infty} x(1-x)^k$ on the interval $[0,1]$. Show that the convergence is not uniform. (Hint: See Example 4.6.2.)

4.6.7. Show that if $\sum_{k=1}^{\infty} f_k$ converges uniformly to f and if each f_k is integrable on $[a,b]$, then f is integrable on $[a,b]$. (Hint: Use the result of problem 4.5.9). This shows that one of the hypotheses of Theorem 4.6.3 was unnecessary.

4.6.8. Use the Weierstrass M-test to show that the following series converge uniformly on the intervals shown.

a) $\displaystyle\sum_{k=0}^{\infty} \frac{2^k x^k}{k!}$ on $[0,10]$ **b)** $\displaystyle\sum_{k=0}^{\infty} \frac{k^2 x^k}{3^k}$ on $[-2,2]$

c) $\displaystyle\sum_{k=0}^{\infty} \frac{kx^2}{k^3 + x^2}$ on $[0,2]$ **d)** $\displaystyle\sum_{k=0}^{\infty} e^{-kx} x^k$ on $[0,\infty)$

(Hint: For **(d)** use calculus to find the maximum value of f_k on $[0,\infty)$.)

4.6.9. **a)** Find the remainder, $r_{n,0}(x)$, for the nth Taylor polynomial for $\cos x$ about $a = 0$.

b) Show that $\lim_{n\to\infty} r_{n,0}(x) = 0$ for *any* x.

c) Find a polynomial series representing $\cos x^2$ and another for $\int \cos x^2\, dx$.

d) Show that the series for $\int \cos x^2\, dx$ converges for each value of x.

4.6.10. **a)** Use the series for $\cos x$ to obtain the series for $\cos\sqrt{x}$ and for $\int_0^x \cos\sqrt{t}\, dt$.

b) Use the first 4 terms of this last series to estimate $\int_0^2 \cos\sqrt{t}\, dt$. Estimate the error by using the fact that the series for $\int_0^2 \cos\sqrt{t}\, dt$ is alternating (problem 4.4.5).

4.6.11. Another integral that is tedious to evaluate is $\int_0^x \frac{1}{1+t^3}\,dt$.

a) Obtain a series for $\frac{1}{1+t^3}$ by substituting $-t^3$ for x into the geometric series $\frac{1}{1-x} = 1 + x + x^2 + x^3 \cdots$. For which values of t does the new series converge?

b) Now obtain a series for $\int_0^x \frac{1}{1+t^3}\,dt$ and use it to estimate the area under the curve $\frac{1}{1+t^3}$ from 0 to $\frac{1}{2}$. (You might wish to compare this estimate with the actual value of $\int_0^x \frac{1}{1+t^3}\,dt$.)

4.6.12. **a)** Find the Taylor series centered at 0 for $\cos 2x$.

b) Find the Taylor series centered at 0 for $\cos^2 x$. (Hint: Use the identity $\cos^2 x = \frac{1}{2}(1 + \cos 2x)$.)

c) Find the Taylor series centered at 0 for $\sin^2 x$.

4.6.13. Find the Taylor series centered at 0 for $(1+x)^{1/2}$. What is its radius of convergence?

4.6.14. The first fundamental theorem cannot be used to compute the definite integral $\int_0^1 e^{-x^2}\,dx$ because there is no elementary antiderivative of e^{-x^2}. However, you can approximate the integral to any degree of accuracy by using Taylor series.

a) Write out the Taylor series for e^{-x^2}. (This can be done simply by substituting $-x^2$ for x in the series expansion for e^x.)

b) By integrating this series expansion term by term show that

$$\int_0^1 e^{-x^2}\,dx = 1 - \frac{1}{3} + \frac{1}{5 \cdot 2!} - \frac{1}{7 \cdot 3!} + \frac{1}{9 \cdot 4!} - \cdots.$$

Why is this valid?

c) Use problem 4.4.5 to show that the approximation

$$\int_0^1 e^{-x^2}\,dx \approx 1 - \frac{1}{3} + \frac{1}{5 \cdot 2!} - \frac{1}{7 \cdot 3!} + \frac{1}{9 \cdot 4!} = \frac{5651}{7560}$$

is accurate to within $1/(11 \cdot 5!) = 1/1320$.

4.6.15. Determine the Taylor series expansion centered at 0 for $\tan^{-1} x$. Begin by recalling that $\int \frac{1}{1+x^2}\,dx = \tan^{-1} x + c$. Find the series expansion for $\frac{1}{1+x^2}$. (See problem 3 for similar questions.) Integrate the series term by term to find an expansion for $\tan^{-1} x + c$. Evaluate c by using $x = 0$ in the expansion. Why is this process valid?

5

Calculus in Two Dimensions

In the first four chapters we have developed a rigorous theory of calculus in one dimension. In the remainder of the book we will explore extensions of the basic definitions and theorems of calculus in two different settings: the vector space $\mathbf{R}^2$ and the complex numbers. Such generalization can be an exciting and enlightening exercise. Seeing the same ideas in a slightly modified context enriches our understanding and leads to a host of interesting questions and comparisons. It tests the strength of our basic ideas and draws attention to the underlying fundamental structures necessary to support those ideas. In the next three chapters we hope you begin to appreciate some of the sweeping unity and power of the mathematical ideas that we worked so hard to develop in the first four chapters.

This chapter explores the basic calculus concepts of limits, continuity, derivatives, and integrals for the vector space $\mathbf{R}^2$. We begin with a look at the properties of the underlying space $\mathbf{R}^2$, concentrating on the fundamental role of the dot product. In a sense the first section is analogous to our study of the real numbers in Chapter 1. Following this, our investigation divides into two parts: First we examine functions of the form $\vec{\mathbf{F}} : \mathbf{R} \to \mathbf{R}^2$, where the range is the vector space $\mathbf{R}^2$. Here our earlier definitions from single-variable calculus extend naturally, leading to componentwise operations. Next we turn to vector functions whose domain is $\mathbf{R}^2$, those of the form $f : \mathbf{R}^2 \to \mathbf{R}$. Limits and derivatives for these functions present certain difficulties and require a rephrasing of the basic concept of the derivative in terms of linear functions. However, with our restructured definition we will be able to prove generalizations of important theorems such as the mean value theorem and the chain rule. The last section of this chapter develops integration for vector functions in a manner parallel to the development of integration in Chapter 3. Although for reasons of simplicity this chapter focuses specifically on $\mathbf{R}^2$, the reader should note that all of the fundamental definitions, theorems, and proofs extend easily to any finite-dimensional vector space, $\mathbf{R}^n$.

5.1 The Dot Product

In this section we present the basic properties of the vector space $\mathbf{R}^2$. Of particular interest in this study is the vector space structure known as the dot product. First, the dot product forms the basis for a distance measure in $\mathbf{R}^2$ upon which the limit definitions of calculus are built. In fact, the ϵ, δ definition of a limit is essentially concerned with how the *distance* between points in the domain ($|x - x_0| < \delta$) is related to the *distance* between their images in the range ($|f(x) - f(x_0)| < \epsilon$). Second, dot products are the foundation for geometric concepts of orthogonality and the measurement of angles via the important Cauchy-Schwarz inequality. Finally, the open and closed sets (i.e., the topology) in $\mathbf{R}^2$ are defined in terms of the metric formed from the dot product. We end the section with the $\mathbf{R}^2$ analogue of the Heine-Borel theorem, which played such a fundamental role in one-dimensional calculus.

5.1.1 The Vector Space $\mathbf{R}^2$

We begin with the definition of the vector space $\mathbf{R}^2$ and its two basic operations. We denote all vectors using boldfaced letters surmounted by arrows.

DEFINITION 5.1.1 *The **vector space** $\mathbf{R}^2$ consists of all ordered pairs (x_1, x_2), where x_1 and x_2 are real numbers. The **sum** of two vectors $\vec{\mathbf{x}} = (x_1, x_2)$ and $\vec{\mathbf{y}} = (y_1, y_2)$ in $\mathbf{R}^2$ is defined componentwise:*

$$\vec{\mathbf{x}} + \vec{\mathbf{y}} \equiv (x_1 + y_1, x_2 + y_2).$$

*The **scalar multiple** of the vector $\vec{\mathbf{x}} = (x_1, x_2)$ by the scalar $k \in \mathbf{R}$ is*

$$k\vec{\mathbf{x}} \equiv (kx_1, kx_2).$$

Notice that since addition of vectors is simply componentwise addition of real numbers, it obeys the same laws of addition as real numbers. Hence addition of vectors is commutative and associative, there is an additive identity $\vec{\mathbf{0}} = (0,0)$, and additive inverses exist. However, there is no provision for multiplication of vectors; therefore *division* by a vector is meaningless. This has important consequences for the definition of the derivative. Consider the expression

$$\lim_{x \to a} \frac{f(x) - f(a)}{x - a}.$$

If the *range* of f is a vector space (i.e., $f(x)$ and $f(a)$ are vectors), the numerator of the derivative still makes sense, since we can add and subtract vectors. But if the *domain* of f is a vector space (i.e., x and a are vectors), the derivative as written is meaningless, since it requires division by a vector. Later on we will see that defining the derivative for functions $f : \mathbf{R} \to \mathbf{R}^2$ requires no alterations, whereas for functions $f : \mathbf{R}^2 \to \mathbf{R}$ we must restructure the definition of the derivative.

Vectors in $\mathbf{R}^2$ are often pictured using arrows in the plane. An arrow from the point (a, b) to (c, d) represents the vector $\vec{\mathbf{v}} = (c - a, d - b)$. Thus to draw an arbitrary vector $\vec{\mathbf{x}} = (x_1, x_2)$ we may start at any initial point (a, b) and end at the point $(a + x_1, b + x_2)$. In this way $\vec{\mathbf{x}}$ can be represented by any of a multitude of parallel arrows beginning at different initial points. Frequently it is drawn as an arrow from the origin to the point (x_1, x_2). Addition of vectors can be carried out geometrically. If $\vec{\mathbf{x}} = (x_1, x_2)$ is drawn with initial point at the origin and $\vec{\mathbf{y}} = (y_1, y_2)$ is drawn with intial point at (x_1, x_2), then the terminal point of $\vec{\mathbf{y}}$ will be $(x_1 + y_1, x_2 + y_2)$, which is precisely the sum of $\vec{\mathbf{x}} + \vec{\mathbf{y}}$. Hence the sum is represented by the arrow from the initial point of $\vec{\mathbf{x}}$ to the terminal point of $\vec{\mathbf{y}}$ (see Figure 5.1.1).

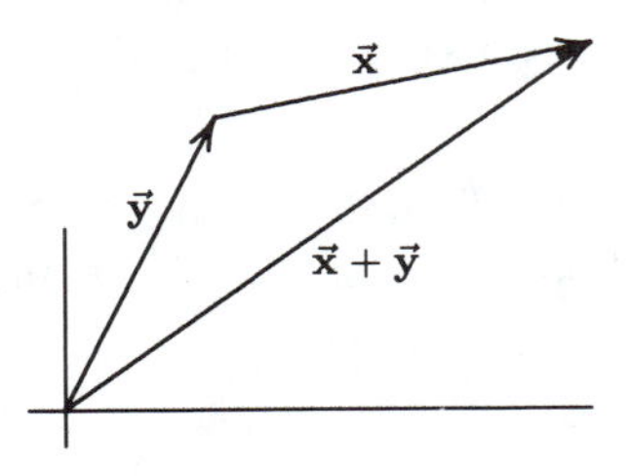

Figure 5.1.1

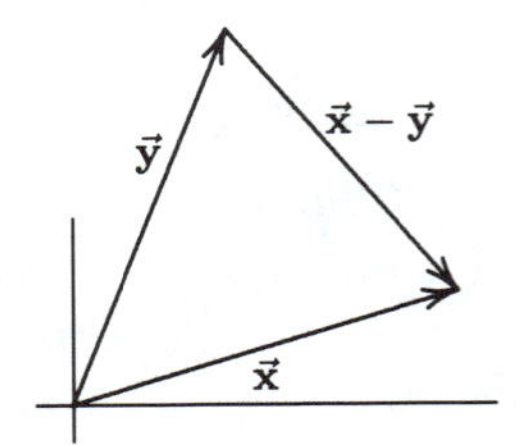

Figure 5.1.2

In a similar manner scalar multipication and subtraction of vectors may be represented geometrically. If $\vec{\mathbf{x}}$ and $\vec{\mathbf{y}}$ share a common initial point, the difference $\vec{\mathbf{x}} - \vec{\mathbf{y}}$ is given by an arrow beginning at the tip of $\vec{\mathbf{y}}$ and ending at the tip of $\vec{\mathbf{x}}$ (Figure 5.1.2). Scalar multiplication by a number k has the effect of stretching or shrinking the vector by a factor of k. If $k < 0$, the direction of the arrow is reversed.

5.1.2 The Dot Product

The measure of distance in $\mathbf{R}^2$ comes directly from the dot product.

DEFINITION 5.1.2 *The **dot product** of $\vec{\mathbf{x}} = (x_1, x_2)$ and $\vec{\mathbf{y}} = (y_1, y_2)$ is defined by*

$$\vec{\mathbf{x}} \cdot \vec{\mathbf{y}} \equiv x_1 y_1 + x_2 y_2 \ .$$

*The **length** of the vector $\vec{\mathbf{x}}$ is defined to be*

$$|\vec{\mathbf{x}}| \equiv \sqrt{\vec{\mathbf{x}} \cdot \vec{\mathbf{x}}} = \sqrt{x_1^2 + x_2^2} \ .$$

*The **distance** between $\vec{\mathbf{x}}$ and $\vec{\mathbf{y}}$ is given by*

$$|\vec{\mathbf{x}} - \vec{\mathbf{y}}| = \sqrt{(\vec{\mathbf{x}} - \vec{\mathbf{y}}) \cdot (\vec{\mathbf{x}} - \vec{\mathbf{y}})} = \sqrt{(x_1 - y_1)^2 + (x_2 - y_2)^2} \ .$$

This definition for the dot product in $\mathbf{R}^2$ leads to the familiar Pythagorean distance formula in the plane. However, the concept of a dot product is frequently defined in more general terms. A **dot product** or **inner product** on a vector space is a function that associates a real number $\vec{\mathbf{x}} \cdot \vec{\mathbf{y}}$ with each pair of vectors $\vec{\mathbf{x}}$ and $\vec{\mathbf{y}}$, and that satisfies the following four axioms:

i.	$\vec{\mathbf{x}} \cdot \vec{\mathbf{x}} \geq 0$ and $\vec{\mathbf{x}} \cdot \vec{\mathbf{x}} = 0 \iff \vec{\mathbf{x}} = \vec{0}$	positivity
ii.	$\vec{\mathbf{x}} \cdot \vec{\mathbf{y}} = \vec{\mathbf{y}} \cdot \vec{\mathbf{x}}$	symmetry
iii.	$(\vec{\mathbf{x}} + \vec{\mathbf{y}}) \cdot \vec{\mathbf{z}} = (\vec{\mathbf{x}} \cdot \vec{\mathbf{z}}) + (\vec{\mathbf{y}} \cdot \vec{\mathbf{z}})$	distributivity
iv.	$k(\vec{\mathbf{x}} \cdot \vec{\mathbf{y}}) = (k\vec{\mathbf{x}}) \cdot \vec{\mathbf{y}}$ for $k \in \mathbf{R}$	homogeneity

In problem 5 you are asked to verify that the dot product in Definition 5.1.2 satisfies each of these properties.

Using the four axioms for a dot product we can show that the distance measure given by $|\vec{\mathbf{x}} - \vec{\mathbf{y}}| = \sqrt{(\vec{\mathbf{x}} - \vec{\mathbf{y}}) \cdot (\vec{\mathbf{x}} - \vec{\mathbf{y}})}$ satisfies the three axioms for a **metric** (see Section 3 of Chapter 1):

1.	$|\vec{\mathbf{x}} - \vec{\mathbf{y}}| \geq 0$ with $|\vec{\mathbf{x}} - \vec{\mathbf{y}}| = 0 \iff \vec{\mathbf{x}} = \vec{\mathbf{y}}$	positivity
2.	$|\vec{\mathbf{x}} - \vec{\mathbf{y}}| = |\vec{\mathbf{y}} - \vec{\mathbf{x}}|$	symmetry
3.	$|\vec{\mathbf{x}} - \vec{\mathbf{z}}| \leq |\vec{\mathbf{x}} - \vec{\mathbf{y}}| + |\vec{\mathbf{y}} - \vec{\mathbf{z}}|$	triangle inequality

The first two of these metric properties follow from the positivity and homogeneity of dot products. The last property, the triangle inequality, is the real workhorse property for limit operations (recall, for example, the proofs in Section 3 of Chapter 2). The triangle inequality is difficult to prove directly but follows easily from another important inequality, which we now present.

THEOREM 5.1.3 ***(The Cauchy-Schwarz Inequality)*** *For any dot product on a vector space we have*

$$|\vec{\mathbf{x}} \cdot \vec{\mathbf{y}}| \leq |\vec{\mathbf{x}}||\vec{\mathbf{y}}|.$$

Futhermore, equality holds if and only if $\vec{\mathbf{y}}$ is a multiple of $\vec{\mathbf{x}}$.

PROOF This proof rests on a fact about quadratic equations: If $At^2+Bt+C \geq 0$ for all t, the graph of $y = At^2 + Bt + C$ does not cross the t-axis. Hence the expression $At^2 + Bt + C$ cannot have two distinct real roots, and its discriminant, $B^2 - 4AC$, must be less than or equal to 0 (if the discriminant were positive there would be two distinct real roots).

Now fix $\vec{\mathbf{x}}$ and $\vec{\mathbf{y}}$ and let $A = (\vec{\mathbf{x}} \cdot \vec{\mathbf{x}}) = |\vec{\mathbf{x}}|^2$, $B = (\vec{\mathbf{x}} \cdot \vec{\mathbf{y}})$, and $C = (\vec{\mathbf{y}} \cdot \vec{\mathbf{y}}) = |\vec{\mathbf{y}}|^2$. Form vectors $t\vec{\mathbf{x}} + \vec{\mathbf{y}}$ using a variable scalar t. For *every* value of t the length squared of this vector is nonnegative. Using the distributive property of dot products we have

$$\begin{aligned} 0 \leq |t\vec{\mathbf{x}} + \vec{\mathbf{y}}|^2 &= (t\vec{\mathbf{x}} + \vec{\mathbf{y}}) \cdot (t\vec{\mathbf{x}} + \vec{\mathbf{y}}) \\ &= t^2(\vec{\mathbf{x}} \cdot \vec{\mathbf{x}}) + 2t(\vec{\mathbf{x}} \cdot \vec{\mathbf{y}}) + (\vec{\mathbf{y}} \cdot \vec{\mathbf{y}}) \\ &= At^2 + 2Bt + C. \end{aligned}$$

Since this quadratic is nonnegative its discriminant is less than or equal to 0.

$$\begin{aligned} 4B^2 - 4AC \leq 0 &\Longrightarrow B^2 \leq AC \\ &\Longrightarrow (\vec{\mathbf{x}} \cdot \vec{\mathbf{y}})^2 \leq |\vec{\mathbf{x}}|^2|\vec{\mathbf{y}}|^2 \\ &\Longrightarrow |\vec{\mathbf{x}} \cdot \vec{\mathbf{y}}| \leq |\vec{\mathbf{x}}||\vec{\mathbf{y}}|. \end{aligned}$$

This proves the Cauchy-Schwarz inequality. Finally, notice that equality holds only if the discriminant is 0, in which case $At^2 + 2Bt + C$ will have two equal roots at $t_0 = -\frac{B}{A}$. This implies that $|t_0\vec{\mathbf{x}} + \vec{\mathbf{y}}| = 0$, or equivalently that $t_0\vec{\mathbf{x}} + \vec{\mathbf{y}} = \vec{\mathbf{0}}$. Hence we have $\vec{\mathbf{y}} = -t_0\vec{\mathbf{x}}$, that is, $\vec{\mathbf{y}}$ is a *multiple* of $\vec{\mathbf{x}}$. ∎

The triangle inequality is now straightforward to prove, as is shown in the exercises. We state the most common version from which the third metric property follows easily.

THEOREM 5.1.4 ***(The Triangle Inequality)*** *For any vectors $\vec{\mathbf{x}}$ and $\vec{\mathbf{y}}$*

$$|\vec{\mathbf{x}} + \vec{\mathbf{y}}| \leq |\vec{\mathbf{x}}| + |\vec{\mathbf{y}}|.$$

The proofs of Theorems 5.1.3 and 5.1.4 rely only on the axioms for dot products. Consequently these theorems hold not only for our particular dot product, $(\vec{\mathbf{x}}\cdot\vec{\mathbf{y}}) = x_1y_1+x_2y_2$, but for *any* vector space with a dot product. This enables us to define a valid distance measure in any vector space with a dot product by the formula $|\vec{\mathbf{x}} - \vec{\mathbf{y}}| = \sqrt{(\vec{\mathbf{x}} - \vec{\mathbf{y}}) \cdot (\vec{\mathbf{x}} - \vec{\mathbf{y}})}$.

Now let's investigate the geometric meaning of the dot product given in Definition 5.1.2.

EXAMPLE 5.1.1 Let $\vec{\mathbf{x}} = (x_1, x_2)$ and $\vec{\mathbf{y}} = (y_1, y_2)$ be nonzero vectors. Show that the dot product is given by

$$\vec{\mathbf{x}} \cdot \vec{\mathbf{y}} = |\vec{\mathbf{x}}||\vec{\mathbf{y}}| \cos\theta,$$

where θ is the angle between the vectors $\vec{\mathbf{x}}$ and $\vec{\mathbf{y}}$.

SOLUTION To verify this look at Figure 5.1.3 and observe that the angle θ is the difference between the angle β of the vector $\vec{\mathbf{y}}$ and the angle α of the vector $\vec{\mathbf{x}}$. We can compute $\cos\theta$ from the cosines and sines of α and β using the formula for the cosine of the difference of two angles:

$$\cos\theta = \cos(\beta - \alpha) = \cos\beta\cos\alpha + \sin\beta\sin\alpha.$$

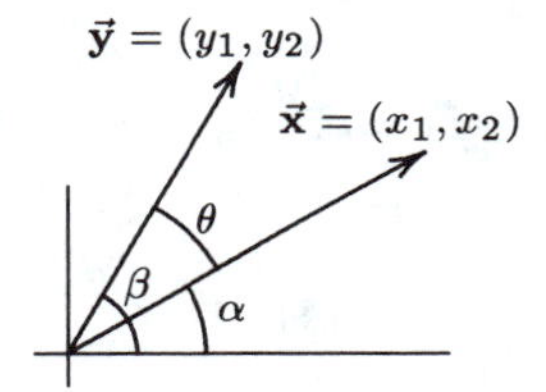

Figure 5.1.3

From our diagram: $\cos\beta = \frac{y_1}{|\vec{\mathbf{y}}|}$, $\sin\beta = \frac{y_2}{|\vec{\mathbf{y}}|}$, $\cos\alpha = \frac{x_1}{|\vec{\mathbf{x}}|}$, $\sin\alpha = \frac{x_2}{|\vec{\mathbf{x}}|}$. Substituting these values in the preceding equation gives

$$\cos\theta = \frac{1}{|\vec{\mathbf{x}}||\vec{\mathbf{y}}|}(x_1y_1 + x_2y_2).$$

Since $x_1y_1 + x_2y_2$ is precisely the dot product of $\vec{\mathbf{x}}$ and $\vec{\mathbf{y}}$ we may write

$$\cos\theta = \frac{\vec{\mathbf{x}} \cdot \vec{\mathbf{y}}}{|\vec{\mathbf{x}}||\vec{\mathbf{y}}|} \quad \text{or} \quad |\vec{\mathbf{x}}||\vec{\mathbf{y}}|\cos\theta = \vec{\mathbf{x}} \cdot \vec{\mathbf{y}}.$$

Notice further that since $|\cos\theta| \leq 1$, we have $|\vec{\mathbf{x}} \cdot \vec{\mathbf{y}}| \leq |\vec{\mathbf{x}}||\vec{\mathbf{y}}|$. This verifies the Cauchy-Schwarz inequality for our dot product. □

Thus in $\mathbf{R}^2$ the expression $\frac{\vec{\mathbf{x}} \cdot \vec{\mathbf{y}}}{|\vec{\mathbf{x}}||\vec{\mathbf{y}}|}$ is the cosine of the angle between $\vec{\mathbf{x}}$ and $\vec{\mathbf{y}}$. But in any vector space with a dot product the Cauchy-Schwarz inequality guarantees that $\frac{\vec{\mathbf{x}} \cdot \vec{\mathbf{y}}}{|\vec{\mathbf{x}}||\vec{\mathbf{y}}|}$ is a number between -1 and $+1$; hence it represents the cosine of some angle θ. We use this fact to extend the notion of angle to any vector space with a dot product as follows: the angle between the vectors $\vec{\mathbf{x}}$ and $\vec{\mathbf{y}}$ is *defined* to be the angle whose cosine equals $\frac{\vec{\mathbf{x}}\cdot\vec{\mathbf{y}}}{|\vec{\mathbf{x}}||\vec{\mathbf{y}}|}$. Thus a dot product on a vector space provides a way of measuring angles as well as distance. In a sense the dot product forms the foundation for the geometry of the space.

The measurement of angles in a vector space enables us to introduce the important concept of orthogonality. Two vectors $\vec{\mathbf{x}}$ and $\vec{\mathbf{y}}$ are **orthogonal** if and only if $\vec{\mathbf{x}}\cdot\vec{\mathbf{y}} = 0$. Here is an illustration of how the dot product may be used to prove a typical geometric fact.

EXAMPLE 5.1.2 Show that the angle inscribed in a semicircle is a right angle.

SOLUTION To put this in vector notation center the circle at the origin and let $\vec{\mathbf{x}}$ be a radius of the circle, so that $-\vec{\mathbf{x}}$ and $\vec{\mathbf{x}}$ form a diameter. Then if $\vec{\mathbf{y}}$ is any vector from the origin to a point on the circle, we have $|\vec{\mathbf{x}}| = |\vec{\mathbf{y}}|$, and the vectors $\vec{\mathbf{x}}+\vec{\mathbf{y}}$ and $\vec{\mathbf{x}}-\vec{\mathbf{y}}$ form the two sides of an angle inscribed in a semicircle (see Figure 5.1.4). To show that this is a right angle we must show that $\vec{\mathbf{x}}-\vec{\mathbf{y}}$ and $\vec{\mathbf{x}}+\vec{\mathbf{y}}$ are orthogonal, or equivalently that their dot product is 0. But

$$\begin{aligned}(\vec{\mathbf{x}}+\vec{\mathbf{y}})\cdot(\vec{\mathbf{x}}-\vec{\mathbf{y}}) &= \vec{\mathbf{x}}\cdot(\vec{\mathbf{x}}-\vec{\mathbf{y}})+\vec{\mathbf{y}}\cdot(\vec{\mathbf{x}}-\vec{\mathbf{y}})\\ &= \vec{\mathbf{x}}\cdot\vec{\mathbf{x}}-\vec{\mathbf{x}}\cdot\vec{\mathbf{y}}+\vec{\mathbf{y}}\cdot\vec{\mathbf{x}}-\vec{\mathbf{y}}\cdot\vec{\mathbf{y}}\\ &= |\vec{\mathbf{x}}|^2-|\vec{\mathbf{y}}|^2\\ &= 0,\end{aligned}$$

since the lengths $|\vec{\mathbf{x}}|$ and $|\vec{\mathbf{y}}|$ are equal. ☐

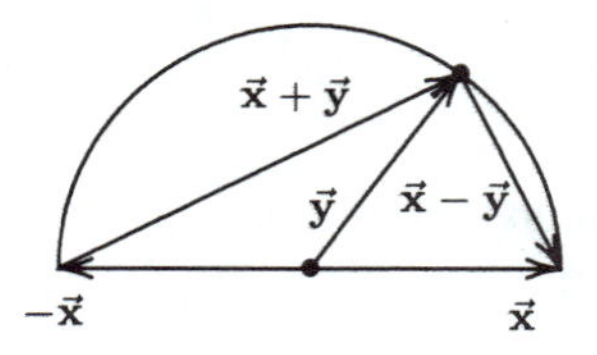

Figure 5.1.4

5.1.3 Open Sets in $\mathbf{R}^2$

In our earlier work with limits, the set of points in an open δ-interval about a given point x_0 (i.e., $\{x : |x - x_0| < \delta\}$) played a central role. In vector spaces the analogous sets, $\{\vec{\mathbf{x}} : |\vec{\mathbf{x}} - \vec{\mathbf{x}}_0| < \delta\}$, are called **open δ-balls** or **open neighborhoods** around a point $\vec{\mathbf{x}}_0$. In $\mathbf{R}^2$ this set is the open disk of radius δ centered at $\vec{\mathbf{x}}_0$ (see Figure 5.1.5). Notice that a δ-ball permits "elbow room" in all directions in the sense that if $|\vec{\mathbf{y}}| < \delta$, then $\vec{\mathbf{x}}_0 + \vec{\mathbf{y}}$ is in the δ-ball about $\vec{\mathbf{x}}_0$ (Figure 5.1.6).

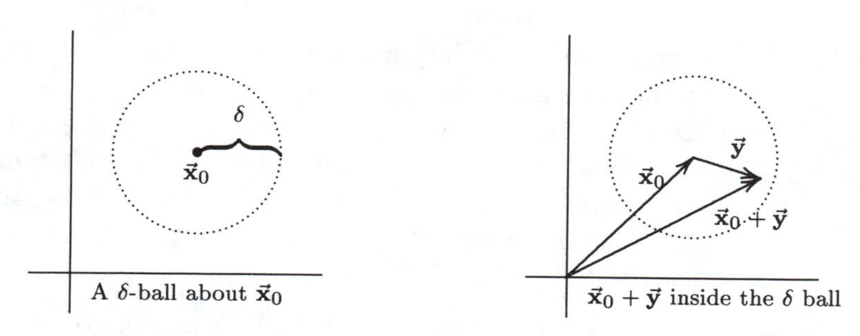

Figure 5.1.5 **Figure 5.1.6**

DEFINITION 5.1.5 *Let $\mathcal{D}$ be a set in $\mathbf{R}^2$. A point $\vec{\mathbf{x}}$ is called an* ***interior point*** *of $\mathcal{D}$ if for some real number $\delta > 0$ the entire δ-ball $\{\vec{\mathbf{z}} : |\vec{\mathbf{z}} - \vec{\mathbf{x}}| < \delta\}$ about $\vec{\mathbf{x}}$ is contained in $\mathcal{D}$. If all points in $\mathcal{D}$ are interior points, the set is called* ***open****.*

A point $\vec{\mathbf{x}}$ is a ***boundary point*** *for the set $\mathcal{D}$ if every δ-ball about $\vec{\mathbf{x}}$ contains some point in $\mathcal{D}$ and some point in the complement, $\mathcal{D}^c$.*

A set is ***closed*** *if it contains all of its boundary points.*

A set is open in $\mathbf{R}^2$ if around each point in the set it is possible to find a δ-ball that is entirely contained in the set. Notice that if one δ-ball is entirely contained in a set, any smaller one is also in the set. At boundary points of a set, any δ-ball, no matter how small, will overlap both inside and outside the set. The boundary points of a set do not necessarily belong to the set, as is illustrated in Figure 5.1.7 where the boundary of $\mathcal{S}$ consists of all points on the unit circle but $\mathcal{S}$ itself contains only the upper semicircle. This set is neither open nor closed, but if we add in the rest of the boundary points the set becomes closed (Figure 5.1.8). From the definition we see that a set is closed if it contains *all* its boundary points, whereas an open set cannot contain *any* of its boundary points. (Why?) There is a certain symmetry between open and closed sets. For example, a set is open if and only if its complement is closed (see problem 5.1.14).

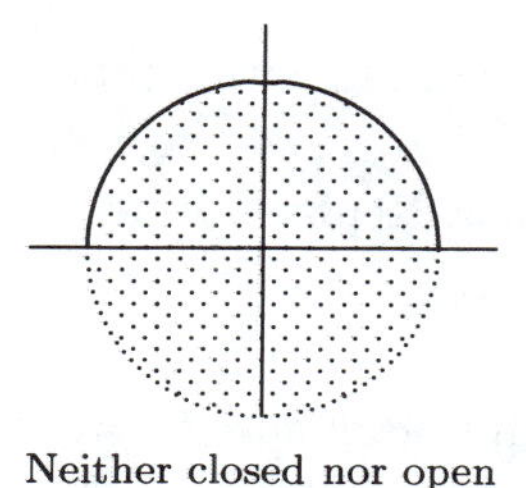

Neither closed nor open

Figure 5.1.7

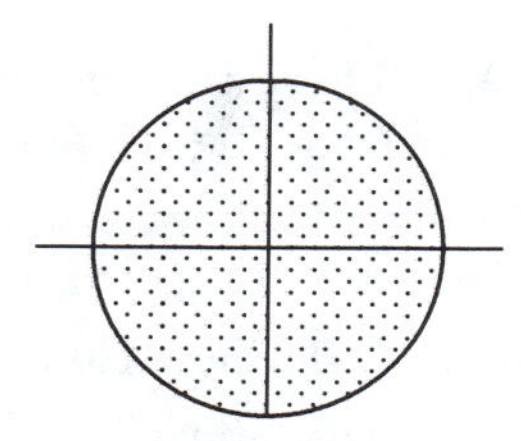

A closed set

Figure 5.1.8

Perhaps the most important result we have seen about closed intervals on the real line is the Heine-Borel theorem. It states that if a *closed, bounded* interval is covered by open intervals, there are a finite number of these open intervals that cover $[a, b]$. This theorem was central to the concept of uniform continuity and enabled us to prove fundamental results in one-dimensional calculus. This theorem can be extended to closed, bounded sets in $\mathbf{R}^2$.

A set $\mathcal{D}$ in $\mathbf{R}^2$ is **bounded** if there is a positive real number M such that for all $\vec{\mathbf{x}} \in \mathcal{D}$ we have $|\vec{\mathbf{x}}| < M$. Intuitively, a set in $\mathbf{R}^2$ is bounded if it can be contained in some large disk.

THEOREM 5.1.6 ***(The Heine-Borel Theorem)*** *Let $\mathcal{D}$ be a closed and bounded set and let $\{O_\alpha\}$ be an open cover for $\mathcal{D}$, that is, $\mathcal{D}$ is contained in $\cup_\alpha O_\alpha$ and each O_α is an open set in $\mathbf{R}^2$. Then there is a finite subset of the collection $\{O_\alpha\}$ that covers $\mathcal{D}$.*

Proofs of this theorem, and of the closely related Bolzano-Weierstrass theorem, are worked out in the exercises at the end of this section. Using the Heine-Borel theorem the concept of uniform continuity and related calculus results can be generalized to $\mathbf{R}^2$. As you may guess, this theorem holds generally in $\mathbf{R}^n$. In fact, the "finite subcover" property, called *compactness*, is a key concept in the study of more abstract topological spaces.

Problems for Section 5.1

•5.1.1. Prove the triangle inequality (Theorem 5.1.4) using the Cauchy-Schwarz inequality. (Hint: First recall from Definition 5.1.2 how the length of a vector is defined in terms of the dot product. Then use the Cauchy-Schwarz inequality to show that $|\vec{\mathbf{x}} + \vec{\mathbf{y}}|^2 \leq (|\vec{\mathbf{x}}| + |\vec{\mathbf{y}}|)^2$. Finally take the positive square root of both sides.)

5.1.2. The triangle inequality as stated in property (3) for a metric says that for any three vectors $\vec{\mathbf{x}}$, $\vec{\mathbf{y}}$, and $\vec{\mathbf{z}}$, we must have $|\vec{\mathbf{x}} - \vec{\mathbf{z}}| \leq |\vec{\mathbf{x}} - \vec{\mathbf{y}}| + |\vec{\mathbf{y}} - \vec{\mathbf{z}}|$. Show that this follows directly from Theorem 5.1.4. (Hint: Let $\vec{\mathbf{x}} - \vec{\mathbf{y}}$ play the role of $\vec{\mathbf{x}}$ and $\vec{\mathbf{y}} - \vec{\mathbf{z}}$ play the role of $\vec{\mathbf{y}}$.)

•5.1.3. There are several useful inequalities concerning norms that are derived directly from the triangle inequality, Theorem 5.1.4.

a) Show that $|\vec{\mathbf{x}}| - |\vec{\mathbf{y}}| \leq |\vec{\mathbf{x}} - \vec{\mathbf{y}}|$. (Hint: Replace $|\vec{\mathbf{x}}|$ by the vector $|\vec{\mathbf{x}} - \vec{\mathbf{y}}|$ in Theorem 5.1.4.)

b) Show that $||\vec{\mathbf{x}}| - |\vec{\mathbf{y}}|| \leq |\vec{\mathbf{x}} - \vec{\mathbf{y}}|$.

c) Show that $|\vec{\mathbf{x}}| - |\vec{\mathbf{y}}| \leq |\vec{\mathbf{x}} + \vec{\mathbf{y}}|$. (Hint: Replace $\vec{\mathbf{y}}$ by $-\vec{\mathbf{y}}$ in **(a)**.)

d) Show that $|\vec{\mathbf{x}} - \vec{\mathbf{y}}| \leq |\vec{\mathbf{x}}| + |\vec{\mathbf{y}}|$.

•5.1.4. Using the triangle inequality we can show that the length of a vector is greater than the length of any coordinate and is less than the sum of the lengths of its coordinates.

The Box Lemma. *Let* $\vec{\mathbf{x}} = (x_1, x_2) \in \mathbf{R}^2$*. Then* $|x_1| \leq |\vec{\mathbf{x}}| \leq |x_1| + |x_2|$*, and similarly,* $|x_2| \leq |\vec{\mathbf{x}}| \leq |x_1| + |x_2|$.

Prove the lemma by writing $\vec{\mathbf{x}}$ as $(x_1, 0) + (0, x_2)$ and using the fact that $|x| = \sqrt{x^2}$.

•5.1.5. Prove that the dot product in Definition 5.1.2 satisfies the four axioms for a dot product listed following Definition 5.1.2.

5.1.6. **a)** Using the properties of a dot product prove that for any scalar k, $|k\vec{\mathbf{x}}| = |k||\vec{\mathbf{x}}|$.

b) Suppose that we define a product in $\mathbf{R}^2$ by

$$(x_1, x_2) \odot (y_1, y_2) = x_1 y_2 + x_2 y_1.$$

Which properties of an inner product does this definition satisfy?

5.1.7. Let $C = \{f(x) \mid f(x) \text{ is continuous on } [0,1]\}$ be the vector space of continuous functions on $[0, 1]$. We define a dot product of two vectors in C by $f \cdot g = \int_0^1 f(x)g(x)\, dx$.

a) Verify that $f \cdot g$ satisfies the four properties of a dot product.

b) Find the length of the vector $f(x) = x$.

c) Find the value of a such that the vector $g(x) = a + x$ is orthogonal to $f(x)$.

5.1.8. The dot product on a vector space determines the geometry of the space. Thus it is not suprising that some theorems of Euclidean geometry can be proved using the (Euclidean) dot product. Prove the following two facts:

a) The sum of the squares of the lengths of the diagonals of a parallelogram equals the sum of the squares of the lengths of the four sides. (Hint: Let $\vec{\mathbf{x}}$ and $\vec{\mathbf{y}}$ be the two sides. Then the diagonals are $\vec{\mathbf{x}} + \vec{\mathbf{y}}$ and $\vec{\mathbf{x}} - \vec{\mathbf{y}}$. Illustrate this with a picture.)

b) If the diagonals of a parallelogram are of equal length, the parallelogram is a rectangle. (Hint: Show that if $|\vec{\mathbf{x}} + \vec{\mathbf{y}}| = |\vec{\mathbf{x}} - \vec{\mathbf{y}}|$, then $\vec{\mathbf{x}}$ and $\vec{\mathbf{y}}$ must be orthogonal.)

5.1.9. Use the definition of dot product and Example 5.1.1 to prove the **law of cosines**:

$$|\vec{\mathbf{x}} - \vec{\mathbf{y}}|^2 = |\vec{\mathbf{x}}|^2 + |\vec{\mathbf{y}}|^2 - 2|\vec{\mathbf{x}}||\vec{\mathbf{y}}| \cos\theta.$$

5.1.10. A common algebra mistake is to say that $a + b = \sqrt{a^2 + b^2}$. Apply the Cauchy-Schwarz inequality to the vectors (a, b) and $(1, 1)$ to obtain a correct relationship between these quantities.

•5.1.11. Let $\vec{\mathbf{u}}$ be a **unit vector**, that is, a vector of length 1. In this problem we will express an arbitrary vector $\vec{\mathbf{x}}$ as a sum of two *orthogonal* vectors. The first part is the vector $(\vec{\mathbf{x}} \cdot \vec{\mathbf{u}})\vec{\mathbf{u}}$, which points in the direction of $\vec{\mathbf{u}}$ and is called **the component of a vector $\vec{\mathbf{x}}$ in the direction $\vec{\mathbf{u}}$.**

a) Show that the vector $\vec{\mathbf{x}} - (\vec{\mathbf{x}} \cdot \vec{\mathbf{u}})\vec{\mathbf{u}}$ is orthogonal to $\vec{\mathbf{u}}$.

b) Conclude that the vector in **(a)** is orthogonal to $(\vec{\mathbf{x}} \cdot \vec{\mathbf{u}})\vec{\mathbf{u}}$ and express $\vec{\mathbf{x}}$ as the sum of two orthogonal pieces.

c) Suppose that $\vec{\mathbf{u}} = (3/5, 4/5)$, then express $\vec{\mathbf{x}} = (3, -4)$ as the sum of two orthogonal components, one of which points in the direction $\vec{\mathbf{u}}$. Illustrate with a diagram.

5.1.12. Suppose that $|\vec{\mathbf{x}} + a\vec{\mathbf{y}}| \geq |\vec{\mathbf{x}}|$ for all $a \in \mathbf{R}$.

a) Show that for every a we must have $a(2\vec{\mathbf{x}} \cdot \vec{\mathbf{y}} + a\vec{\mathbf{y}} \cdot \vec{\mathbf{y}}) \geq 0$. (Hint: Square both sides of $|\vec{\mathbf{x}} + a\vec{\mathbf{y}}| \geq |\vec{\mathbf{x}}|$ and expand as a dot product.)

b) Suppose that $\vec{\mathbf{x}} \cdot \vec{\mathbf{y}} < 0$. Show that this violates the inequality $a(2\vec{\mathbf{x}} \cdot \vec{\mathbf{y}} + a\vec{\mathbf{y}} \cdot \vec{\mathbf{y}}) \geq 0$ by considering the sign of both terms in the product when a is very small and positive.

c) Similarly show that $\vec{\mathbf{x}} \cdot \vec{\mathbf{y}}$ cannot be positive. What does this mean about the vectors $\vec{\mathbf{x}}$ and $\vec{\mathbf{y}}$?

5.1.13. Determine whether each of the following sets is open, closed, or neither. A graphic illustration is sufficient.

a) $\mathcal{D}$ is the set of points $(x, y) \in \mathbf{R}^2$ such that $xy > 0$.

b) $\mathcal{D}$ is the set of points $(x, y) \in \mathbf{R}^2$ such that $x^2 y \geq 0$.

c) $\mathcal{D}$ is the set of points $(x, y) \in \mathbf{R}^2$ such that $x^2 - y^2 \leq 0$.

d) $\mathcal{D}$ is the set of points $(x, y) \in \mathbf{R}^2$ such that $x^{1/2} y^{-1/2} > 0$.

5.1.14. For this problem you will need to review Definition 5.1.5. Recall that the **complement** of a set $\mathcal{D}$ in $\mathbf{R}^2$ is denoted by $\mathcal{D}^c$ and consists of all the points in $\mathbf{R}^2$ that are not in $\mathcal{D}$. Prove that a set $\mathcal{D}$ is closed $\iff$ its complement, $\mathcal{D}^c$, is open.

a) Begin by showing that the boundary of any set $\mathcal{D}$ (not necessarily closed) is the same as the boundary of its complement $\mathcal{D}^c$.

b) Now show that if $\mathcal{D}$ is closed, $\mathcal{D}^c$ is open.

c) Finally show that if $\mathcal{D}^c$ is open, $\mathcal{D}$ is closed.

⋆5.1.15. Prove the Bolzano-Weierstrass theorem for $\mathbf{R}^2$. Any infinite set of vectors $\{\vec{\mathbf{x}}_\alpha \mid \alpha \in A\}$ that is bounded has an accumulation point $\vec{\mathbf{x}}^*$, that is, a point $\vec{\mathbf{x}}^*$ such that for any $\epsilon > 0$ there is some α such that $0 < |\vec{\mathbf{x}}_\alpha - \vec{\mathbf{x}}^*| < \epsilon$.

a) Let $\vec{\mathbf{x}}_\alpha = (x_\alpha, y_\alpha)$ and show that the sets of first coordinates $\{x_\alpha \mid \alpha \in A\}$ and second coordinates $\{y_\alpha \mid \alpha \in A\}$ are both bounded, and that at least one is infinite. (Hint: If $|\vec{\mathbf{x}}_\alpha| < M$ for all $\alpha \in A$, what must be true about the x_α? If both sets of first and second coordinates are finite, how many different $\vec{\mathbf{x}}_a$ are possible?)

b) Assume without loss of generality that the set of first coordinates is infinite. Use the Bolzano-Weierstrass theorem (Theorem 1.4.4) to find an accumulation point x^* for $\{x_\alpha \mid \alpha \in A\}$.

c) Use the fact that x^* is an accumulation point for the first coordinates to find an infinite sequence of distinct vectors $B = \{(x_n, y_n) \mid n \in \mathbf{N}\}$ such that $0 < |x_n - x^*| < \frac{1}{2^n}$. (Hint: See Lemma 1.4.5 and pick a different x_α for each n.)

d) Suppose that there are only finitely many distinct y coordinates in the foregoing sequence (x_n, y_n) (How can this happen?). Argue that at least one of the y values, say y_k, must appear in infinitely many pairs (x_n, y_n). Show that (x^*, y^*) for $y^* = y_k$ is an accumulation point for the set (x_α, y_α) (Hint: Given $\epsilon > 0$ find an N such that $\frac{1}{2^N} < \epsilon$. Next find an $n > N$ such that (x_n, y^*) is in the set B (how do you know that such a point exists?) and compute its distance from (x^*, y^*).)

e) If the set $\{y_n\}$ is infinite, show that it must have an accumulation point y^* and show that $\vec{\mathbf{x}} = (x^*, y^*)$ is an accumulation point for A.

⋆5.1.16. This problem provides a brief sketch showing how to prove the Heine-Borel theorem using the Bolzano-Weierstrass theorem. Let $\mathcal{D}$ be a closed, bounded set in $\mathbf{R}^2$ with an open cover $\{O_\alpha\}$ and assume that there is no finite subcover. The proof is by contradiction using a sequence of points in $\mathcal{D}$ generated as follows:

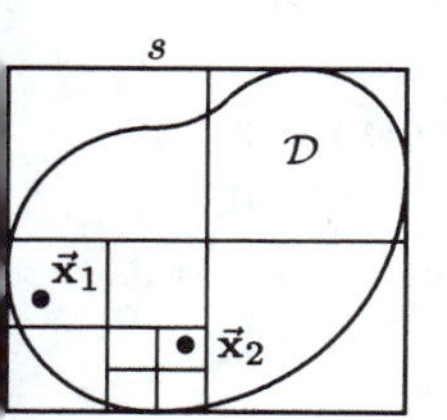

Figure 5.1.9

Since $\mathcal{D}$ is bounded, it may be enclosed in some large square of side s. Divide the square into four smaller squares of side $s/2$. At least one portion of $\mathcal{D}$ lying in one of these smaller squares must have no finite subcover from $\{O_\alpha\}$. (Why?) From such a subsquare choose a point $\vec{\mathbf{x}}_1 \in \mathcal{D}$. Now subdivide this smaller square into four squares of side $s/2^2$. One of these smaller squares must in turn have no finite subcover; from it choose a second point $\vec{\mathbf{x}}_2 \in \mathcal{D}$. Continue this process to generate a sequence of distinct points $\{\vec{\mathbf{x}}_n\} \in \mathcal{D}$ such that $\vec{\mathbf{x}}_n$ lies in a nested square of side $s/2^n$. By construction, the portion of $\mathcal{D}$ in each one of these nested squares has no finite subcover from $\{O_\alpha\}$.

a) Show that the set $\{\vec{\mathbf{x}}_n\}$ has an accumulation point $\vec{\mathbf{x}}^*$ which must lie inside $\mathcal{D}$. (Hint: Since $\mathcal{D}$ is closed, $\mathcal{D}^c$ is open. If $\vec{\mathbf{x}}^* \in \mathcal{D}^c$, there would be an open ball about $\vec{\mathbf{x}}^*$ in $\mathcal{D}^c$. How does this contradict the fact that $\vec{\mathbf{x}}^*$ is an accumulation point $\{\vec{\mathbf{x}}_n\}$?)

b) Since $\vec{\mathbf{x}}^* \in \mathcal{D}$, then $\vec{\mathbf{x}}^* \in O_{\alpha_i}$ for some α_i. (Why?) Thus there is a δ-ball about $\vec{\mathbf{x}}^*$ that is entirely contained in O_{α_i}. Show that this δ-ball (and consequently O_{α_i}) must entirely cover one of the subsquares which presumably had no finite subcover. (Hint: This requires a bit of geometry. Show first there exists an n such that $|\vec{\mathbf{x}}_n - \vec{\mathbf{x}}^*| < \delta/2$. Now show that if n is large enough so that $s/2^n < \delta/4$, then the entire square for $\vec{\mathbf{x}}_n$ is within δ of $\vec{\mathbf{x}}^*$. Draw a picture.)

5.2 Vector-Valued Functions

This section examines the calculus of functions $\vec{\mathbf{F}} : \mathbf{R} \to \mathbf{R}^2$ whose range is the vector space $\mathbf{R}^2$. For these *vector-valued functions* each real number, t, is mapped into an ordered pair of real numbers $(x(t), y(t))$. Therefore functions $\vec{\mathbf{F}} : \mathbf{R} \to \mathbf{R}^2$ are described by a pair of real-valued **coordinate functions**: $\vec{\mathbf{F}}(t) = (x(t), y(t))$.

Vector-valued functions are most frequently interpreted in terms of the motion of a point. If $\vec{\mathbf{F}}(t)$ represents the position of a particle in the plane at time t, then as time progresses $\vec{\mathbf{F}}(t)$ traces out the trajectory of the moving particle.

EXAMPLE 5.2.1 Sketch the trajectory of a particle whose position at time $t \in [-2, 2]$ is given by $\vec{\mathbf{F}}(t) = (t - t^3, t^2 - 1)$.

SOLUTION In Figure 5.2.1 several points of $\vec{\mathbf{F}}(t)$ are plotted for increasing values of t and connected by a smooth curve. The resulting trajectory shows the set of image points traced out by $\vec{\mathbf{F}}$ over time. A trajectory is not exactly a graph, since it does not show which points t in the domain correspond to which image points on the trajectory. (A complete graph would be three-dimensional.) In fact, the same image path can be traced out by many different functions moving at different speeds and directions along the path. For example, the function $\vec{\mathbf{G}}(t) = (t^3 - t, t^2 - 1)$ traces out the same points as $\vec{\mathbf{F}}$ but in the opposite direction. Functions that trace out the same image path are called **parameterizations** of the path. ☐

t	$\vec{\mathbf{F}}(t)$
-2	$(6, 3)$
-1	$(0, 0)$
$-\frac{1}{2}$	$(-\frac{3}{8}, -\frac{3}{4})$
0	$(0, -1)$
$\frac{1}{2}$	$(\frac{3}{8}, -\frac{3}{4})$
1	$(0, 0)$
2	$(6, 3)$

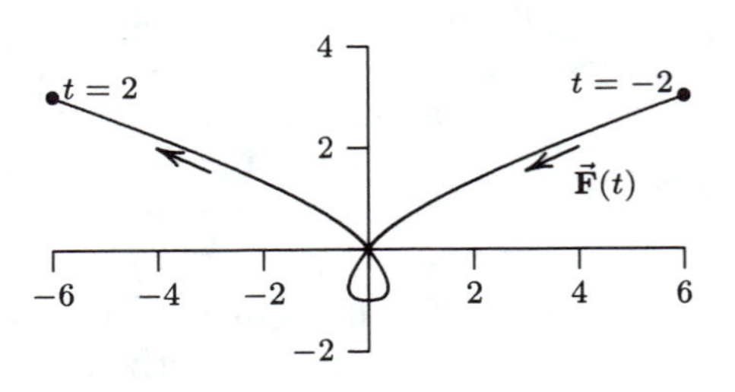

Figure 5.2.1

5.2.1 Limits and Derivatives for $\vec{\mathbf{F}}$

Whenever distances can be measured in both the domain and range of a function, the definition of the limit can be framed in the standard ϵ, δ manner. The Euclidean distance for $\mathbf{R}^2$ leads to the following limit definition:

DEFINITION 5.2.1 *Let $\vec{\mathbf{F}}(t) = (x(t), y(t))$ and $\vec{\mathbf{L}} = (L_1, L_2)$. Then $\lim_{t\to t_0} \vec{\mathbf{F}}(t) = \vec{\mathbf{L}}$ if for any $\epsilon > 0$ there is some $\delta > 0$ such that*

$$\text{if } 0 < |t - t_0| < \delta, \text{ then } |\vec{\mathbf{F}}(t) - \vec{\mathbf{L}}| = \sqrt{(x(t) - L_1)^2 + (y(t) - L_2)^2} < \epsilon.$$

You might guess that working with the square root in this definition could lead to problems. But the next theorem reduces the limit for a vector function to the individual limits of its coordinate functions and enables us to apply the entire limit theory developed for ordinary real-valued functions to limits of vector-valued functions.

THEOREM 5.2.2 *Let $\vec{\mathbf{F}}(t) = (x(t), y(t))$ and $\vec{\mathbf{L}} = (L_1, L_2)$. Then $\lim_{t\to t_0} \vec{\mathbf{F}}(t) = \vec{\mathbf{L}}$ if and only if $\lim_{t\to t_0} x(t) = L_1$ and $\lim_{t\to t_0} y(t) = L_2$.*

PROOF We prove this using the box lemma of problem 5.1.4. ($\Leftarrow$) Assume that $\lim_{t\to t_0} x(t) = L_1$ and $\lim_{t\to t_0} y(t) = L_2$. Then given any $\epsilon > 0$, there are positive numbers δ_1 and δ_2 such that if $0 < |t - t_0| < \delta_1$, then $|x(t) - L_1| < \epsilon/2$ and if $0 < |t - t_0| < \delta_2$, then $|y(t) - y(t_0)| < \epsilon/2$. Let $\delta = \min\{d_1, \delta_2\}$. If $0 < |t - t_0| < \delta$, then by the box lemma

$$|\vec{\mathbf{F}}(t) - \vec{\mathbf{L}}| \leq |x(t) - L_1| + |y(t) - L_2| < \epsilon/2 + \epsilon/2 = \epsilon,$$

so $\lim_{t\to t_0} \vec{\mathbf{F}}(t) = \vec{\mathbf{L}}$. (The proof of the converse is left as an exercise.) ∎

With this theorem all the limit definitions for the calculus of vector functions can be translated into the corresponding limits of the coordinate functions. For example, consider the case of continuity and differentiability.

DEFINITION 5.2.3

a) *$\vec{\mathbf{F}}$ is **continuous** at t_0 if and only if $\lim_{t\to t_0} \vec{\mathbf{F}}(t) = \vec{\mathbf{F}}(t_0)$.*

b) *The **derivative** of $\vec{\mathbf{F}}$ at t is $\vec{\mathbf{F}}'(t) = \lim_{h\to 0} \frac{1}{h}[\vec{\mathbf{F}}(t+h) - \vec{\mathbf{F}}(t)]$. If this limit exists the function is called **differentiable** at t.*

THEOREM 5.2.4

a) *$\vec{\mathbf{F}}(t) = (x(t), y(t))$ is continuous if and only if both $x(t)$ and $y(t)$ are continuous.*

b) *The derivative $\vec{\mathbf{F}}'(t)$ exists if and only if both $x'(t)$ and $y'(t)$ exist, in which case the derivative is given by $\vec{\mathbf{F}}'(t) = (x'(t), y'(t))$.*

The proof of the preceding theorem is immediate using Definition 5.2.3 and Theorem 5.2.2 (see the exercises).

The derivative $\vec{\mathbf{F}}'(t)$ is often called the **velocity vector** of the particle at time t. Let's intuitively see how this interpretation arises from the definition of $\vec{\mathbf{F}}'$. First we examine the direction of $\vec{\mathbf{F}}'$. The vector $\vec{\mathbf{F}}(t+h) - \vec{\mathbf{F}}(t)$ points between the two locations of the particle at time t and $t+h$ (see Figure 5.2.2). Assuming that the motion is smooth, then as $h \to 0$ the vector $\vec{\mathbf{F}}(t+h)$ gets close to $\vec{\mathbf{F}}(t)$ and the difference $(\vec{\mathbf{F}}(t+h) - \vec{\mathbf{F}}(t))$ visually approaches a tangent to the trajectory. Hence in the definition of $\vec{\mathbf{F}}'$ the scalar multiples $\frac{1}{h}[\vec{\mathbf{F}}(t+h) - \vec{\mathbf{F}}(t)]$ will also approach tangency. Notice that whether h is positive or negative the vector $\frac{1}{h}[\vec{\mathbf{F}}(t+h) - \vec{\mathbf{F}}(t)]$ always points in the forward direction of motion. Taking the limit we see that $\vec{\mathbf{F}}'(t) = (x'(t), y'(t))$ *represents a vector tangent to the trajectory, pointing in the direction of movement.*

What about the length of $\vec{\mathbf{F}}'$? When h is small the distance along the curve is nearly equal to the straight line distance given by the length of $\vec{\mathbf{F}}(t+h) - \vec{\mathbf{F}}(t)$; therefore the length of $\frac{1}{h}[\vec{\mathbf{F}}(t+h) - \vec{\mathbf{F}}(t)]$ approximates the average speed (distance traveled over elapsed time). In the limit *the length* $|\vec{\mathbf{F}}'(t)| = \sqrt{(x'(t))^2 + (y'(t))^2}$ *represents the instantaneous speed of the particle.* (We will see a more rigorous justification of this interpretation when we discuss arc length.)

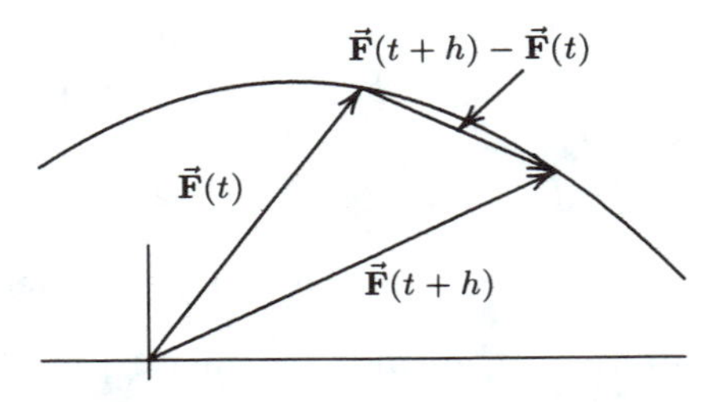

Figure 5.2.2

Example 5.2.2 Sketch the trajectories and the velocity vectors at the indicated points for each of these functions:

a) $\vec{\mathbf{K}}(t) = (\cos t, \sin t)$ for $t = 0, \frac{\pi}{2}$ **b)** $\vec{\mathbf{F}}(t) = (t - t^3, t^2 - 1)$ for $t = 0$

c) $\vec{\mathbf{G}}(t) = (t^3 - t, t^2 - 1)$ for $t = 0$

SOLUTION **a)** The function $\vec{\mathbf{K}}$ is the common parameterization of a circle. $\vec{\mathbf{K}}'(t) = (-\sin t, \cos t)$, so $\vec{\mathbf{K}}'(0) = (0, 1)$ and $\vec{\mathbf{K}}'(\frac{\pi}{2}) = (-1, 0)$. These vectors are shown in Figure 5.2.3

b, c) The function $\vec{\mathbf{F}}$ was sketched in Example 5.2.1. $\vec{\mathbf{F}}'(t) = (1 - 3t^2, 2t)$, so $\vec{\mathbf{F}}'(0) = (1, 0)$. The parameterization $\vec{\mathbf{G}}$ has the same image as $\vec{\mathbf{F}}$ in the plane, but it is traced out in the opposite direction. Notice that

$\vec{\mathbf{G}}'(0) = (-1, 0)$ is also tangent but points in the opposite direction from $\vec{\mathbf{F}}'(0)$, reflecting the difference in the direction of motion. (See Figure 5.2.4.) □

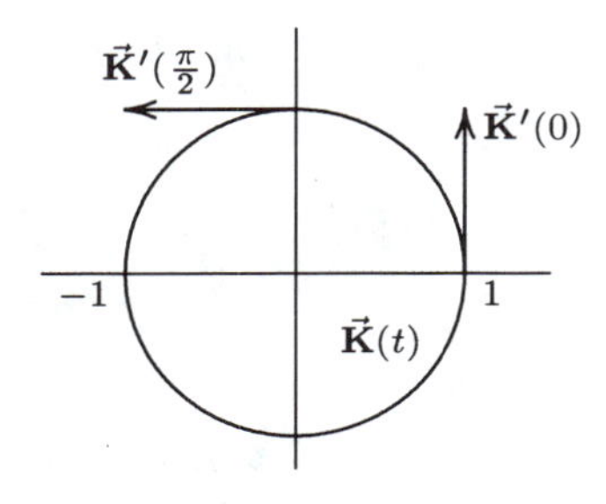

Figure 5.2.3

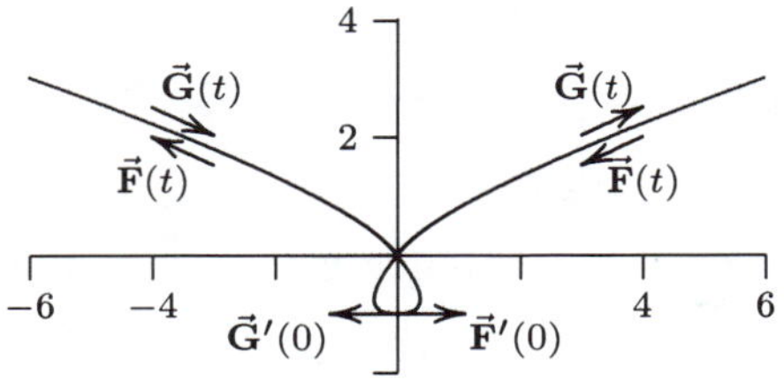

Figure 5.2.4

5.2.2 Differentiation Rules

The differentiation rules for vector functions follow directly from the standard differentiation rules applied to the coordinate functions.

THEOREM 5.2.5 *Let $\vec{\mathbf{F}}$ and $\vec{\mathbf{G}}$ be differentiable functions from $\mathbf{R}$ to $\mathbf{R}^2$ and let $h(t)$ be a differentiable real-valued (scalar) function. Then*

a) $\frac{d}{dt}(\vec{\mathbf{F}} + \vec{\mathbf{G}})(t) = \vec{\mathbf{F}}'(t) + \vec{\mathbf{G}}'(t)$;

b) $\frac{d}{dt}(\vec{\mathbf{F}} \cdot \vec{\mathbf{G}})(t) = \vec{\mathbf{F}}'(t) \cdot \vec{\mathbf{G}}(t) + \vec{\mathbf{F}}(t) \cdot \vec{\mathbf{G}}'(t)$;

c) $\frac{d}{dt}[h(t)\vec{\mathbf{F}}(t)] = h(t)\vec{\mathbf{F}}'(t) + h'(t)\vec{\mathbf{F}}(t)$;

d) $\frac{d}{dt}\vec{\mathbf{F}}(h(t)) = \vec{\mathbf{F}}'(h(t))h'(t)$.

PROOF We prove **(b)** and leave the others as exercises. Let $\vec{\mathbf{F}}(t) = (x(t), y(t))$ and $\vec{\mathbf{G}}(t) = (w(t), z(t))$. Then $(\vec{\mathbf{F}} \cdot \vec{\mathbf{G}})(t) = x(t)w(t) + y(t)z(t)$, and by the product rule for real-valued functions,

$$\begin{aligned}
\frac{d}{dt}(\vec{\mathbf{F}} \cdot \vec{\mathbf{G}})(t) &= \frac{d}{dt}[x(t)w(t) + y(t)z(t)] \\
&= x'(t)w(t) + x(t)w'(t) + y'(t)z(t) + y(t)z'(t) \\
&= x'(t)w(t) + y'(t)z(t) + x(t)w'(t) + y(t)z'(t) \\
&= (x'(t), y'(t)) \cdot (w(t), z(t)) + (x(t), y(t)) \cdot (w'(t), z'(t)) \\
&= \vec{\mathbf{F}}'(t) \cdot \vec{\mathbf{G}}(t) + \vec{\mathbf{F}}(t) \cdot \vec{\mathbf{G}}'(t).
\end{aligned}$$
■

The following corollary illustrates the relationship that exists between differentiation and the geometry of vector functions.

COROLLARY 5.2.6 *If $|\vec{\mathbf{F}}(t)|$ is constant, then $\vec{\mathbf{F}}'(t)$ is orthogonal to $\vec{\mathbf{F}}(t)$.*

PROOF Suppose that $|\vec{\mathbf{F}}(t)| = \sqrt{\vec{\mathbf{F}}(t) \cdot \vec{\mathbf{F}}(t)} = k$. Then $\vec{\mathbf{F}}(t) \cdot \vec{\mathbf{F}}(t) = k^2$ is also constant and so has a derivative equal to 0. Using **(b)** of Theorem 5.2.5 we have

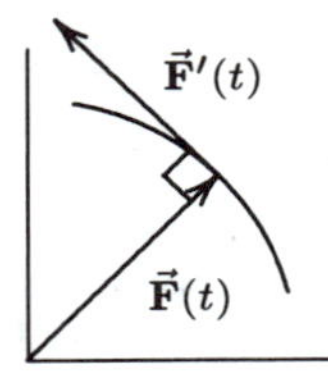

$|\vec{\mathbf{F}}(t)|$ constant.

Figure 5.2.5

$$0 = \frac{d}{dt}(\vec{\mathbf{F}} \cdot \vec{\mathbf{F}})(t) = 2(\vec{\mathbf{F}}'(t) \cdot \vec{\mathbf{F}}(t)).$$

Hence $\vec{\mathbf{F}}'(t) \cdot \vec{\mathbf{F}}(t) = 0$. ∎

The length of $\vec{\mathbf{F}}(t)$ measures the distance from the origin. If the length is constant, the particle moves on a circle with radius $|\vec{\mathbf{F}}|$. This corollary says that under these circumstances the velocity vector $\vec{\mathbf{F}}'(t)$ will be perpendicular to the position vector $\vec{\mathbf{F}}(t)$.

Corollary 5.2.6 has important consequences for the study of acceleration. The **acceleration vector**, denoted $\vec{\mathbf{a}}(t)$, is the derivative of the velocity vector:

$$\vec{\mathbf{a}}(t) = \frac{d}{dt}\vec{\mathbf{F}}'(t) = \lim_{h \to 0} \frac{1}{h}[\vec{\mathbf{F}}'(t+h) - \vec{\mathbf{F}}'(t)].$$

This vector describes the direction and rate of change in the velocity vector. Since $\vec{\mathbf{F}}'(t) : \mathbf{R} \to \mathbf{R}^2$, Theorem 5.2.4 shows that the next derivative, $\vec{\mathbf{F}}''$, is also computed componentwise; hence $\vec{\mathbf{a}}(t) = \vec{\mathbf{F}}''(t) = (x''(t), y''(t))$. But acceleration may also be computed in a different way, which reveals the geometry involved. Assuming $\vec{\mathbf{F}}' \neq \vec{\mathbf{0}}$, write the velocity vector as

$$\vec{\mathbf{F}}'(t) = v(t)\vec{\mathbf{t}}(t),$$

where the scalar function $v(t) = |\vec{\mathbf{F}}'(t)|$ is the speed of the particle and $\vec{\mathbf{t}}(t) = \frac{\vec{\mathbf{F}}'(t)}{|\vec{\mathbf{F}}'(t)|}$ is the **unit tangent vector** to the trajectory. To find the acceleration vector of $\vec{\mathbf{F}}$ using this representation, we apply **(c)** of Theorem 5.2.5:

$$\vec{\mathbf{a}}(t) = \frac{d}{dt}\vec{\mathbf{F}}'(t) = \frac{d}{dt}v(t)\vec{\mathbf{t}}(t) = v'(t)\vec{\mathbf{t}}(t) + v(t)\vec{\mathbf{t}}'(t). \tag{1}$$

Since $\vec{\mathbf{t}}(t)$ is of constant unit length, Corollary 5.2.6 shows that its derivative $\vec{\mathbf{t}}'(t)$ is *orthogonal* to $\vec{\mathbf{t}}(t)$. That is, $\vec{\mathbf{t}}'(t)$ is *normal* to the trajectory. Thus equation (1) decomposes acceleration into two perpendicular components.

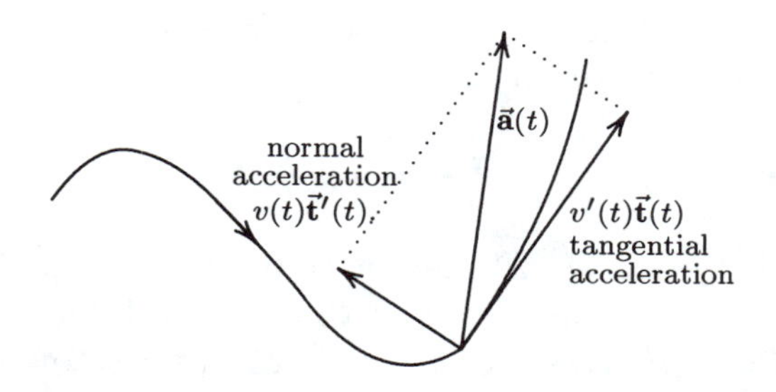

Figure 5.2.6

The two components represent two distinct aspects of acceleration. The first term, $v'(t)\vec{\mathbf{t}}(t)$, is called the **tangential component** of acceleration and is directed *along* the path of motion. This term describes how the object is speeding up (or slowing down) in the direction of motion. The second term, $v(t)\vec{\mathbf{t}}'(t)$, is the **normal component** of acceleration and describes the rate at which the *direction* of motion is changing as the path bends or turns. Thus an object accelerates through some combination of speeding up and changing direction. (See Figure 5.2.6.)

EXAMPLE 5.2.3 Compute the tangential and normal components of acceleration at $t = 1$ for the functions $\vec{\mathbf{K}}$ and $\vec{\mathbf{F}}$ in Example 5.2.2. Sketch the acceleration vector and its two components.

SOLUTION **a)** For $\vec{\mathbf{K}}$ the speed $v(t) = |\vec{\mathbf{K}}'(t)| = 1$. Therefore $\vec{\mathbf{t}} = (-\sin t, \cos t)$ and $\vec{\mathbf{t}}'(t) = (-\cos t, -\sin t)$. Since the speed is constant, its derivative is 0 and there is no tangential component of acceleration. The normal component of the acceleration is $v(t)\vec{\mathbf{t}}' = 1\vec{\mathbf{t}}'(t) = (-\cos t, -\sin t)$. (See Figure 5.2.7.)

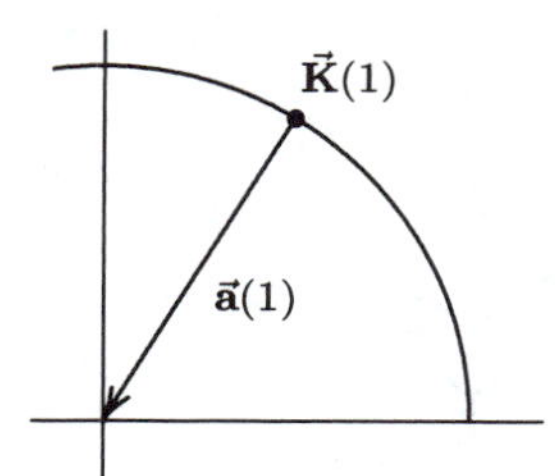

The tangential component of $\vec{\mathbf{K}}$ is 0.

Figure 5.2.7

b) The two components of the acceleration of $\vec{\mathbf{F}}$ may be computed directly from equation (1) as in **(a)**, but this leads to complicated derivatives, particularly for $v'(t)$ and $\vec{\mathbf{t}}'(t)$. A simpler method relies on a clever use of dot products.

Suppose we wish to express a vector $\vec{\mathbf{a}}$ as the sum of scalar multiples of two orthogonal vectors $\vec{\mathbf{u}}$ and $\vec{\mathbf{c}}$, where $\vec{\mathbf{u}}$ is a unit vector:

$$\vec{\mathbf{a}} = k_1\vec{\mathbf{u}} + k_2\vec{\mathbf{c}}.$$

Then the scalar coefficient k_1 may be computed by simply taking the dot product of $\vec{\mathbf{a}}$ with $\vec{\mathbf{u}}$. Using the fact that $\vec{\mathbf{c}} \cdot \vec{\mathbf{u}} = 0$ and $\vec{\mathbf{u}}$ is a unit vector we have $\vec{\mathbf{a}} \cdot \vec{\mathbf{u}} = k_1\vec{\mathbf{u}} \cdot \vec{\mathbf{u}} + k_2\vec{\mathbf{c}} \cdot \vec{\mathbf{u}} = k_1$. In the case of equation (1), this means that we may compute the scalar $v'(t)$ by taking the dot product of $\vec{\mathbf{t}}$ and $\vec{\mathbf{a}}$. In this way we can avoid the complications of computing $v'(t)$ directly. For $\vec{\mathbf{F}}(t)$ we have

$$\vec{\mathbf{a}}(t) = (-6t, 2) \qquad \text{and} \qquad \vec{\mathbf{t}}(t) = \frac{1}{\sqrt{(1-3t^2)^2 + 4t^2}}(1 - 3t^2, 2t),$$

so

$$v'(t) = \frac{1}{\sqrt{(1-3t^2)^2 + 4t^2}}(1 - 3t^2, 2t) \cdot (-6t, 2) = \frac{-2t + 18t^3}{\sqrt{(1-3t^2)^2 + 4t^2}}.$$

Thus

$$v'(t)\vec{\mathbf{t}}(t) = \frac{-2t + 18t^3}{(1-3t^2)^2 + 4t^2}(1 - 3t^2, 2t).$$

At $t = 1$ this vector is $(-4, 4)$. The normal component can be found by subtracting the tangential component from the total acceleration $\vec{\mathbf{a}}(t)$ (see Figure 5.2.8):

$$v(t)\vec{\mathbf{t}}'(t)\Big|_{t=1} = \vec{\mathbf{a}}(t) - v'(t)\vec{\mathbf{t}}(t)\Big|_{t=1} = (-6, 2) - (-4, 4) = (-2, -2). \quad \square$$

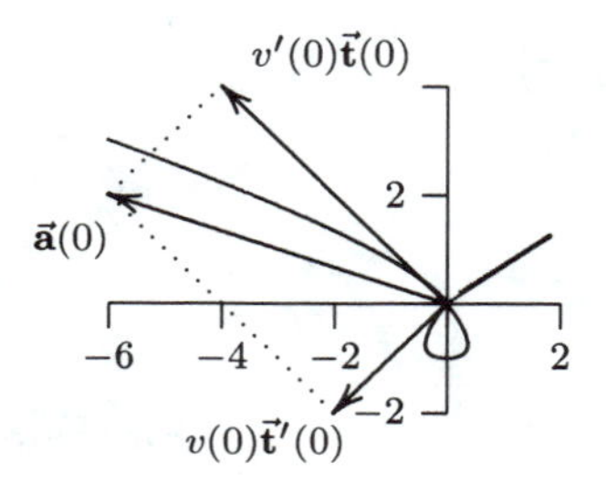

Figure 5.2.8

5.2.3 Arc Length

In this section we extend the definition of length to encompass curved arcs traced out by vector functions $\vec{\mathbf{F}}$. Recall from our study of area in Chapter 3 that we began with the simple formula for the area of rectangles. We then proceeded to *define* the area for more irregularly shaped regions as the limit of approximations formed by using basic rectangles.

We are in a similar situation with the concept of length. The standard Euclidian distance formula gives the length of straight lines between points; we would like to extend this to *define* length along a curve. We begin by forming successive **polygonal approximations** to the curve using straight lines (see Figure 5.2.9). Intuitively, as more points are added the approximations approach the curve. The length of any polygonal approximation can be computed by adding up the lengths of the straight line pieces. The basic idea is to *define* the length of a curve as the limit of the lengths of these polygonal approximations as we add more points. We now make these ideas rigorous.

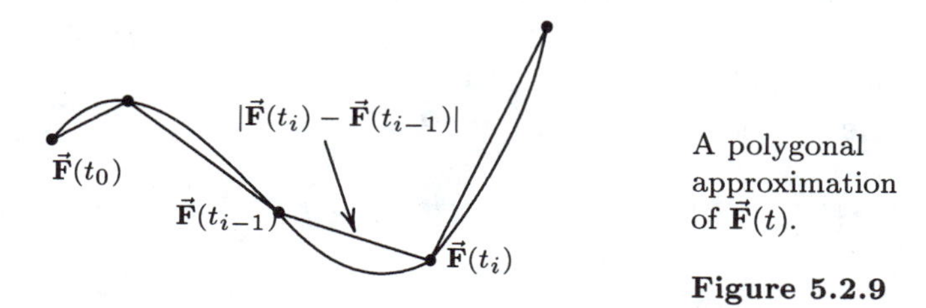

A polygonal approximation of $\vec{\mathbf{F}}(t)$.

Figure 5.2.9

Consider the trajectory of $\vec{\mathbf{F}}(t) = (x(t), y(t))$ for $t \in [a, b]$. To construct a polygonal approximation we form a partition of the interval $[a, b]$ given by $P = \{a = t_0, t_1, t_2, \ldots, t_n = b\}$. The images $\vec{\mathbf{F}}(t_0), \vec{\mathbf{F}}(t_1), \ldots, \vec{\mathbf{F}}(t_n)$ form points for the polygonal approximation. The length of this approximation, denoted by $\mathcal{L}(P, \vec{\mathbf{F}})$, is simply the sum of the lengths of each straight line segment:

$$\mathcal{L}(P, \vec{\mathbf{F}}) = \sum_{i=1}^{n} |\vec{\mathbf{F}}(t_i) - \vec{\mathbf{F}}(t_{i-1})|. \tag{2}$$

Now suppose that a single point, say t^*, is added to the partition between t_{i-1} and t_i. Then the term $|\vec{\mathbf{F}}(t_i) - \vec{\mathbf{F}}(t_{i-1})|$ in the sum (2) for the polygonal approximation is replaced by $|\vec{\mathbf{F}}(t^*) - \vec{\mathbf{F}}(t_{i-1})| + |\vec{\mathbf{F}}(t_i) - \vec{\mathbf{F}}(t^*)|$. By the triangle inequality the replacement term is larger; hence the length of the polygonal approximation increases. As more points are added to the partition the sum continues to increase. This proves the following theorem:

THEOREM 5.2.7 *If P' is a refinement of P, then $\mathcal{L}(P', \vec{\mathbf{F}}) \geq \mathcal{L}(P, \vec{\mathbf{F}})$.*

The definition of arc length is motivated by the fact that the length of polygonal approximations always *increases* as the partition is refined.

DEFINITION 5.2.8 *The* ***arc length*** *of $\vec{\mathbf{F}}(t)$ for $t \in [a, b]$ is the least upper bound of the lengths of the polygonal approximations, $\mathcal{L}(P, \vec{\mathbf{F}})$, over all partitions P of the interval $[a, b]$. If the least upper bound exists, then $\vec{\mathbf{F}}$ is said to be* ***rectifiable*** *on $[a, b]$.*

When does arc length exist and how can we compute it? Consider a single term in the length of the polygonal approximation from the sum in (2):

$$|\vec{\mathbf{F}}(t_i) - \vec{\mathbf{F}}(t_{i-1})| = \big|(x(t_i) - x(t_{i-1}), y(t_i) - y(t_{i-1}))\big|.$$

If $x(t)$ and $y(t)$ are differentiable, then by the mean value theorem there are points t^* and $t^{**} \in [t_{i-1}, t_i]$ such that $x(t_i) - x(t_{i-1}) = x'(t_i^*)\Delta t_i$ and $y(t_i) - y(t_{i-1}) = y'(t_i^{**})\Delta t_i$, where $\Delta t_i = t_i - t_{i-1}$. Substituting this into the preceding equation gives

$$|\vec{\mathbf{F}}(t_i) - \vec{\mathbf{F}}(t_{i-1})| = |(x'(t_i^*)\Delta t_i, y'(t_i^{**})\Delta t_i)| = |(x'(t_i^*), y'(t_i^{**}))|\Delta t_i.$$

Summing over all the pieces in the polygonal approximations yields

$$\mathcal{L}(P, \vec{\mathbf{F}}) = \sum_{i=1}^{n} |\vec{\mathbf{F}}(t_i) - \vec{\mathbf{F}}(t_{i-1})| = \sum_{i=1}^{n} |(x'(t_i^*), y'(t_i^{**}))|\Delta t_i. \tag{3}$$

This resembles a Riemann sum for $\int_a^b |(x'(t), y'(t))|\, dt$, which leads us to suspect that $\mathcal{L}(P, \vec{\mathbf{F}})$ will approach this integral as the partition is refined. Unfortunately, the sum in (3) is not precisely a Riemann sum, since x' and y' in the integrand are evaluated at different intermediate points, t_i^* and t_i^{**}. This causes the proof to be somewhat delicate.

We can compare the expression for $\mathcal{L}(P, \vec{\mathbf{F}})$ to a true Riemann sum for $|(x'(t), y'(t)| = |\vec{\mathbf{F}}'(t)|$, which has the general form

$$R(P, |\vec{\mathbf{F}}'|) = \sum_{i=1}^{n} |x'(\tau_i), y'(\tau_i)|\Delta t_i.$$

When the subintervals in the partition are small, the *uniform continuity* of x' can be used to guarantee that $x'(t_i^*)$ is close to $x'(\tau_i)$ in each subinterval and similarly that $y'(t_1^{**})$ is close to $y'(\tau_i)$. This allows us to show that the sums $\mathcal{L}(P, \vec{\mathbf{F}})$ and $R(P, |\vec{\mathbf{F}}'|)$ are also close when the subintervals are sufficiently small. This is the key idea in the following proof of the arc length formula.

THEOREM 5.2.9 *If $\vec{\mathbf{F}}(t) = (x(t), y(t))$ has a continuous derivative $\vec{\mathbf{F}}'(t) = (x'(t), y'(t))$, then the arc length of the trajectory of $\vec{\mathbf{F}}(t)$ for $t \in [a, b]$ is*

$$\int_a^b |\vec{\mathbf{F}}'(t)|\, dt = \int_a^b \sqrt{(x'(t))^2 + (y'(t))^2}\, dt.$$

PROOF Fix ϵ and P. The functions $x'(t)$ and $y'(t)$ are continuous and hence uniformly continuous on $[a, b]$. Let $\epsilon' = \epsilon/2(b-a)$. There exists a $\delta > 0$ such that when $|t-\tau| < \delta$, then $|x'(t)-x'(\tau)| < \epsilon'$ and $|y'(t)-y'(\tau)| < \epsilon'$. Let P' be any refinement of the partition P such that the length of each subinterval in P' is less than δ. Consequently if t_i and τ_i belong to the same subinterval of P', then $|x'(t_i)-x'(\tau_i)| < \epsilon'$ and $|y'(t_i)-y'(\tau_i)| < \epsilon'$.

Now we show that the length of $(x'(\tau_i), y'(\tau_i))$ in a Riemann sum $R(P', |\vec{\mathbf{F}}|)$ and the length of $(x'(t_i^*), y'(t_i^{**}))$ in $\mathcal{L}(P', \vec{\mathbf{F}})$ (see equation (3)) and are nearly the same. First note that, by the mean value theorem, t_i^*, t_i^{**}, and τ_i all belong to the same subinterval. By using the triangle inequality and then the box lemma (problem 5.1.4) we have

$$\begin{aligned} \big|\,|(x'(t_i^*), y'(t_i^{**}))| - |(x'(\tau_i), y'(\tau_i))|\,\big| &\le |(x'(t_i^*) - x'(\tau_i), y'(t_i^{**}) - y'(\tau_i))| \\ &\le |x'(t_i^*) - x'(\tau_i)| + |y'(t_i^{**}) - y'(\tau_i)| \\ &< 2\epsilon'. \end{aligned}$$

Using this inequality together with the triangle inequality we are able to compare $\mathcal{L}(P', \vec{\mathbf{F}})$ with an arbitrary Riemann sum $R(P', |\vec{\mathbf{F}}'|)$.

$$\begin{aligned} |\mathcal{L}(P', \vec{\mathbf{F}}) - R(P', |\vec{\mathbf{F}}'|)| &= \left| \sum_{i=1}^n |(x'(t_i^*), y'(t_i^{**}))|\Delta t_i - \sum_{i=1}^n |(x'(\tau_i), y'(\tau_i))|\Delta t_i \right| \\ &\le \sum_{i=1}^n \Big|\, |(x'(t_i^*), y'(t_i^{**}))| - |(x'(\tau_i), y'(\tau_i))| \,\Big| \Delta t_i \\ &< \sum_{i=1}^n 2\epsilon' \Delta t_i = 2\epsilon' \sum_{i=1}^n \Delta t_i = 2\epsilon'(b-a) = \epsilon. \end{aligned}$$

This inequality ($|\mathcal{L} - R| < \epsilon$) is valid for *any* Riemann sum with partition P', including the upper and lower sum. For the lower sum $L(P', |\vec{\mathbf{F}}'|)$ we have $|\mathcal{L}(P', \vec{\mathbf{F}}) - L(P', |\vec{\mathbf{F}}'|)| < \epsilon$. Because the lower sum is a lower bound for the integral,

$$\mathcal{L}(P', \vec{\mathbf{F}}) < L(P', |\vec{\mathbf{F}}'|) + \epsilon \le \int_a^b |\vec{\mathbf{F}}'(t)|\, dt + \epsilon.$$

Finally, since P' is a refinement of P,

$$\mathcal{L}(P, \vec{\mathbf{F}}) \le \mathcal{L}(P', \vec{\mathbf{F}}) < \int_a^b |\vec{\mathbf{F}}'(t)|\, dt + \epsilon.$$

This argument holds for *any* partition P. Since $\epsilon > 0$ is arbitrary, $\int_a^b |\vec{\mathbf{F}}'(t)|\, dt$ is an upper bound for $\mathcal{L}(P, \vec{\mathbf{F}})$ over all partitions P. (Why?)

To show that $\int_a^b |\vec{\mathbf{F}}'(t)|\, dt$ is the least upper bound, we apply the inequality $|\mathcal{L} - R| < \epsilon$ using the upper Riemann sum: $|\mathcal{L}(P', \vec{\mathbf{F}}) - U(P', |\vec{\mathbf{F}}'|)| < \epsilon$. Hence

$$\mathcal{L}(P', \vec{\mathbf{F}}) > U(P', |\vec{\mathbf{F}}'|) - \epsilon \ge \int_a^b |\vec{\mathbf{F}}'(t)|\, dt - \epsilon.$$

Since ϵ is arbitrary, $\int_a^b |\vec{\mathbf{F}}'(t)|\, dt$ is the *least* upper bound. ∎

EXAMPLE 5.2.4 Let $\vec{\mathbf{H}}(t) = (t^2, t^3)$. Find the length of the path traced out from $t = 0$ to $t = 1$.

SOLUTION $\vec{\mathbf{H}}'(t) = (2t, 3t^2)$, so the arc length is

$$\int_0^1 |\vec{\mathbf{H}}'(t)|\, dt = \int_0^1 \sqrt{4t^2 + 9t^4}\, dt = \int_0^1 t\sqrt{4 + 9t^2}\, dt.$$

A substitution using $u = 4 + 9t^2$ and $du = 18t\, dt$ gives

$$\int_0^1 |\vec{\mathbf{H}}'(t)|\, dt = \int_4^{13} \frac{1}{18}\sqrt{u}\, du = \frac{1}{27} u^{3/2} \Big|_4^{13} = \frac{13^{3/2} - 8}{27}. \quad \square$$

Now that we can find the exact length traveled by a particle, we can rigorously compute the instantaneous speed. The average speed between times t and $t + h$ is the total distance traveled divided by the elapsed time h. Using the arc length formula, this yields $\frac{1}{h}\int_t^{t+h} |\vec{\mathbf{F}}'(\tau)|\, d\tau$. The **instantaneous speed** is the limit of this expression as h goes to 0. By the second fundamental theorem of calculus this limit is simply the integrand evaluated at t, provided that the integrand $|\vec{\mathbf{F}}'|$ is continuous:

$$\lim_{h \to 0} \frac{1}{h} \int_t^{t+h} |\vec{\mathbf{F}}'(\tau)|\, d\tau = |\vec{\mathbf{F}}'(t)|.$$

Hence $|\vec{\mathbf{F}}'(t)|$ represents the speed of a particle moving along $\vec{\mathbf{F}}$ at time t.

5.2.4 Kepler's Laws and Gravitation

Perhaps the greatest single step in science was Newton's discovery of the universal law of gravitation. With this simple formula Newton swept away centuries of confusion and mystery that had surrounded the motion of heavenly bodies and replaced it with an easily understood idea that applied to all objects. Perhaps more than any other single discovery this law gave humankind the confidence to believe that nature could be understood in mathematical terms (recall the Pythagoreans' philosophy).

How did Newton discover this? Not by a falling apple but on the "shoulders of giants." Newton could not have hoped to explain the motion of the planets without the struggles of earlier scientists to formulate an orderly and reasonably correct description of that motion. This description was the life's work of Johannes Kepler, who distilled three laws of planetary motion from the volumes of observations carefully made by the astronomer Tycho Brache.

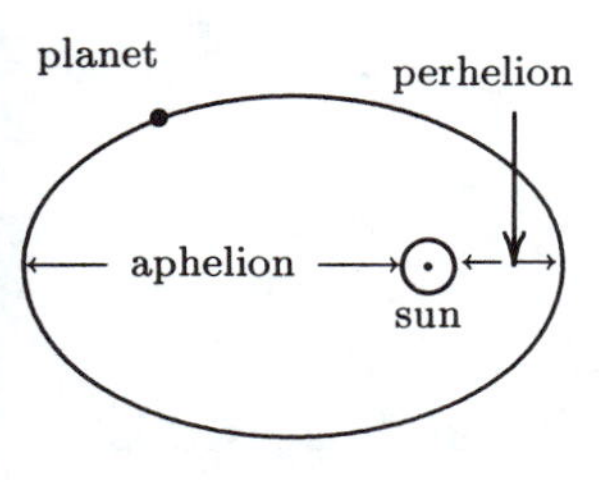

Figure 5.2.10

Kepler's Laws

1. Each planet moves in an elliptical orbit with the sun at one focus of the ellipse.
2. Each planet moves so that the arm from the sun to the planet sweeps out area at a constant rate.
3. The ratio of the cube of the mean distance a to the square of the period T is the same for all planets. That is, $\frac{a^3}{T^2}$ is constant. (The mean distance is $a = 1/2$(aphelion + perhelion). (See Figure 5.2.10.)

Newton's contribution was to show that these three laws implied two properties for the acceleration of planets.

Newton's Law of Gravitation

a. The acceleration is always directed toward the sun.
b. The magnitude of the acceleration is inversely proportional to the square of the distance between the planet and the sun, and the constant of proportionality is the same (universal) for all planets.

For Newton force, mass, and acceleration were intimately related by the equation $F = ma$, and his conclusions implied that there was some force acting between the sun and the planets whose magnitude varied inversely with the square of the distance. In this section we will show

how Kepler's three laws lead to the conclusions **a** and **b**. To accomplish this we will work with polar coordinates $r(t)$ and $\theta(t)$. To utilize Kepler's laws we must first determine how to express both the equation for an ellipse and the measure of "area swept out" in polar coordinates. The common formula for an ellipse is $\frac{x^2}{a^2} + \frac{y^2}{b^2} = 1$.

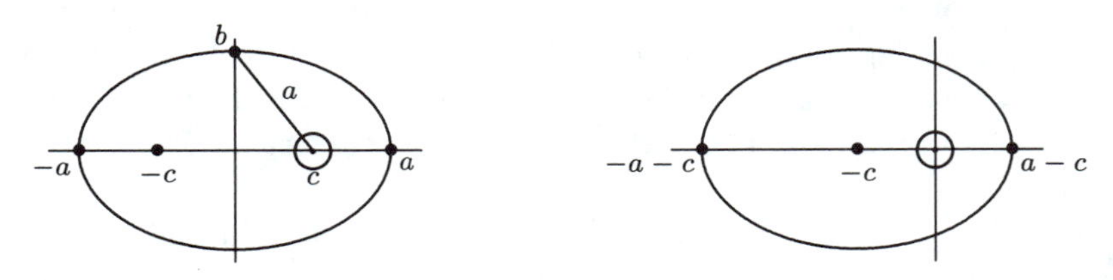

Figure 5.2.11

An ellipse has two foci on its major axis at c and $-c$, where c is given by $c^2 = a^2 - b^2$. To locate the sun at the origin of our figure we translate the ellipse c units to the left (see Figure 5.2.11). This gives a new equation for the ellipse $\frac{(x+c)^2}{a^2} + \frac{y^2}{b^2} = 1$. In polar coordinates this equation becomes

$$\frac{b^2}{r} = a + c\cos\theta, \tag{4}$$

where $x = r\cos\theta$ and $y = r\sin\theta$ (see problem 22).

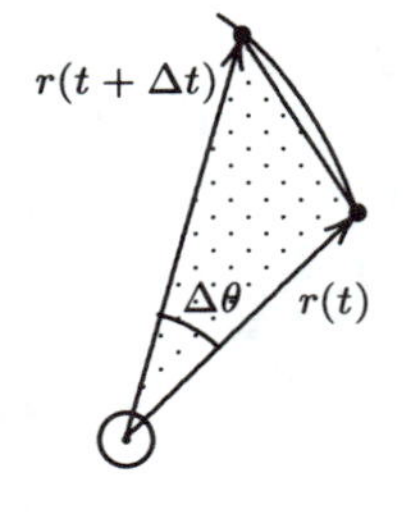

Figure 5.2.12

Next examine the area swept out in a short time Δt, shown in Figure 5.2.12. Observe that it is very nearly equal to a triangle with area $1/2\, r(t)r(t+\Delta t)\sin(\Delta\theta)$. To compute the *rate* at which area is swept out we should divide this quantity by Δt and take the limit. The limit is a bit easier to see if we multiply and divide the expression by $\Delta\theta$. According to Kepler's second law the resulting limit should be a constant rate K;

$$\lim_{\Delta t\to 0} 1/2\, r(t)r(t+\Delta t)\frac{\sin(\Delta\theta)}{\Delta\theta}\frac{\Delta\theta}{\Delta t} = 1/2(r(t))^2\theta'(t) = K\,. \tag{5}$$

This limit exists since as Δt approaches 0 the term $\frac{\sin(\Delta\theta)}{\Delta\theta}$ approaches 1, the term $\frac{\Delta\theta}{\Delta t}$ approaches $\theta'(t)$, and $r(t+\Delta t)$ approaches $r(t)$.

Now we have the necessary tools to apply Kepler's law to the motion of an object. First write the position of the planet in terms of polar coordinates:

$$\vec{\mathbf{F}}(t) = r(t)(\cos\theta(t), \sin\theta(t)).$$

Note that $(\cos\theta(t), \sin\theta(t))$ is a unit vector pointing in the direction of the planet and $r(t)$ is the distance from the sun (see Figure 5.2.13). The velocity is computed using **(c)** and **(d)** of Theorem 5.2.5:

$$\vec{\mathbf{F}}'(t) = r'(t)(\cos\theta(t), \sin\theta(t)) + r(t)\theta'(t)(-\sin\theta(t), \cos\theta(t)).$$

Thus the velocity is decomposed into two orthogonal components, one in the direction of the sun and one perpendicular to that direction.

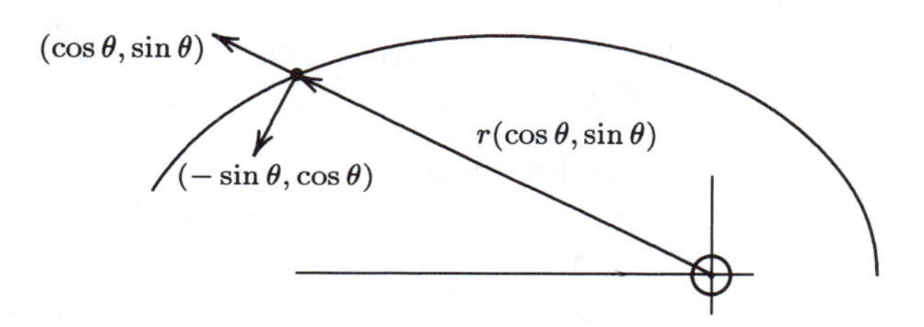

Figure 5.2.13

Next we compute the acceleration. In the computation we suppress the argument t in $r(t)$ and $\theta(t)$ for clarity.

$$\begin{aligned}\vec{\mathbf{a}} = \vec{\mathbf{F}}'' &= r''(\cos\theta, \sin\theta) + r'\theta'(-\sin\theta, \cos\theta) \\ &\qquad + r(\theta')^2(-\cos\theta, -\sin\theta) + (r'\theta' + r\theta'')(-\sin\theta, \cos\theta) \\ &= (r'' - r'(\theta')^2)(\cos\theta, \sin\theta) + (2r'\theta' + r\theta'')(-\sin\theta, \cos\theta).\end{aligned}$$

Hence the acceleration has also been decomposed into a radial component $\vec{\mathbf{a}}_r$ with length $r'' - r(\theta')^2$ and a normal component $\vec{\mathbf{a}}_n$ with length $2r'\theta' + r\theta''$. But Kepler's second law implies that *this normal component is* 0! To see this recall from (5) that area is swept out at a constant rate, $1/2(r)^2\theta' = K$. Therefore differentiating this gives 0.

$$\frac{d}{dt}\,1/2\,r^2\theta' = rr'\theta' + 1/2\,r^2\theta'' = 1/2\,r(2r'\theta' + r\theta'') = 0\ .$$

As r is always positive, the only way for this to be 0 is for $2r'\theta' + r\theta'' = 0$. But this is exactly the normal component of the acceleration. Since the normal component is 0, *all of the acceleration of the planets is directed radially* along the line connecting to the sun. For Newton this meant that there was some force acting between the planet and the sun.

What about the magnitude of this radial acceleration? To show that this varies inversely with the planet's distance r from the sun we need to express $r'' - r(\theta')^2$ in terms of r. We begin by finding r''. To do this we will differentiate the polar form of the ellipse (4), and use the fact that $r^2\theta' = 2K$ from (5). Differentiating both sides of (4) gives

$$\frac{-b^2}{r^2} r' = -c(\sin\theta)\,\theta' \quad \text{or} \quad r' = \frac{r^2\theta' c}{b^2}\sin\theta = \frac{2Kc}{b^2}\sin\theta\,.$$

Differentiating a second time gives

$$r'' = \frac{2Kc}{b^2}(\cos\theta)\,\theta'\,.$$

From (4) $c\cos\theta = \frac{b^2}{r} - a$ and from (5) $\theta' = \frac{2K}{r^2}$. Therefore the first part of the radial component is given by

$$r'' = \frac{2K}{b^2}\left(\frac{b^2}{r} - a\right)\frac{2K}{r^2} = \frac{4K^2}{r^3} - \frac{4K^2 a}{b^2 r^2}\,.$$

The second term in the radial component is easy to find using $\theta' = \frac{2K}{r^2}$:

$$r(\theta')^2 = r\left(\frac{2K}{r^2}\right)^2 = \frac{4K^2}{r^3}\,.$$

Altogether the complete radial component is

$$r'' - r(\theta')^2 = \left(\frac{4K^2}{r^3} - \frac{4K^2 a}{b^2 r^2}\right) - \frac{4K^2}{r^3} = -\frac{4K^2 a}{b^2 r^2}\,.$$

This shows that the *acceleration is inversely proportional to the square of the distance* r *from the planet* with constant of proportionality equal to $\frac{-4K^2 a}{b^2}$. The negative constant indicates that the acceleration is directed *toward* the sun.

On the face of it the constant $\frac{-4K^2 a}{b^2}$ would seem to depend on the particular parameters a and b of the elliptical orbit of the planet. But using Kepler's third law we can show that this constant is the same for all planets. It is this observation that leads to a universal law of gravitation. To begin, first note that the area of an ellipse is given by πab. Now the planet sweeps out area at a constant rate K. Therefore during a full period the planet should sweep out the entire area of the elliptical orbit, that is,

$$KT = \pi ab \qquad \text{or} \qquad K = \frac{\pi ab}{T}\,.$$

If we substitute this expression for K into the constant $-\frac{4K^2 a}{b^2}$, we obtain

$$-\frac{4K^2 a}{b^2} = -\frac{4\pi^2 a^2 b^2 a}{T^2 b^2} = -4\pi^2\frac{a^3}{T^2}\,.$$

Kepler's third law states that the ratio $\frac{a^3}{T^2}$ is the same for each planet. Therefore the motion of *all* the planets can be described as arising from a radial acceleration directed toward the sun with magnitude equal to $\frac{G}{r^2}$, where G is a constant.

Problems for Section 5.2

5.2.1. For each of the following functions sketch the trajectory of $\vec{\mathbf{F}}$ on the interval shown and draw in the velocity vectors at the indicated points t.

a) $\vec{\mathbf{F}}(t) = (t^2, t^3 - t)$ on $[-2, 2]$ at $t = -1,\ 0,\ 1$.

b) $\vec{\mathbf{F}}(t) = (4t - t^2, \frac{4}{t})$ on $[1, 4]$ at $t = 2, 3$.

c) $\vec{\mathbf{F}}(t) = (\sin^2 t, \cos^2 t)$ on $[0, \pi]$ at $t = 0,\ \pi/4,\ \pi/2$.

d) $\vec{\mathbf{G}}(t) = (\cos t, \sin 2t)$ for $t \in [0, 2\pi]$ at $t = 0, \frac{\pi}{4}, \frac{3\pi}{4}$.

5.2.2. Prove the other half of Theorem 5.2.2. That is, show that if $\lim_{t\to t_0} \vec{\mathbf{F}}(t) = \vec{\mathbf{L}}$, then $\lim_{t\to t_0} x(t) = L_1$ and $\lim_{t\to t_0} y(t) = L_2$.

5.2.3. Prove Theorem 5.2.4 using Theorem 5.2.2.

5.2.4. Let $\vec{\mathbf{F}}(t) = (t^2, t^3 - t)$ and $\vec{\mathbf{G}}(t) = (t^4 + t, t)$. Calculate $\frac{d}{dt}(\vec{\mathbf{F}} \cdot \vec{\mathbf{G}})(t)$ in two ways. First use Theorem 5.2.5. Second, begin by computing the dot product and then take its derivative.

5.2.5. Complete the proof of Theorem 5.2.5 **a**, **c**, and **d**.

5.2.6. Prove the converse of Corollary 5.2.6. That is, if $\vec{\mathbf{F}}(t)$ is orthogonal to $\vec{\mathbf{F}}'(t)$ for all t, then $|\vec{\mathbf{F}}|$ is a constant.

5.2.7. Suppose that the polar representation of $\vec{\mathbf{F}}(t)$ is $r(t)(\cos\theta(t), \sin\theta(t))$.

a) Find $\vec{\mathbf{F}}'(t)$ using the product rule for scalar products in Theorem 5.2.5.

b) Show that the two pieces of $\vec{\mathbf{F}}'(t)$ in **(a)** are always orthogonal.

c) Prove that this orthogonality will always occur when $\vec{\mathbf{F}}$ is expressed as a scalar function times a vector function of unit length, that is, $\vec{\mathbf{F}}(t) = g(t)\vec{\mathbf{u}}(t)$, where $|\vec{\mathbf{u}}| = 1$.

5.2.8. Assume that $\vec{\mathbf{F}}(t)$ is differentiable for all t and does not pass through the origin. Suppose that at time t_0 the trajectory of $\vec{\mathbf{F}}$ is at its closest point to the origin (i.e., $|\vec{\mathbf{F}}(t)|$ is a minimum at t_0). Prove that at t_0 the velocity vector is orthogonal to $\vec{\mathbf{F}}(t)$. What about when $|\vec{\mathbf{F}}|$ is a maximum?

5.2.9. The function $\vec{\mathbf{F}}(t) = (x(t), y(t)) = (3\cos t, 2\sin t)$ for $0 \le t \le 2\pi$ traces out an ellipse because $4(x(t))^2 + 9(y(t))^2 = 36$. Sketch the trajectory and geometrically determine the points at which $\vec{\mathbf{F}}(t) \perp \vec{\mathbf{F}}'(t)$. Verify analytically that at these points we really have $\vec{\mathbf{F}}(t) \perp \vec{\mathbf{F}}'(t)$.

5.2.10. For each of the following functions sketch the trajectory and the acceleration vector at the indicated point. Then decompose the vector into tangential and normal parts and sketch in these components on your graph.

a) $\vec{\mathbf{F}}(t) = (t^2 - 2t, t^2)$ at $t = 2$.

b) $\vec{\mathbf{F}}(t) = (4t - t^2, \frac{4}{t})$ at $t = 2$.

c) $\vec{\mathbf{G}}(t) = (\cos t, \sin 2t)$ at $t = \frac{\pi}{4}$.

5.2.11. Suppose that for all t we have $\vec{\mathbf{F}}(t) \cdot \vec{\mathbf{F}}'(t) = k$.

a) Write $|\vec{\mathbf{F}}(t)|^2$ as a dot product and apply the formula for differentiating dot products to determine the value of $\frac{d}{dt}|\vec{\mathbf{F}}(t)|^2$.

b) Use your result to find a formula for $|\vec{\mathbf{F}}(t)|$. (Hint: First determine what $|\vec{\mathbf{F}}(t)|^2$ must be; it will involve an arbitrary constant.)

5.2.12. **a)** Write $|\vec{\mathbf{F}}(t)|$ in terms of its coordinate functions and then compute $\frac{d}{dt}|\vec{\mathbf{F}}(t)|$ in terms of coordinate functions.

b) Show that your expression in **(a)** is just the component of the velocity vector in the direction of $\vec{\mathbf{F}}(t)$. (Note: The component of $\vec{\mathbf{x}}$ in the direction of $\vec{\mathbf{y}}$ is $\vec{\mathbf{x}} \cdot \frac{\vec{\mathbf{y}}}{|\vec{\mathbf{y}}|}$.) Can you describe why this should be so?

c) Now compute $\frac{d}{dt}|\vec{\mathbf{F}}'(t)|$ in terms of components. What dot product is involved and how does this relate to tangential and normal acceleration?

5.2.13. In this problem we develop general formulas for the tangential and normal components of acceleration. Let $\vec{\mathbf{F}}(t) = (x(t), y(t))$.

a) Show that the tangential component of acceleration can be written as

$$\vec{\mathbf{a}}_T = \frac{x''x' + y''y'}{(x')^2 + (y')^2}(x', y').$$

b) Compute the magnitude of the tangential component. (Note: For any vector $\vec{\mathbf{z}}$ we have $|a\vec{\mathbf{z}}| = |a||\vec{\mathbf{z}}|$. You should get some cancellation.)

c) Now subtract this tangential component from the acceleration to get

$$\vec{\mathbf{a}}_n = \frac{1}{(x')^2 + (y')^2}(y'(x''y' - y''x'), x'(y''x' - x''y'))$$

d) Compute the magnitude of the normal component. (Hint: It is a mess! Let $a = x''y' - y''x'$ so that the vector part becomes $(y'a, -x'a) = a(y', x')$. Now compute the magnitude in terms of a and resubstitute.)

5.2.14. A curve has **unit speed** if $|\vec{\mathbf{F}}'(t)| = 1$. For such a curve the number $|\vec{\mathbf{a}}(t)|$ is called the **curvature** of the trajectory at t.

a) Show that $\vec{\mathbf{a}}(t)$ is orthogonal to $\vec{\mathbf{F}}'(t)$ if $\vec{\mathbf{F}}(t)$ has unit speed.

b) Verify that $\vec{\mathbf{F}}(t) = (r\cos(\frac{t}{r}), r\sin(\frac{t}{r}))$ is a unit speed curve and compute its curvature at t.

5.2.15. When the curve is not parameterized as a unit speed curve the **curvature** can still be computed without reparameterizing. It is given by the magnitude of the normal component of acceleration, $\vec{\mathbf{a}}_n$, divided by the square of the speed,

$$\kappa = \frac{|\vec{\mathbf{a}}_n|}{|\vec{\mathbf{F}}'(t)|^2}.$$

a) Sketch a graph of $(t, \sin t)$ and compute the curvature at $t = 0, \frac{\pi}{4}, \frac{\pi}{2}$.

b) Compute the curvature of the parabola (t, t^2). Where is it largest? Does this make geometric sense?

c) Show that the two curves given by $\vec{\mathbf{F}}(t) = (x(t), y(t))$ for $t \in [a, b]$ and $\vec{\mathbf{G}}(s) = (x(2s), y(2s))$ for $s \in [\frac{1}{2}a, \frac{1}{2}b]$ trace out the same points in the plane. How do these functions differ?

d) Show that although the velocity and accelerations of the two curves $\vec{\mathbf{F}}$ and $\vec{\mathbf{G}}$ differ their curvatures agree. This illustrates that the curvature is measuring a geometric aspect of the image curve independent of its parameterization.

5.2.16. Write down the integral expressing the arc length from $t \in [a, b]$ for each of the curves in problem 1.

5.2.17. Find the arc length of each of the given functions on the indicated interval.

a) $\vec{\mathbf{F}}(t) = (t, 2t^{3/2})$ on $[0, 1]$.

b) $\vec{\mathbf{F}}(t) = (\sin^2 t, -\cos^2 t)$ on $[0, \pi/2]$.

c) $\vec{\mathbf{F}}(t) = (e^t(\cos t + \sin t), e^t(\cos t - \sin t))$ on $[1, 4]$.

5.2.18. Determine a formula for the arc length of the graph of a function $y = f(t)$ over an interval $[a, b]$. (Hint: Think of the graph as the trajectory of the vector-valued function $\vec{\mathbf{F}}(t) = (t, f(t))$ for t in the interval $[a, b]$.)

5.2.19. Suppose that $\vec{\mathbf{F}}(t) = (f(t)\cos t, f(t)\sin t)$, where $f(t)$ is a continuously differentiable real-valued function of t on the interval $[a, b]$.

a) Show that the arc length of $\vec{\mathbf{F}}$ is given by the formula

$$\int_a^b \sqrt{(f(t))^2 + (f'(t))^2}\, dt.$$

b) Use this to find the circumference of the circle of radius r centered at the origin that is parameterized by $\vec{\mathbf{F}}(t) = (r\cos t, r\sin t)$ for $t \in [0, 2\pi]$.

c) Use the formula in **(a)** to find the length of the exponential spiral given by $\vec{\mathbf{F}}(t) = (e^{-t}\cos t, e^{-t}\sin t)$ from $t = 0$ to $t = \infty$.

5.2.20. Suppose that $|\vec{\mathbf{F}}(t) - \vec{\mathbf{F}}(t^*)| < M|t - t^*|$ for all $t, t^* \in [a, b]$. Prove that $\vec{\mathbf{F}}(t)$ is rectifiable, (i.e., that its arc length exists). (Note: You cannot assume that $\vec{\mathbf{F}}$ is differentiable. The problem is a matter of showing that a certain least upper bound exists.)

5.2.21. **a)** Sketch the graph of $\vec{\mathbf{F}}(t) = (t - \sin t, 1 - \cos t))$ for $0 < t < 2\pi$.

b) Sketch the velocity and acceleration vectors at $t = \frac{\pi}{2}$ and $t = \pi$. What is happening to the motion of the particle at each of these points?

c) Find a general expression (for any t) for the normal and tangential components of acceleration $\vec{\mathbf{F}}$.

d) Find the arc length of $\vec{\mathbf{F}}$ on $[0, 2\pi]$. (Hint: Use the trigonmetric identity $\frac{1}{2} - \frac{1}{2}\cos\theta = \sin^2(\theta/2)$ to eliminate the square root.)

5.2.22. Derive the rectangular form for an ellipse $\frac{(x+c)^2}{a^2} + \frac{y^2}{b^2} = 1$ from the polar equation $b^2/r = a + c\cos\theta$. (Hint: Rewrite the polar form as $b^2 - cr\cos\theta = ar$ and substitute $r\cos\theta = x$ on the left and $r = \sqrt{x^2+y^2}$ on the right. Square both sides and use the relation $b^2 = a^2 - c^2$ to algebraically manipulate the equation into the form $b^2(x+c)^2 + a^2y^2 = a^2b^2$.)

5.2.23. Newton's law of gravitation says that the force of gravity on an object with mass M_2 exerted by another object with mass M_1 has a magnitude proportional to the product of the two masses and inversely proportional to the distance between them, and is directed along the line joining the two bodies:

$$\vec{\mathbf{F}} = -\frac{\gamma M_1 M_2}{r^2}\vec{\mathbf{u}},$$

where γ is the *universal* gravitational constant of proportionality, r is the distance between the objects and $\vec{\mathbf{u}}$ is the unit vector from M_1 in the direction of M_2. Newton's second law of motion says that the acceleration of an object with mass M is determined by the force on the object via $\vec{\mathbf{F}} = M\vec{\mathbf{a}}$.

a) Combine these two laws to show that the acceleration of the earth (mass M_e) due to the sun (mass M_s) is given by $\vec{\mathbf{a}} = \frac{\gamma M_s}{r^2}\vec{\mathbf{u}}$.

b) In the text, the magnitude of the acceleration of a planet was computed from Kepler's laws to be $4\pi^2\frac{a^3}{T^2r^2}$, where a was the mean (average) distance and T was the period of the planet. Assume that the orbit is nearly circular so that a is just the radius r. Equate this with the magnitude for the acceleration in **(a)** to get a formula for γM_s in terms of r and T.

c) Assuming that the orbit of the earth is circular with a radius of 1.5×10^{11} meters, a period of 3.15×10^7 seconds (365 days), and that $\gamma = 6.67 \times 10^{-11}\frac{\text{m}^3}{\text{sec}^2\text{kg}^2}$, find the mass of the sun in kilograms.

5.2.24. Suppose that $|\vec{\mathbf{F}}(t)| = R$, where R is a constant for all t and that the acceleration is given by $\vec{\mathbf{F}}''(t) = -a^2\vec{\mathbf{F}}(t)$.

a) Draw a picture representing this type of motion. Display a couple of sample velocity and acceleration vectors on your diagram.

b) Why must $\vec{\mathbf{F}}(t) \cdot \vec{\mathbf{F}}'(t) = 0$?

c) Now take the derivative of $\vec{\mathbf{F}} \cdot \vec{\mathbf{F}}'$ using Theorem 5.2.5 and show that the speed must be the constant $v(t) = |\vec{\mathbf{F}}'(t)| = aR$.

5.3 Functions of a Vector Variable

This section begins our study of functions of the form $f : \mathbf{R}^2 \to \mathbf{R}$ whose *domain* is the vector space $\mathbf{R}^2$. After a brief look at graphical representations of these functions, we examine the concept of limits and continuity. The limit behavior of functions of a vector variable is somewhat subtle, and interesting pathologies can arise. The section ends with a discussion of the preliminary concept of a partial derivative. This prepares the way for Section 4 in which the definition of differentiability is presented and the calculus of these functions is explored.

A function $f : \mathbf{R}^2 \to \mathbf{R}$ associates to each point $\vec{\mathbf{x}} = (x, y)$, a real number written $f(\vec{\mathbf{x}})$ or $f(x, y)$. The *graph* of $f(x, y)$ consists of a surface over the xy plane generated by plotting the points $(x, y, f(x, y))$ in three dimensions for all values of (x, y) in the domain. Three-dimensional pictures of surfaces frequently have hidden sections, which make them difficult to comprehend and awkward to draw. Often a more accurate understanding of $f(x, y)$ can be obtained from a contour map of the surface. To draw such a map we first find the **level curves** of the surface. These are curves in the xy plane for which $f(x, y)$ is a constant. The level curve at height c is found by sketching the curve $f(x, y) = c$. If an adequate number of level curves are drawn for different values of c, then just as with a geographical contour map we begin to develop the impression of a surface with peaks, valleys, ridges, and saddle points. Let's illustrate these ideas.

EXAMPLE 5.3.1 Sketch some level curves for $f(x, y) = 4x^2 + y^2$ and then draw the graph of f.

SOLUTION The level curves $4x^2 + y^2 = c$ are all ellipses centered at the origin. We can easily sketch these for any particular c by finding the x and y intercepts. For example, when $c = 4$ the contour $4x^2 + y^2 = 4$ has x intercepts at ± 1 and y intercepts at ± 2. Several level curves and a sketch of the surface are shown in Figures 5.3.1 and 5.3.2. Notice that the height of the surface increases in all directions as we move from the origin. Hence the origin is a minimum point for the surface. □

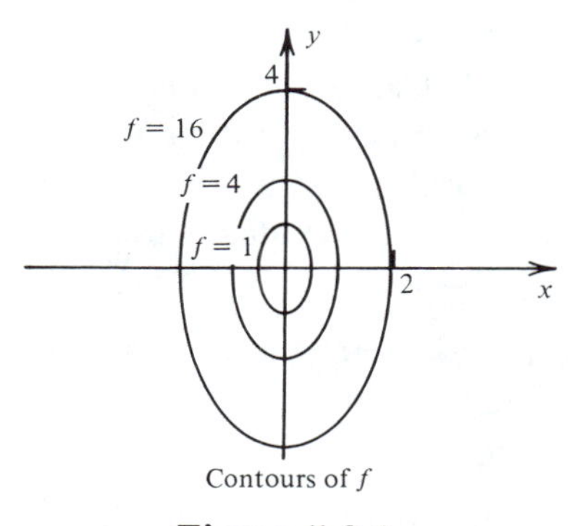

Figure 5.3.1

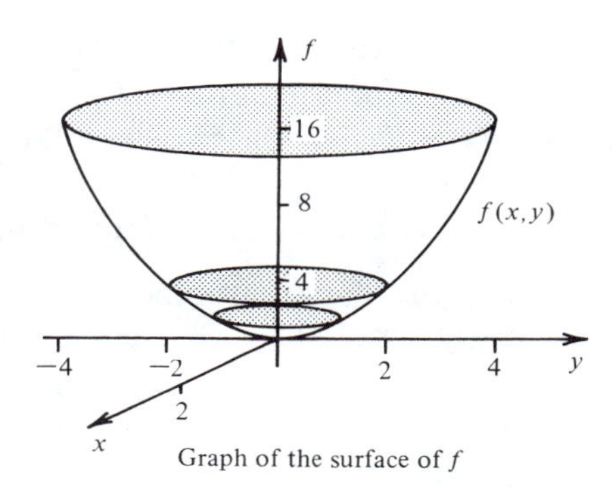

Figure 5.3.2

5.3.1 Limits and Continuity for $f : \mathbf{R}^2 \to \mathbf{R}$

The definition of limits for $f : \mathbf{R}^2 \to \mathbf{R}$ takes the standard ϵ,δ form using the metrics defined on the domain $\mathbf{R}^2$ and range $\mathbf{R}$.

DEFINITION 5.3.1 *The* $\lim_{\vec{\mathbf{x}} \to \vec{\mathbf{a}}} f(\vec{\mathbf{x}}) = L$ *if for any* $\epsilon > 0$ *there is a* $\delta > 0$ *such that*

$$\text{if} \quad 0 < |\vec{\mathbf{x}} - \vec{\mathbf{a}}| < \delta, \quad \text{then} \quad |f(\vec{\mathbf{x}}) - L| < \epsilon.$$

If we write $\vec{\mathbf{x}} = (x, y)$ *and* $\vec{\mathbf{a}} = (a, b)$, *this becomes*

$$\text{if} \quad 0 < \sqrt{(x-a)^2 + (y-b)^2} < \delta, \quad \text{then} \quad |f(x, y) - L| < \epsilon,$$

and is written $\lim_{(x,y) \to (a,b)} f(x, y) = L$.

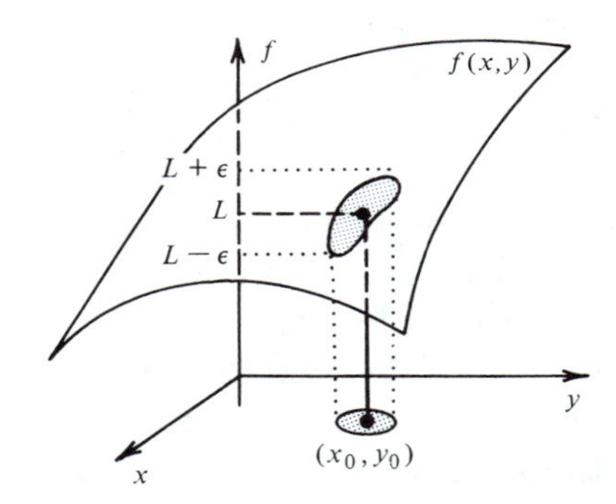

Figure 5.3.3

In terms of the graph of f this definition says that for any $\epsilon > 0$ we must be able to find an open δ-disk about the point (a, b) such that for all points (x, y) inside this disk the height $f(x, y)$ is within ϵ of L (with the possible exception of the point (a, b) itself). This is portrayed in Figure 5.3.3. Notice that by this definition a limit can exist only when a function is *defined in some little open disk about the point.*

EXAMPLE 5.3.2 The functions $P_1(x, y) = x$ and $P_2(x, y) = y$ are called **projection functions**, since they take any point (x, y) and "project" onto first and second coordinate respectively. Using the ϵ, δ definition show that $\lim_{(x,y) \to (a,b)} P_1(x, y) = a$ and $\lim_{(x,y) \to (a,b)} P_2(x, y) = b$.

SOLUTION For the function $P_1(x, y) = x$ we must find a δ such that

$$\text{if} \quad 0 < |\vec{\mathbf{x}} - \vec{\mathbf{a}}| < \delta, \quad \text{then} \quad |x - a| < \epsilon.$$

By the box lemma, $|x - a| \leq |\vec{\mathbf{x}} - \vec{\mathbf{a}}|$, so we may take $\delta = \epsilon$. A similar argument shows that $\lim_{(x,y) \to (a,b)} y = b$. ◻

Once the concept of limit is in hand the definition of continuity follows immediately.

DEFINITION 5.3.2 *$f(x, y)$ is* ***continuous*** *at (a, b) if and only if*

$$\lim_{(x,y)\to(a,b)} f(x, y) = f(a, b).$$

According to this definition Example 5.3.2 shows that the two projection functions x and y are continuous everywhere. The continuity of constant functions is just as easy to show. Most other commonly encountered functions are built up from these three basic functions (x, y, and constants) using addition, subtraction, multiplication, division, and composition with other standard functions. For example, the function $f(x, y) = \sin(x^2y + 2y)$ is formed by first combining the x, y, and 2 using addition and multiplication to form $x^2y + 2y$ and then composing the result with the sine function. To show that expressions like $f(x, y)$ are continuous we need only prove that continuity is preserved under the operations of arithmetic and under composition. This in turn rests on understanding how limits behave under these operations.

THEOREM 5.3.3 *If $\lim_{(x,y)\to(a,b)} f(x, y) = L$ and $\lim_{(x,y)\to(a,b)} g(x, y) = M$, then*

a) $\displaystyle\lim_{(x,y)\to(a,b)} (f \pm g)(x, y) = L \pm M;$

b) $\displaystyle\lim_{(x,y)\to(a,b)} (f \cdot g)(x, y) = LM;$

c) $\displaystyle\lim_{(x,y)\to(a,b)} \left(\frac{f}{g}\right)(x, y) = \frac{L}{M}$, *provided* $M \neq 0$.

Theorem 5.3.3 is the direct analogue of the sum, product, and quotient theorems for limits that were covered in Chapter 2. The proofs for those theorems used only the basic distance axioms for the domain variable x. From Section 5.1 we know that the metric in $\mathbf{R}^2$ (indeed in any vector space with a dot product) satisfies the same axioms. In fact, to prove Theorem 5.3.3 we can simply carry over each step in the proofs for the sum, product, and quotient of limits presented in Chapter 2, replacing x by $\vec{\mathbf{x}}$ and a by $\vec{\mathbf{a}}$. (The reader should check this broad assertion.)

Theorem 5.3.3 may be immediately applied to show that sums, products, and quotients of continuous functions are again continuous. These rules, together with the continuity of compositions (proven in the following theorem), enable us to quickly verify the continuity of most common expressions except, of course, at problem points such as when a denominator is 0.

THEOREM 5.3.4 *Assume that $f(x, y)$ and $g(x, y)$ are continuous at (a, b).*

a) *Then $f + g$ and $f \cdot g$ are continuous at (a, b).*

b) *If $g(a, b) \neq 0$, then $\frac{f}{g}$ is continuous at (a, b) .*

c) *Further, if $h : \mathbf{R} \to \mathbf{R}$ is continuous at $t = f(a, b)$, then the composition $h(f(x, y)) : \mathbf{R}^2 \to \mathbf{R}$ is continuous at (a, b).*

PROOF Continuity for the sum, product, and quotient follows immediately from Theorem 5.3.3. To prove continuity for compositions, given any $\epsilon > 0$ we must exhibit a δ such that when $|(x, y) - (a, b)| < \delta$, then $|h(f(x, y)) - h(f(a, b))| < \epsilon$. (Why can we eliminate the 0 in the expression $0 < |(x, y) - (a, b)| < \delta$?). The proof requires two steps:

First notice that by assumption the function $h(t)$ is continuous at $f(a, b)$; therefore given any $\epsilon > 0$ there is a δ_1 such that

$$\text{if} \quad |t - f(a, b)| < \delta_1, \quad \text{then} \quad |h(t) - h(f(a, b))| < \epsilon.$$

Now since $f(x, y)$ is continuous at (a, b), for this δ_1 there is some $\delta > 0$ such that

$$\text{if} \quad |(x, y) - (a, b)| < \delta, \quad \text{then} \quad |f(x, y) - f(a, b)| < \delta_1.$$

But since $|f(x, y) - f(a, b)| < \delta_1$ we let $f(x, y)$ play the role of t and conclude that

$$|h(f(x, y)) - h(f(a, b))| < \epsilon.$$

Hence $h(f(x, y))$ is continuous at (a, b). ∎

EXAMPLE 5.3.3 Find the points where these functions are continuous:

$$\textbf{a)} \quad f(x, y) = \frac{x^2}{x^2 + y^2}; \qquad \textbf{b)} \quad g(x, y) = \frac{1}{x^2 - y^2} + \ln(xy).$$

SOLUTION **a)** The numerator and denominator of f are obtained from the projection functions x and y by multiplication and addition. Since by Example 5.3.2 these basic functions are continuous everywhere, Theorem 5.3.4 shows that both the numerator and the denominator of f are continuous at all points. By Theorem 5.3.4, the quotient is continuous everywhere except where the denominator is 0. The expression $x^2 + y^2 = 0$ only when $(x, y) = (0, 0)$, so $f(x, y)$ is continuous everywhere except at the origin where it is undefined.

b) As in **(a)**, the first term in the function g is continuous except where $x^2 - y^2 = 0$. Factoring this expression yields $x^2 - y^2 = (x - y)(x + y) = 0$, which implies that $x-y=0$ or $x+y=0$. Therefore $\frac{1}{x^2-y^2}$ is continuous everywhere except along the two lines $y=x$ and $y=-x$. The second term is a composition, $\ln(xy)$. The natural log is defined and continuous only for positive numbers. The function xy is positive only at points in the first and third quadrants. Since xy is continuous at these points we conclude that the composition $\ln(xy)$ is continuous for (x, y) in quadrants 1 and 3. Therefore $g(x, y)$ is continuous in these quadrants except on the diagonal line $y=x$. ◻

Although the expression $\frac{x^2}{x^2+y^2}$ in the preceding example was not continuous at $(0,0)$, we may still ask if the *limit* of this function exists at the origin.

EXAMPLE 5.3.4 Show that $\lim\limits_{(x,y)\to(0,0)} \dfrac{x^2}{x^2+y^2}$ does not exist.

SOLUTION One standard way to prove the nonexistence of a limit is to show that the expression gets close to different values along different paths of approach. In this example, if we approach the origin along the x-axis (i.e., where $y=0$), then $f(x, y)=\frac{x^2}{x^2+y^2}=\frac{x^2}{x^2}=1$. But if we approach (0,0) along the y-axis (i.e., where $x=0$), the expression is $\frac{x^2}{x^2+y^2}=\frac{0}{y^2}=0$. This implies that *any* δ-disk about the origin, no matter how small, will contain some points at which $f(x, y)=1$ and other points at which $f(x, y)=0$. Therefore there can be no limit. ◻

From Example 5.3.4 we see that for f to have a limit L at a point (a, b), the values of f must approach L along *all* possible routes that converge to (a, b). It is important to be aware that *it is not enough to check that f approachs the same limit along the x direction and the y direction*, since the behavior of f along other routes can be entirely different. This subtle behavior is illustrated in the next example.

EXAMPLE 5.3.5 Show that along *every* straight line through the origin, the function $f(x, y)=\frac{x^2 y}{x^4+y^2}$ approaches 0, but along the parabolic path $y=x^2$ the function approaches 1/2.

SOLUTION To show that f approaches 0 along any straight-line path, we first show that $f \rightarrow 0$ along the x-axis and the y-axis. If $x=0$, then $f(x, y)=\frac{0}{y^2}$; and if $y=0$, then $f(x, y)=\frac{0}{x^4}$. Hence along both axes f is always 0. Now let $y=ax$ be the equation of a line through the origin. On this line, as $x \rightarrow 0$ we have

$$f(x, y)=\frac{x^2 y}{x^4+y^2}=\frac{x^2 ax}{x^4+(ax)^2}=\frac{ax}{x^2+a^2} \rightarrow 0.$$

However, if we are on the parabola $y = x^2$, then

$$f(x,y) = \frac{x^2 y}{x^4 + y^2} = \frac{x^2 x^2}{x^4 + (x^2)^2} = \frac{1}{2}.$$

Although f approaches 0 along every straight line, notice that in any δ-disk about the origin, no matter how tiny, there will be a little piece of the parabola $y = x^2$ where $f(x,y) = \frac{1}{2}$. Bizarre behavior indeed! □

This is a discouraging example. It shows that we cannot hope to know if a limit exists by checking what happens along certain paths of approach. Hence *the problem of finding the limit* $\lim_{(x,y)\to(a,b)} f(x,y)$ *cannot simply be broken into the individual limits* $\lim_{x\to a} f(x,b)$ *and* $\lim_{y\to b} f(a,y)$. But this should be no surprise, since the two individual limits only examine the behavior along the two curves $f(x,b)$ and $f(a,y)$ and this doesn't necessarily tell us anything about what is happening to $f(x,y)$ at the many other points near (a,b).

How, then, can we show that a limit does exist? Except in cases in which Theorems 5.3.3 and 5.3.4 can be employed, there is no general method. The next example illustrates the type of argument that can sometimes be made to show that a limit exists.

EXAMPLE 5.3.6 Show that $\lim_{(x,y)\to(0,0)} \frac{x^2 y - y^2 x}{x^2 + y^2} = 0$.

SOLUTION We wish to find a δ such that

$$\text{if} \quad 0 < |\vec{\mathbf{x}} - \vec{\mathbf{0}}| < \delta, \quad \text{then} \quad \left| \frac{x^2 y - y^2 x}{x^2 + y^2} - 0 \right| < \epsilon.$$

By the triangle inequality and the box lemma,

$$\left| \frac{x^2 y - y^2 x}{x^2 + y^2} \right| = \frac{|x||y|(|x - y|)}{|x^2 + y^2|} \le \frac{|x||y|(|x| + |y|)}{|x^2 + y^2|} \le \frac{|\vec{\mathbf{x}}||\vec{\mathbf{x}}|(|\vec{\mathbf{x}}| + |\vec{\mathbf{x}}|)}{|\vec{\mathbf{x}}|^2} = 2|\vec{\mathbf{x}}|.$$

Thus let $\delta = \epsilon/2$. □

5.3.2 Partial Derivatives

Perhaps as a first attempt to define the notion of derivative for functions $f : \mathbf{R}^2 \to \mathbf{R}$, we might try the analogue of the one-dimensional derivative:

$$\lim_{\vec{\mathbf{h}}\to\vec{\mathbf{0}}} \frac{f(\vec{\mathbf{x}} + \vec{\mathbf{h}}) - f(\vec{\mathbf{x}})}{\vec{\mathbf{h}}}.$$

But this expression is clearly meaningless, since it involves division by a vector. Mathematicians of the early eighteenth century were not detered by this problem. They simply carried over the concept of an ordinary derivative to functions of two variables by considering change in only one variable at a time. This led to the idea of a partial derivative, which is the subject of this section. Partial derivatives are a simple and very useful tool to describe the behavior of functions. However, we will see that partial derivatives alone don't really capture the full idea of the derivative. The problem is that the limits defining partial derivatives examine the behavior in only the x and y directions, which we have seen is insufficient for describing two-dimensional limits. In Section 4 a restructuring of the definition of the derivative will enable us to arrive at the modern concept of differentiation for functions of a vector variable.

DEFINITION 5.3.5 *The* ***partial derivative*** *of $f(x, y)$ with respect to x at the point (x_0, y_0) is*

$$f_x(x_0, y_0) = \frac{\partial f}{\partial x}(x_0, y_0) = \lim_{h \to 0} \frac{f(x_0 + h,\, y_0) - f(x_0, y_0)}{h},$$

if the limit exists. Similarly the partial derivative with respect to y is

$$f_y(x_0, y_0) = \frac{\partial f}{\partial y}(x_0, y_0) = \lim_{h \to 0} \frac{f(x_0,\, y_0 + h) - f(x_0, y_0)}{h},$$

if the limit exists.

Let $f(x, y_0)$ be the function of the single variable x that is obtained from $f(x, y)$ by keeping y constant at y_0. Notice that the partial derivative $f_x(x_0, y_0)$ is defined in exactly the same way as you would define the ordinary derivative of the function $f(x, y_0)$ at the point x_0. This means that *partial differentiation with respect to x is just ordinary differentiation with the x's treated as variables and the y's held constant.* Therefore, for computational purposes all the familiar rules for differentiation from Chapter 2 apply. Similarly f_y is computed by treating the x's as constants and differentiating with respect to the variable y.

EXAMPLE 5.3.7 Compute the partial derivatives of the following functions: **(a)** $\dfrac{x^2 y}{2y + 1}$ and **(b)** $y \sin(xy)$.

SOLUTION **a)** First, treating all y's as constant and taking the derivative with respect to x gives

$$f_x(x, y) = \frac{\partial f}{\partial x}(x, y) = \frac{2xy}{2y + 1}.$$

Then treating x as constant and using the quotient rule for y gives

$$f_y(x,y) = \frac{\partial f}{\partial y}(x,y) = \frac{x^2(2y+1) - 2x^2 y}{(2y+1)^2} \frac{x^2}{(2y+1)^2}.$$

b) For g_x we treat y as constant, and apply the chain rule to get $g_x(x,y) = y^2\cos(xy)$. For g_y we must apply the product rule and the chain rule. We obtain $g_y(x,y) = \sin(xy) + xy\cos(xy)$. □

We can interpret the partial derivatives at a point in terms of slopes on the surface of $f(x,y)$. The partial derivative $f_x(x_0, y_0)$ is just the derivative of the single variable function $f(x, y_0)$ at the point x_0. The graph of $f(x, y_0)$ is given by the curve lying on the surface of $f(x,y)$ above the line $y = y_0$. (See Figure 5.3.4.) Similarly $f(x_0, y)$ is a curve on the surface in the y direction, above the line $x = x_0$. Since the ordinary derivatives of these functions at x_0 and y_0 correspond to slopes of their respective graphs, we see that the partial derivatives f_x and f_y *represent the slopes of the surface of $f(x,y)$ along curves in the x direction and y direction respectively* (Figure 5.3.4). Therefore f_x and f_y tell us the rate of change in $f(x,y)$ along these two curves.

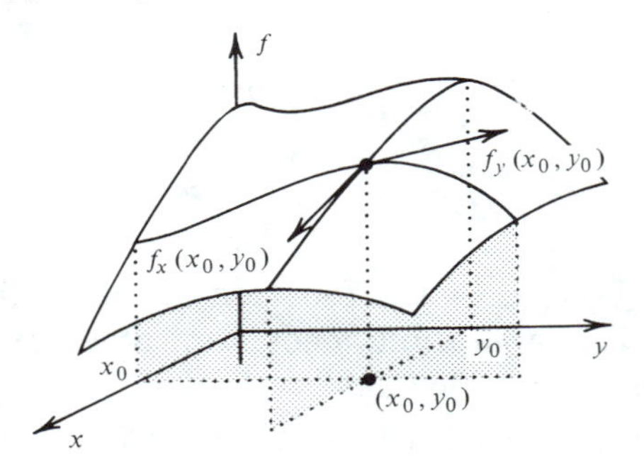

Figure 5.3.4

EXAMPLE 5.3.8 Compute the partial derivatives of $f(x,y) = 4x^2 + y^2$ at the point $(1,2)$ and determine how the surface is changing in the x and y directions.

SOLUTION $f_x(x,y) = 8x$, so $f_x(1,2) = 8$; and $f_y(x,y) = 2y$, so $f_y(1,2) = 4$. Thus at $(1,2)$ the height of the surface of f is increasing at a rate of 8 units for every unit in the x direction and 4 units for every unit in the y direction. □

Partial derivatives are easy to compute and have a nice geometric interpretation. However, from the discussion of limits earlier in this section, we know that limits in the x and y directions alone don't suffice to tell us what is happening in the nearby region. In fact, partial derivatives can exist at points where the function is discontinuous!

EXAMPLE 5.3.9 Let

$$f(x,y) = \begin{cases} \frac{xy}{x^2+y^2}, & \text{if } (x,y) \neq (0,0) \\ 0, & \text{if } (x,y) = (0,0). \end{cases}$$

Show that the partial derivatives of f exist at the origin even though f is discontinuous there.

SOLUTION The function $f(x,0)$ is defined at all points and equals 0 everywhere. Hence the partial derivative $f_x(0,0) = \frac{d}{dx}f(x,0) = 0$ and similarly $f_y(0,0) = 0$. Although the limit of f at the origin along the x and y directions is 0, approaching the origin along the line $y = x$ yields

$$f(x,y) = \frac{x^2}{x^2+x^2} = 1/2.$$

Therefore the function is not continuous at the origin. □

One of the most interesting and useful properties of partial derivatives is the equality of mixed partials. This means that if the partial derivative of a function is taken first with respect to x and then the result is differentiated with respect to y the answer is the same as starting with y and then differentiating with respect to x. (This holds provided that the resulting expressions are continuous.) Thus the *order* of differentiation is irrelevant. Such iterated partial derivatives are called **mixed partials**. If differentiation with respect to x is to be carried out first, the mixed partial is written

$$f_{xy} = \frac{\partial^2 f}{\partial y \partial x} = \frac{\partial}{\partial y}\left(\frac{\partial f}{\partial x}\right).$$

If the derivative with respect to y is first, we write

$$f_{yx} = \frac{\partial^2 f}{\partial x \partial y} = \frac{\partial}{\partial x}\left(\frac{\partial f}{\partial y}\right).$$

EXAMPLE 5.3.10 If $f(x,y) = xe^{xy^2}$, then $\frac{\partial f}{\partial x} = e^{xy^2} + xy^2e^{xy^2}$; therefore

$$\frac{\partial^2 f}{\partial y \partial x} = \frac{\partial}{\partial y}\left(\frac{\partial f}{\partial x}\right) = 4xye^{xy^2} + 2x^2y^3e^{xy^2}.$$

In the other direction we have $\frac{\partial f}{\partial y} = 2x^2ye^{xy^2}$, so

$$\frac{\partial^2 f}{\partial x \partial y} = \frac{\partial}{\partial x}\left(\frac{\partial f}{\partial y}\right) = 4xye^{xy^2} + 2x^2y^3e^{xy^2}. \quad \square$$

On the surface there seems to be no reason why these two very different computational procedures should necessarily give the same result. To understand why begin by writing the mixed partials as a limit:

$$\frac{\partial^2 f}{\partial y \partial x} = \frac{\partial}{\partial y}\left(\frac{\partial f}{\partial x}\right) = \lim_{k \to 0} \frac{\frac{\partial f}{\partial x}(x, y+k) - \frac{\partial f}{\partial x}(x, y)}{k}.$$

Next rewrite the inside partials as limits to get

$$\frac{\partial^2 f}{\partial y \partial x} = \lim_{k \to 0} \lim_{h \to 0} \frac{f(x+h, y+k) - f(x, y+k) - f(x+h, y) + f(x, y)}{hk}. \tag{1}$$

But if we began with the partials in the reversed order, we would obtain almost the same expression:

$$\frac{\partial^2 f}{\partial x \partial y} = \lim_{h \to 0} \lim_{k \to 0} \frac{f(x+h, y+k) - f(x+h, y) - f(x, y+k) + f(x, y)}{hk}. \tag{2}$$

The numerators and denominators in both (1) and (2) are exactly the same; the only difference is in *the order of the limits.* In general changing the order of these *iterated* limits can make a difference, but with the added assumption of continuity things work nicely.

THEOREM 5.3.6 ***(Equality of Mixed Partials)*** *If the functions $\frac{\partial^2 f}{\partial y \partial x}$ and $\frac{\partial^2 f}{\partial x \partial y}$ exist in an open region about (x, y) and are continuous at (x, y), then the mixed partials are equal:*

$$\frac{\partial^2 f}{\partial y \partial x}(x, y) = \frac{\partial^2 f}{\partial x \partial y}(x, y).$$

PROOF Fix x and y and denote the common numerator appearing in (1) and (2) by

$$N(h, k) = f(x+h, y+k) - f(x+h, y) - f(x, y+k) + f(x, y)$$

Observe that if $g(x) = f(x, y+k) - f(x, y)$, then $N(h, k)$ can be expressed as $g(x + h) - g(x)$. The mean value theorem applied to the $g(x)$ on the interval $[x, x + h]$ allows us to express $N(h, k)$ in terms of derivatives. Noting that $g'(x) = \frac{\partial f}{\partial x}(x, y + k) - \frac{\partial f}{\partial x}(x, y)$ we have (for some $x^* \in [x, x + h]$)

$$\begin{aligned} N(h, k) &= g(x + h) - g(x) = hg'(x^*) \\ &= h\left(\frac{\partial f}{\partial x}(x^*, y+k) - \frac{\partial f}{\partial x}(x^*, y)\right). \end{aligned} \tag{3}$$

Applying the mean value theorem again to the function $\frac{\partial f}{\partial x}(x^*, y)$ on the interval $[y, y+k]$ enables us to rewrite the last difference in (3) as $\frac{\partial f}{\partial x}(x^*, y+k) - \frac{\partial f}{\partial x}(x^*, y) = k\frac{\partial^2 f}{\partial y \partial x}(x^*, y^*)$ for some $y^* \in [y, y+k]$. Substituting this in (3) and dividing through by hk gives

$$\frac{N(h,k)}{hk} = \frac{\partial^2 f}{\partial y \partial x}(x^*, y^*).$$

Since $x^* \in [x, x+h]$ and $y^* \in [y, y+k]$, as $(h,k) \to (0,0)$ we have $(x^*, y^*) \to (x, y)$. Recalling that the mixed partials are assumed to be continuous at (x, y), we can conclude that

$$\lim_{(h,k)\to(0,0)} \frac{N(h,k)}{hk} = \lim_{(h,k)\to(0,0)} \frac{\partial^2 f}{\partial y \partial x}(x^*, y^*) = \frac{\partial^2 f}{\partial y \partial x}(x, y).$$

In exactly the same way you can show that $\lim_{(h,k)\to(0,0)} \frac{N(h,k)}{hk}$ is also equal to $\frac{\partial^2 f}{\partial x \partial y}(x, y)$ by starting with $r(y) = f(x+h, y) - f(x, y)$. Since both mixed partials are equal to the same limit they must be equal to each other. ∎

Problems for Section 5.3

5.3.1. Let $P_2(x, y) = y$ be the projection function on the second coordinate. Prove that P_2 is continuous at any point (a, b).

5.3.2. **a)** Prove that if $\lim_{(x,y)\to(a,b)} f(x, y) = L$, then both of the one-dimensional limits exist: $\lim_{x\to a} f(x, b) = L$ and $\lim_{y\to b} f(a, y) = L$.

b) Now consider the function $f(x, y) = \frac{2xy}{x^2+y^2}$ near $(0,0)$. Show that both one-dimensional limits $\lim_{x\to 0} f(x, 0) = 0$ and $\lim_{y\to 0} f(0, y) = 0$ exist.

c) Show that $\lim_{(x,y)\to(0,0)} f(x, y)$ does not exist. Hence the converse of **(a)** is not true in general.

5.3.3. After Theorem 5.3.3 we stated that the proofs of the sum, product, and quotient rules for limits of vector functions are completely analogous to the proofs for scalar functions found in Chapter 2. Look back at these theorems in Chapter 2 and make the necessary changes to prove Theorem 5.3.3.

5.3.4. Prove that $f(x, y) = \frac{x+y}{x-y}$ does not have a limit at $(0,0)$.

5.3.5. Prove that $g(x, y) = \frac{(x+y)^2}{x^2+y^2}$ does not have a limit at $(0,0)$.

5.3.6. Consider again the function in Example 5.3.9 and show that on every line through the origin, $y = ax$, the function has a constant value (excluding (0,0)). On which lines does the function have the largest and smallest values? Can you draw a sketch explaining the behavior of this function around the origin?

5.3.7. A **linear function** is any function satisfying two properties:
(i) $L(\vec{\mathbf{x}} + \vec{\mathbf{y}}) = L(\vec{\mathbf{x}}) + L(\vec{\mathbf{y}})$ and
(ii) $L(k\vec{\mathbf{x}}) = kL(\vec{\mathbf{x}})$ for any constant k.

a) Show that any linear function on $\mathbf{R}^2$ can be written as $L(x, y) = m_1 x + m_2 y$ for some values of m_1 and m_2. (Hint: Let $L(1, 0) = m_1$ and $L(0, 1) = m_2$. Use properties (i) and (ii) to show that for any (x, y) we can write $L(x, y) = xL(1, 0) + yL(0, 1)$.)

b) Show that $|L(x, y)| \leq |(m_1, m_2)| \cdot |(x, y)|$. (Hint: Look at the expression for L in **(a)** and think about dot products and the Cauchy-Schwarz inequality.)

c) Use **(b)** to prove that any linear function is continuous at any point (a, b). (Hint: First use property (i) to write $|L(x, y) - L(a, b)| = |L(x - a, y - b)|$, then apply the inequality **(b)**.)

5.3.8. Show that $g(x, y)$ is continuous everywhere if

$$g(x, y) = \begin{cases} \frac{6x^2 y}{x^2 + y^2}, & \text{if } (x, y) \neq (0, 0) \\ 0, & \text{if } (x, y) = (0, 0). \end{cases}$$

5.3.9. Suppose that f is continuous at (a, b) and that $f(a, b) > 0$. Show that there exists a sufficiently small δ such that if $|(x, y) - (a, b)| < \delta$, then $f(x, y) > 0$.

5.3.10. Let $f(x, y)$ be continuous on a closed, bounded set $\mathcal{D}$ in $\mathbf{R}^2$. We will prove that f is bounded using the Heine-Borel theorem from Section 5.1.

a) For every point (a, b) in $\mathcal{D}$, select a δ-ball such that $|(x, y) - (a, b)| < \delta$ implies $|f(x, y) - f(a, b)| < 1$. What guarantees that we can find such a δ for any point in $\mathcal{D}$?

b) Using the Heine-Borel theorem obtain a finite cover of the form

$$B_{\delta_1}(a_1, b_1), \ldots, B_{\delta_n}(a_n, b_n)$$

and show that $\max\{f(a_i, b_i)\} + 1$ is an upper bound for f on $\mathcal{D}$.

5.3.11. Show that $f(x, y)$ in the previous problem must actually achieve a maximum value on $\mathcal{D}$. (Hint: Just as in Theorem 2.6.8 use the boundedness of f to obtain a least upper bound, M, and then consider $\frac{1}{M - f(x,y)}$ on $\mathcal{D}$.)

5.3.12. Suppose that we tried to define the derivative of $f(x, y)$ as

$$f'(x, y) = \lim_{(h_1, h_2) \to (0,0)} \frac{f(x + h_1, y + h_2) - f(x, y)}{|(h_1, h_2)|}.$$

Show that even the simple function $f(x, y) = x$ would not be differentiable by this definition. (Hint: Consider the limits from the right and left in the x direction, that is, for $(h_1, 0)$ where h_1 is first positive and then negative.)

5.3.13. Find the slope of the surface in the x and y directions at the indicated points for each of the surfaces defined by the following functions:

a) $f(x, y) = 4 - x^2 - y^2 + 2xy$ at $(1, 1)$

b) $g(x, y) = \sin(xy)$ at $(1, \frac{\pi}{2})$

c) $h(x, y) = \frac{y^2}{x+y}$ at $(2, 2)$

5.3.14. For each of the functions in the preceding problem, compute both $\frac{\partial^2}{\partial x \partial y}$ and $\frac{\partial^2}{\partial y \partial x}$.

5.3.15. Let $g(x, y) = \begin{cases} \frac{(x-y)^2}{x^2+y^2}, & \text{if } (x, y) \neq (0, 0) \\ 1, & \text{if } (x, y) = (0, 0) \end{cases}$

a) Show that g has a constant value on the x- and y-axes and compute the partial derivatives at the origin.

b) Show that g is not continuous at the origin.

5.3.16. Complete the proof of Theorem 5.3.6 by showing that

$$\lim_{(h,k)\to(0,0)} \frac{N(h, k)}{hk} = \frac{\partial^2 f}{\partial x \partial y}(x, y).$$

5.3.17. We encountered iterated limits in the definition of mixed partials. In general the iterated limit $\lim_{x\to a} \lim_{y\to b}$ is not the same as the two-dimensional limit $\lim_{(x,y)\to(a,b)}$.

a) Consider the function in Example 5.3.9 and show that the iterated limit given by $\lim_{x\to 0} \lim_{y\to 0} f(x, y)$ exists even though the two-dimensional limit fails to exist.

b) Let $f(x, y) = \begin{cases} x \sin \frac{1}{y} & \text{if } y \neq 0 \\ 0 & \text{if } y = 0 \end{cases}$. Prove that $\lim_{(x,y)\to(0,0)} f(x, y) = 0$ and then carefully explain why $\lim_{x\to 0} \lim_{y\to 0} f(x, y)$ fails to exist. (It is true, however, that if both the two-dimensional and iterated limits exist, then they give the same value.)

5.3.18. When the mixed partials are not continuous, Theorem 5.3.6 can fail. Consider the function

$$f(x, y) = \begin{cases} \frac{x^3 y - y^3 x}{x^2+y^2}, & \text{if } (x, y) \neq (0, 0) \\ 0, & \text{if } (x, y) = (0, 0). \end{cases}$$

This is an example of a function for which the mixed partial derivatives are not equal at (0,0).

a) Consider $\frac{\partial^2 f}{\partial y \partial x}$ as defined by the limit in equation (1) preceding Theorem 5.3.6. For our particular function at the point $x = 0, y = 0$, show that (1) collapses to $\lim_{k\to 0} \lim_{h\to 0} \frac{f(h,k)}{hk}$.

b) Show $\frac{\partial^2 f}{\partial y \partial x}(0, 0) = -1$.

c) Similarly show that $\frac{\partial^2 f}{\partial x \partial y}(0, 0) = 1$. (Use (2) before Theorem 5.3.6.)

5.4 The Derivative for Vector Functions

In the last section we saw that partial derivatives can exist even at points at which the surface is discontinuous (see Example 5.3.9). This is disturbing, since we have come to expect that the existence of derivatives represents a degree of smoothness at least sufficient to gaurantee continuity. Partial derivatives, which only depend on two particular curves through a point, are clearly not enough. In this section we restructure our definition of the derivative in a way that will premit us to generalize derivatives to vector functions in a more satisfactory way. Our definition of the derivative will involve approximating functions using linear functions.

5.4.1 Tangent Lines and Planes

Return for a moment to the basic geometric notion that gave birth to the derivative. Newton knew nothing of our formal definitions of limits. Instead he worked geometrically with tangents and velocities. Newton thought of curves as being formed from infinitesimally small linear segments that were smoothly joined to make a sort of "infinitesimal" polygonal approximation. The tangent line at any point was simply a matter of extending the infinitesimal linear piece of the curve into a line.[1] In Newton's conception the tangent line coincided with the curve in an infinitesimally small region. Today we express these ideas by saying that the tangent line is a **local approximation** to the curve. The following limit definition makes this idea rigorous:

> *The function $f(x)$ has a tangent at $x = x_0$ if there is a straight line $T(x) = m(x - x_0) + f(x_0)$ through $(x_0, f(x_0))$ such that*
>
> $$\lim_{x \to x_0} \frac{f(x) - T(x)}{|x - x_0|} = 0.$$

Look at this limit more closely. It implies that for x's close to x_0 the curve $f(x)$ must be much closer to the tangent line $T(x)$ than x is to x_0 (since the ratio of these distances goes to 0). For example, in a δ-region corresponding to $\epsilon = 1/10$, the distance between $f(x)$ and $T(x)$ must be less than $1/10$ of the distance between x and x_0. For $\epsilon = 1/100$ it cannot exceed a hundredth the distance. Thus the limit forces the curve

[1] Carl B. Boyer, *The History of the Calculus and Its Conceptual Development* (New York: Dover Publications, Inc., 1959), pp. 189–190.

$f(x)$ to flatten out and adhere more and more tightly to its tangent line as x gets close to x_0. This can be illustrated effectively on a computer by magnifying the graph of a function about a particular point. Under successively higher magnifications the graph of the function appears more and more like a straight line with slope equal to the slope of the tangent line at that point. (Try it!)

How does this "tangent line" limit relate to our earlier definition of the derivative? The next theorem shows that these two ideas are really the same.

THEOREM 5.4.1 *$f(x)$ is differentiable at $x = x_0$ if and only if there is a straight line $T(x) = m(x - x_0) + f(x_0)$ through $(x_0, f(x_0))$ such that*

$$\lim_{x \to x_0} \frac{f(x) - T(x)}{|x - x_0|} = 0. \tag{1}$$

Furthermore, the slope m is the derivative $f'(x_0)$.

PROOF We prove that the limit (1) implies that the function f is differentiable with $f'(x_0) = m$. The implication in the other direction is left as an exercise.

Assume that for some line $T(x)$ limit (1) holds. Hence for some m

$$\lim_{x \to x_0} \frac{f(x) - (m(x - x_0) + f(x_0))}{|x - x_0|} = 0.$$

In terms of the ϵ, δ definition this says that for any $\epsilon > 0$ there is a $\delta > 0$ such that

$$\text{if} \quad 0 < |x - x_0| < \delta, \quad \text{then} \quad \left| \frac{f(x) - (m(x - x_0) + f(x_0))}{|x - x_0|} - 0 \right| < \epsilon.$$

Equivalently we may say that $\left| \frac{f(x) - (m(x - x_0) + f(x_0))}{x - x_0} \right| < \epsilon$. Simplifying further we conclude that there is a $\delta > 0$ such that

$$\text{if} \quad 0 < |x - x_0| < \delta, \quad \text{then} \quad \left| \frac{f(x) - f(x_0)}{x - x_0} - m \right| < \epsilon.$$

But this is the meaning of $\lim_{x \to x_0} \frac{f(x) - f(x_0)}{x - x_0} = m$ according to the ϵ, δ definition of limit. Hence f is differentiable with derivative $f'(x_0) = m$. ∎

Theorem 5.4.1 shows that our original definition of differentiability is entirely equivalent to the existence of a local approximation by a tangent line expressed by the limit (1). Therefore this limit can serve as an *alternative definition for the derivative.* The advantage of this lies in the fact that the concept of local approximation by linear functions underlying this approach can be extended to vector spaces where it can serve as the *definition* of the derivative.

In the same way as Newton conceived of curves as chains of infinitesimally short lines, differentiable surfaces described by $f : \mathbf{R}^2 \to \mathbf{R}$ might be thought of as being composed of infinitesimally small flat planar pieces sewn smoothly together. Intuitively, differentiability at a point on this surface means that in a small region about the point the surface is flat enough to be "almost" a plane. The tangent plane is that plane which locally approximates the surface at the point.

The geometric notion of flatness and planes is captured algebraically by the concept of **linearity**. In one dimension, linear functions are lines of the form $l(x) = mx$, where m is the slope. In $\mathbf{R}^2$, linear functions describe planes. Every linear function on $\mathbf{R}^2$ can be written as

$$l(x, y) = (m_1, m_2) \cdot (x, y) = m_1 x + m_2 y$$

for some pair (m_1, m_2). We might think of (m_1, m_2) as being the slopes of the plane $l(x, y)$. For our purposes linear functions are too restrictive, since they describe only planes through the origin. In general tangent planes need not pass through the origin; therefore we broaden our category of functions to include translations of linear functions. These are known as affine functions.

Definition 5.4.2 *An **affine function** $A(x, y)$ defined on $\mathbf{R}^2$ is a function of the form*

$$A(x, y) = m_1 x + m_2 y + c,$$

where c is a constant.

Notice that if $A(x, y)$ passes through the point $(x_0, y_0, f(x_0, y_0))$, then it must be true that $f(x_0, y_0) = m_1 x_0 + m_2 y_0 + c$. Hence we conclude $c = f(x_0, y_0) - m_1 x_0 - m_2 y_0$. Therefore any affine function $A(x, y)$ that passes through $(x_0, y_0, f(x_0, y_0))$ may be expressed as follows:

$$\begin{aligned} A(x, y) &= m_1(x - x_0) + m_2(y - y_0) + f(x_0, y_0) \\ &= (m_1, m_2) \cdot (x - x_0, y - y_0) + f(x_0, y_0). \end{aligned}$$

This is analogous to the point-slope form of the equation for a straight line with slope m through the point $(x_0, f(x_0))$.

The heart of differentiability lies in being able to *locally approximate a function f by these affine functions.* The formal definition of this for $\mathbf{R}^2$ is the direct generalization of limit (1) in Theorem 5.4.1: instead of approximating curves by tangent lines we approximate surfaces by tangent planes.

DEFINITION 5.4.3 *A function $f(x, y) : \mathbf{R}^2 \to \mathbf{R}$ is* ***differentiable*** *at (x_0, y_0) if and only if f can be locally approximated by an affine function*

$$T(x, y) = (m_1, m_2) \cdot (x - x_0, y - y_0) + f(x_0, y_0)$$

in the sense that

$$\lim_{(x,y)\to(x_0,y_0)} \frac{f(x, y) - T(x, y)}{|(x, y) - (x_0, y_0)|} = 0.$$

The pair of slopes (m_1, m_2) is called the ***(total) derivative*** *of f at (x_0, y_0) and the function $T(x, y)$ is the* ***tangent plane****.*

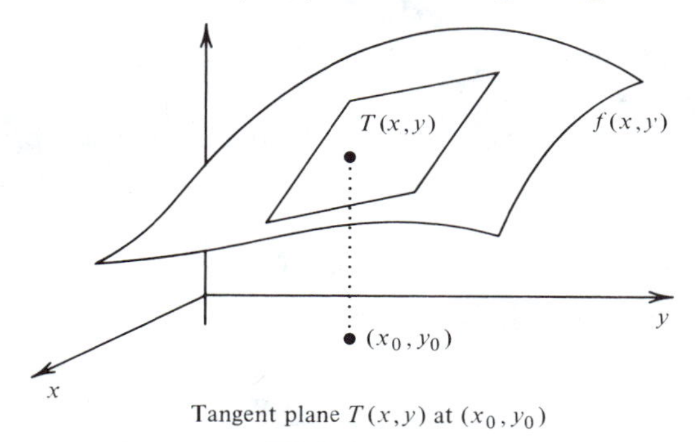

Tangent plane $T(x, y)$ at (x_0, y_0)

Figure 5.4.1

The modifier "total" is used to distinguish this derivative from our earlier notion of partial derivative. Geometrically this limit says that as a δ-disk closes down on the point (x_0, y_0), the surface of $f(x, y)$ flattens out and cleaves more and more closely to the plane. Thus in a small enough region about any point the surface of a differentiable function resembles a flat affine plane. In the spirit of Newton we may think of such a surface as composed of infinitesimally small planar pieces sewn smoothly together.

Consider this derivative (m_1, m_2) in Definition 5.4.3. Since the surface is approximated by the tangent plane near (x_0, y_0), we expect that at (x_0, y_0) the slope of f in any direction should coincide with the slope of the tangent plane in the same direction. In particular, the rates of change of f in the x and y directions (given by $\frac{\partial f}{\partial x}$ and $\frac{\partial f}{\partial y}$ at (x_0, y_0)) should be the same as the rates of change for the tangent plane T in the x and y directions, which are given by $\frac{\partial T}{\partial x} = m_1$ and $\frac{\partial T}{\partial y} = m_2$, respectively.

THEOREM 5.4.4 *If $f(x, y)$ is differentiable at (x_0, y_0), then the partial derivatives of f exist at (x_0, y_0) and*

$$\frac{\partial f}{\partial x}(x_0, y_0) = m_1 \quad \text{and} \quad \frac{\partial f}{\partial y}(x_0, y_0) = m_2.$$

Therefore when f is differentiable, its total derivative is $(\frac{\partial f}{\partial x}, \frac{\partial f}{\partial y})$.

PROOF The partial derivative $\frac{\partial f}{\partial x}(x_0, y_0)$ is simply the derivative of $f(x, y_0)$ at x_0, treating y_0 as constant. We must show that this derivative exists and equals m_1. By Theorem 5.4.1 this is equivalent to showing the limit

$$\lim_{x \to x_0} \frac{f(x, y_0) - f(x_0, y_0) - m_1(x - x_0)}{|x - x_0|} = 0.$$

Now assume that $f(x, y)$ is differentiable. By Definition 5.4.3

$$\lim_{(x,y) \to (x_0,y_0)} \frac{f(x, y) - (m_1, m_2) \cdot (x - x_0, y - y_0) - f(x_0, y_0)}{|(x, y) - (x_0, y_0)|} = 0$$

for some m_1 and m_2. Hence for any $\epsilon > 0$ there is a $\delta > 0$ such that if $|(x, y) - (x_0, y_0)| < \delta$, then

$$\left| \frac{f(x, y) - (m_1, m_2) \cdot (x - x_0, y - y_0) - f(x_0, y_0)}{|(x, y) - (x_0, y_0)|} - 0 \right| < \epsilon.$$

In particular, if $y = y_0$ is kept constant, then whenever we have $|(x, y_0) - (x_0, y_0)| = |x - x_0| < \delta$ it follows that

$$\left| \frac{f(x, y_0) - f(x_0, y_0) - m_1(x - x_0)}{|x - x_0|} \right| < \epsilon.$$

But this is the meaning of $\lim_{x \to x_0} \frac{f(x, y_0) - f(x_0, y_0) - m_1(x - x_0)}{|x - x_0|} = 0$ according to the ϵ, δ definition of the limit. (A similar argument works for y.) ∎

This is nice. When the (total) derivative exists we can compute it easily using partial derivatives. Consequently the ordered pair of partials is important and is given a special name:

DEFINITION 5.4.5 *The **gradient** of a function $f(x, y) : \mathbf{R}^2 \to \mathbf{R}$ at the point (x_0, y_0) is the vector*

$$\nabla f(x_0, y_0) \equiv \left(\frac{\partial f}{\partial x}(x_0, y_0), \frac{\partial f}{\partial y}(x_0, y_0) \right).$$

Notice by the previous theorem that *if f is differentiable, its total derivative is the gradient $\nabla f(x, y)$.* But at this point we do not know if there is *any* function at all which satisfies the criteria for differentiability. One way to remedy this is to show that the simple functions x, y and constants are differentiable, and then show that differentiablity is preserved by arithmetic operations (as we did with continuity). In this way a class of differentiable functions could be constructed. However, we will follow another more direct approach, which shows that if the resulting partial derivatives are continuous, the function is differentiable.

THEOREM 5.4.6 *If f_x and f_y exist in an open region containing (x_0, y_0) and are continuous at (x_0, y_0), then $f(x, y)$ is differentiable at (x_0, y_0).*

PROOF From Theorem 5.4.4 the only possible candidates for m_1 and m_2 in Definition 5.4.3 are $m_1 = f_x(x_0, y_0)$ and $m_2 = f_y(x_0, y_0)$. Thus to prove that f is differentiable we must show that

$$\lim_{(x,y)\to(x_0,y_0)} \frac{f(x,y) - (f_x(x_0,y_0), f_y(x_0,y_0)) \cdot (x - x_0, y - y_0) - f(x_0,y_0)}{|(x,y) - (x_0,y_0)|} = 0.$$

That is, given an arbitrary $\epsilon > 0$ we must find a δ such that whenever $0 < |(x, y) - (x_0, y_0)| < \delta$, then

$$\left| \frac{f(x,y) - (f_x(x_0,y_0), f_y(x_0,y_0)) \cdot (x - x_0, y - y_0) - f(x_0,y_0)}{|(x - x_0, y - y_0)|} \right| < \epsilon. \quad (2)$$

To show there is such a δ we first rewrite the numerator by expressing $f(x, y) - f(x_0, y_0)$ as the sum of two differences, $[f(x, y) - f(x_0, y)]$ and $[f(x_0, y) - f(x_0, y_0)]$. Now we apply the mean value theorem to each difference (first making sure to choose the δ-region small enough so that f_x and f_y exist). Thus for some $x^* \in [x, x_0]$ and $y^* \in [y, y_0]$ we have

$$\begin{aligned} f(x,y) - f(x_0,y_0) &= [f(x,y) - f(x_0,y)] + [f(x_0,y) - f(x_0,y_0)] \\ &= f_x(x^*,y)(x - x_0) + f_y(x_0,y^*)(y - y_0) \\ &= \Big(f_x(x^*,y), f_y(x_0,y^*)\Big) \cdot \Big(x - x_0, y - y_0\Big). \end{aligned}$$

Substituting this dot product for $f(x, y) - f(x_0, y_0)$ in the numerator of expression (2) and combining it with the remaining dot product yields

$$\left| \frac{\Big(f_x(x^*,y) - f_x(x_0,y_0), f_y(x_0,y^*) - f_y(x_0,y_0)\Big) \cdot \Big(x - x_0, y - y_0\Big)}{|(x - x_0, y - y_0)|} \right| < \epsilon. \quad (3)$$

Thus the theorem will be proven if we can find a δ region about (x_0, y_0) such that (3) holds.

Consider the vector $(f_x(x^*, y) - f_x(x_0, y_0), f_y(x_0, y^*) - f_y(x_0, y_0))$. Since both partials are continuous at (x_0, y_0), there exists a $\delta > 0$ such that, $0 < |(x, y) - (x_0, y_0)| < \delta$ implies

$$|f_x(x, y) - f_x(x_0, y_0)| < \epsilon/2 \quad \text{and} \quad |f_y(x, y) - f_y(x_0, y_0)| < \epsilon/2.$$

Since $x^* \in [x_0, x]$ and $y^* \in [y_0, y]$, we know that $|(x^*, y) - (x_0, y_0)| < \delta$ and $|(x_0, y^*) - (x_0, y_0)| < \delta$; thus the preceding inequalities hold at the points (x^*, y) and (x_0, y^*). Therefore, using the box lemma, the length of our vector is bounded as follows:

$$\Big|\Big(f_x(x^*, y) - f_x(x_0, y_0), f_y(x_0, y^*) - f_y(x_0, y_0)\Big)\Big| < \epsilon/2 + \epsilon/2 = \epsilon.$$

Therefore, if $0 < |(x, y) - (x_0, y_0)| < \delta$, then

$$\frac{\Big|\Big(f_x(x^*, y) - f_x(x_0, y_0), f_y(x_0, y^*) - f_y(x_0, y_0)\Big)\Big| \cdot \Big|\Big(x - x_0, y - y_0\Big)\Big|}{|(x - x_0, y - y_0)|} < \epsilon. \tag{4}$$

By the Cauchy-Schwarz inequality, (4) implies (3), which proves the theorem. ∎

It is usually easy to verify that the expressions for f_x and f_y are continuous. Theorem 5.4.6 then assures us that f is differentiable and hence a tangent affine plane approximation to f exists.

EXAMPLE 5.4.1 The tangent plane approximation to $f(x, y) = 3x^2\sqrt{y}$ at $(1, 4)$ exists, since the partials are continuous there. To find the plane we compute both partials at $(1, 4)$. First, $f_x(x, y) = 6x\sqrt{y}$, so $f_x(1, 4) = 12$. Next, $f_y(x, y) = 3x^2/2\sqrt{y}$, so $f_y(1, 4) = 3/4$. The equation of the tangent affine approximation is then

$$T(x, y) = (12, 3/4) \cdot (x - 1, y - 4) + 6.$$

Since the plane approximates f near $(1, 4)$, we may use it to estimate the value of f at nearby points. For instance, at $(1.1, 3.8)$ f will be very nearly equal to $T(1.1, 3.8) = 12(.1) + 3/4(-.2) + 6 = 7.05$. The actual value of f is closer to 7.076. □

In Example 5.3.9 we saw that the existence of partial derivatives alone is not sufficient to guarantee that the function is continuous. However, differentiability as defined in Definition 5.4.3 *does* imply continuity.

THEOREM 5.4.7 *If $f(x, y)$ is differentiable at (x_0, y_0), then it is continuous at (x_0, y_0).*

PROOF Because f is differentiable we know that

$$\lim_{(x,y)\to(x_0,y_0)} \frac{f(x,y) - T(x,y)}{|(x,y) - (x_0,y_0)|} = 0.$$

We also have $\lim_{(x,y)\to(x_0,y_0)} |(x,y) - (x_0,y_0)| = 0$. Multiplying these two limits yields

$$\lim_{(x,y)\to(x_0,y_0)} f(x,y) - T(x,y) = 0,$$

which says that

$$\lim_{(x,y)\to(x_0,y_0)} f(x,y) - m_1(x - x_0) - m_2(y - y_0) - f(x_0, y_0) = 0.$$

But

$$\lim_{(x,y)\to(x_0,y_0)} m_1(x - x_0) + m_2(y - y_0) = 0.$$

Therefore

$$\lim_{(x,y)\to(x_0,y_0)} f(x,y) - f(x_0,y_0) = 0.$$

This shows that f is continuous. ∎

The essence of differentiability lies in being able to locally approximate a function with affine functions. In one dimension these affine functions are straight lines and their slope is the derivative $m = f'(x_0)$. In $\mathbf{R}^2$, affine functions are planes and the pair of slopes (m_1, m_2) is given by the (total) derivative $(f_x(x_0, y_0), f_y(x_0, y_0))$. Thus the key to understanding differentiability is to first thoroughly understand how linear (affine) functions work. Then using limits we may exploit the principle that in any small region a differentiable function is "almost" equal to one of these affine functions.

5.4.2 Directional Derivatives

In Section 5.3 we saw that the partial derivatives of a function represented the rate of change in f in the x and y directions. There is nothing sacred about the x and y directions, and we could equally well measure the rate of change in f along any direction. This leads to the concept of a directional derivative.

In directional derivatives we specify the direction by using a unit vector $\vec{\mathbf{u}} = (u_1, u_2)$. For example, $\vec{\mathbf{u}} = (1, 0)$ indicates the x direction, whereas $\vec{\mathbf{u}} = (\frac{\sqrt{2}}{2}, \frac{\sqrt{2}}{2})$ is the diagonal direction at 45°. In general $\vec{\mathbf{u}} = (\cos\theta, \sin\theta)$ is the direction given by the angle θ measured counterclockwise from the x-axis. (See Figure 5.4.2.) To measure the rate of change in f at $\vec{\mathbf{x}}_0$ in the direction of $\vec{\mathbf{u}}$ we first move a small distance h in the $\vec{\mathbf{u}}$ direction from the point $\vec{\mathbf{x}}_0$ to the point $\vec{\mathbf{x}}_0 + h\vec{\mathbf{u}}$. Then we compare the change in f between these points (Δf) with the distance moved (h). The limit as h tends to 0 then gives the directional derivative (Figure 5.4.3).

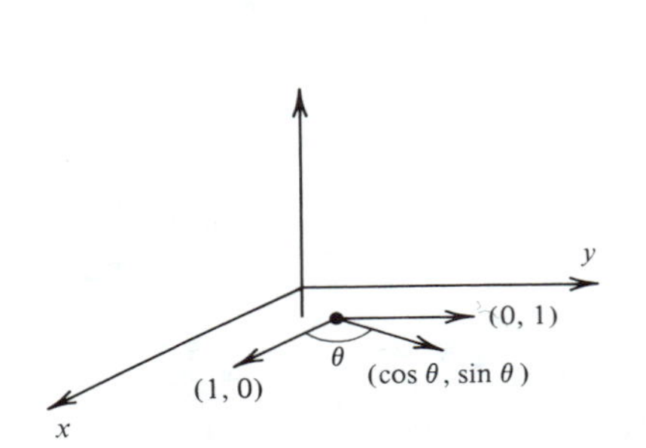

Figure 5.4.2

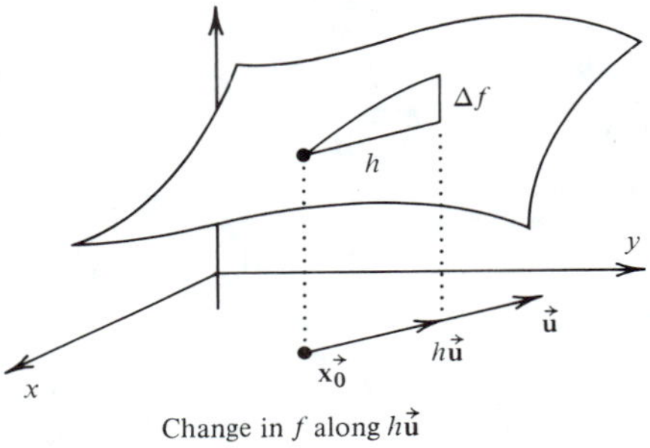

Change in f along $h\vec{u}$

Figure 5.4.3

Definition 5.4.8 *The* ***directional derivative*** *of f at the point $\vec{\mathbf{x}}_0 = (x_0, y_0)$ in the direction of the unit vector $\vec{\mathbf{u}} = (u_1, u_2)$ is*

$$\frac{\partial f}{\partial \vec{\mathbf{u}}} \equiv \lim_{h\to 0} \frac{f(\vec{\mathbf{x}}_0 + h\vec{\mathbf{u}}) - f(\vec{\mathbf{x}}_0)}{h} = \lim_{h\to 0} \frac{f(x_0+hu_1, y_0+hu_2) - f(x_0, y_0)}{h}.$$

Example 5.4.2

a) Find the directional derivative of $f(x, y) = e^{x+xy}$ at $(0, 0)$ in the direction $\vec{\mathbf{u}} = (\frac{\sqrt{2}}{2}, -\frac{\sqrt{2}}{2})$.

b) Find the directional derivative of $A(x, y) = m_1(x - x_0) + m_2(y - y_0) + k$ at (a, b) in the direction $\vec{\mathbf{u}} = (u_1, u_2)$.

SOLUTION **a)** From the definition,

$$\frac{\partial f}{\partial \vec{\mathbf{u}}}(0,0) = \lim_{h\to 0} \frac{f(h\frac{\sqrt{2}}{2}, -h\frac{\sqrt{2}}{2}) - f(0,0)}{h} = \lim_{h\to 0} \frac{e^{h\frac{\sqrt{2}}{2} - \frac{1}{2}h^2} - e^0}{h}.$$

The denominator and numerator both approach 0, and we may therefore use L'Hôpital's rule to evaluate this limit:

$$\frac{\partial f}{\partial \vec{\mathbf{u}}}(0,0) = \lim_{h\to 0} \left(\frac{\sqrt{2}}{2} - h\right) e^{h\frac{\sqrt{2}}{2} - \frac{1}{2}h^2} = \frac{\sqrt{2}}{2}.$$

b) To compute the directional derivative for the affine function A at (a,b) first verify that $A(a+hu_1, b+hu_2) - A(a,b) = m_1hu_1 + m_2hu_2$. Thus

$$\frac{\partial A}{\partial \vec{\mathbf{u}}} = \lim_{h\to 0} \frac{m_1hu_1 + m_2hu_2}{h} = m_1u_1 + m_2u_2 = (m_1, m_2)\cdot(u_1, u_2). \quad \square$$

Part **(b)** of the above example shows that the rate of change of an affine function at any point (a,b) in any direction is given as the dot product of the slopes (m_1, m_2) with the direction $\vec{\mathbf{u}} = (u_1, u_2)$. The idea of differentiability at a point (x_0, y_0) is that a function is locally approximated by the tangent affine plane that has slopes given by the gradient $\nabla f(x_0, y_0) = \left(\frac{\partial f}{\partial x}(x_0, y_0), \frac{\partial f}{\partial y}(x_0, y_0)\right)$. As we close in on (x_0, y_0) the function resembles this tangent plane more and more. Thus we expect that right at (x_0, y_0) the rate of change in f in *any* direction should be equal to the rate of change in the approximating tangent plane in that same direction.

THEOREM 5.4.9 *If f is differentiable at (x_0, y_0), then the directional derivative in any direction $\vec{\mathbf{u}}$ exists and is given by*

$$\frac{\partial f}{\partial \vec{\mathbf{u}}}(x_0, y_0) = \nabla f(x_0, y_0)\cdot \vec{\mathbf{u}} = \left(\frac{\partial f}{\partial x}(x_0, y_0), \frac{\partial f}{\partial y}(x_0, y_0)\right)\cdot(u_1, u_2).$$

PROOF Since f is differentiable at (x_0, y_0), the limit in Definition 5.4.3 exists, so for any $\epsilon > 0$ there is a $\delta > 0$ such that if $|(x,y) - (x_0, y_0)| < \delta$, then

$$\left|\frac{f(x,y) - T(x,y)}{|(x,y) - (x_0, y_0)|}\right| < \epsilon, \tag{5}$$

where $T(x,y) = (f_x(x_0, y_0), f_y(x_0, y_0))\cdot(x - x_0, y - y_0) + f(x_0, y_0)$. In particular, let $(x,y) = (x_0, y_0) + h(u_1, u_2)$ with $|h| < \delta$. Then $|(x,y) - (x_0, y_0)| = |h||\vec{\mathbf{u}}| < \delta$, so (5) holds. Substituting for (x,y) in (5) we have

$$\left|\frac{f(x_0+hu_1, y_0+hu_2) - (f_x(x_0, y_0), f_y(x_0, y_0))\cdot(hu_1, hu_2) - f(x_0, y_0)}{h}\right| < \epsilon.$$

or equivalently,

$$\left|\frac{f(x_0+hu_1, y_0+hu_2) - f(x_0, y_0)}{h} - (f_x(x_0, y_0), f_y(x_0, y_0))\cdot(u_1, u_2)\right| < \epsilon.$$

This holds whenever $h < \delta$. But this is precisely the required ϵ, δ proof that the directional derivative $\frac{\partial f}{\partial \vec{\mathbf{u}}}$ exists and equals

$$\lim_{h\to 0} \frac{f(x_0 + hu_1, y_0 + hu_2) - f(x_0, y_0)}{h} = (f_x(x_0, y_0), f_y(x_0, y_0))\cdot(u_1, u_2). \quad \blacksquare$$

In the discussion of limits in Section 5.3 we saw that limits along the x and y directions were not sufficient to characterize the behavior of a function near a point (see Examples 5.3.5 and 5.3.9). However, *if f is differentiable*, Theorem 5.4.9 shows that the directional derivative of f is determined by $\nabla f(x_0, y_0)$. Hence in this case the rate of change of f in any direction *is determined* by the slopes of f along only two curves, $f(x, y_0)$ and $f(x_0, y)$. The earlier pathologies of limit behavior do not arise here because the definition of differentiability implies that the function is nearly equal to a flat tangent plane in a sufficiently small region near (x_0, y_0). This forces f to behave nicely near the point, like its affine approximation. Thus in the case of differentiable functions the rates of change in any direction are determined by the two rates $\frac{\partial f}{\partial x}$ and $\frac{\partial f}{\partial y}$.

Now that the rate of change in any direction can be easily computed, we may ask which direction yields the maximum rate of increase and decrease in f.

COROLLARY 5.4.10 *Let f be differentiable at (x_0, y_0). Then the gradient $\nabla f(x_0, y_0)$ points in the direction of maximum increase in f at the point (x_0, y_0), and this rate of increase is $|\nabla f(x_0, y_0)|$. The maximum rate of decrease occurs in the direction $-\nabla f(x_0, y_0)$ with magnitude $|\nabla f(x_0, y_0)|$.*

PROOF By Theorem 5.4.9 the rate of change in the direction $\vec{\mathbf{u}}$ is $\nabla f(x_0, y_0) \cdot \vec{\mathbf{u}}$. From Example 5.1.1 we may write this dot product as

$$\frac{\partial f}{\partial \vec{\mathbf{u}}} = \nabla f(x_0, y_0) \cdot \vec{\mathbf{u}} = |\nabla f(x_0, y_0)||\vec{\mathbf{u}}| \cos \theta = |\nabla f(x_0, y_0)| \cos \theta,$$

where θ is the angle between the unit vector $\vec{\mathbf{u}}$ and the vector $\nabla f(x_0, y_0)$. Since $|\nabla f(x_0, y_0)|$ is a constant, this expression is maximized when $\cos \theta$ is a maximum. This occurs when $\cos \theta = 1$ at $\theta = 0$, in which case $\vec{\mathbf{u}}$ is pointing in the direction of ∇f. Similarly the minimum value of this expression is attained when $\cos \theta = -1$ at $\theta = \pi$, in which case $\vec{\mathbf{u}}$ points in the direction of $-\nabla f$. ∎

EXAMPLE 5.4.3 Heat tends to flow toward colder regions at a rate that is proportional to the drop in temperature. Corollary 5.4.10 allows us to express this rate of flow in terms of the gradient of the temperature. If $T(x, y)$ gives the temperature at the point (x, y), then $\nabla T(x, y)$ is a vector pointing in the direction of greatest increase in temperature. Similarly, the coldest direction is given by $-\nabla T(x, y)$ and the rate of change in temperature in this direction is $-|\nabla T(x, y)|$. Therefore the vector describing the flow of heat energy is given by

$$\text{heat flow vector } = \mathbf{J}(x, y) = -\kappa \nabla T(x, y).$$

The constant κ is called the conductivity of the material in which the heat is flowing. For example, if the temperature is $T(x, y) = 3x^2y^2$ and $\kappa = 2$, we have $\mathbf{J}(x, y) = -2(6xy^2, 6yx^2)$. Thus at the point $(1, 2)$, the heat energy will flow in the direction of $(-\frac{2\sqrt{5}}{5}, -\frac{\sqrt{5}}{5})$ at a rate of $|(48, 24)| = 24\sqrt{5}$. □

Local maxima and minima are among the most important points for a function. Graphically these points correspond to peaks and valleys on the surface of f. Corollary 5.4.10 shows that at any point (x_0, y_0), f is increasing at a rate of $|\nabla f(x_0, y_0)|$ along the direction given by the gradient. However, if f is to have a local maximum, f should not increase in any direction. This can only happen when $\nabla f(x_0, y_0) = (0, 0)$. The proof of this is left as an exercise. Such points where the gradient is the zero vector are called **critical points** of f.

COROLLARY 5.4.11 *If (x_0, y_0) is a local maximum or mimimum for f and if f is differentiable at (x_0, y_0), then $\nabla f(x_0, y_0) = (0, 0)$.*

EXAMPLE 5.4.4 To find the critical points $f(x, y) = (x + y)^2 + (x + 2)^2$, we must locate all points at which both partials are zero:

$$\frac{\partial f}{\partial x}(x, y) = 4x + 2y + 4 = 0, \qquad \frac{\partial f}{\partial y}(x, y) = 2x + 2y = 0.$$

Solving these two equations simultaneously gives $x = -2$ and $y = 2$. (In this particular case it is easy to check that this point is a minimum since $f(-2, 2) = 0$ and at other points $f(x, y) \geq 0$.) □

The simultaneous equations $\frac{\partial f}{\partial x} = 0$ and $\frac{\partial f}{\partial y} = 0$ are not always linear as they are in the preceding example, and it may be impossible to solve them analytically. Nevertheless, the geometry of Corollary 5.4.10 suggests a way to proceed. Since the gradient always points in the direction of maximum increase in the surface, starting at any point $\vec{\mathbf{a}}$ and following the gradient direction, the value of f will increase. However, since f and its gradient are always changing, we should stop frequently after short distances to recompute the gradient and find the new ascending direction. If the function is fairly well behaved and we reevaluate after sufficiently short steps, this method will always increase f, bringing us to the locally highest point if such a point exists. This procedure is similar to the determined hiker who, wishing to reach the top of a mountain, begins by taking the steepest direction and every few yards reevaluates that direction. This iterative numerical method of locating a maximum is called the **gradient search**. The next point $\vec{\mathbf{a}}_1$ is always found from the previous point $\vec{\mathbf{a}}$ by the simple algorithm

$$\vec{\mathbf{a}}_1 = \vec{\mathbf{a}} + h\nabla f(\vec{\mathbf{a}})$$

As we approach a local maximum the values of $|\nabla f|$ tend to 0.

5.4.3 The Mean Value Theorem and The Chain Rule

The mean value theorem for vector functions is derived by applying the one-dimensional mean value theorem to a carefully constructed auxiliary function. Consider the function $g(t)$ of a single variable obtained from a function $f(x, y)$ by

$$g(t) = f(\vec{\mathbf{a}} + t\vec{\mathbf{u}}) = f(a + tu_1, b + tu_2) \tag{6}$$

where $\vec{\mathbf{a}} = (a, b)$ is a fixed point and $\vec{\mathbf{u}} = (u_1, u_2)$ is a unit vector. The function $g(t)$ is just f evaluated at a point t units along a line from $\vec{\mathbf{a}}$ in the direction of $\vec{\mathbf{u}}$. Thus the graph of $g(t)$ is simply the curve on the surface of f where the vertical plane through this line intersects the surface. (See Figure 5.4.4.)

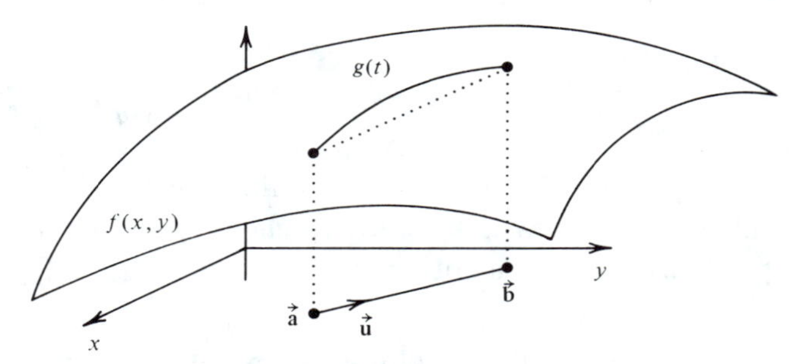

The Mean Value Theorem applied to $g(t)$.

Figure 5.4.4

The next theorem shows that if $f(x, y)$ is differentiable, then the auxiliary function $g(t)$ is differentiable. Moreover, as the picture suggests, the derivative of g turns out to be the same as the derivative of f in the direction of $\vec{\mathbf{u}}$. This enables us to apply the mean value theorem to $g(t)$ and prove the following:

THEOREM 5.4.12 ***(The Mean Value Theorem)*** *Suppose that $f : \mathbf{R}^2 \to \mathbf{R}$ is differentiable in an open set containing the line segment between $\vec{\mathbf{a}}$ and $\vec{\mathbf{b}}$. Then there is a point $\vec{\mathbf{c}}$ on this segment such that*

$$\nabla f(\vec{\mathbf{c}}) \cdot (\vec{\mathbf{b}} - \vec{\mathbf{a}}) = f(\vec{\mathbf{b}}) - f(\vec{\mathbf{a}}).$$

Notice that since ∇f is the derivative of f, this theorem is the exact analogue of the one-dimensional mean value theorem.

PROOF Let $g(t)$ be defined as in (6) with $\vec{\mathbf{u}} = \frac{(\vec{\mathbf{b}}-\vec{\mathbf{a}})}{|\vec{\mathbf{b}}-\vec{\mathbf{a}}|}$. We will show that g is differentiable for $t \in [0, |\vec{\mathbf{b}} - \vec{\mathbf{a}}|]$. Consider the limit

$$\lim_{h\to 0} \frac{g(t+h)-g(t)}{h} = \lim_{h\to 0} \frac{f(\vec{\mathbf{a}}+t\vec{\mathbf{u}}+h\vec{\mathbf{u}})-f(\vec{\mathbf{a}}+t\vec{\mathbf{u}})}{h}.$$

The limit on the right is the derivative of f in the direction of $\vec{\mathbf{u}}$ at the point $\vec{\mathbf{x}} = \vec{\mathbf{a}} + t\vec{\mathbf{u}}$. Since f is differentiable, this limit must exist (Theorem 5.4.9) and therefore

$$g'(t) = \nabla f(\vec{\mathbf{x}}) \cdot \vec{\mathbf{u}} = \nabla f(\vec{\mathbf{a}}+t\vec{\mathbf{u}}) \cdot \frac{(\vec{\mathbf{b}}-\vec{\mathbf{a}})}{|\vec{\mathbf{b}}-\vec{\mathbf{a}}|}. \tag{7}$$

This is valid for any $t \in [0, |\vec{\mathbf{b}} - \vec{\mathbf{a}}|]$; hence we can apply Theorem 3.3.2 to g on this interval. Thus for some $t^* \in [0, |\vec{\mathbf{b}} - \vec{\mathbf{a}}|]$,

$$g'(t^*)(|\vec{\mathbf{b}}-\vec{\mathbf{a}}|) = g(|\vec{\mathbf{b}}-\vec{\mathbf{a}}|) - g(0).$$

But by definition $g(0) = f(\vec{\mathbf{a}})$ and $g(|\vec{\mathbf{b}}-\vec{\mathbf{a}}|) = f(\vec{\mathbf{a}} + |\vec{\mathbf{b}}-\vec{\mathbf{a}}|\vec{\mathbf{u}}) = f(\vec{\mathbf{b}})$. Thus the right side of the preceding equation is simply $f(\vec{\mathbf{b}}) - f(\vec{\mathbf{a}})$. Using (7) to translate the left side in terms of f, and letting $\vec{\mathbf{c}} = \vec{\mathbf{a}} + t^*\vec{\mathbf{u}}$, we have

$$\nabla f(\vec{\mathbf{c}}) \cdot (\vec{\mathbf{b}}-\vec{\mathbf{a}}) = f(\vec{\mathbf{b}}) - f(\vec{\mathbf{a}}). \quad ∎$$

As in Chapter 3, this mean value theorem is the tool that allows us to utilize local "infinitesimal" information given by the derivative to extract global information about the function. A region R is called **convex** if any two points in R can be connected by a line segment that lies entirely in the region. The proof of the following analogue to Theorem 3.3.4 is left as an exercise.

COROLLARY 5.4.13 *If f is differentiable in a convex region and $\nabla f(x,y) = 0$ everywhere in the region, then f is constant there.*

The mean value theorem is also used to extend the chain rule. Suppose that we have two functions

$$\vec{\mathbf{F}}(t) = (x(t), y(t)) : \mathbf{R} \to \mathbf{R}^2 \quad \text{and} \quad g(x,y) : \mathbf{R}^2 \to \mathbf{R}.$$

Then the composition $g \circ \vec{\mathbf{F}}(t) = g(x(t), y(t))$ maps $\mathbf{R}$ to $\mathbf{R}$. The chain rule states that the derivative of this composition can be computed as the dot product of the separate derivatives ∇g and $\vec{\mathbf{F}}'(t)$.

THEOREM 5.4.14 ***(The Chain Rule)*** *If $\vec{\mathbf{F}}(t) = (x(t), y(t))$ is differentiable at the point t and if $g(x, y)$ is continuously differentiable in some δ-disk containing $(x(t), y(t))$, then*

$$\frac{d}{dt} g \circ \vec{\mathbf{F}}(t) = \frac{d}{dt} g(\vec{\mathbf{F}}(t)) = \nabla g(\vec{\mathbf{F}}(t)) \cdot \vec{\mathbf{F}}'(t),$$

or equivalently,

$$\frac{d}{dt} g(x(t), y(t)) = \nabla g(x(t), y(t)) \cdot (x'(t), y'(t)).$$

PROOF We begin the proof of the chain rule by noting that the composition is a map from $\mathbf{R}$ to $\mathbf{R}$, so its derivative is defined by

$$\frac{d}{dt} g(x(t), y(t)) = \lim_{h \to 0} \frac{g(x(t+h), y(t+h)) - g(x(t), y(t))}{h}. \tag{8}$$

Since $\vec{\mathbf{F}}$ is continuous, we can find a sufficiently small h such that the point $(x(t+h), y(t+h))$ will be in the δ-disk where g is differentiable with a continuous derivative ∇g. We may then use Theorem 5.4.12 to write the difference in the numerator of (8) as

$$g(x(t+h), y(t+h)) - g(x(t), y(t)) = \nabla g(\vec{\mathbf{c}}) \cdot [(x(t+h), y(t+h)) - (x(t), y(t))],$$

where $\vec{\mathbf{c}}$ is on the segment joining $(x(t), y(t))$ and $(x(t+h), y(t+h))$. Using the product theorem for limits, we may write the limit in (8) as

$$\frac{d}{dt} g(x(t), y(t)) = \lim_{h \to 0} \nabla g(\vec{\mathbf{c}}) \cdot \lim_{h \to 0} \frac{(x(t+h), y(t+h)) - (x(t), y(t))}{h},$$

provided that both of these limits exist. The second limit is the definition of the derivative $\vec{\mathbf{F}}'(t)$, which exists by assumption. The first limit is more delicate. Notice that since since $\vec{\mathbf{F}}(t)$ is continuous, as h goes to 0 the point $(x(t+h), y(t+h))$ approaches $(x(t), y(t))$; hence $\vec{\mathbf{c}}$ also approaches $(x(t), y(t))$. (Why?) By the continuity of ∇g, we now conclude that $\lim_{h \to 0} \nabla g(\vec{\mathbf{c}}) = \nabla g(x(t), y(t))$, which proves the theorem. ∎

EXAMPLE 5.4.5 Suppose that $g(x, y) = x^2 y - \sin y$ and $\vec{\mathbf{F}}(t) = (2t, t^2)$, then

$$g \circ \vec{\mathbf{F}}(t) = g(2t, t^2) = 4t^4 - \sin(t^2).$$

Therefore $\frac{d(g \circ \vec{\mathbf{F}})}{dt} = 16t^3 - 2t \cos(t^2)$.

The chain rule states that the derivative $\frac{d(g \circ \vec{\mathbf{F}})}{dt}$ can also be computed as $\nabla g = (2xy, x^2 - \cos y)$ evaluated at $\vec{\mathbf{F}} = (2t, t^2)$ dotted with $\vec{\mathbf{F}}'(t) = (x'(t), y'(t)) = (2, 2t)$. This computation gives

$$\begin{aligned}\nabla g(x(t), y(t)) \cdot (x'(t), y'(t)) &= (4t^3, 4t^2 - \cos(t^2)) \cdot (2, 2t) \\ &= 16t^3 - 2t\cos(t^2). \quad \square\end{aligned}$$

There is a graphical interpretation of the chain rule. Since $\vec{\mathbf{F}}(t)$ traces out a curve in the plane, the composition $g(\vec{\mathbf{F}}(t))$ traces a curve on the surface of g (see Figure 5.4.5). The derivative of the composition measures how fast the height of this curve is changing. The chain rule says that the rate of change in height along the curve is the product of how quickly the surface is increasing in the direction of the curve (directional derivative of g in the direction of $\vec{\mathbf{F}}$) times the speed $|\vec{\mathbf{F}}|$ at which the curve is traced out by $\vec{\mathbf{F}}$.

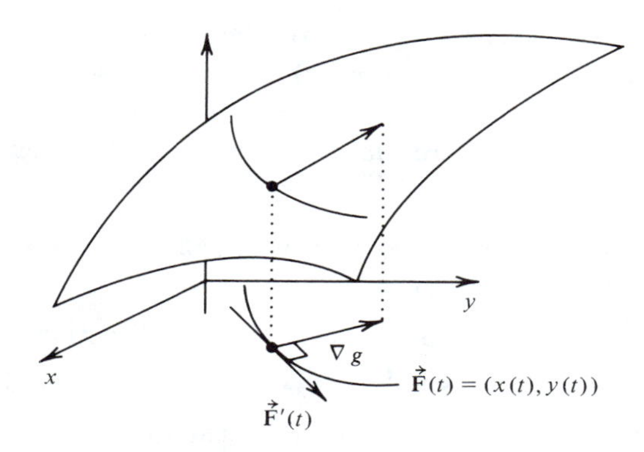

Figure 5.4.5

Example 5.4.6 If $\vec{\mathbf{F}}(t)$ parameterizes some level curve in the plane for the function g, then $g(\vec{\mathbf{F}}(t)) = c$, so its derivative must be 0. Theorem 5.4.14 implies that at any point $(x(t), y(t))$ on the curve we have

$$\nabla g(x(t), y(t)) \cdot (x'(t), y'(t)) = 0.$$

Since the dot product is 0, these two vectors are perpendicular. Now $(x'(t), y'(t))$ is a vector in the direction of a level curve $\vec{\mathbf{F}}$ and ∇g is the gradient. Hence *the gradient always points perpendicularly to the level curves.* $\square$

Problems for Section 5.4

5.4.1. Complete the proof of Theorem 5.4.1 by proving that if f is differentiable at x_0, then the limit (1) holds where $m = f'(x_0)$.

5.4.2. Write out the tangent plane approximation for $f(x,y) = \ln(x^2 + y)$ at $(2,-3)$. Use this plane to estimate $f(1.9,-2.8)$.

5.4.3. Using Theorem 5.4.6 verify that $f(x,y) = \frac{x^2}{y}$ is differentiable at $(-1,2)$ and write down the equation of its tangent plane at $(-1,2)$.

5.4.4. Let $T_f(x,y) = (m_1,m_2)\cdot(x-x_0,y-y_0)+c_1$ be the tangent plane for f at (x_0,y_0) and $T_g(x,y) = (n_1,n_2)\cdot(x-x_0,y-y_0)+c_2$ be the tangent plane for g at (x_0,y_0). What is the tangent plane at (x_0,y_0) for $f+g$? For fg?

5.4.5. If $f(x,y) = ax+by+c$, what does the expression in the limit in Definition 5.4.3 become? What does this say about the tangent plane approximation to f?

5.4.6.

a) Verify that the tangent plane to $f(x,y) = xy$ at the origin is $T(x,y) = 0$.

b) Write out the limit in Definition 5.4.3 and verify that it is indeed equal to 0.

c) Now write the equation for the tangent plane of f at (x_0,y_0) and verify that the limit in Definition 5.4.3 reduces to $\frac{(x-x_0)(y-y_0)}{\sqrt{(x-x_0)^2+(y-y_0)^2}}$.

5.4.7. Prove the sum and product rules for the (total) derivative of f, that is, $\nabla(f+g) = \nabla f + \nabla g$ and $\nabla(fg) = f\nabla g + g\nabla f$.

5.4.8. Compute the directional derivatives of the following functions. (Be sure to make $\vec{\mathbf{v}}$ a unit vector.)

a) $f(x,y) = 2x^2 - 3y^2$ at (1,-1) in the direction of $\vec{\mathbf{v}} = (-1,2)$.

b) $g(x,y) = \ln(x+3y)$ at (2,4) in the direction of $\vec{\mathbf{v}} = (1,1)$.

c) $h(x,y) = x\cos(xy)$ at $(1,\pi)$ in the direction of $\vec{\mathbf{v}} = (1,-2)$.

5.4.9. Compute the directional derivative of $x^2 + xy$ at $(2,-1)$ in the direction of $(-3,4)$ in two ways: first, directly from the definition; and second, using Theorem 5.4.9. (Be sure to convert $(3,4)$ to a unit vector).

5.4.10.

a) Prove that $\lim_{h\to 0} g(h) = L$ if and only if $\lim_{-h\to 0} g(h) = L$.

b) Prove directly from the definition of a directional derivative that $\frac{\partial f}{\partial \vec{\mathbf{u}}} = -\frac{\partial f}{\partial(-\vec{\mathbf{u}})}$. What is this saying in terms of movement on the surface of f? (Hint: Use **(a)** to write the definition of $\frac{\partial f}{\partial \vec{\mathbf{u}}}$ using $\lim_{-h\to 0}$ and then compare this to the definition $\frac{\partial f}{\partial(-\vec{\mathbf{u}})}$.)

5.4.11. The height of a mountain above the point (x,y) is given by $h(x,y) = 3000 - 2x^2 - y^2$. In what direction above the point $(x,y) = (30,-20)$ is the mountain's slope steepest? How steep is it?

5.4.12. The temperature at the point (x, y) is given by $T(x, y) = e^{x+y^2}$. If a heat seeking missile always heads in the direction of maximum increase in temperature, what is the heading of the missile at the point (1,1)?

5.4.13. Assume that f is differentiable everywhere and that $|\nabla f| \leq M$ at all points (x, y). Prove that $|f(x, y) - f(w, z)| \leq M|(x, y) - (w, z)|$. (Hint: Start with the mean value theorem.)

5.4.14. Prove Corollary 5.4.11.

5.4.15. For each of the following functions verify the chain rule. First use Theorem 5.4.14 and then take the derivative of $g(\vec{\mathbf{F}}(t))$ directly.

a) $g(x, y) = xy^2$ and $\vec{\mathbf{F}}(t) = (\cos t, \sin t)$

b) $g(x, y) = y/x$ and $\vec{\mathbf{F}}(t) = (t^2, t^3)$

5.4.16. **a)** Prove Corollary 5.4.13.

b) Extend the result to **polygonally connected regions**, that is, domains in which any two points can be connected by a finite number of straight line segments.

5.4.17. Prove each of the following chain rules:

a) If $h(t) : \mathbf{R} \to \mathbf{R}$ is differentiable at t_0 and $\vec{\mathbf{F}}$ is differentiable at $h(t_0)$, then $\vec{\mathbf{F}}(h(t))$ is differentiable at t_0 with derivative

$$\frac{d}{dt}\vec{\mathbf{F}}(h(t)) = \vec{\mathbf{F}}'(h(t))h'(t).$$

b) If $g(x, y)$ is differentiable at (x_0, y_0) and $h(t)$ is continuously differentiable at $g(x_0, y_0)$, then $h(g(x, y))$ is differentiable at (x_0, y_0) with derivative

$$\nabla h(g(x, y)) = h'(g(x_0, y_0))\nabla g(x_0, y_0).$$

5.4.18. At the end of the proof of Theorem 5.4.14 we claimed that $\lim_{h\to 0} g(\vec{\mathbf{c}}) = g(x(t), y(t))$, where $\vec{\mathbf{c}}$ was a point on the line between the points $(x(t), y(t))$ and $(x(t + h), y(t + h))$. To justify this we must find a δ such that if $|h| < \delta$, then

$$|g(\vec{\mathbf{c}}) - g(x(t), y(t))| < \epsilon.$$

a) First think of $(x(t), y(t))$ as fixed. Show that there exists a δ_1 such that $|(a, b) - (x(t), y(t))| < \delta_1$ implies $|g(a, b) - g(x(t), y(t))| < \epsilon$.

b) Using the fact that $\vec{\mathbf{F}}$ is continuous, show that there is a δ such that when $|h| < \delta$, we have $|(x(t + h), y(t + h)) - (x(t), y(t))| < \delta_1$.

c) Finally, show that if $|h| < \delta$, then $|g(\vec{\mathbf{c}}) - g(x(t), y(t))| < \epsilon$.

5.4.19. Let $g(x, y) = x^2 + 4y^2$.

a) First show that $\vec{\mathbf{F}}(t) = (2a \cos t, a \sin t)$ describes a level curve of g.

b) Verify that $\nabla g(\vec{\mathbf{F}})$ is orthogonal to $\vec{\mathbf{F}}'$.

5.4.20. **a)** Locate a level curve for the surface $g(x, y) = yf(x)$. (Hint: Let $g(x, y) = c$ and solve for y.)

b) Parameterize the curve in **(a)** with a function $\vec{\mathbf{F}}(t)$ and show that $\nabla g(\vec{\mathbf{F}})$ is orthogonal to $\vec{\mathbf{F}}'$. (Hint: Parameterize the curve as you did in problem 5.2.18.)

5.4.21. **Lagrange Multipliers.** Suppose that we have a function $g(x, y)$ and we wish to find the points (x, y) that maximize g subject to a constraint given by $h(x, y) = 0$.

a) Let $\vec{\mathbf{F}}(t)$ describe the level curve of $h(x, y) = 0$. What is the relationship between the direction of ∇h and $\vec{\mathbf{F}}'$ at any point $(x(t), y(t))$? (See Example 5.4.6.)

b) Since $\vec{\mathbf{F}}(t)$ describes the points at which $h(x, y) = 0$, our problem becomes one of *maximizing* $g(\vec{\mathbf{F}}(t))$ *as a function of* t. Use the chain rule to find the derivative of this function with respect to t.

c) If this is to be a maximum, then the derivative must be 0. What does this say about the direction of ∇g and $\vec{\mathbf{F}}'$?

d) Show that $\nabla g = \lambda \nabla h$ for some value of λ. (The scalar λ is called a Lagrange multiplier.)

e) Apply this principle to the case in which $g(x, y) = xy$ and $h(x, y) = x + 2y - 4$ and conclude that $x = 2\lambda$ and $y = \lambda$. Plug this into $h(x, y) = 0$ to solve for λ and so determine x and y.

5.5 Integration

Integration is the limit of a summation process. To implement this process for functions $f : \mathbf{R}^2 \to \mathbf{R}$ we begin by dividing the region of integration in the plane into small pieces. This may be accomplished, for example, by using a grid, as in Figure 5.5.1.

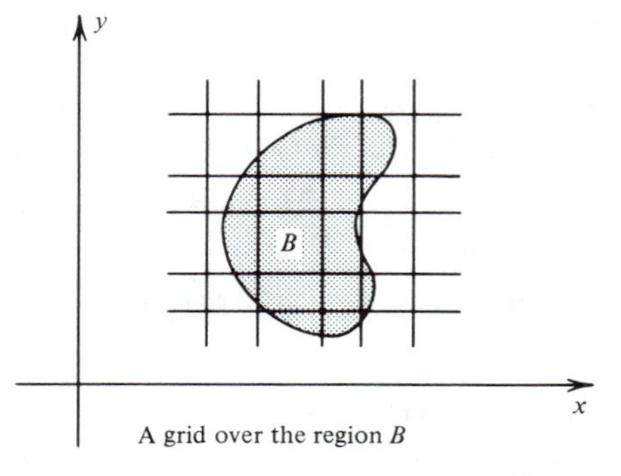

A grid over the region B

Figure 5.5.1

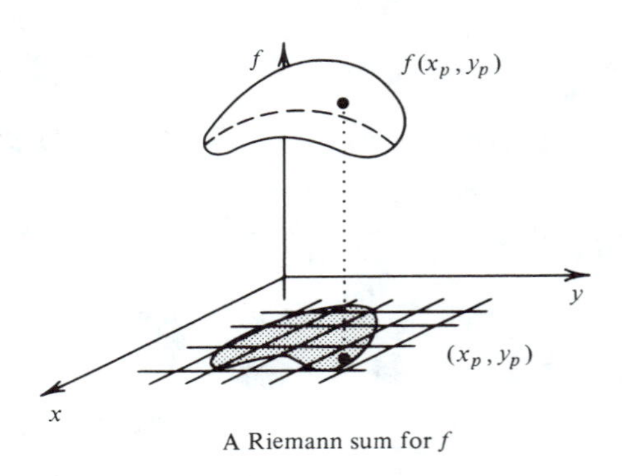

A Riemann sum for f

Figure 5.5.2

For each piece p in the partition the function is evaluated at some point (x_p, y_p) on the piece, and this value is multiplied by the area of the piece. Finally, all the terms are summed:

$$\sum_{\text{pieces}} f(x_p, y_p)(\text{area of piece } p). \tag{1}$$

This sum forms a preliminary approximation for the integral, which is then refined by partitioning the region into smaller pieces. If these refinements lead to a limit, we call this limit the integral.

What does the sum (1) represent? Geometrically, each term in the sum represents the volume of a column whose base is formed from the piece p of the partition and whose height is $f(x_p, y_p)$. Together these columns approximate the *volume of the region lying under the surface* $f(x, y)$ *above the region* B *in the plane* (Figure 5.5.2). In the limit the integral will represent the exact volume of this region.

5.5.1 Definitions

Intuitively this description of integration over planar regions is fine, but from a formal viewpoint it presents many problems. Which types of sets in $\mathbf{R}^2$ can be used as regions of integration? What can be done about places where the grid overlaps the boundary of B (Figure 5.5.1)? Finally what is the precise meaning of "limit" in this summation process and for which functions will this limit exist? These are complex questions, and to develop a sound theory of integration we must start with simpler cases.

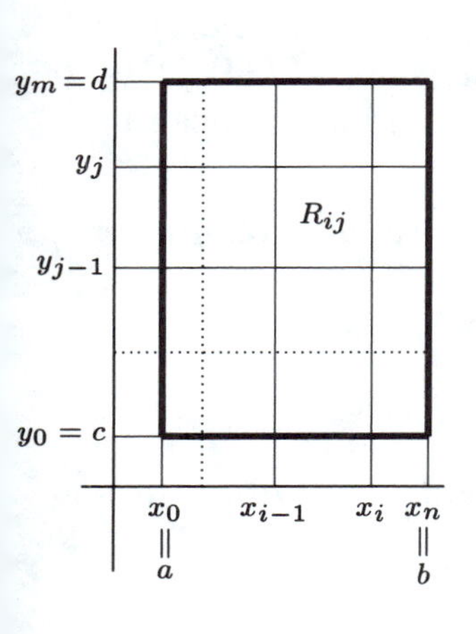

Figure 5.5.3

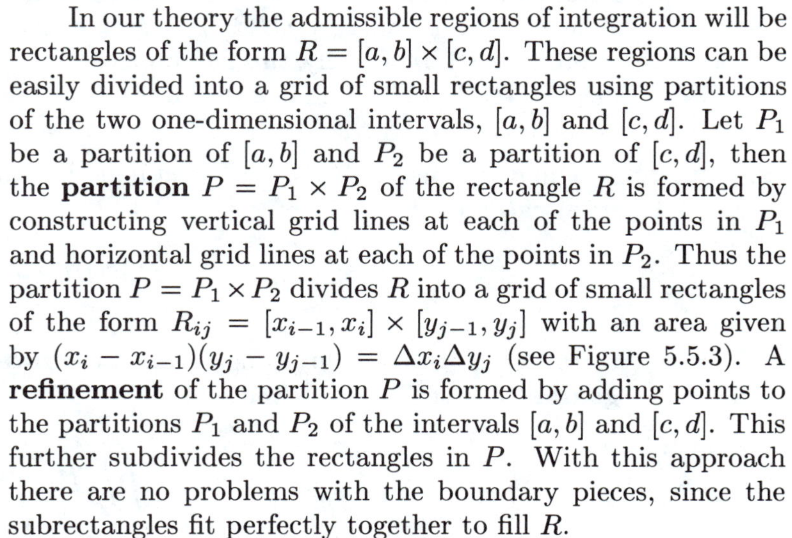

In our theory the admissible regions of integration will be rectangles of the form $R = [a, b] \times [c, d]$. These regions can be easily divided into a grid of small rectangles using partitions of the two one-dimensional intervals, $[a, b]$ and $[c, d]$. Let P_1 be a partition of $[a, b]$ and P_2 be a partition of $[c, d]$, then the **partition** $P = P_1 \times P_2$ of the rectangle R is formed by constructing vertical grid lines at each of the points in P_1 and horizontal grid lines at each of the points in P_2. Thus the partition $P = P_1 \times P_2$ divides R into a grid of small rectangles of the form $R_{ij} = [x_{i-1}, x_i] \times [y_{j-1}, y_j]$ with an area given by $(x_i - x_{i-1})(y_j - y_{j-1}) = \Delta x_i \Delta y_j$ (see Figure 5.5.3). A **refinement** of the partition P is formed by adding points to the partitions P_1 and P_2 of the intervals $[a, b]$ and $[c, d]$. This further subdivides the rectangles in P. With this approach there are no problems with the boundary pieces, since the subrectangles fit perfectly together to fill R.

The sum in (1) can now be expressed more explicitly. We must evaluate f at an arbitrary point (x_{ij}^*, y_{ij}^*) in each subrectangle R_{ij}, multiply by the area $\Delta x_i \Delta y_j$, and sum all of the terms. Suppose that P_1 divides $[a, b]$ into n pieces and P_2 divides $[c, d]$ into m pieces. Then the partition P breaks R into nm subrectangles. Substituting into (1) we obtain the following sum, which is called a **Riemann sum** for f over the rectangle R with partition P:

$$\sum_{i,j} f(x_{ij}^*, y_{ij}^*)\Delta x_i \Delta y_j = \sum_{i=1}^{n}\sum_{j=1}^{m} f(x_{ij}^*, y_{ij}^*)\Delta x_i \Delta y_j. \qquad (2)$$

The integral is obtained as a "limit" of these Riemann sum approximations. As in Chapter 3, to make this concept of limit precise we consider upper and lower sums. For the function f defined on R with partition P the **upper sum** $U(P, f)$ and the **lower sum** $L(P, f)$ are respectively

$$U(P, f) = \sum_{i,j} M_{ij}\Delta x_i \Delta y_j \quad \text{and} \quad L(P, f) = \sum_{i,j} m_{ij}\Delta x_i \Delta y_j,$$

where M_{ij} is the supremum and m_{ij} is the infimum of f on the rectangle R_{ij}. (If f is continuous, these values will be the maximum and minimum values of f on R_{ij}). Upper and lower sums represent the largest and smallest possible Riemann sums for a particular partition, and all other Riemann sums must lie between these values.

The same facts that were proven in Chapter 3 about upper and lower sums are true in this setting as well. Let's review these briefly. First, refinements of a partition always decrease upper sums and increase lower sums (Theorem 3.4.4). Then by using a common refinement it is shown that *any* upper sum $U(P, f)$ is greater than *any* lower sum $L(Q, f)$ (Corollary 3.4.5). Finally, this fact implies that we have $\sup_P\{L(P, f)\} \leq \inf_P\{U(P, f)\}$ (Theorem 3.4.6). Thus integrability and integrals can be defined as before.

DEFINITION 5.5.1 *A bounded function $f(x, y)$ on a rectangle $R = [a, b] \times [c, d]$ is **Riemann integrable** on R if*

$$\sup_P\{L(P, f)\} = \inf_P\{U(P, f)\}.$$

*In this case the common value is called the **integral** (or **double integral**) of f over R and is denoted*

$$\iint_R f \qquad \text{or} \qquad \iint_R f(x, y)\, dA.$$

To prove that a function is integrable by this definition we must show that the upper and lower sums can be made arbitrarily close (Theorem 3.4.9).

EXAMPLE 5.5.1 Prove that the function $f(x, y) = x + y$ is integrable on $R = [0, 1] \times [0, 2]$ and compute its integral.

SOLUTION Given an arbitrary $\epsilon > 0$, we must exhibit a partition such that the upper and lower sums are within ϵ. To do this choose partitions P_1 of $[0, 1]$ and P_2 of $[0, 2]$ into n even pieces. Then $P = P_1 \times P_2$ divides R into n^2 subrectangles of the same size with sides $\Delta x_i = \frac{1}{n}$ and $\Delta y_i = \frac{2}{n}$. On the rectangle $R_{ij} = [x_{i-1}, x_i] \times [y_{j-1}, y_j]$ the maximum of $x + y$ is $M_{ij} = x_i + y_j$ and the minimum is $m_{ij} = x_{i-1} + y_{j-i}$. Therefore

$$\begin{aligned}
|U(P, f) - L(P, f)| &= \left|\sum_{i,j}(x_i + y_j)\Delta x_i \Delta y_j - \sum_{i,j}(x_{i-1} + y_{j-1})\Delta x_i \Delta y_j\right| \\
&= \left|\sum_{i,j}((x_i - x_{i-1}) + (y_j - y_{j-1}))\Delta x_i \Delta y_j\right| \\
&= \left|\sum_{i,j}(\Delta x_i)^2 \Delta y_j + \Delta x_i (\Delta y_j)^2\right| \\
&= \sum_{i,j}\frac{6}{n^3} = n^2\frac{6}{n^3} = \frac{6}{n}.
\end{aligned}$$

The last step follows since there are n^2 subrectangles R_{ij}. If we choose $n > \frac{6}{\epsilon}$ so that $\frac{6}{n} < \epsilon$, the upper and lower sums for this partition will differ by less than ϵ.

To compute the integral we evaluate the upper sum. Notice that by the even spacing of the partition, $x_i = \frac{i}{n}$ and $y_j = \frac{2j}{n}$. Recalling that $\sum_{i=1}^{n} i = \frac{n(n+1)}{2}$ we have

$$U(P, f) = \sum_{i,j}(x_i + y_j)\Delta x_i \Delta y_j = \frac{2}{n^2}\sum_{i=1}^{n}\left[\sum_{j=1}^{n}\left(\frac{i}{n} + \frac{2j}{n}\right)\right]$$

In the inner sum indexed on j the first term sums to i and the second term sums to $\left(\frac{2}{n}\right)\left(\frac{n(n+1)}{2}\right) = n + 1$. Thus

$$\begin{aligned}
U(P, f) = \frac{2}{n^2}\sum_{i=1}^{n}(i + n + 1) &= \frac{2}{n^2}\left(\frac{n(n+1)}{2} + n(n+1)\right) \\
&= \frac{n + 1 + 2(n+1)}{n} = 3 + \frac{2}{n}.
\end{aligned}$$

For sufficiently large n this is arbitrarily close to 3, so $\iint_R f = 3$. Geometrically this number represents the volume of the region under the surface $f(x, y) = x + y$ above the rectangle R. □

5.5.2 Iterated Integration

Computing an integral directly from its definition involves complex sums that are difficult to manipulate. In the one-dimensional case, the fundamental theorem of calculus allowed us to compute integrals using antiderivatives. In $\mathbf{R}^2$ double integrals are computed by a two-stage antidifferentiation process called iterated integration.

DEFINITION 5.5.2 *The* ***iterated integrals*** *of $f(x,y)$ over the rectangle $R = [a,b] \times [c,d]$ are*

$$\int_c^d \int_a^b f(x,y)\,dx\,dy = \int_c^d \left(\int_a^b f(x,y)\,dx \right) dy$$

and

$$\int_a^b \int_c^d f(x,y)\,dy\,dx = \int_a^b \left(\int_c^d f(x,y)\,dy \right) dx.$$

Iterated integration is ordinary one-dimensional integration applied iteratively. For example, to compute $\int_c^d \left(\int_a^b f(x,y)\,dx \right) dy$ we begin with the interior integral and locate an antiderivative for $f(x,y)$ with respect to x, treating y as a constant. (This is the inverse process of partial differentiation with respect to x.) If this antiderivative is $F(x,y)$, for example, then evaluating the interior integral yields $F(b,y) - F(a,y)$, which is a function of y alone. This reduces the iterated integral to $\int_c^d F(b,y) - F(a,y)\,dy$. This in turn is computed by finding an antiderivative with respect to y for the expression $F(b,y) - F(a,y)$.

EXAMPLE 5.5.2 Compute both iterated integrals of $x + y$ over $R = [0,1] \times [0,2]$,

$$\int_0^2 \int_0^1 x + y\,dx\,dy \qquad \text{and} \qquad \int_0^1 \int_0^2 x + y\,dy\,dx.$$

SOLUTION To begin, an antiderivative of $x + y$ with respect to x is $1/2x^2 + yx$. Thus the first iterated integral is

$$\int_0^2 \left(\int_0^1 x + y\,dx \right) dy = \int_0^2 \left(1/2x^2 + yx \Big|_0^1 \right) dy$$
$$= \int_0^2 (1/2 + y)\,dy = 1/2y + 1/2y^2 \Big|_0^2 = 3.$$

The second is done similarly:

$$\int_0^1 \left(\int_0^2 x + y\,dy \right) dx = \int_0^1 \left(xy + 1/2y^2 \Big|_0^2 \right) dy$$

$$= \int_0^1 (2x + 2)\, dy = x^2 + 2x \Big|_0^2 = 3. \quad \square$$

Observe that in Examples 5.5.1 and 5.5.2 we have obtained identical results for the double integral and both iterated integrals of $x+y$ over $R = [0,1] \times [0,2]$. The equality of double integrals and iterated integration is an important general result.

THEOREM 5.5.3 ***(Fubini's Theorem)*** *Suppose that $f(x,y)$ is integrable on $R = [a,b] \times [c,d]$. Then if the interated integrals exist,*

$$\iint_R f = \int_c^d \int_a^b f(x,y)\, dx\, dy = \int_a^b \int_c^d f(x,y)\, dy\, dx.$$

A discussion of Fubini's theorem is delayed until after the more general Theorem 5.5.7 is proven.

Fubini's theorem is the central result for integration over regions in the plane. Provided that we know f is integrable, Fubini's theorem allows us to compute the integral using iterated integration. Effectively, we have reduced a two-dimensional integration to a two-stage process of one-dimensional integrations. These one-dimensional integrations are generally handled by finding antiderivatives.

Fubini's theorem also states that both iterated integrals are equal. Therefore *the order of integration in an interated integral may be reversed without affecting the result.* This can be very useful, since frequently one iterated integral is easier to compute than the other, as is illustrated in the following example.

EXAMPLE 5.5.3 Assume that xe^{xy} is integrable over $R = [1,2] \times [0,1]$. Compute $\iint_R xe^{xy}$.

SOLUTION By Fubini's theorem the integral can be evaluated via iterated integration. Integration first with respect to x would require integration by parts. This may be avoided by reversing the order of integration and starting with y.

$$\int_0^1 \int_1^2 xe^{xy}\, dx\, dy = \int_1^2 \int_0^1 xe^{xy}\, dy\, dx = \int_1^2 \left(\frac{x}{x} e^{xy} \Big|_0^1 \right) dx$$

$$= \int_1^2 e^x - 1\, dx = e^2 - e - 1. \quad \square$$

Examining the geometric significance of the iterated integral suggests why Fubini's theorem should be true. Let's consider the iterated integral $\int_a^b \left(\int_c^d f(x,y)\,dy\right)dx$. Approximating the outer integral by a Riemann sum gives

$$\sum_{i=1}^{n} \left(\int_c^d f(x_i^*, y)\,dy\right) \Delta x_i.$$

For any fixed x_i^* the interior integral $\int_c^d f(x_i^*, y)\,dy$ represents the area under the curve $f(x_i^*, y)$. Multiplying this by Δx_i gives us the volume of the slab shown in Figure 5.5.4. Provided that the surface is sufficiently smooth, the sum of the volumes of these slabs is approximately the total volume under $f(x,y)$. The approximation improves as the Δx_i shrink, and in the limit the iterated integral should represent the exact volume under the surface f. But this volume is exactly what the double integral represents. A similar picture shows that the iterated integral in the opposite order will also be the volume under the surface. Thus we may think of the integrals in Fubini's theorem as being three different ways of computing the same volume.

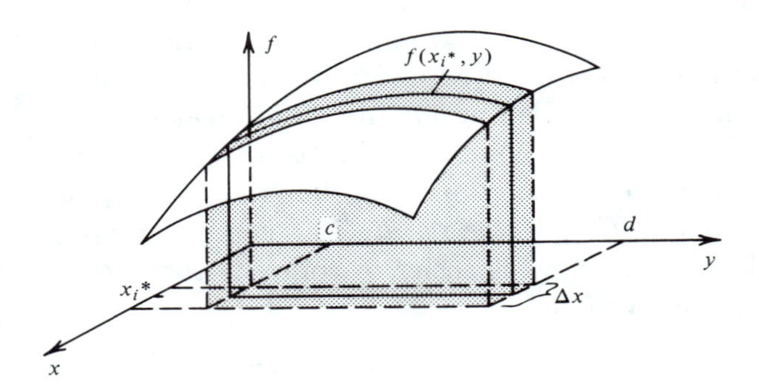

Figure 5.5.4

5.5.3 Continuity Implies Integrability

Fubini's theorem states that iterated integration may be used to compute double integrals *provided that the function is integrable.* However, so far we have established integrability for only one function: $f(x,y) = x + y$. In this section we show that all continuous functions on rectangles are integrable. Since it is often easy to verify the continuity of expressions by inspection, this result will enable us to invoke Fubini's theorem routinely in the computation of double integrals.

Integrability of continuous functions in the one-dimensional case was proven in Theorem 3.5.1. The crucial step in the proof made use of the property of uniform continuity. To extend this result we need to define uniform continuity for functions on $\mathbf{R}^2$. Ordinary continuity is defined in terms of a limit at a single point: given an ϵ we must find a δ that will work for the particular point. But uniform continuity applies to whole regions: for a given ϵ we must find a single δ that works *simultaneously for all points in the region.*

DEFINITION 5.5.4 *A function $f(x,y)$ defined on a region B in the plane is* ***uniformly continuous*** *on B if given any $\epsilon > 0$ there is a $\delta > 0$ such that for any two points (x,y) and (x',y') in the region B*

$$\text{if} \quad |(x,y)-(x',y')| < \delta, \quad \text{then} \quad |f(x,y)-f(x',y')| < \epsilon.$$

If a function is uniformly continuous on a region, we may cover the region with disks of the same radius ($\delta/2$), and within any of these disks the function will vary less than ϵ. One way to guarantee such uniform behavior is to control how quickly f changes. For example, if f is differentiable and its derivative is bounded on a region B (i.e., $|\nabla f| \le M$), then the function cannot increase or decrease faster than M. This limit on the rate of change in f applies to all points in B simultaneously and so forces f to be uniformly continuous on B.

In Theorem 2.6.4 we saw that *any* continuous function was uniformly continuous when restricted to a closed, bounded interval $[a,b]$. This result extends directly to $\mathbf{R}^2$, since the definition of uniform continuity in $\mathbf{R}^2$ is exactly parallel to Definition 2.6.3 of uniform continuity in $\mathbf{R}^1$.

THEOREM 5.5.5 *Any continuous function $f(x,y)$ on a closed and bounded set is uniformly continuous.*

The proof of Theorem 2.6.4 hinged on the special properties of closed, bounded sets expressed in the Heine-Borel theorem. As we saw in Section 5.1, the Heine-Borel theorem extends to $\mathbf{R}^2$ and allows us to extract a finite subcover from any open cover. The proof of Theorem 5.5.5 then follows by an argument similar to Theorem 2.6.4, and we leave this as an exercise for the reader. Theorem 5.5.5 points again to the deep relationship that exists between closed and open sets and continuous functions.

Since rectangles are closed and bounded, Theorem 5.5.5 shows that a continuous function on R must be uniformly continuous. This uniformity is exactly the key needed to prove integrability.

THEOREM 5.5.6 *If $f(x, y)$ is a continuous function on the rectangle $R = [a, b] \times [c, d]$, then $f(x, y)$ is integrable on R.*

PROOF We show that for a sufficiently fine partition the upper and lower sums will be within ϵ.

Let A be the area of the rectangle. By Theorem 5.5.5, f is uniformly continuous on R, so we may find a δ such that whenever two points in R are closer than δ, we have

$$|f(x, y) - f(x', y')| < \frac{\epsilon}{A}. \tag{3}$$

Now choose a partition $P = P_1 \times P_2$ such that within any subrectangle R_{ij} any two points are within δ. (This can be accomplished if all Δx_i in P_1 and all Δy_j in P_2 are smaller than $\frac{\delta}{\sqrt{2}}$). By (3) the values of f on any R_{ij} in this partition are within $\frac{\epsilon}{A}$; in particular $(M_{ij} - m_{ij}) < \frac{\epsilon}{A}$. Thus

$$\begin{aligned}
|U(P, f) - L(P, f)| &= \sum_{i,j} M_{ij}\Delta x_i \Delta y_j - \sum_{i,j} m_{ij}\Delta x_i \Delta y_j \\
&= \sum_{i,j} (M_{ij} - m_{ij})\Delta x_i \Delta y_j \\
&< \sum_{i,j} \frac{\epsilon}{A}\Delta x_i \Delta y_j \\
&= \frac{\epsilon}{A}\sum_{i,j} \Delta x_i \Delta y_j = \frac{\epsilon}{A}A = \epsilon. \quad \blacksquare
\end{aligned}$$

5.5.4 Properties of Integrals

Integrals enjoy three important characteristic properties. The first is linearity, and it is proven using upper and lower sums in the same way as in Theorems 3.5.6 and 3.5.7 for the one-dimensional case.

Property 1. ***(Linearity)*** *If f and g are integrable on R and c is a constant, then the functions $f + g$ and cf are integrable and*

$$\iint_R f + g = \iint_R f + \iint_R g \quad \text{and} \quad \iint_R cf = c\iint_R f.$$

The next property is the analogue of Theorem 3.5.3. It permits us to break up the region of integration and compute the integral separately over each part.

Property 2. ***(Additivity)*** *If f is integrable on R and P is a partition of R, then f is integrable on each subrectangle R_{ij} of P and*

$$\iint_R f = \sum_{i,j} \iint_{R_{ij}} f.$$

PROOF First we show that f is integrable on R_{ij}. We must locate a partition Q_{ij} of R_{ij} such that $|U(Q_{ij}, f) - L(Q_{ij}, f)| < \epsilon$.

Since f is integrable on all of R we can find a partition for R that includes the points of the partition P and for which the upper and lower sums are within ϵ (use a common refinement if necessary). Call this partition Q and let Q_{ij} be the restriction of Q to R_{ij}. The upper sum on Q is the sum of the upper sums on the Q_{ij}, and similarly for the lower sum:

$$\sum_{i,j} U(Q_{ij}, f) = U(Q, f) \qquad \text{and} \qquad \sum_{i,j} L(Q_{ij}, f) = L(Q, f). \tag{4}$$

Thus we may write

$$|U(Q, f) - L(Q, f)| = \sum_{i,j} (U(Q_{ij}, f) - L(Q_{ij}, f)) < \epsilon.$$

All the terms in this sum are nonnegative, so each one must certainly be less than ϵ. In particular, for rectangle R_{ij}, $|U(Q_{ij}, f) - L(Q_{ij}, f)| < \epsilon$; therefore f is integrable on R_{ij}.

Now we show that $\iint_R f = \sum_{i,j} \iint_{R_{ij}} f$. Let S be *any* partition of R. Let Q be the common refinement of P and S. Then (4) holds for this particular partition Q, so

$$\begin{aligned} L(S, f) \le L(Q, f) &= \sum_{i,j} L(Q_{ij}, f) \\ &\le \sum_{i,j} \iint_{R_{ij}} f \\ &\le \sum_{i,j} U(Q_{ij}, f) = U(Q, f) \le U(S, f). \end{aligned}$$

That is, for every partition S of R, $\sum_{i,j} \iint_{R_{ij}} f$ lies between $L(S, f)$ and $U(S, f)$. As in problem 3.4.11, this means $\iint_R f = \sum_{i,j} \iint_{R_{ij}} f$. ∎

The last property states that any integral is bounded above by the maximum value of f times the area of the rectangle and below by its minimum value times the area. This is the two-dimensional generalization of problem 3.4.4. We leave its proof as an exercise.

Property 3. (Boundedness) *If f is integrable on $R = [a,b] \times [c,d]$ and $m \le f(x,y) \le M$ for all $(x,y) \in R$, then*

$$m(b-a)(d-c) \le \iint_R f \le M(b-a)(d-c).$$

These three properties actually *characterize* the double integral in the sense that any other formulation or definition of an integral which satisfies these properties must agree with the double integral. Thus these may be regarded as the defining properties or *axioms for integration.*

THEOREM 5.5.7 *Let $I_R f$ be an integral defined on the rectangle R that satisfies Properties 1 to 3. Then, if the double integral of f also exists on R we have*

$$I_R f = \iint_R f.$$

PROOF The proof is similar to the proof of the second half of Property 2. Let P be any partition of R. As usual, let M_{ij} and m_{ij} denote the supremum and infimum of f on R_{ij}. Thus $m_{ij} \le f(x,y) \le M_{ij}$; consequently by Property 3

$$m_{ij}\Delta x_i \Delta y_j \le I_{R_{ij}} f \le M_{ij}\Delta x_i \Delta y_j.$$

Summing over all rectangles gives

$$\sum_{i,j} m_{ij}\Delta x_i \Delta y_j \le \sum_{i,j} I_{R_{ij}} f = I_R f \le \sum_{i,j} M_{ij}\Delta x_i \Delta y_j,$$

where the equality in the middle comes from Property 2. But the summations on the left and right sides of the preceding inequality are just $L(P,f)$ and $U(P,f)$, respectively. So for any partition P of R,

$$L(P,f) \le I_R f \le U(P,f).$$

Since f is integrable, this means that $I_R f = \iint_R f$. ∎

We have seen that integration over rectangles in the plane may also be defined using iterated integrals. Further, iterated integration satisfies each of these three characteristic properties (see the exercises). Therefore by Theorem 5.5.7, when both the double integral and the iterated integrals exist, they must agree. Thus Fubini's theorem is a corollary to this more general result. Note, however, that the hypothesis of existence is necessary, since it is possible for one type of integral to exist over R while another does not, as the next example shows.

EXAMPLE 5.5.4 Let f be defined on $R = [.1, 1] \times [.1, 1]$ in the following tricky manner using inverted decimal strings.

$$f(x, y) = \begin{cases} 1 & \text{if } x = .a_1 a_2 \ldots a_n \text{ and } y = .a_n a_{n-1} \ldots a_1 \text{ where } a_n \neq 0 \\ 0 & \text{otherwise.} \end{cases}$$

Show that the iterated integrals exist and are 0, but that the double integral fails to exist.

SOLUTION (Iterated integrals) For every fixed x_0 the function $f(x_0, y)$ is 0 except possibly at one point. In fact, if x_0 is a nonterminating decimal, the function $f(x_0, y) \equiv 0$; hence $\int_{.1}^{1} f(x_0, y)\, dy = 0$. If x_0 terminates, then $f(x_0, y) = 0$ except at one particular value of y. For example, $f(.1083, y)$ is 0 except at $y = .3801$. From problem 3.5.11, the single discontinuity doesn't affect the integral, which is also 0. Thus

$$\int_{.1}^{1}\int_{.1}^{1} f(x, y)\, dy\, dx = \int_{.1}^{1}\Big(\int_{.1}^{1} f(x, y)\, dy\Big)\, dx = \int_{.1}^{1} 0\, dx = 0$$

(Double integral) No matter how fine the partition we will show that in each subrectangle there are points for which f is 1 and points for which f is 0. Thus for every partition the upper sum will be $1 \times \text{area}(R) = .81$, while the lower sum will be 0. Hence $\sup L(P, f) \neq \inf U(P, f)$ and f is not integrable.

To see this consider an arbitrary rectangle, $[x_{i-1}, x_i] \times [y_{j-1}, y_j]$. It is easy to verify that f will be 0 at some points in this rectangle (e.g., when x is irrational). To find a point at which f is 1 take a subinterval of $[y_{j-1}, y_j]$ of the form

$$[0.a_1 a_2 \ldots a_n,\ 0.a_1 a_2 \ldots (a_n + 1)], \tag{6}$$

taking care that $1 \leq a_n \leq 8$. You should be able to convince yourself that such an interval always exists in $[y_{j-1}, y_j]$. Now returning to the interval $[x_{i-1}, x_i]$, locate a decimal of the form $d = .b_1 b_2 \ldots b_m a_n a_{n-1} \ldots a_2 a_1$. That is, find any decimal terminating in the string $a_n a_{n-1} \ldots a_2 a_1$. Reversing this string gives $d^* = .a_1 a_2 \ldots a_n b_m \ldots b_2 b_1$ which lands in the interval (6). Thus the point (d, d^*) is in the little rectangle, and at that point f is 1 by definition. Hence the double integral does not exist. ☐

5.5.5 Integrals over General Regions

The initial intuitive idea of integration presented at the beginning of this section seemed to apply to general regions. But in the definition of double integrals our theory was restricted to rectangles in order to treat partitions precisely. Is it possible to extend this theory to more general regions bounded by curves?

With iterated integration the restriction to rectangles is unnecessary. For example, consider the region B in Figure 5.5.5. For each fixed x value $f(x, y)$ is a function of y with limits of integration $h(x)$ and $g(x)$. If at each stage we obtain integrable expressions (in the ordinary one-dimensional sense), then there is no particular conceptual problem with carrying out an iterated integral over the region B

$$\int_a^b \int_{h(x)}^{g(x)} f(x, y)\, dy\, dx.$$

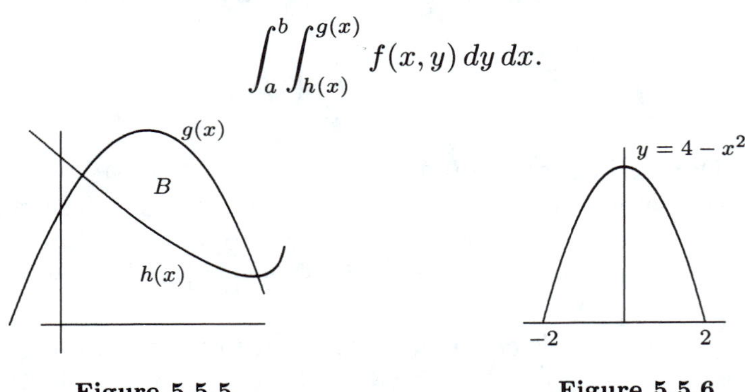

Figure 5.5.5 Figure 5.5.6

EXAMPLE 5.5.5 Integrate $2x^2y$ over the region bounded by $y = 4 - x^2$ and $y = 0$.

SOLUTION The region is shown in Figure 5.5.6. Notice that for each value of x between -2 and 2 the y values go from 0 to $4 - x^2$. Therefore

$$\begin{aligned}\int_{-2}^{2} \left(\int_0^{4-x^2} 2x^2 y\, dy \right) dx &= \int_{-2}^{2} x^2 y^2 \Big|_0^{4-x^2} dx \\ &= \int_{-2}^{2} x^2 (4 - x^2)^2\, dx \\ &= {}^{16}\!/_{3}x^3 - {}^{8}\!/_{5}x^5 + {}^{1}\!/_{7}x^7 \Big|_{-2}^{2} \approx 19.5 \quad \square\end{aligned}$$

EXAMPLE 5.5.6 Compute the integral of $e^{y^2} + x$ over the region bounded by the lines $x = 0$, $y = 2$, and $y = x$.

SOLUTION This region is shown in Figure 5.5.7. If we try to integrate with respect to y first, we find that no elementary antiderivative is available for $e^{y^2} + x$

and our computations are blocked. However, integration in the other order is tractable.

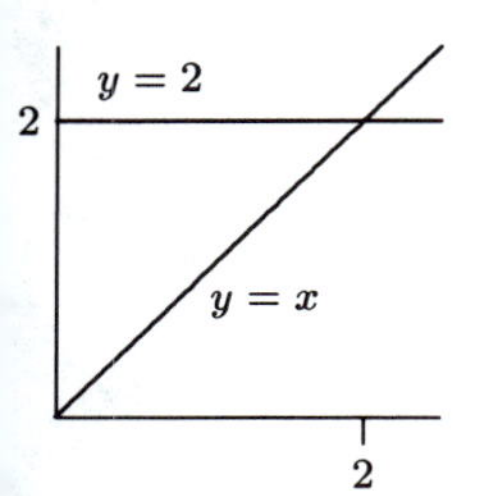

Figure 5.5.7

$$\int_0^2 \int_0^y e^{y^2} + x\,dx\,dy = \int_0^2 \left(xe^{y^2} + \frac{x^2}{2}\Big|_0^y \right) dy$$
$$= \int_0^2 \left(ye^{y^2} + \frac{y^2}{2} \right) dy$$
$$= \frac{e^{y^2}}{2} + \frac{y^3}{6}\Big|_0^2 = \frac{5}{6} + \frac{e^4}{2}. \ \square$$

Sometimes for purposes of computation it is necessary to divide the region of integration into disjoint pieces and compute the integral for each part separately.

EXAMPLE 5.5.7 Integrate $f(x, y) = 2x + y$ over the region bounded by $x = y^2$, $y = x - 2$, and $y = \frac{3}{x}$.

SOLUTION After sketching these curves and solving for the points of intersection, the region in Figure 5.5.8 is obtained. If in the iterated integral we began with the y's, then it would be necessary to break the region into three parts: the first subregion has x going from 0 to 1, the second subregion goes from 1 to $3^{2/3}$, and the third goes from $3^{2/3}$ to 3. This is because at each of these points the function that defines either the upper limit or the lower limit for y changes. In the other direction it is necessary to break the integral into only two parts.

$$\iint_B 2x + y\,dx\,dy = \int_{-1}^{1} \int_{y^2}^{y+2} 2x + y\,dx\,dy + \int_1^{\sqrt[3]{3}} \int_{y^2}^{3/y} 2x + y\,dx\,dy.$$

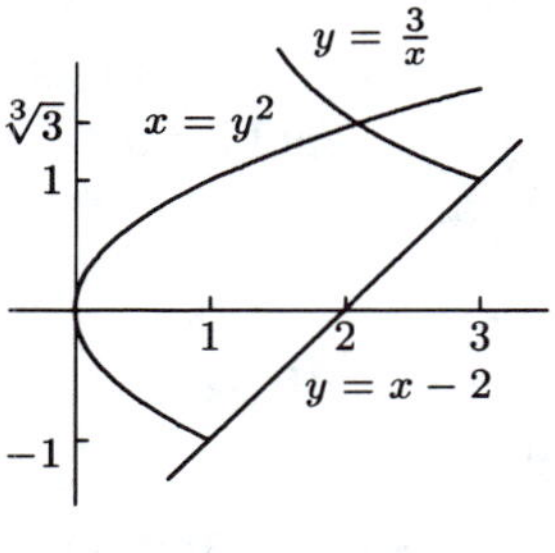

The first of these is

$$\int_{-1}^{1} \int_{y^2}^{y+2} 2x + y\,dx\,dy = \int_{-1}^{1} x^2 + xy\Big|_{y^2}^{y+2} dy$$
$$= \int_{-1}^{1} (y+2)^2 + y(y+2) - y^4 - y^3\,dy$$
$$= \int_{-1}^{1} -y^4 - y^3 + 2y^2 + 6y + 4\,dy$$
$$= 8.9\overline{3}.$$

Figure 5.5.8

The second term is

$$\int_1^{\sqrt[3]{3}}\int_{y^2}^{3/y} 2x + y\,dx\,dy = \int_1^{\sqrt[3]{3}} x^2 + xy\Big|_{y^2}^{3/y} dy$$
$$= \int_1^{\sqrt[3]{3}} \frac{9}{y^2} + 3 - y^4 - y^3\,dy$$
$$= -\frac{9}{y} + 3y - \frac{y^5}{5} - \frac{y^4}{4}\Big|_1^{\sqrt[3]{3}} \approx 2.21.$$

The sum of the two parts gives the value $\iint_B 2x + y\,dx\,dy \approx 11.14$ □

Now we ask whether there is a way to extend the concept of the double integral to these more general regions bounded by curves. The difficulty with a direct approach comes in trying to define a partition for a general region B. A way to circumvent this problem is to extend the region of integration, B, to a rectangle R and simply let the function be 0 on points outside of B. In this way to integrate f over B we will form a partition on a larger rectangular region, but the subrectangles that are outside B contribute nothing to the upper and lower sums. Let's call the function which agrees with f on B and is 0 at other points f_B:

$$f_B(x,y) = \begin{cases} f(x,y), & \text{if } (x,y) \in B \\ 0, & \text{otherwise.} \end{cases}$$

The function f_B is no longer continuous on the rectangle R that encloses B, so integrability does not follow from Theorem 5.5.6. But if f is continuous on B, then f_B will be continuous except along the *boundary curves* of B. Will these discontinuities along the boundary of B destroy the integrability of f_B? Intuitively, what happens on this thin set of boundary points should not influence the integral.

In the case of integration on the real line, the discontinuities at a finite number of points made no difference in the integrability of a bounded function. This was because these points could be enclosed in such small intervals that the contribution to the upper or lower sums was negligible. The generalization of this idea in $\mathbf{R}^2$ is to enclose a set of points with subrectangles whose total area is arbitrarily small.

DEFINITION 5.5.8 *A set S contained in a rectangle R is of* **content zero** *if for any $\epsilon > 0$ there is some partition P of R such that the sum of the areas of the subrectangles R_{ij} containing points of S is less than ϵ.*

For the purposes of integration, sets of content zero can be ignored, as long as f is bounded. In particular, if the discontinuities of f form a set of content zero, f will remain integrable.

THEOREM 5.5.9 *If f is bounded and continuous on $R = [a, b] \times [c, d]$ except on a set of content zero, then it is integrable.*

PROOF Assume that f is bounded by $|f(x, y)| < M$. This implies that for any upper and lower sums, $(M_{ij} - m_{ij}) < 2M$. Let S be the set of discontinuities. By assumption, S is of content zero. Choose a partition P that will enclose S with subrectangles whose total area is less than $\frac{\epsilon}{4M}$. The difference between the upper and lower sums on these pieces of the partition is

$$\sum_{R_{ij} \cap S \neq \phi} (M_{ij} - m_{ij}) \Delta x_i \Delta y_j \leq 2M \sum_{R_{ij} \cap S \neq \phi} \Delta x_i \Delta y_j \leq 2M \frac{\epsilon}{4M} = \frac{\epsilon}{2}.$$

Now f is continuous at all points in the remaining rectangles. Together these form a closed set bounded by R, so by Theorem 5.5.5 f is uniformly continuous on this set. In the same manner as in the proof of Theorem 5.5.6, we can find a refinement of P such that within each subrectangle (excluding those which enclose the set of content zero) the function varies less than $\frac{\epsilon}{2A}$, where A is the area of R. The difference in the upper and lower sums on this set is

$$\sum_{R_{ij} \cap S = \phi} (M_{ij} - m_{ij}) \Delta x_i \Delta y_j \leq \frac{\epsilon}{2A} \sum_{R_{ij} \cap S = \phi} \Delta x_i \Delta y_j \leq \frac{\epsilon}{2A} A = \frac{\epsilon}{2}.$$

Adding together both parts of the upper and lower sums we see that the total difference is less than ϵ. ∎

To extend integration to regions with curved boundaries, it only remains to show that reasonably behaved boundary curves are of content zero.

THEOREM 5.5.10 *If $\vec{\mathbf{F}}(t) = (x(t), y(t))$ is a continuously differentiable curve for $t \in [a, b]$, then its image is of content zero.*

PROOF The first step is to divide $[a, b]$ into pieces $[t_{i-1}, t_i]$ that are small enough so that all points in the image of this subinterval are within ϵ of each other (see Figure 5.5.9). To see how small these pieces must be, we begin by noting that since $\vec{\mathbf{F}}'(t) = (x'(t), y'(t))$ is continuous on $[a, b]$, then by the max-min theorem there is an M such that $|x'(t)| < M$ and $|y'(t)| < M$. Now we may apply the mean value theorem to any two arbitrary points s and t to obtain

$$|x(s) - x(t)| = |x'(t^*)(s - t)| < M|s - t|$$

and

$$|y(s) - y(t)| = |y'(s^*)(s - t)| < M|s - t|.$$

Together these imply that

$$|(x(s), y(s)) - (x(t), y(t))| < 2M|s - t|.$$

Thus whenever $|s - t| \leq \frac{\epsilon}{2M}$, the two image points are within ϵ. Hence, to guarantee that all the image points of the subinterval $[t_{i-1}, t_i]$ will be within ϵ of each other, we should divide $[a, b]$ into pieces of length $\frac{\epsilon}{2M}$. The number of such pieces will be $\frac{2M(b-a)}{\epsilon}$. (If this quantity is not an integer, simply choose a smaller ϵ that evenly divides $2M(b - a)$.)

Consider a rectangle R containing the image of $\vec{\mathbf{F}}$ that is partitioned into subrectangles of size $\epsilon \times \epsilon$. The image of any subinterval, $[t_{i-1}, t_i]$, can intersect *at most four* of these subrectangles, since image points in this subinterval are all closer than ϵ (Figure 5.5.9). There are $\frac{2M(b-a)}{\epsilon}$ such subintervals, so the entire image is enclosed by at most four times this number of subrectangles. Each subrectangle has area ϵ^2, thus the total area is

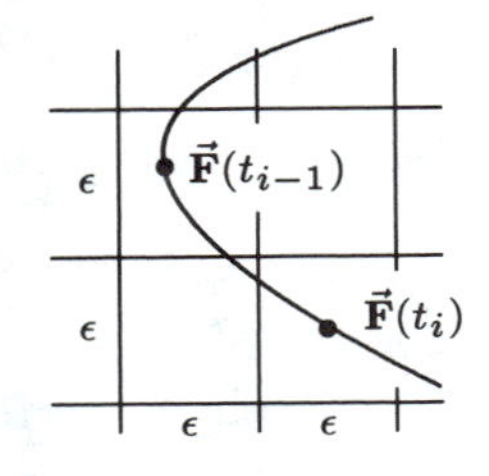

Figure 5.5.9

$$4\epsilon^2 \frac{2M(b-a)}{\epsilon} = 8M(b-a)\epsilon.$$

By choosing ϵ close enough to 0 this can be made arbitrarily small. ∎

This theorem together with Theorem 5.5.9 shows that continuous functions on regions with continuously differentiable boundaries are integrable. Fubini's theorem now assures us that we can compute these double integrals with iterated integrals. One useful consequence of this is Leibniz's rule, which says that differentiation with respect to a parameter can be brought under the integral sign. The proof is left as an exercise.

COROLLARY 5.5.11 ***(Leibniz's Rule)*** *If the function $\frac{\partial f}{\partial y}(x, y)$ is continuous on $R = [a, b] \times [c, d]$, then $\int_a^b f(x, y)\,dx$ is differentiable for $y \in [c, d]$ and*

$$\frac{d}{dy}\int_a^b f(x, y)\,dx = \int_a^b \frac{\partial f}{\partial y}(x, y)\,dx.$$

EXAMPLE 5.5.8 The integral $\int_1^3 \frac{e^{xy}}{x}\,dx$ cannot be evaluated easily, since an antiderivative of $\frac{e^{xy}}{x}$ with respect to x is not available. However, by this theorem we know that the result is a differentiable function even if we have no explicit formula, and furthermore we may even compute its derivative:

$$\frac{d}{dy}\int_1^3 \frac{e^{xy}}{x}\,dx = \int_1^3 \frac{\partial}{\partial y}\frac{e^{xy}}{x}\,dx = \int_1^3 e^{xy}\,dx = \left.\frac{e^{xy}}{y}\right|_1^3 = \frac{e^{3y} - e^y}{y}. \quad \square$$

Problems for Section 5.5

5.5.1. Let $R = [0,1] \times [0,1]$ and the partition $P = P_1 \times P_2$, where $P_1 = P_2 = \{0, \frac{1}{n}, \frac{2}{n}, \ldots, \frac{n}{n}\}$.

a) Compute the upper and lower sums for $\iint_R 2x + y \, dx\, dy$ for the partition P.

b) Show as $n \to \infty$ that the upper and lower sums approach the same number.

c) Repeat **(a)** and **(b)** for $\iint_R x^2 y \, dx\, dy$. (Note: $\sum_{i=1}^n i^2 = \frac{n(n+1)(2n+1)}{6}$.)

5.5.2. In each case sketch the region of integration and compute the integral of f over B using iterated integration.

a) $f(x,y) = x + y^2$ over the region $2 \le x \le 3$ and $-2 \le y \le 3$.

b) $f(x,y) = x + y$ over the region $0 \le x \le 3$ and $\sqrt{x} \le y \le x^2$.

c) $f(x,y) = \frac{1}{x+y}$ over the triangle with vertices $(1,1)$, $(1,2)$, and $(2,2)$.

5.5.3. Sketch the region of integration and then write these iterated integrals in reverse order.

a) $\displaystyle\int_0^1 \int_x^1 f(x,y)\, dy\, dx$ **b)** $\displaystyle\int_0^1 \int_1^{e^x} f(x,y)\, dy\, dx$

c) $\displaystyle\int_0^1 \int_0^{4-y^2} f(x,y)\, dx\, dy$

5.5.4. Here are some integrated integrals that are difficult to compute as written but are easy when the order of integration is reversed. Sketch the domain of integration and then reverse the order of integration to evaluate these integrals.

a) $\displaystyle\int_0^1 \int_0^{\pi} x\cos(xy)\, dx\, dy$ **b)** $\displaystyle\int_0^1 \int_x^1 \sqrt{1+y^2}\, dy\, dx$

c) $\displaystyle\int_1^2 \int_1^{\frac{2}{x}} x^2 y\, dx\, dy$ **d)** $\displaystyle\int_0^1 \int_y^1 y e^{x^3}\, dx\, dy$

e) $\displaystyle\int_0^{\frac{\pi}{2}} \int_x^{\frac{\pi}{2}} \frac{\sin y}{y}\, dy\, dx$

•5.5.5. Recall that the volume of a cylinder is the area of the base times the height. Consequently we should be able to compute the area of certain plane regions using iterated integrals where the height function is 1.

a) Suppose that R is a region in the plane bounded above by $y = f(x)$, below by $y = g(x)$, on the left by $x = a$, and on the right by $x = b$. Show that area $(R) = \iint_R 1\, dy\, dx$. (Hint: Show that the iterated integral reduces to the familiar integral for the area between two curves.)

b) Suppose instead that R is a region in the plane bounded on the right by $x = h(y)$, on the left by $x = k(y)$, above by $y = d$, and below by $y = c$. Show that area $(R) = \iint_R 1\, dx\, dy$.

5.5.6. **The Mean Value Theorem for Double Integrals:** *If f is continuous on the rectangle R, then*

$$\iint_R f(x,y) = f(x_0, y_0)A(R)$$

where $A(R)$ is the area of the rectangle and (x_0, y_0) is a point in R. Mimic the ideas of the one-dimensional integral mean value theorem (Theorem 3.6.4) to try to prove this. What theorem do you need to make your proof work? Does that theorem hold?

5.5.7. Let $f(x) > 0$ and let B be the region bounded by $f(x)$, the x-axis, and the lines $x = a$ and $x = b$. Show that $2\pi \iint_B y\,dx\,dy$ gives the volume generated by revolving B about the x-axis.

5.5.8.
a) Sketch the rectangle R with vertices at (a,b), (a,d), (c,d), and (c,b).

b) Prove that

$$\iint_R f_{xy}(x,y)\,dx\,dy = f(a,b) - f(a,d) + f(c,d) - f(b,d).$$

c) If $f_{xy}(x,y) = 0$ everywhere, what must be true about the values of f on the vertices of any rectangle?

5.5.9. Suppose that f is continuous. Prove that if $\iint_R f = 0$ over every rectangle, then $f = 0$ everywhere.

•5.5.10.
a) Show that the double integral on rectangles satisfies the linearity and boundedness properties.

b) Prove that iterated integration satisfies the three characteristic properties for integrals: linearity, additivity, and boundedness.

5.5.11. Suppose that $\int_a^b f(x)\,dx = A$ and $\int_c^d g(x)\,dx = B$.

a) Compute $\iint_R f(x)g(y)\,dy\,dx$, where $R = [a,b] \times [c,d]$.

b) Compute $\iint_R f(x) + g(y)\,dy\,dx$.

5.5.12. Using Theorem 5.1.6 prove Theorem 5.5.5. (Hint: Review the proof of Theorem 2.6.4.)

5.5.13. Consider the iterated integral $\int_c^y \int_a^b f_t(x,t)\,dx\,dt$, where the partial f_t is continuous on the rectangle of integration.

a) Use the second fundamental theorem of calculus to find the derivative of this with respect to y.

b) By Fubini's theorem this should give the same result if we first change the order of integration and then differentiate. Reverse the order of integration, carry out the interior integral, and differentiate the result to prove Leibniz's rule, Corollary 5.5.11.

5.5.14. Suppose that f_x, f_y, f_{xy}, and f_{yx} are all continuous. Show that $f_{xy} = f_{yx}$. (This is Theorem 5.3.6 again.)

a) Use Leibniz's rule to differentiate $\int_a^x f_t(t, y)\, dt$ with respect to y. Then differentiate the result to obtain $f_{yx}(x, y)$.

b) Now integrate $\int_a^x f_t(t, y)\, dt$ using the fundamental theorem of calculus . Then differentiate first with respect to y and then x to obtain $f_{yx}(x, y)$.

5.5.15. Find a formula for the point (a, b) that minimizes the integral of the squared distance from (x, y) to (a, b). That is, minimize $\iint_B (x-a)^2 + (y-b)^2\, dx\, dy$. This is analogous to *least squares*, which minimizes the sum of squared distances. (Hint: Apply Leibniz's rule to differentiate under the integral sign. Your formula will be in terms of an integral.)

5.5.16. The **average values** of x and of y over a region B are respectively,

$$\bar{x} = \frac{1}{\text{area}(B)} \iint_B x\, dx\, dy \quad \text{and} \quad \bar{y} = \frac{1}{\text{area}(B)} \iint_B y\, dx\, dy$$

The point $(\bar{x}, \bar{y})$ is sometimes called the **centroid** of B.

a) Find the centroid of a triangle with vertices at $(0, 0)$, $(0, 2)$, and $(2, 0)$.

b) Find the centroid of a region bounded by the parabola $y = 4x - x^2$ and the x-axis.

c) Suppose that B_1 and B_2 are disjoint and that $\text{area}(B_1) = \text{area}(B_2)$. Prove that the centroid of $B_1 \cup B_2$ is the midpoint of the centroids for B_1 and B_2.

d) Prove that the centroid of the union of two disjoint regions B_1 and B_2 lies on the line between the centroids of B_1 and B_2.

6

Line Integrals and Green's Theorem

The central theme of this chapter is the generalization of the fundamental theorem of calculus. In each higher dimension there is actually a *series* of extensions that are linked to each other geometrically in a lovely way. In this chapter we present the two extensions for $\mathbf{R}^2$ and examine how they are linked. The first of these involves the idea of integrating a vector function $\vec{\mathbf{F}}(x, y) = (P(x, y), Q(x, y))$ along a curve C in the plane, a concept known as a *line integral.* It turns out that integrating a vector function given by the derivative $\nabla f = \left(\frac{\partial f}{\partial x}, \frac{\partial f}{\partial y}\right)$ along a curve, gives the same result as the difference in $f(x, y)$ at the two endpoints of the curve. This is the content of the fundamental theorem of calculus for line integrals. The second generalization, known as Green's theorem, involves two dimensional regions and their boundary curves. It equates the line integral of $\vec{\mathbf{F}}(x, y)$ around the boundary curve of a region with the double integral of a kind of "derivative" over the interior enclosed by the curve. In Sections 6.1 and 6.3 we attempt to "discover" these two extensions of the fundamental theorem of calculus in $\mathbf{R}^2$ by generalizing the proof from one dimension. Both theorems arise from simple geometric ideas about regions, boundaries, and canceling sums and differences, all mediated by applications of the mean value theorem. These are the basic topological ideas behind the general form of the fundamental theorem of calculus for n dimensions. Sections 6.1 and 6.3 may be read independently and are intended only to reveal the unity behind the different forms of the fundamental theorem of calculus. The standard presentation of line integrals and Green's theorem is developed in Sections 6.2, 6.4, and 6.5.

This is a beautiful topic in the study of calculus. First, it reveals that calculus is not just about δ's and ϵ's but involves deep geometric and topological ideas concerning relationships between regions and their boundaries. This "geometric" side of calculus leads eventually to the study of differential geometry and topology in which the tools of calculus are used to investigate the geometry of higher dimensional surfaces. Second, the notion of a line integral is a central concept in physics used to define such basic ideas as work, potentials, and flows across boundaries. We explore a few of these applications in some of our examples. The interaction between the mathematical analysis and our geometric and physical intuition is one of the highlights of this subject.

6.1 The Fundamental Theorem of Calculus: Part I

In one dimension, the fundamental theorem of calculus states that

$$f(b) - f(a) = \int_a^b f'(x)\,dx.$$

Suppose that we seek to generalize this theorem to curves in the plane. As a first step recall the central ideas in the proof to see if they might be extended. The proof began by partitioning the interval $[a, b]$ into small pieces. This enabled us to write the difference of f on the entire interval $[a, b]$ in the following way, as the sum of the differences on each of the interior segments $[x_{i-1}, x_i]$:

$$\begin{aligned} f(b) - f(a) &= f(x_n) - f(x_0) \\ &= [f(x_n) - f(x_{n-1})] + [f(x_{n-1}) - f(x_{n-2})] + \cdots \\ &\qquad + [f(x_2) - f(x_1)] + [f(x_1) - f(x_0)] \\ &= \sum_{i=1}^{n} [f(x_i) - f(x_{i-1})]. \end{aligned} \tag{1}$$

This simple idea is a purely combinatorial-geometrical fact about telescoping sums that is valid no matter how the interval is divided, provided that the differences on the interior segments $[x_{i-1}, x_i]$ are all aligned in the same direction, so that they collapse when summed.

Next, using the mean value theorem, each of the differences of f is expressed in terms of a derivative: $f(x_i) - f(x_{i-1}) = f'(x_i^*)\Delta x_i$ (where $x_i^* \in [x_{i-1}, x_i]$). Hence the sum in (1) becomes

$$f(b) - f(a) = \sum_{i=1}^{n} f'(x_i^*)\Delta x_i. \tag{2}$$

Finally, integrals enter the picture as the interval is divided into smaller and smaller pieces. Thus the right-hand side of (2), which is a Riemann sum, approaches $\int_a^b f'(x)\,dx$, giving us the fundamental theorem of calculus.

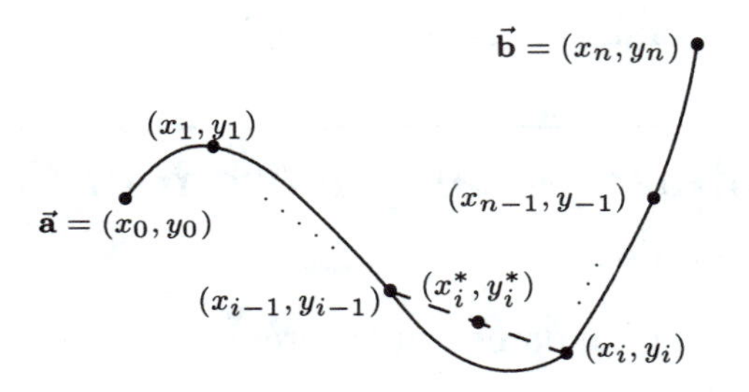

Partitioning the curve C.

Figure 6.1.1

Now let's attempt to mimic this canceling technique for a function $f(x, y)$ on a curve C in the plane. First partition the curve C into little pieces (see Figure 6.1.1). Provided that the interior segments in this partition are all oriented in the same direction, the sum of the differences of $f(x, y)$ across the interior segments will again cancel in a telescoping manner, leaving us with the overall difference across the boundary points of the curve $f(\vec{\mathbf{b}}) - f(\vec{\mathbf{a}})$. Hence, exactly as in equation (1), we have

$$f(\vec{\mathbf{b}}) - f(\vec{\mathbf{a}}) = \sum_{i=1}^{n} [f(x_i, y_i) - f(x_{i-1}, y_{i-1})]. \tag{3}$$

Next we look for a way to express the differences across each of the line segments given by $[f(x_i, y_i) - f(x_{i-1}, y_{i-1})]$ in terms of derivatives. The mean value theorem for functions $f : \mathbf{R}^2 \to \mathbf{R}$ accomplishes this nicely, saying that

$$f(x_i, y_i) - f(x_{i-1}, y_{i-1}) = \nabla f(x_i^*, y_i^*) \cdot (\Delta x_i, \Delta y_i),$$

where $\Delta x_i = x_i - x_{i-1}$, $\Delta y_i = y_i - y_{i-1}$, and (x_i^*, y_i^*) is some point on the line segment between (x_{i-1}, y_{i-1}) and (x_i, y_i). The telescoping sum in (3) can now be written as

$$f(\vec{\mathbf{b}}) - f(\vec{\mathbf{a}}) = \sum_{i=1}^{n} \nabla f(x_i^*, y_i^*) \cdot (\Delta x_i, \Delta y_i)$$
$$= \sum_{i=1}^{n} \left(\frac{\partial f}{\partial x}(x_i^*, y_i^*), \frac{\partial f}{\partial y}(x_i^*, y_i^*) \right) \cdot (\Delta x_i, \Delta y_i). \tag{4}$$

So far we have been able to mimic all the steps in the proof of the fundamental theorem of calculus; all that remains is to recognize this as a Riemann sum of some integral. But in fact, the construction of the preceding sum is just like forming a kind of Riemann sum on a curve instead of an interval.

STEP 1 First the curve C is divided into little vector pieces $(\Delta x_i, \Delta y_i)$ by choosing points on C.

STEP 2 The vector function pair $(P(x,y), Q(x,y)) = \left(\frac{\partial f}{\partial x}, \frac{\partial f}{\partial y}\right)$ is then evaluated at an intermediate point of each piece.

STEP 3 The dot product of each pair is taken with the corresponding piece $(\Delta x_i, \Delta y_i)$, and the results are summed.

As the curve is divided into smaller and smaller pieces, this sum approaches a limit called the *line integral* of the vector function $\vec{\mathbf{F}}(x,y) = (P(x,y), Q(x,y))$ over the curve C. A natural notation for such a line integral is

$$\int_C (P(x,y), Q(x,y)) \cdot (dx, dy) \quad \text{or} \quad \int_C \vec{\mathbf{F}} \cdot \mathbf{d}\vec{\mathbf{x}},$$

where $\mathbf{d}\vec{\mathbf{x}} = (dx, dy)$. In our sum (4) the vector function pair is $\vec{\mathbf{F}} = \nabla f$. As the curve is divided into smaller and smaller pieces the right-hand side of (4) approaches the line integral $\int_C \nabla f \cdot \mathbf{d}\vec{\mathbf{x}}$. Hence we are led to conjecture the following *fundamental theorem of calculus for line integrals:*

$$f(\vec{\mathbf{b}}) - f(\vec{\mathbf{a}}) = \int_C \nabla f \cdot \mathbf{d}\vec{\mathbf{x}}.$$

We hasten to note that this result has not been proven and that the concept of a line integral has not been made rigorous; yet we have seen some very suggestive ideas. In the next section the notion of a line integral will be rigorously defined and a proof of this theorem will be presented. In Section 6.3 we will see that these simple geometric ideas about cancellation of the differences across interior boundaries can be advanced one dimension and applied to line integrals considered as boundaries of two-dimensional regions. This leads us to another form of the fundamental theorem of calculus in $\mathbf{R}^2$, called Green's theorem.

6.2 Line Integrals

We begin by presenting the intuitive concept of a line integral of a vector function along a curve C. The vector functions that we will be considering are of the form $\vec{\mathbf{F}}(x,y) = (P(x,y), Q(x,y))$, which map $\mathbf{R}^2$ to $\mathbf{R}^2$. These functions define a **vector field** in $\mathbf{R}^2$. To visualize a vector field think of attaching the vector $(P(x,y), Q(x,y))$ at each point (x,y) in the plane.

EXAMPLE 6.2.1 Consider the vector field given by $\vec{\mathbf{F}}(x,y) = (-y+1, x)$. Then at the point $(1,4)$, we have a vector $\vec{\mathbf{F}}(1,4) = (-3,1)$, at $(2,1)$ we have $\vec{\mathbf{F}}(2,1) = (0,2)$, and so on.

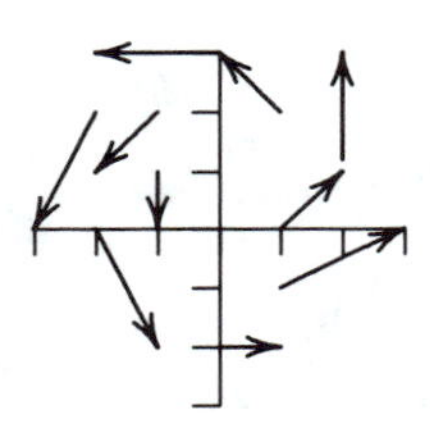

Figure 6.2.1

A few of the vectors for this vector field are shown in Figure 6.2.1. As more vectors are drawn in, the sense of a flow in the plane begins to develop. We may imagine that the vector field is describing the flow of some liquid in the plane, where the vectors represent the *velocity* of the liquid at each point. Vector fields are also employed to represent *force fields*, where the vector $\vec{\mathbf{F}}(x,y)$ describes the force acting on the point (x,y) in the plane. □

Integrals usually arise from a Riemann sum process, and the concept of a line integrals can be motivated in a similar way. Given a curve C in the plane and a vector field $\vec{\mathbf{F}} = (P,Q)$, we first partition the curve into tiny segments using points (x_i, y_i) on the curve. The vector function is then evaluated at an intermediate point on each of the segments, $\vec{\mathbf{F}}(x_i^*, y_i^*) = (P(x_i^*, y_i^*), Q(x_i^*, y_i^*))$, and we take dot products with the little pieces of our curve $(\Delta x_i, \Delta y_i)$. Finally, the terms are summed to yield an expression resembling a two-dimensional Riemann sum:[1]

$$\sum_{i=1}^{n} \vec{\mathbf{F}}(x_i^*, y_i^*) \cdot (\Delta x_i, \Delta y_i) = \sum_{i=1}^{n} P(x_i^*, y_i^*)\Delta x_i + Q(x_i^*, y_i^*)\Delta y_i. \qquad (1)$$

[1] This procedure differs slightly from the approach in Section 6.1 equation (4). In that section the mean value theorem supplied the intermediate point (x_i^*, y_i^*) that lay on the straight line between the two points of the curve, that is, on the polygonal approximation to the curve and not on the curve itself. However, as the partition of C becomes finer, the polygonal approximations approach C. If the functions P and Q are reasonably smooth, it can be shown that both procedures lead to the same limit as the partition is refined.

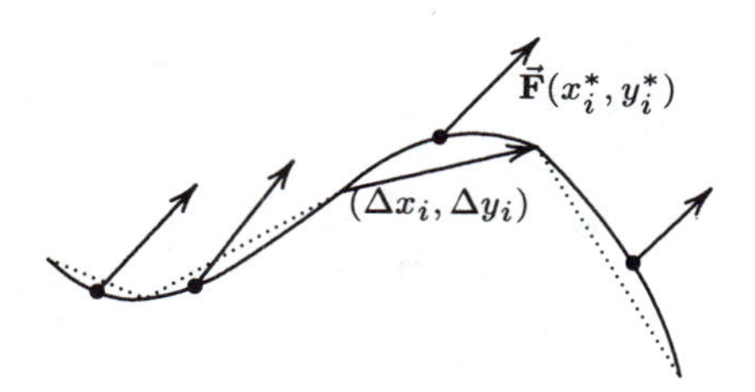

The line integral of the vector field $\vec{\mathbf{F}}$ taken along a curve.

Figure 6.2.2

The Riemann sum for an ordinary integral represents an approximation of the area under a curve. To see what quantity the sum in (1) represents, first recall that $\vec{\mathbf{F}} \cdot (\Delta x_i, \Delta y_i) = |\vec{\mathbf{F}}||(\Delta x_i, \Delta y_i)| \cos\theta$. Observe that $|\vec{\mathbf{F}}|\cos\theta$ measures the magnitude of $\vec{\mathbf{F}}$ in the direction of the vector $(\Delta x_i, \Delta y_i)$, and $|(\Delta x_i, \Delta y_i)|$ is the amount of displacement along the curve (Figure 6.2.3). Thus the sum in (1) *represents the amount of the vector field in the direction of the curve times the distance along the curve accumulated over the curve.*

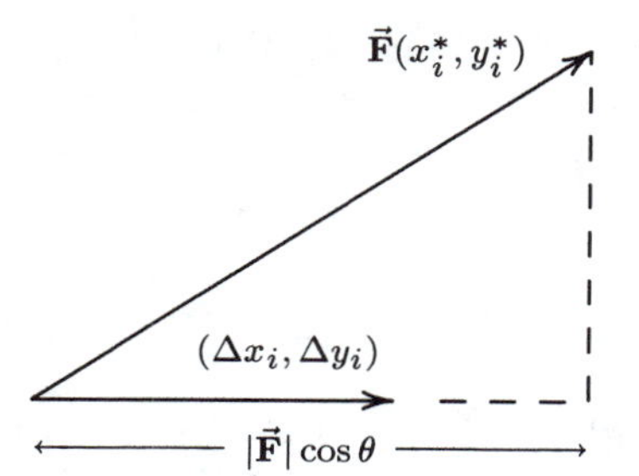

The component of $\vec{\mathbf{F}}$ along $(\Delta x_i, \Delta y_i)$ is given by $\vec{\mathbf{F}} \cos\Theta$.

Figure 6.2.3

6.2.1 Definitions and Properties

To make the intuitive concept of a line integral rigorous and to actually compute them we first need a precise definition for curves.

Definition 6.2.1 *A* ***(smooth) curve*** *C is a continuously differentiable function of the form $(x(t), y(t))$ that maps an interval $[a, b]$ in $\mathbf{R}$ to the plane $\mathbf{R}^2$.*

By definition, a curve *is* its parameterization. Consequently, two parameterizations that trace out the same set of points are regarded as different curves. This may seem somewhat strange, since we often view curves as sets of points lying in the plane, but there are deep problems with this way of thinking. For example, the set of points traced out by a space-filling curve is a two-dimensional region. This image set alone makes no sense as a curve. It can be understood as a curve only in terms of the way in which a map traces out points, that is, in terms of its parameterization. Thus the essence of a curve lies in how it traces out points and not in the set of image points that are traced out.

Let's apply the Riemann sum concept of a line integral to the curve C given by $(x(t), y(t))$ for $t \in [a, b]$. To divide C into small pieces, we partition the parameter space $[a, b]$ with points t_0, t_1, $t_2, \ldots$, t_n into pieces of length Δt_i. This induces a partition on C given by $(x(t_i), y(t_i))$ (see Figure 6.2.4).

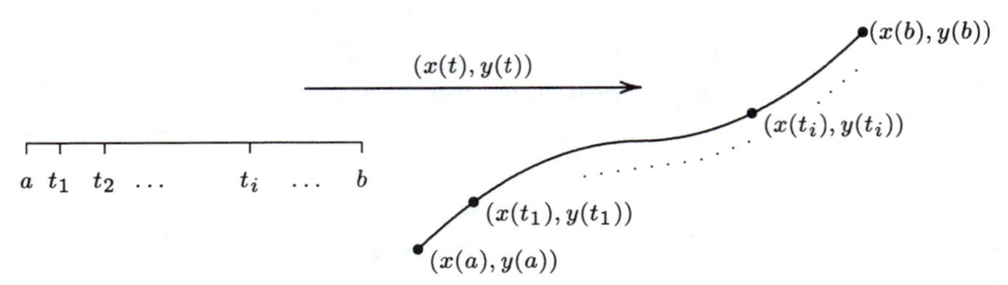

The partition of $[a, b]$ induces a partition of C under $(x(t), y(t))$.

Figure 6.2.4

With this partition, the segments $(\Delta x_i, \Delta y_i)$ in equation (1) are given by $(x(t_{i+1}) - x(t_i), y(t_{i+1}) - y(t_i))$. By using the mean value theorem twice, once for $x(t)$ and again for $y(t)$, this pair of differences can be expressed as

$$(x(t_{i+1}) - x(t_i), y(t_{i+1}) - y(t_i)) = (x'(t_i^*), y'(t_i^{**}))\Delta t_i,$$

where $\Delta t_i = t_i - t_{i-1}$ and both points t_i^*, t_i^{**} belong to $[t_{i-1}, t_i]$. Next we evaluate the vector function $\vec{\mathbf{F}} = (P, Q)$ at an intermediate point on each segment of the partitioned curve, say at $x(t_i^*)$ and $y(t_i^*)$. Then the sum in (1) becomes

$$\sum_{i=1}^{n} (P(x(t_i^*), y(t_i^*)), Q(x(t_i^*), y(t_i^*))) \cdot (x'(t_i^*), y'(t_i^{**}))\, \Delta t_i.$$

Multiplying out the dot product we see that this sum is almost exactly a Riemann sum for the function

$$f(t) = P(x(t), y(t))x'(t) + Q(x(t), y(t))y'(t)$$

over the interval $t \in [a, b]$. The only fact preventing it from being an exact Riemann sum is that t_i^* and t_i^{**} are not necessarily the same point in the subinterval. The reader may recall that in our development of arc length we encountered a similar discrepency in evaluation points. There the problem was handled with an argument using uniform continuity (see Theorem 5.2.9). A similar argument works here to show that when the partition of $[a,b]$ is sufficiently fine, the foregoing sum differs from a Riemann sum by an arbitrarily small amount. Thus the Riemann sum concept of a line integral leads naturally to an ordinary one-dimensional integral over the parameterization interval on the t-axis. Therefore, we *define* line integrals as follows:

DEFINITION 6.2.2 *Let $\vec{\mathbf{F}}(x,y) = (P(x,y), Q(x,y)) : \mathbf{R}^2 \to \mathbf{R}^2$ be a continuous vector field and let the curve C be given by $(x(t), y(t))$ for $t \in [a,b]$. The* ***line integral*** *of $\vec{\mathbf{F}}$ along C is denoted by $\int_C \vec{\mathbf{F}} \cdot \mathbf{d}\vec{\mathbf{x}}$ or $\int_C P\,dx + Q\,dy$ and is given by*

$$\begin{aligned}\int_C \vec{\mathbf{F}} \cdot \mathbf{d}\vec{\mathbf{x}} &= \int_a^b (P(x(t),y(t)), Q(x(t),y(t))) \cdot (x'(t), y'(t))\,dt \\ &= \int_a^b P(x(t),y(t))x'(t) + Q(x(t),y(t))y'(t)\,dt.\end{aligned}$$

Now we are in familiar territory. These are integrals of a single variable, so the computation of line integrals is quite straightforward.

EXAMPLE 6.2.2 Compute the line integral of $(P(x,y), Q(x,y)) = (x+y, xy)$ along the parabola C in Figure 6.2.5 parameterized by $x(t) = t$ and $y(t) = t^2$, where $t \in [0,2]$.

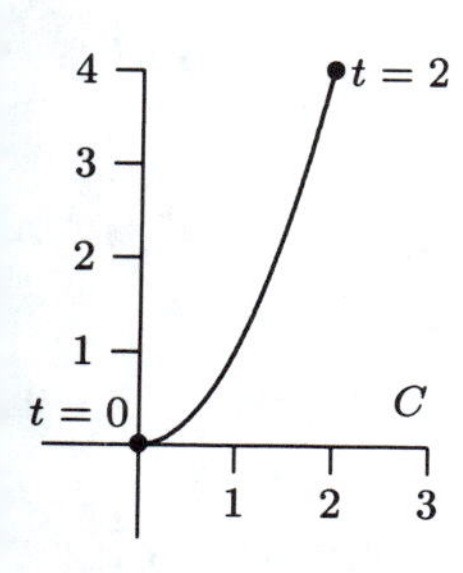

Figure 6.2.5

We have $x'(t) = 1$ and $y'(t) = 2t$, so

$$\begin{aligned}\int_C P\,dx + Q\,dy &= \int_0^2 (t+t^2, t^3) \cdot (1, 2t)\,dt \\ &= \int_0^2 t + t^2 + 2t^4\,dt \\ &= \frac{t^2}{2} + \frac{t^3}{3} + \frac{2t^5}{5}\Bigg|_0^2 \\ &= 17\tfrac{7}{15}. \quad \square\end{aligned}$$

EXAMPLE 6.2.3 Consider the vector field $\vec{\mathbf{F}}(x,y) = (-y+1, x)$ along the semicircle, Γ, parameterized by $(x(t), y(t)) = (2\cos t, 2\sin t)$ for $t \in [0,\pi]$.

Computing the flow of $\vec{\mathbf{F}}$ along the semicircle counterclockwise according to the parameterization gives

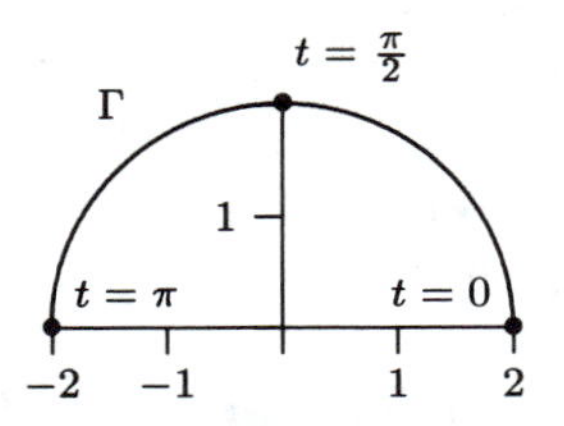

Figure 6.2.6

$$\begin{aligned}\int_\Gamma \vec{\mathbf{F}} \cdot \mathbf{d}\vec{\mathbf{x}} &= \int_0^\pi (-2\sin t + 1, 2\cos t) \cdot (-2\sin t, 2\cos t)\, dt \\ &= \int_0^\pi 4\sin^2 t - 2\sin t + 4\cos^2 t\, dt \\ &= \int_0^\pi 4 - 2\sin t\, dt = 4\pi - 4. \quad \square\end{aligned}$$

Line integrals inherit the important characteristic properties of integrals (linearity, additivity, and boundedness) from their definition in terms of dot products and ordinary integrals.

THEOREM 6.2.3 ***Linearity***: *If $\vec{\mathbf{F}}$ and $\vec{\mathbf{G}}$ are two continuous vector fields, C a smooth curve, and a and b scalars, then*

$$\int_C (a\vec{\mathbf{F}} + b\vec{\mathbf{G}}) \cdot \mathbf{d}\vec{\mathbf{x}} = a\int_C \vec{\mathbf{F}} \cdot \mathbf{d}\vec{\mathbf{x}} + b\int_C \vec{\mathbf{G}} \cdot \mathbf{d}\vec{\mathbf{x}}.$$

Additivity: *If C, C_1, and C_2 are parameterized by $(x(t), y(t))$ for the respective intervals $t \in [a, b]$, $t \in [a, c]$, and $t \in [c, b]$, then*

$$\int_C \vec{\mathbf{F}} \cdot \mathbf{d}\vec{\mathbf{x}} = \int_{C_1} \vec{\mathbf{F}} \cdot \mathbf{d}\vec{\mathbf{x}} + \int_{C_2} \vec{\mathbf{F}} \cdot \mathbf{d}\vec{\mathbf{x}}.$$

Boundedness: *If $\vec{\mathbf{F}}$ satisfies $|\vec{\mathbf{F}}| \le M$ for all points on the curve C and the length of C is bounded by L, then*

$$\left|\int_C \vec{\mathbf{F}} \cdot \mathbf{d}\vec{\mathbf{x}}\right| \le ML.$$

PROOF The first two properties are immediate consequences of the definition of line integrals and the analogous property for ordinary integrals and are left as exercises. To show boundedness, let the curve C be parameterized by $\vec{\alpha}(t) = (x(t), y(t))$ for $t \in [a, b]$. Recall from Theorem 5.2.9 that the length of C is given by $L = \int_a^b |\vec{\alpha}'(t)|\, dt$. Consider the line integral $\int_C \vec{\mathbf{F}} \cdot \mathbf{d}\vec{\mathbf{x}} = \int_a^b \vec{\mathbf{F}}(\vec{\alpha}(t)) \cdot \vec{\alpha}'(t)\, dt$. By the Cauchy-Schwarz inequality, the integrand satisfies

$$|\vec{\mathbf{F}}(\vec{\alpha}) \cdot \vec{\alpha}'(t)| \le |\vec{\mathbf{F}}(\vec{\alpha})||\vec{\alpha}'(t)| \le M|\vec{\alpha}'(t)|.$$

From problem 3.5.15, we know that $|\int_a^b f(t)\,dt| \leq \int_a^b |f(t)|\,dt$ for any integrable function $f(t)$. Therefore

$$\left|\int_C \vec{\mathbf{F}} \cdot \mathbf{d}\vec{\mathbf{x}}\right| \leq \int_a^b |\vec{\mathbf{F}}(\vec{\alpha}(t)) \cdot \vec{\alpha}'(t)|\,dt$$
$$\leq \int_a^b |\vec{\mathbf{F}}(\vec{\alpha}(t))|\,|\vec{\alpha}'(t)|\,dt \leq M \int_a^b |\vec{\alpha}'(t)|\,dt = ML. \blacksquare$$

Another common notation for the line integral is obtained by writing

$$\int_C \vec{\mathbf{F}} \cdot \mathbf{d}\vec{\mathbf{x}} = \int_a^b \vec{\mathbf{F}}(\vec{\alpha}(t)) \cdot \vec{\alpha}'(t)\,dt = \int_a^b \vec{\mathbf{F}}(\vec{\alpha}(t)) \cdot \vec{\tau}(t)|\vec{\alpha}'(t)|\,dt,$$

where $\vec{\tau}(t) = \frac{\vec{\alpha}'(t)}{|\vec{\alpha}'(t)|}$ is the unit tangent to the curve. If we denote the element of arc length for the curve by $ds = |\vec{\alpha}'(t)|\,dt$, this expression for the line integral becomes

$$\int_C \vec{\mathbf{F}} \cdot \mathbf{d}\vec{\mathbf{x}} = \int_C \vec{\mathbf{F}} \cdot \vec{\tau}\,ds. \tag{2}$$

Since $\vec{\mathbf{F}} \cdot \vec{\tau}$ is exactly the magnitude of $\vec{\mathbf{F}}$ in the direction tangent to the curve and ds represents an infinitesimal length of arc, this notation makes explicit the interpretation of the line integral developed at the start of this section.

6.2.2 Independence of Parameterization

The motivating idea behind the line integral depended only on the image of the curve and not on the way those points are traced out. However, the line integral is defined and computed in terms of a particular parameterization. What happens if we compute the line integral using two different parameterizations of the same image curve? Will we get the same answer?

Consider for example, the parabolic segment in Figure 6.2.5. This trajectory is also traced by the parameterization C' given by

$$\phi(s) = 2s - 2, \qquad \psi(s) = (2s-2)^2, \qquad s \in [1, 2].$$

Does $\vec{\mathbf{F}}(x, y) = (x+y, xy)$ along C' give us the same answer as was found in Example 6.2.2? (Compute the line integral along C' and verify that the values of both line integrals are equal.) Of course, we may generate many other parameterizations of the form $\phi(s) = h(s)$, $\psi(s) = (h(s))^2$ for different functions $h(s)$. Will they all yield the same value?

Let's examine another example. The parameterization

$$\phi(u) = u, \qquad \psi(u) = \sqrt{4 - u^2}, \qquad u \in [-2, 2]$$

traces out the semicircle in Figure 6.2.6, but as u goes from -2 to 2 the curve is traced in the opposite direction. How does this change affect the line integral? Notice that in the Riemann sum concept of the line integral, $\sum_{i=1}^{n} \vec{\mathbf{F}}(x_i^*, y_i^*) \cdot (\Delta x_i, \Delta y_i)$, reversing the direction of the curve changes the sign of $(\Delta x_i, \Delta y_i)$. Therefore we might expect that *the line integral along a path in the reverse direction should be the negative of the line integral in the original direction.* (Verify that this is true for Example 6.2.4 in the preceding parameterization.[2])

With an appropriate definition we can prove generally that line integrals depend only on the direction in which the path is traced and *not* on the particular parameterization of the path.

DEFINITION 6.2.4 *A curve C' given by $(\phi(s), \psi(s))$ for $s \in [c, d]$ is said to be* **equivalent** *to a curve C given by $(x(t), y(t))$ for $t \in [a, b]$ if there is a continuously differentiable function h that maps $[c, d]$ onto $[a, b]$ with $h' \neq 0$ such that*

$$\phi(s) = x(h(s)) \quad \text{and} \quad \psi(s) = y(h(s)). \tag{3}$$

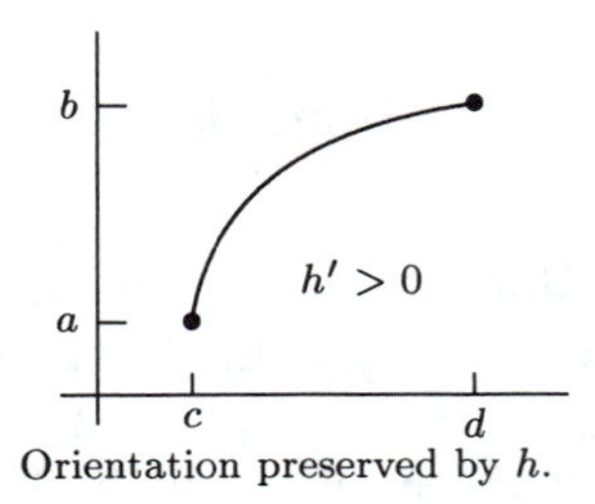

Orientation preserved by h.

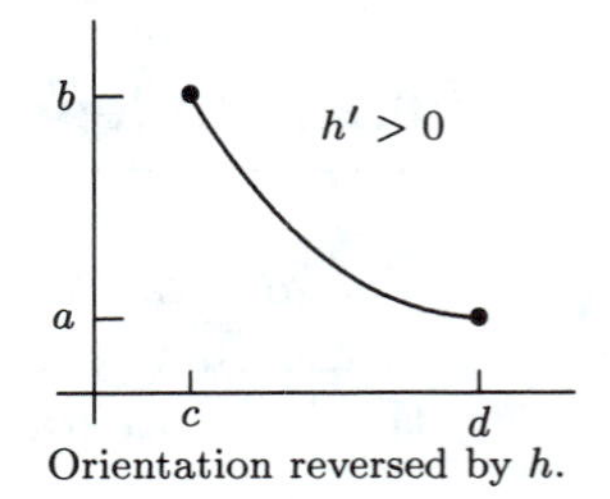

Orientation reversed by h.

Figure 6.2.7

Intuitively this definition says that the parameterization (ϕ, ψ) may be obtained from the original parameterization (x, y) by a function h that translates, stretches, squeezes, and possibly reverses the direction of the interval $[a, b]$. The alternative parameterization for the path in Figure 6.2.5 was generated by the function h, whrere

$$h(s) = 2s - 2 : [1, 2] \to [0, 2].$$

---

[2] Hint: the antiderivative of $1/\sqrt{4 - x^2}$ is $\sin^{-1}(\frac{x}{2})$.

For Figure 6.2.6 the function h is given by

$$h(u) = \cos^{-1}\left(\frac{u}{2}\right) : [-2, 2] \to [0, \pi].$$

In our definition h' is continuous and $h' \neq 0$; therefore h' must be either always positive or always negative. (What theorem guarantees this?) If h' is always positive, $h(s)$ is increasing and the orientation of $[a, b]$ is preserved. So as we move from c to d, $h(s)$ moves from a to b. If $h' < 0$, the function is decreasing and the orientation is reversed (see Figure 6.2.7). In either case h will be a one-to-one mapping and therefore invertible. When h is not a one-to-one function, the line integrals can differ (see the exercises).

THEOREM 6.2.5 *Let C and C' be two equivalent curves parameterized by $(x(t), y(t))$ for $t \in [a, b]$ and $(\phi(s), \psi(s))$ for $s \in [c, d]$, respectively. If the orientations of C and C' are the same, then*

$$\int_{C'} \vec{\mathbf{F}} \cdot \mathbf{d}\vec{\mathbf{x}} = \int_C \vec{\mathbf{F}} \cdot \mathbf{d}\vec{\mathbf{x}}.$$

If the orientations are opposite, then

$$\int_{C'} \vec{\mathbf{F}} \cdot \mathbf{d}\vec{\mathbf{x}} = -\int_C \vec{\mathbf{F}} \cdot \mathbf{d}\vec{\mathbf{x}}. \tag{4}$$

PROOF Since C and C' are equivalent, there is a function $h(s)\colon [c, d] \to [a, b]$ with $\phi(s) = x(h(s))$ and $\psi(s) = y(h(s))$. From Definition 6.2.2,

$$\begin{aligned}
\int_{C'} \vec{\mathbf{F}} \cdot \mathbf{d}\vec{\mathbf{x}} &= \int_c^d \vec{\mathbf{F}}(\phi(s), \psi(s)) \cdot (\phi'(s), \psi'(s))\, ds \\
&= \int_c^d \vec{\mathbf{F}}(x(h(s)), y(h(s))) \cdot (x'(h(s))h'(s), y'(h(s))h'(s))\, ds \\
&= \int_c^d \vec{\mathbf{F}}(x(h(s)), y(h(s))) \cdot (x'(h(s)), y'(h(s)))h'(s)\, ds.
\end{aligned}$$

Now use the substitution $t = h(s)$ and $dt = h'(s)\, ds$. If the orientation is preserved, h is increasing. Thus $h(c) = a$ and $h(d) = b$ and the substitution gives

$$\int_{C'} \vec{\mathbf{F}} \cdot \mathbf{d}\vec{\mathbf{x}} = \int_a^b \vec{\mathbf{F}}(x(t), y(t)) \cdot (x'(t), y'(t))\, dt = \int_C \vec{\mathbf{F}} \cdot \mathbf{d}\vec{\mathbf{x}}.$$

If the orientation is reversed, h is decreasing with $h(c) = b$ and $h(d) = a$, so

$$\int_{C'} \vec{\mathbf{F}} \cdot \mathbf{d}\vec{\mathbf{x}} = \int_b^a \vec{\mathbf{F}}(x(t), y(t)) \cdot (x'(t), y'(t))\, dt = -\int_C \vec{\mathbf{F}} \cdot \mathbf{d}\vec{\mathbf{x}}. \quad \blacksquare$$

At least for equivalent curves this theorem assures us that line integrals depend only on the path traced and the direction of tracing. For this reason the curve C traced in the reversed direction is denoted by C^- and we write

$$\int_{C^-} \vec{\mathbf{F}} \cdot \mathbf{d}\vec{\mathbf{x}} = -\int_C \vec{\mathbf{F}} \cdot \mathbf{d}\vec{\mathbf{x}}.$$

EXAMPLE 6.2.4 Consider the segment of the hyperbola $y = 2/x$ shown in Figure 6.2.8. The easiest way to parameterize any curve given by $y = f(x)$ is to let $x(t) = t$ and $y(t) = f(t)$. In our case this gives the parameterization $(t, 2/t)$ for $t \in [1, 3]$. Unfortunately $(t, 2/t)$ traces this curve in a direction opposite to the one indicated in the picture. There are two approaches:

a) (Reparameterization) If a curve C is given by $(x(t), y(t))$ for $t \in [a, b]$, a simple reparameterization that will reverse the direction is given by the function $h(s) = a + b - s$ which maps $[a, b]$ onto $[a, b]$. Since $h'(s) < 0$, we see that the orientation is reversed. Thus the curve C^- will be traced by $(\phi(s), \psi(s)) = (x(a+b-s), y(a+b-s))$ for $s \in [a, b]$. For our path this gives the parameterization $\phi(s) = 4 - s$, $\psi(s) = \frac{2}{4-s}$ for $t \in [1, 3]$. The line integral of $\vec{\mathbf{F}}(x, y)$ may be computed as

$$\int_1^3 \vec{\mathbf{F}}\left(4 - s, \frac{2}{4-s}\right) \cdot \left(-1, \frac{2}{(4-s)^2}\right) ds.$$

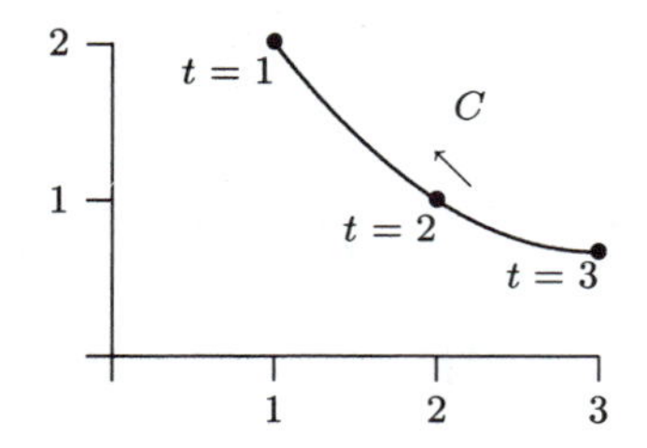

Parameterization $(t, \frac{2}{t})$ reverses the direction of C.

Figure 6.2.8

b) (Reorientation) Alternatively by Theorem 6.2.5, we may simply use the given parameterization and multiply the answer by -1 to account for the opposite orientation.

$$\int_C \vec{\mathbf{F}} \cdot \mathbf{d}\vec{\mathbf{x}} = -\int_1^3 \vec{\mathbf{F}}\left(s, \frac{2}{s}\right) \cdot \left(1, -\frac{2}{s^2}\right) ds. \quad \square$$

Paths that are composed of several curves $C_1, C_2, \ldots, C_n$, that are joined so that the endpoint of C_i is the initial point of C_{i+1} are called **contours** (Figure 6.2.9). We can extend our definition of line integrals to encompass contours in the natural way: simply define the integral of $\vec{\mathbf{F}}$ along a contour as the sum of the integrals along each curve. Thus if C is a contour composed of curves $C_1, C_2, \ldots, C_n$, we define the **contour integral** of $\vec{\mathbf{F}}$ on C as

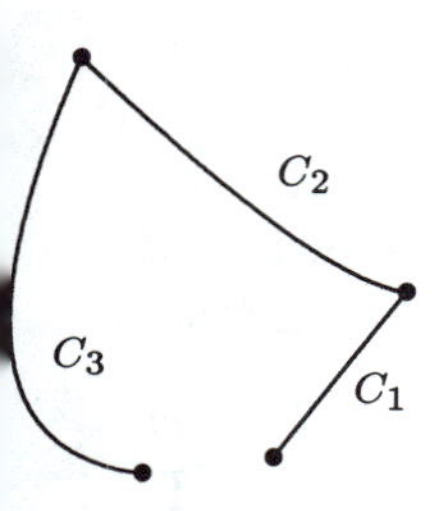

A contour C.

Figure 6.2.9

$$\int_C \vec{\mathbf{F}} \cdot \mathbf{d}\vec{\mathbf{x}} = \int_{C_1} \vec{\mathbf{F}} \cdot \mathbf{d}\vec{\mathbf{x}} + \int_{C_2} \vec{\mathbf{F}} \cdot \mathbf{d}\vec{\mathbf{x}} + \cdots + \int_{C_n} \vec{\mathbf{F}} \cdot \mathbf{d}\vec{\mathbf{x}}.$$

When the path's endpoint is the same as the beginning point, the contour is said to be **closed**. Integrals around closed contours are often written with a special integral sign: $\oint_C \vec{\mathbf{F}} \cdot \mathbf{d}\vec{\mathbf{x}}$.

Example 6.2.5 Integrate $\vec{\mathbf{F}}(x, y) = (y, x^2)$ over the closed contour $\Lambda = \Lambda_1 + \Lambda_2 + \Lambda_3$, the union of three distinct line segments oriented in the counterclockwise direction in Figure 6.2.10.

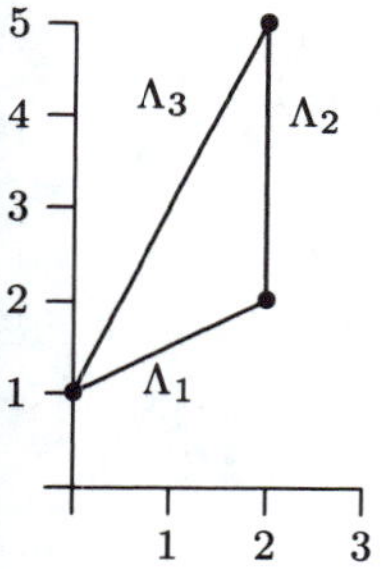

Figure 6.2.10

To integrate $\vec{\mathbf{F}}$ over Λ will require a parameterization for each line segment. To parameterize a segment, first determine the equation of the line containing it. For Λ_1 we have $y = \frac{1}{2}x + 1$. Now let $x(t) = t$ and $y(t) = \frac{1}{2}t + 1$. Since x begins at 0 and goes to 2, we let $t \in [0, 2]$. This gives

$$\int_{\Lambda_1} \vec{\mathbf{F}} \cdot \mathbf{d}\vec{\mathbf{x}} = \int_0^2 \left(\frac{t}{2} + 1, t^2\right) \cdot \left(1, \frac{1}{2}\right) dt = \frac{13}{3}.$$

The next segment is the vertical line $x = 2$. Here x is constant and y moves from 2 to 5, so our parameterization is $x(t) = 2$ and $y(t) = t$ for $t \in [2, 5]$. This yields

$$\int_{\Lambda_2} \vec{\mathbf{F}} \cdot \mathbf{d}\vec{\mathbf{x}} = \int_2^5 (t, 4) \cdot (0, 1)\, dt = 12.$$

Finally, the last segment goes from $(2, 5)$ to $(0, 1)$. The equation for this line is $y = 2x + 1$. Therefore the parameterization is $x(t) = t$ and $y(t) = 2t + 1$. Note that as t goes from 0 to 2 the curve is traced in the reverse direction, so

$$\int_{\Lambda_3} \vec{\mathbf{F}} \cdot \mathbf{d}\vec{\mathbf{x}} = -\int_0^2 (2t + 1, t^2) \cdot (1, 2)\, dt = -11\frac{1}{3}.$$

The integral around the complete contour is therefore

$$\oint_\Lambda \vec{\mathbf{F}} \cdot \mathbf{d}\vec{\mathbf{x}} = \int_{\Lambda_1} \vec{\mathbf{F}} \cdot \mathbf{d}\vec{\mathbf{x}} + \int_{\Lambda_2} \vec{\mathbf{F}} \cdot \mathbf{d}\vec{\mathbf{x}} + \int_{\Lambda_3} \vec{\mathbf{F}} \cdot \mathbf{d}\vec{\mathbf{x}} = 5. \quad \square$$

6.2.3 The Fundamental Theorem of Calculus

The idea of line integrals as developed in Section 6.1 was motivated by trying to extend the fundamental theorem of calculus. Our defintion of line integrals makes this theorem a simple consequence of the chain rule.

THEOREM 6.2.6 ***(The Fundamental Theorem of Calculus for Line Integrals)*** *Let C be a contour beginning at $\vec{\mathbf{a}}$ and ending at $\vec{\mathbf{b}}$. Let $f(x,y) : \mathbf{R}^2 \to \mathbf{R}$ be continuously differentiable in an open set containing C. Then*

$$\int_C \nabla f \cdot \mathbf{d}\vec{\mathbf{x}} = \int_C \frac{\partial f}{\partial x}\, dx + \frac{\partial f}{\partial y}\, dy = f(\vec{\mathbf{b}}) - f(\vec{\mathbf{a}}).$$

PROOF Suppose that C is a curve parameterized by $(x(t), y(t))$ for $t \in [a, b]$. Then

$$\int_C \nabla f \cdot \mathbf{d}\vec{\mathbf{x}} = \int_a^b \nabla f(x(t), y(t)) \cdot (x'(t), y'(t))\, dt. \tag{5}$$

Since f and C are differentiable functions, the composition $f(x(t), y(t))$ is also differentiable. Applying the chain rule for $\mathbf{R}^2$ (Theorem 5.1.14), we have

$$\frac{d}{dt} f(x(t), y(t)) = \nabla f(\, x(t), y(t)) \cdot (x'(t), y'(t)).$$

Substituting this in (5) yields

$$\begin{aligned}\int_C \nabla f \cdot \mathbf{d}\vec{\mathbf{x}} &= \int_a^b \frac{d}{dt} f(x(t), y(t))\, dt \\ &= f(x(b), y(b)) - f(x(a), y(a)) = f(\vec{\mathbf{a}}) - f(\vec{\mathbf{b}}),\end{aligned}$$

since C starts at $\vec{\mathbf{a}} = (x(a), y(a))$ and ends at $\vec{\mathbf{b}} = (x(b), y(b))$. This proves the result for single curves.

Next suppose that C is a contour composed of curves $C_1, C_2, \ldots C_n$, where C_1 goes from $\vec{\mathbf{a}} = \vec{\mathbf{a}}_0$ to $\vec{\mathbf{a}}_1$, C_2 goes from $\vec{\mathbf{a}}_1$ to $\vec{\mathbf{a}}_2$, ..., and C_n goes from $\vec{\mathbf{a}}_{n-1}$ to $\vec{\mathbf{a}}_n = \vec{\mathbf{b}}$. Then by the definition of a line integral along a contour and by employing the foregoing result on each curve C_i we have

$$\int_C \nabla f \cdot \mathbf{d}\vec{\mathbf{x}} = \sum_{i=1}^{n} \int_{C_i} \nabla f \cdot \mathbf{d}\vec{\mathbf{x}} = \sum_{i=1}^{n} f(\vec{\mathbf{a}}_i) - f(\vec{\mathbf{a}}_{i-1}).$$

This sum collapses to give $f(\vec{\mathbf{a}}_n) - f(\vec{\mathbf{a}}_0) = f(\vec{\mathbf{b}}) - f(\vec{\mathbf{a}})$. ∎

One startling consequence of this theorem is that since the value of any line integral of ∇f depends only on the values of f at the initial and final points of the path, it is *completely independent of the path used to connect the two points!* This property of **path independence** has many interesting consequences and will be explored in more detail in Section 6.5.

EXAMPLE 6.2.6 Let $f(x,y) = xy^2 - 2x$. Then $\nabla f(x,y) = (y^2-2, 2xy)$. Since f is continuous everywhere, we can compute $\int_C \nabla f \cdot \mathbf{d\vec{x}}$ for paths between any two points. For example, to find the line integral over the curve C given by $\left(\frac{5t}{t^2+4}, \sqrt{t^2+9}\right)$ for $t \in [0,4]$ we need only compute the beginning and ending points of C, $\vec{\mathbf{a}} = (0,3)$ and $\vec{\mathbf{b}} = (1,5)$. Hence $\int_C \nabla f \cdot \mathbf{d\vec{x}} = f(1,5) - f(0,3) = 23$. □

If the contour is closed, the beginning and ending point is the same, and Theorem 6.2.6 shows that the integral of a gradient will be 0:

$$\oint_C \nabla f \cdot \mathbf{d\vec{x}} = 0.$$

It is remarkably easy to compute line integrals of gradients, since there is no need to work with any parameterization. It would be useful to be able to identify whether a given vector field is the gradient of some function. A function $f(x,y)$ is called a **potential function** for the vector field $(P(x,y), Q(x,y))$ if $\nabla f(x,y) = (P(x,y), Q(x,y))$. Potential functions are like antiderivatives in that they allow us to compute line integrals by simply evaluating the potential function at the endpoints of the curve. But not all vector fields have a potential function, so it is useful to check the following preliminary criterion:

THEOREM 6.2.7 *Suppose that $P(x,y)$ and $Q(x,y)$ are continuously differentiable functions. If $(P(x,y), Q(x,y)) = \nabla f(x,y)$ for some function f, then*

$$\frac{\partial P}{\partial y} = \frac{\partial Q}{\partial x}.$$

PROOF If $(P(x,y), Q(x,y)) = \nabla f(x,y)$, then $P(x,y) = \frac{\partial f}{\partial x}$ and $Q(x,y) = \frac{\partial f}{\partial y}$. Therefore

$$\frac{\partial P(x,y)}{\partial y} = \frac{\partial^2 f}{\partial y \partial x} \quad \text{and} \quad \frac{\partial Q(x,y)}{\partial x} = \frac{\partial^2 f}{\partial x \partial y}.$$

But if the mixed partials of f are continuous, by Theorem 5.3.6 they must be equal. Hence $\frac{\partial P}{\partial y} = \frac{\partial Q}{\partial x}$. ∎

This is a convenient check to see if a vector field has no potential function.

Example 6.2.7 Suppose $(P(x,y), Q(x,y)) = (2xy+1, x^2y)$; then

$$\frac{\partial P}{\partial y} = 2x \qquad \text{and} \qquad \frac{\partial Q}{\partial x} = 2xy.$$

Since these are unequal, there is no hope of finding a potential function and to compute line integrals we must parameterize our curves and use Definition 6.2.2. □

We see from this example that the situation for potential functions is different from antiderivatives. In one-dimensional calculus if f is continuous, an antiderivative for f always exists. In fact, the second fundamental theorem of calculus states that $\int_a^x f(t)\,dt$ is an antiderivative for f. But potential functions for a vector field often do not exist. Even knowing that $\frac{\partial P}{\partial y} = \frac{\partial Q}{\partial x}$ does not necessarily guarantee that a potential function will exist (see problem 15). However, in most instances, when this criterion is satisfied we can construct a potential function f that will work, as in the following example:

Example 6.2.8 Consider the vector field $(P(x,y), Q(x,y)) = (2x + 4xy^3, 6x^2y^2 + y^{-2})$. Both P and Q are continuously differentiable except along the x-axis, where $y = 0$. Furthermore,

$$\frac{\partial P}{\partial y} = 12xy^2 = \frac{\partial Q}{\partial x},$$

so there is some hope that we can find a potential function for (P, Q).

To construct a potential function we work backward, first integrating P with respect to x and then Q with respect to y. When we integrate with respect to x, the y's are considered constant; thus the arbitrary constant of integration is an *arbitrary function of* y. In our case we obtain

$$\int P\,dx = \int 2x + 4xy^3\,dx = x^2 + 2x^2y^3 + h(y).$$

Similarly, in the integration with respect to y we get an *arbitrary function of* x:

$$\int Q\,dy = \int 6x^2y^2 + y^{-2}\,dy = 2x^2y^3 - y^{-1} + g(x).$$

To obtain a potential function we must find arbitrary functions, $h(y)$ and $g(x)$ so that $\int P\,dx = f = \int Q\,dy$. Looking back at $\int P\,dx$ and

$\int Q\,dy$, we see that this is possible by letting $h(y) = -y^{-1}$ and $g(x) = x^2$; this gives us $f(x,y) = x^2 + 2x^2y^3 - y^{-1}$ as a potential function. Now that we have a potential function, we can integrate along even the most complex curve by simply evaluating at the endpoints. However, we must be careful that our curve does not intersect points where $f(x,y)$ is undefined (i.e., along the x-axis). This is an important stipulation, since Theorem 6.1.4 is valid only for *curves lying entirely in a region where $f(x,y)$ is continuously differentiable.* For instance, along any curve going from $(1,2)$ to $(3,4)$ that avoids the x-axis,

$$\int_C P\,dx + Q\,dy = \int_C \nabla f \cdot \mathbf{d}\vec{\mathbf{x}} = f(3,4) - f(1,2) = 1144\tfrac{1}{4}. \quad \square$$

6.2.4 Work

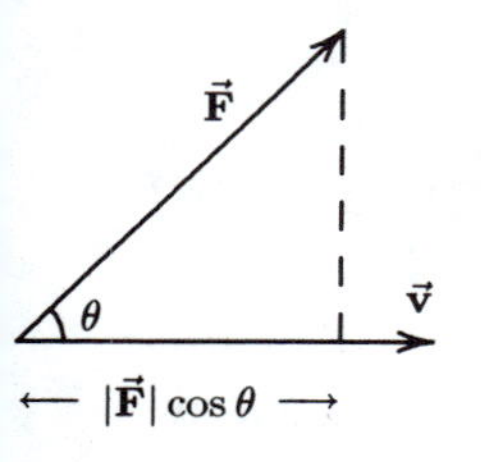

Figure 6.2.11

One of the most important applications of line integrals occurs in the definition of work in physics. In the simplest case, for a constant force moving an object along a straight path the work is

work = (force in the direction of motion) × (distance moved).

If $\vec{\mathbf{F}}$ represents the force vector and $\vec{\mathbf{v}}$ is a displacement vector, this quantity can be computed by $|\vec{\mathbf{F}}||\vec{\mathbf{v}}|\cos\theta$, or more simply $\vec{\mathbf{F}} \cdot \vec{\mathbf{v}}$ (Figure 6.2.11).

Suppose now that the force $\vec{\mathbf{F}}$ changes continuously from point to point and that the path of motion is a curve. A reasonable procedure to estimate the work done by $\vec{\mathbf{F}}$ is to break the curve into small straight line pieces $(\Delta x_i, \Delta y_i)$, approximate the work on each of these pieces, and sum the results. Assuming that $\vec{\mathbf{F}}$ varies gradually and that the pieces are short enough, the work along each piece can be estimated as the dot product of $\vec{\mathbf{F}}(x_i^*, y_i^*)$ with the segment $(\Delta x_i, \Delta y_i)$, where (x_i^*, y_i^*) is an intermediate point on the segment of the curve. Summing these terms we estimate the total work to be

$$\sum_{i=1}^{n} \vec{\mathbf{F}}(x_i^*, y_i^*) \cdot (\Delta x_i, \Delta y_i).$$

Finer divisions of the curve should lead to increasingly accurate estimates. Observe that this basic idea was exactly our starting point for the concept of a line integral; therefore work is defined as a line integral.

DEFINITION 6.2.8 *The **work** done by a force $\vec{\mathbf{F}}$ in moving an object along a curve C is*

$$\text{work} = \int_C \vec{\mathbf{F}} \cdot \mathbf{d}\vec{\mathbf{x}}.$$

The importance of the physical concept of work lies in its relation to energy. The **kinetic energy** of a particle is defined to be K.E. $= \frac{1}{2}mv^2$, where $v = |\vec{\mathbf{v}}|$ is the particle's speed. Kinetic energy and work are intimately related.

THEOREM 6.2.9 ***(The Work-Energy Principle)*** *If a particle obeys Newton's second law of motion, $\vec{\mathbf{F}} = m\vec{\mathbf{a}}(t)$, as it moves along a curve C for $t \in [t_0, t_1]$, then the work done by $\vec{\mathbf{F}}$ along C is equal to the change in kinetic energy, ΔK.E., experienced by the particle.*

$$\int_C \vec{\mathbf{F}} \cdot \mathbf{d}\vec{\mathbf{x}} = \frac{1}{2}mv^2(t_1) - \frac{1}{2}mv^2(t_0) = \Delta\text{K.E.} \tag{6}$$

PROOF The function $(x(t), y(t))$ represents the particle's position at time t and provides a parameterization of the trajectory C for $t \in [t_0, t_1]$. Thus $v^2 = |\vec{\mathbf{v}}|^2 = (x'(t))^2 + (y'(t))^2$. Newton's second law gives the relationship between the force $\vec{\mathbf{F}}$ and this curve C. Since acceleration is the second derivative of position,

$$\vec{\mathbf{F}}(t) = m\vec{\mathbf{a}}(t) = m(x''(t), y''(t)).$$

The work done by $\vec{\mathbf{F}}$ over the time interval $[t_0, t_1]$ is therefore

$$\begin{aligned}\int_C \vec{\mathbf{F}} \cdot \mathbf{d}\vec{\mathbf{x}} &= \int_{t_0}^{t_1} m(x''(t), y''(t)) \cdot (x'(t), y'(t))\, dt \\ &= \int_{t_0}^{t_1} mx''(t)x'(t)\, dt + \int_{t_0}^{t_1} my''(t)y'(t)\, dt.\end{aligned}$$

Using integration by substitution, it is easy to see that $\int x'(t)x''(t)\, dt = \frac{1}{2}(x'(t))^2$ and $\int y'(t)y''(t)\, dt = \frac{1}{2}(y'(t))^2$. Thus

$$\begin{aligned}\int_C \vec{\mathbf{F}} \cdot \mathbf{d}\vec{\mathbf{x}} &= \tfrac{1}{2}m\left((x'(t))^2 + (y'(t))^2\right)\Big|_{t_0}^{t_1} \\ &= \tfrac{1}{2}m\left((x'(t_1))^2 + (y'(t_1))^2\right) - \left((x'(t_0))^2 + (y'(t_0))^2\right) \\ &= \tfrac{1}{2}mv^2(t_1) - \tfrac{1}{2}mv^2(t_0) = \Delta\text{K.E.} \quad \blacksquare\end{aligned}$$

If we are fortunate enough to have a potential function for $\vec{\mathbf{F}}$, say $f(x, y)$, then by Theorem 6.2.6 the work $\int_C \vec{\mathbf{F}} \cdot \mathbf{d}\vec{\mathbf{x}}$ is given by

$$f(x(t_1), y(t_1)) - f(x(t_0), y(t_0)) = f(\text{final}) - f(\text{initial}) = \Delta f$$

and is independent of the path C connecting the points. Physicists think of this type of field as having an energy associated with position called **potential energy**, $U(x, y)$, defined by $U(x, y) = -f(x, y)$. If $\vec{\mathbf{F}}$ has a potential function, then by the preceding theorem

$$-\Delta U = \Delta f = \int_C \vec{\mathbf{F}} \cdot \mathbf{d}\vec{\mathbf{x}} = \Delta\text{K.E.}$$

This implies

$$\Delta U + \Delta\text{K.E.} = 0.$$

According to this equation, the net change in the total energy, potential and kinetic, always remains 0. Thus the *total amount of energy is conserved.* Vector fields that have potential functions are therefore called **conservative** fields.[3] This conservation property means that every change in potential energy is exactly compensated by an opposite change in kinetic energy and vice versa. Hence the work done by a force is often viewed as converting potential into kinetic energy (when the work in positive) or as storing kinetic energy as potential energy (when the work is negative).

Example 6.2.9 **(Central Force Fields are Conservative)** A **central force field** is a field in which the magnitude of the force at any position (x, y) depends only on the distance from the origin, $r = \sqrt{x^2 + y^2}$, and the direction of the force is either toward (attractive force) or away (repulsive force) from the origin.

At a point (x, y) the direction of a central force is given by the unit vector $\pm\left(\frac{x}{\sqrt{x^2+y^2}}, \frac{y}{\sqrt{x^2+y^2}}\right) = \pm\frac{1}{r}(x, y)$. Therefore central force fields can be described by

$$\vec{\mathbf{F}} = \lambda(r)\frac{1}{r}(x, y), \tag{7}$$

where λ is a scalar function of the distance r which describes the magnitude of the force. If the force is attractive λ is negative; if it is repulsive λ is positive.

[3] A potential function is defined only up to an arbitrary constant. Consequently the total energy of a system can be arbitrary. However, this never affects our computations, which only deal with changes between potential and kinetic energy.

Earth's gravity is an example of such a force. It acts on a straight line between the center of the earth (origin) and the object in question, and is proportional to the product of the masses and inversely proportional to the square of the distance. We have

$$\vec{\mathbf{F}}_G(x,y) = \frac{-GMm}{r^2}\frac{1}{r}(x,y).$$

To show that any force field of the form (7) is conservative we must exhibit a potential. Since $\lambda(r)$ is just a scalar function, it has some antiderivative, $\Lambda(r)$, assuming that $\lambda(r)$ is continuous. We claim that a potential for $\vec{\mathbf{F}}$ is given by $\Lambda(r) = \Lambda(\sqrt{x^2+y^2})$. By the chain rule (problem 5.4.17b), since $\Lambda'(r) = \lambda(r)$,

$$\nabla\Lambda(\sqrt{x^2+y^2}) = \lambda(\sqrt{x^2+y^2})\left(\frac{x}{\sqrt{x^2+y^2}}, \frac{x}{\sqrt{x^2+y^2}}\right) = \lambda(r)\frac{1}{r}(x,y).$$

For example, we may compute the gravitational potential function as the antiderivative of $\frac{-GMm}{r^2}$, which is

$$\text{gravitational potential} = \frac{GMm}{r}. \quad \square$$

Problems for Section 6.2

6.2.1. Suppose that we do not have a specific formula for a curve but we know that it passes through the points $(0,0)$, (2,1), (3,2), (4,3), and $(5,5)$.

a) If the vector field $\vec{\mathbf{F}}(x,y)$ is a constant $(1,-2)$ at all points in the plane, estimate the line integral of $\vec{\mathbf{F}}$ along the curve. (See the sum in equation (1) that generated the basic notion of a line integral.)

b) If $\vec{\mathbf{G}}(x,y) = (y+x,y)$, estimate the line integral of $\vec{\mathbf{G}}$ along the curve. (Hint: Estimate $\vec{\mathbf{G}}$ at the midpoint of each segment of the curve and use equation (1).)

6.2.2. Sketch the vector field $\vec{\mathbf{F}}(x,y) = (-y/2, x/2)$ at the following points: $(\pm1,0)$, $(0,\pm1)$, $(\pm2,0)$, $(0,\pm2)$, $(\pm1,\pm2)$, $(\pm2,\pm1)$, and $(\pm2,\pm2)$. From the look of the vector field would you expect the line integral of $\vec{\mathbf{F}}$ about the unit circle centered at the origin in the counterclockwise direction to be positive or negative? Verify your conjecture by computing the line integral.

6.2.3. **a)** Sketch the vector field $\vec{\mathbf{F}}(x,y) = (x,y)$ at the following points on the upper semicircle of radius 2: $(\pm2,0)$, $(\pm\sqrt{3},1)$, $(\pm1,\sqrt{3})$, and $(0,2)$.

b) Now compute the line integral $\int_C \vec{\mathbf{F}}(x,y)\cdot \vec{\mathbf{dx}}$, where C is the upper semicircle given by $(x(t),y(t)) = (2\cos t, 2\sin t)$ for $t\in[0,\pi]$.

c) Looking at the picture in **(a)**, why would you expect the answer in **(b)**?

6.2.4. Let $\vec{\mathbf{F}}(x, y) = (P(x, y), Q(x, y))$ be continuous and $(x(t), y(t))$ be continuously differentiable for $t \in [a, b]$.

a) Show that the difference between $\sum_{i=1}^{n} \vec{\mathbf{F}}(x(t_i^*), y(t_i^*)) \cdot (x'(t_i^*), y'(t_i^{**}))\Delta t_i$ and the Riemann sum $\sum_{i=1}^{n} \vec{\mathbf{F}}(x(t_i^*), y(t_i^*)) \cdot (x'(t_i^*), y'(t_i^*))\Delta t_i$ can be written as $\sum_{i=1}^{n} Q(x(t_i^*), y(t_i^*))(y'(t_i^*) - y'(t_i^{**}))\Delta t_i$.

b) Explain why $Q(x(t), y(t))$ is bounded by some M for $t \in [a, b]$ and why $y'(t)$ is uniformly continuous.

c) Using the uniform continuity prove for any $\epsilon > 0$ that there is a partition of $[a, b]$ sufficiently fine so that when t^* and t^{**} belong to the same interval, then $\frac{y'(t^*) - y'(t^{**})}{M(b-a)} < \epsilon$. Use this to show that the sum in **(a)** is less than ϵ for such a partition.

6.2.5. Let C be given by $(t, \sqrt{t^2 + 9})$ for $t \in [0, 4]$. Draw some vectors from the vector field $\vec{\mathbf{G}}(x, y) = (\frac{x}{2}, -\frac{y}{2})$ at various points on C. Geometrically, what do you expect the value of $\int_C \vec{\mathbf{G}} \cdot \mathbf{d}\vec{\mathbf{x}}$ should be and why? Verify your claim with a computation.

6.2.6. Sketch each of the following curves and then compute the line integral of $\vec{\mathbf{F}}(x, y) = (y, x + 1)$ along each of the curves.

a) C is $(x(t), y(t)) = (\cos t, \sin t)$ for $t \in [0, \frac{\pi}{2}]$.

b) C is $(x(t), y(t)) = (1 - t, \frac{2}{2-t} - 1)$ for $t \in [0, 1]$.

c) The straight line segment going from (1,0) to (0,0) followed by the segment from (0,0) to (0,1).

6.2.7. Sketch the curve and then compute the line integral for the given vector field along the curve.

a) $\vec{\mathbf{F}}(x, y) = (x^2, y^2)$ along the straight line from $(0, 0)$ to $(2, -4)$.

b) $\vec{\mathbf{G}}(x, y) = (xy, y - x)$ around the unit circle in a *clockwise direction.*

c) $\vec{\mathbf{G}}(x, y) = (xy, y - x)$ around the arc of the parabola $y = 4 - x^2$ from $(2, 0)$ to $(-2, 0)$ and then back to the point $(2, 0)$ along the x–axis.

6.2.8. Prove the linearity and additivity properties for line integrals in Theorem 6.2.3.

6.2.9. Prove that if at each point of a curve C the vector field $\vec{\mathbf{F}} = \tau$, where τ is the unit tangent to the curve, then $\int_C \vec{\mathbf{F}} \cdot \vec{\mathbf{x}} = L$, where L is the length of C.

6.2.10. Suppose that the vector field is of the form $\vec{\mathbf{F}}(x, y) = (P(x), Q(y))$; that is, P depends only on x and Q depends only on y. Prove that the line integral for any such vector field around a closed curve C is 0.

•6.2.11. **a)** Show that if C is a horizontal line, then $\int_C (P(x, y), Q(x, y)) \cdot \mathbf{d}\vec{\mathbf{x}}$ is completely independent of $Q(x, y)$. (Hint: Assume that the line goes from (a, c) to (b, c). Find a parameterization of this line and write down the line integral.)

b) Restate the problem for vertical line segments and prove it.

6.2.12. Let $\alpha(t)$, $\beta(t)$, and $\gamma(t)$ be curves. Using Definition 6.2.4 prove that equivalent curves satisfy the following three properties. (These are the three basic properties of equivalence relations.)

a) Reflexivity: Any curve $\alpha(t)$ is equivalent to itself.

b) Symmetry: If $\alpha(t)$ is equivalent to $\beta(t)$, then $\beta(t)$ is equivalent to $\alpha(t)$.

c) Transitivity: If $\alpha(t)$ is equivalent to $\beta(t)$ and $\beta(t)$ is equivalent to $\gamma(t)$, then $\alpha(t)$ is equivalent to $\gamma(t)$.

6.2.13. For each of the following vector fields determine a potential function, if it exists.

a) $\vec{\mathbf{F}}(x, y) = (3x^2y, x^3)$.

b) $\vec{\mathbf{G}}(x, y) = (2xe^y + y, x^2e^y + x - 2y)$.

c) $\vec{\mathbf{H}}(x, y) = (2xy, y)$.

d) $\vec{\mathbf{F}}(x, y) = (\sin y - y \sin x + x, \cos x + x \cos y + y)$.

e) Now integrate each of these vector fields along each of the following curves: (i) the straight line from (1,1) to (4,2), (ii) the parabola $y = x^2$ from (1,1) to (4,2), and (iii) clockwise about the unit circle centered at the origin starting at $(1, 0)$.

6.2.14. **a)** Prove that $\vec{\mathbf{F}}(x, y) = (x^2 - y^2, x^2y)$ cannot have a potential function.

b) Find a potential function for $\vec{\mathbf{G}}(x, y) = (xy^4 - 2x, 2x^2y^3 + 1)$ and use it to compute $\int_\Gamma \vec{\mathbf{G}} \cdot \mathbf{d}\vec{\mathbf{x}}$, where $\Gamma(t) = (2\cos t, 1 + \sin t)$ for $t \in [0, \frac{\pi}{2}]$.

•6.2.15. Let the vector field $\vec{\mathbf{F}}(x, y) = (\frac{-y}{x^2+y^2}, \frac{x}{x^2+y^2})$.

a) Show that $\frac{\partial P}{\partial y}$ is equal to $\frac{\partial Q}{\partial x}$.

b) From the definition, compute the line integral of $\vec{\mathbf{F}}$ about the unit circle that is parameterized by $(\cos t, \sin t)$ for $t \in [0, 2\pi]$.

c) Conclude that the converse of Theorem 6.2.7 does not necessarily hold.

6.2.16. A force field is given by $\vec{\mathbf{F}}(x, y) = \alpha(x^2 + y^2)^{-3/2}(x, y)$, where α is a constant.

a) Compute the work done by this force in moving a particle from $(1, 1)$ to $(4, 4)$ along a straight line.

b) Compute the work done in moving the particle about a circle of radius 2 in the counterclockwise direction.

6.2.17. Prove that any constant force field $\vec{\mathbf{F}} = (k_1, k_2)$ is conservative. How much work is done by such a field in moving from (x_0, y_0) to (x_1, y_1)?

6.2.18. In Example 6.2.9 we saw that all central force fields had a potential function. In this problem we investigate force fields that are perpendicular to (x, y). Suppose that $\vec{\mathbf{F}}(x, y)$ is always perpendicular to the position vector (x, y). Show that if $\vec{\mathbf{F}}$ has a potential function f that exists for all points (x, y) in the plane, then the potential function f must be constant and $\vec{\mathbf{F}}$ must be 0. (Hint: Show that $f(\vec{\mathbf{b}}) - f(\vec{\mathbf{a}}) = 0$ by considering the value of the line integral from $\vec{\mathbf{a}}$ to $\vec{\mathbf{b}}$ that goes first from $\vec{\mathbf{a}}$ to the origin along the radial direction and then out to $\vec{\mathbf{b}}$ along the radius.)

6.3 The Fundamental Theorem of Calculus: Part II

In Section 6.1 we extended the proof of the fundamental theorem of calculus to curves using four basic steps:

STEP 1 The curve was divided into many little pieces.

STEP 2 The difference $f(\vec{\mathbf{b}}) - f(\vec{\mathbf{a}})$ of the function across the boundary of the curve was written as a collapsing sum of differences across each of the interior pieces $f(x_i, y_i) - f(x_{i-1}, y_{i-1})$.

STEP 3 We used the mean value theorem to express these interior differences in terms of a derivative.

STEP 4 Finally, by summing the derivative expressions we obtained the Riemann sum for an integral. This gave us the fundamental theorem of calculus for line integrals, which expressed the difference of f at the two boundary points of the curve as a (line) integral of the derivative of f over the interior of the curve.

6.3.1 Extending to Two Dimensional Regions

We now wish to apply the same process to two-dimensional regions in the hope of discovering a further generalization of the fundamental theorem of calculus. However, the boundary of a region in the plane doesn't consist of simple isolated points but rather of an entire boundary curve. Therefore in Step 2, instead of expressing the difference of the *value of* f at boundary points as a collapsing sum of *values of* f across interior segments, we will be expressing the *line integral of* $\vec{\mathbf{F}} = (P, Q)$ around the outer boundary of the entire region as a collapsing sum of *line integrals* around little interior pieces. As before, we must make sure that the interior pieces are oriented in the same way so that they will collapse when added. In the case of two-dimensional regions, to achieve the desired cancellation we will give all of the boundaries a counterclockwise orientation.[1]

Thus let us begin with a two-dimensional region R in the plane and a vector function $\vec{\mathbf{F}} = (P, Q)$.

STEP 1 Using a grid, the region R is decomposed into tiny pieces, R_{ij}, which all possess the same counterclockwise orientation on their boundaries (see Figure 6.3.1).

STEP 2 Now take the line integral of $\vec{\mathbf{F}} = (P, Q)$ counterclockwise around the boundary of each interior region. Looking at the entire figure, notice that *every interior boundary is traced twice, once in each direction. Thus all the line integrals on all the interior boundaries cancel, leaving us with the line integral around the outer boundary of R traced in a counterclockwise direction.* This is the direct analogue of the telescoping sum in equation (1) of Section 6.1, but here the canceling is of one-dimensional curves instead of zero-dimensional boundary points. The boundary of a region R is usually denoted by ∂R. Using this notation the cancellation of interior boundaries then yields

$$\oint_{\partial R} (P, Q) \cdot \mathbf{d}\vec{\mathbf{x}} = \sum_{i,j} \oint_{\partial R_{ij}} (P, Q) \cdot \mathbf{d}\vec{\mathbf{x}}. \tag{1}$$

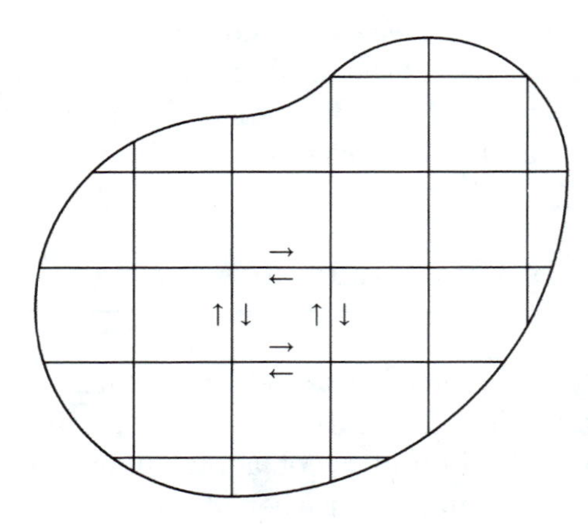

Region R decomposed.

Figure 6.3.1

[1] For highly convoluted regions it might appear difficult to determine a counterclockwise direction. However, at any point we can always find the correct direction, since as the boundary is traversed counterclockwise the region being circumscribed always stays on the left.

STEP 3 Next we seek to express these pieces $\oint_{\partial R_{ij}} (P,Q) \cdot \mathbf{d\vec{x}}$ in terms of derivatives using the mean value theorem. To do this, we divide the boundary curve into four separate paths (see Figure 6.3.2). First we deal with the horizontal line segments C_1 and C_3. These line integrals reduce as follows (see the exercises):

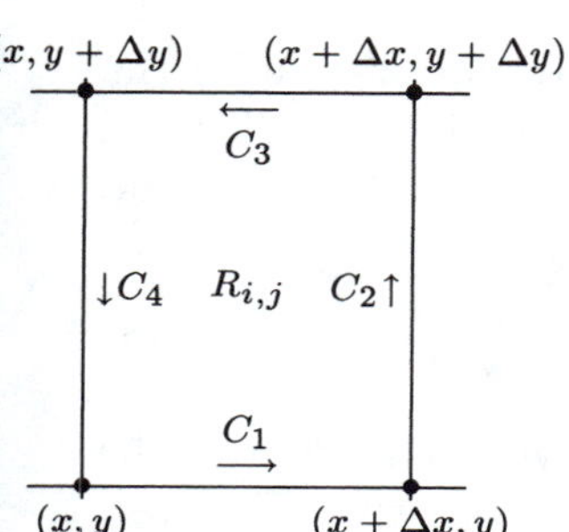

Figure 6.3.2

$$\int_{C_1} (P,Q) \cdot \mathbf{d\vec{x}} = \int_x^{x+\Delta x} P(x,y)\, dx$$

and

$$\int_{C_3} (P,Q) \cdot \mathbf{d\vec{x}} = -\int_x^{x+\Delta x} P(x, y+\Delta y)\, dx,$$

where the negative sign accounts for the direction of C_3. Combining these two integrals yields

$$-\int_x^{x+\Delta x} P(x,y+\Delta y) - P(x,y)\, dx.$$

Since both y and $y+\Delta y$ are constants for the integration, using the integral mean value theorem

$$-\int_x^{x+\Delta x} P(x,y+\Delta y) - P(x,y)\, dx = -[P(x^*, y+\Delta y) - P(x^*, y)]\Delta x$$

for some $x^* \in [x, x+\Delta x]$. A final use of the mean value theorem allows us to express this difference as

$$[P(x^*, y+\Delta y) - P(x^*, y)] = \frac{\partial P}{\partial y}(x^*, y^*)\Delta y$$

for some $y^* \in [y, y+\Delta y]$. Thus we may write

$$\int_{C_1} (P,Q) \cdot \mathbf{d\vec{x}} + \int_{C_3} (P,Q) \cdot \mathbf{d\vec{x}} = -\frac{\partial P}{\partial y}(x^*, y^*)\Delta y \Delta x.$$

A similar computation for the other curves, C_2 and C_4, yields

$$\int_{C_2} (P,Q) \cdot \mathbf{d\vec{x}} + \int_{C_4} (P,Q) \cdot \mathbf{d\vec{x}} = \frac{\partial Q}{\partial x}(x^{**}, y^{**})\Delta y \Delta x.$$

Altogether the integral of (P,Q) around the boundary of R_{ij} is

$$\oint_{\partial R_{ij}} (P,Q) \cdot \mathbf{d\vec{x}} = \left(\frac{\partial Q}{\partial x}(x^{**}, y^{**}) - \frac{\partial P}{\partial y}(x^*, y^*)\right) \Delta y \Delta x.$$

Thus we have expressed the integral of each interior piece R_{ij} using the derivative expression $\frac{\partial Q}{\partial x} - \frac{\partial P}{\partial y}$ evaluated at an intermediate point multiplied by $\Delta x \Delta y$.

STEP 4 From Step 2, the sum of the interior integrals $\oint_{\partial R_{ij}} (P,Q) \cdot \mathbf{d\vec{x}}$ gives the integral on the entire boundary. Substituting the preceding expression for the line integral around $R_{i,j}$ into the sum (1) yields

$$\oint_{\partial R} (P,Q) \cdot \mathbf{d\vec{x}} = \sum_{i,j} \left(\frac{\partial Q}{\partial x}(x_{ij}^{**}, y_{ij}^{**}) - \frac{\partial P}{\partial x}(x_{ij}^{*}, y_{ij}^{*}) \right) \Delta x \Delta y.$$

The sum on the right-hand side of this equation is a Riemann sum for the *double integral* of $\frac{\partial Q}{\partial x} - \frac{\partial P}{\partial y}$ over R. As the grid is divided into smaller and smaller pieces, we obtain our second generalization of the fundamental theorem of calculus, known as Green's theorem:

$$\oint_{\partial R} (P,Q) \cdot \mathbf{d\vec{x}} = \iint_R \frac{\partial Q}{\partial x} - \frac{\partial P}{\partial y}\, dx\, dy \tag{2}$$

This argument is somewhat incomplete, since we have ignored the irregular pieces of the region R next to the outer boundary. However, as the region is divided more finely, the total area of the irregular pieces shrinks and so becomes negligible for the double integral.

The derivative expression $\frac{\partial Q}{\partial x} - \frac{\partial P}{\partial y}$ appearing in Green's theorem is called the **curl** of the vector field (P,Q). This term is derived from the physical interpretation associated with line integrals around a boundary. The line integral measures the accumulated magnitude of the vector field in the direction of motion as we circumnavigate a region. It represents a sort of net tendency of the vector field to "circulate" or "curl" around the boundary. In Step 3 we showed that the counterclockwise circulation or curling around a small region about (x,y) can be written as $\frac{\partial Q}{\partial x} - \frac{\partial P}{\partial y}$ times the area of the region $\Delta x \Delta y$; therefore this derivative expression is called the curl.

EXAMPLE 6.3.1 Let $\vec{\mathbf{F}}(x,y) = (-y, x)$ and let R be the circular disk of radius $r = 1$ centered at the origin. The curl of $\vec{\mathbf{F}}$ is

$$\frac{\partial Q}{\partial x} - \frac{\partial P}{\partial y} = 1 - (-1) = 2.$$

According to our interpretation, the curl represents the counterclockwise circulation per unit area for $\vec{\mathbf{F}}$. Since the curl happens to be constant over the disk, this sum is just curl $\times$ area, which equals $2 \times \pi r^2 = 2\pi$. We check Green's theorem by computing the counterclockwise integral of $\vec{\mathbf{F}}$ around the unit circle that bounds the disk using the parameterization of $x(t) = \cos t,\ y(t) = \sin t$ for $t \in [0, 2\pi]$. This gives

$$\int_0^{2\pi} (-\sin t, \cos t) \cdot (-\sin t, \cos t)\, dt = \int_0^{2\pi} 1\, dt = 2\pi. \quad \square$$

6.3.2 Further Generalizations

We now have two different generalizations of the fundamental theorem of calculus for $\mathbf{R}^2$. The first is the fundamental theorem of calculus for line integrals, which applies to one-dimensional curves in $\mathbf{R}^2$ and their boundary points. The second is Green's theorem, which relates to two-dimensional regions in $\mathbf{R}^2$ and their one-dimensional boundary curves. But why should we stop at two dimensions? In fact, if we were to apply our methods in $\mathbf{R}^3$ we would discover three generalizations of the fundamental theorem of calculus:

1. one for curves in $\mathbf{R}^3$ and their boundary points (the fundamental theorem of calculus for line integrals),
2. a second for two-dimensional surfaces in $\mathbf{R}^3$ and their boundary curves (known as Stokes' theorem), and
3. a third for solid regions and their boundary surfaces (known as Gauss's theorem).

The curious reader can find presentations and proofs of these lovely theorems in most calculus texts.

At this point you may have guessed that some sort of universal form of the fundamental theorem of calculus might be lurking behind these many different expressions. In its most general form this theorem equates the *integral of a function over the (oriented) boundary of a region* to the *integral of a certain kind of derivative of the function over the interior of the region.* It is known as the general form of Stokes' theorem, and is expressed symbolically as

$$\int_{\partial R} F = \int_R dF, \tag{3}$$

where ∂R denotes the (oriented) boundary of R and dF denotes the special derivative of F (called an **exterior** derivative).

Let's review what these symbols mean for our two versions of this theorem in $\mathbf{R}^2$. Consider first the fundamental theorem of calculus for line integrals. Here R represents a curve and the boundary, ∂R, consists of (oriented) isolated points. In this case the integral of the left side of (3) is interpreted as the oriented *sum* of F at the boundary points $F(\vec{\mathbf{b}}) - F(\vec{\mathbf{a}})$. The exterior derivative in this case is just the gradient of F, that is, $dF = \nabla F$. Thus (3) reads:

$$\int_{\partial R} F = F(\vec{\mathbf{b}}) - F(\vec{\mathbf{a}}) = \int_R \nabla F \cdot \mathbf{d}\vec{\mathbf{x}} = \int_R dF.$$

Turning to Green's theorem we see that R stands for a two-dimensional region and that ∂R represents the boundary curve of R traced counterclockwise. Our function is $\vec{\mathbf{F}} = (P, Q)$ and the exterior derivative is given by the curl $d\vec{\mathbf{F}} = (\frac{\partial Q}{\partial x} - \frac{\partial P}{\partial y})$. In this case (3) becomes

$$\int_{\partial R} \vec{\mathbf{F}} = \oint_{\partial R} (P, Q) \cdot \mathbf{d}\vec{\mathbf{x}} = \iint_R \frac{\partial Q}{\partial x} - \frac{\partial P}{\partial y}\, dx\, dy = \int_R d\vec{\mathbf{F}}.$$

Further development of the methods we have used to extend the fundamental theorem of calculus lead to general algorithms known as *exterior algebra* for determining these exterior derivatives.

6.3.3 Linking the Theorems

Finally, let's observe how these two versions of Stokes' theorem in $\mathbf{R}^2$ are linked. We begin with a curious property of exterior derivatives dF. If we start with a scalar function $F(x, y)$ and apply the fundamental theorem of calculus for line integrals, then $dF = \nabla F$. Since $\nabla F = (\frac{\partial F}{\partial x}, \frac{\partial F}{\partial y})$ is a vector function, we may apply Green's theorem, in which the exterior derivative is given by the curl. Thus

$$d(dF) = \operatorname{curl}(\nabla F) = \operatorname{curl}\left(\frac{\partial F}{\partial x}, \frac{\partial F}{\partial y}\right) = \frac{\partial^2 F}{\partial x \partial y} - \frac{\partial^2 F}{\partial y \partial x} = 0.$$

Here we have used the fact that continuous mixed partials are equal. Hence, applying the special exterior derivative twice we obtain

$$d(dF) = 0. \tag{4}$$

This property of the iterated exterior derivative is neatly mirrored in the geometry of the regions of integration involved. If we start with a two-dimensional region R and take its boundary ∂R, we obtain a closed curve. Now the boundary of a curve is just its beginning point minus its ending point. Since a closed curve begins and ends at the same point, the boundary is 0. We obtain

$$\partial(\partial R) = 0. \tag{5}$$

Equation (4) is a purely analytic statement about our derivatives; equation (5) is purely geometric. They are beautifully linked through the general Stokes' theorem.

At this point we are in a position to understand an important question from combinatorial topology. We have just seen that $d(dF) = 0$

always holds. Suppose now that we have a vector function $\vec{\mathbf{G}}$ such that $d(\vec{\mathbf{G}}) = 0$. When is it true that $\vec{\mathbf{G}} = dF$ for some F? Let's translate this question to see exactly which derivatives we are referring to. If $\vec{\mathbf{G}} = (P, Q)$, then

$$d(\vec{\mathbf{G}}) = \operatorname{curl}(P, Q) = \frac{\partial Q}{\partial x} - \frac{\partial P}{\partial y}.$$

If $d\vec{\mathbf{G}} = 0$, this means that $\frac{\partial Q}{\partial x} = \frac{\partial P}{\partial y}$. Our question now becomes: if $\frac{\partial Q}{\partial x} = \frac{\partial P}{\partial y}$, can we say that $(P, Q) = \nabla F$ for some F? That is, does (P, Q) necessarily have a potential function? Returning to Theorem 6.2.7 we see that our question asks when the *converse* of this proposition holds. Interestingly, the answer to this question depends deeply on the geometry, or more precisely on the topology of the region in which (P, Q) is defined.

We may glimpse a little of what is involved by asking the parallel question for the geometric operator ∂. Suppose that we know that $\partial C = 0$ (i.e., C is a closed curve). Then is it true that $C = \partial R$? That is, is C the boundary of some two-dimensional region in our domain? The answer depends on what we consider our domain to be. For example, let C be the unit circle centered at the origin. If our domain is all of $\mathbf{R}^2$, then C is the boundary of the unit disk D and we can write $C = \partial D$. But suppose that the domain under consideration had a "hole" in it, say, all points such that $|\vec{\mathbf{x}}| < \frac{1}{2}$ are excluded. Then C would form only part of the boundary for an annular region R (see Figure 6.3.3); the other portion of the boundary of R is the inner circle of radius $\frac{1}{2}$. Thus, although C is a closed curve, we can't say that $C = \partial R$. In general, if there are "holes" in the domain, a closed curve surrounding a hole will not be the entire boundary of any region in the domain. Investigations along these lines lead to a kind of algebra connected to the geometry of different domains, which is the subject of a branch of mathematics called *combinatorial* or *algebraic topology*.

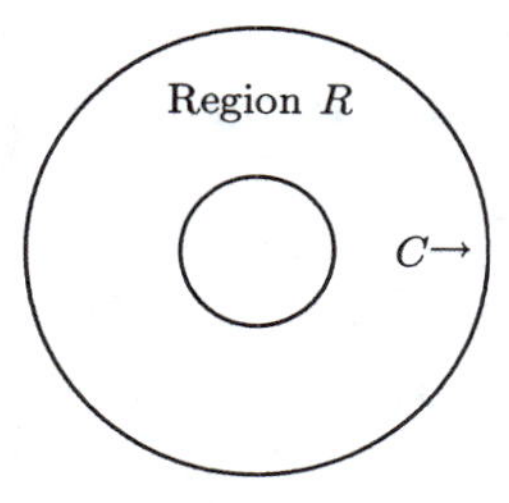

Figure 6.3.3

Problems for Section 6.3

6.3.1. Show that $\int_{C_1} (P, Q) \cdot \mathbf{d}\vec{\mathbf{x}} = \int_x^{x+\Delta x} P(x, y)\, dx$, and similarly that $\int_{C_3} (P, Q) \cdot \mathbf{d}\vec{\mathbf{x}} = -\int_x^{x+\Delta x} P(x, y + \Delta y)\, dx$.

6.3.2. In Figure 6.3.2 verify that the line integral of $\vec{\mathbf{F}} = (P, Q)$ along the vertical segments C_2 and C_4 can be written as

$$\int_{C_2} (P, Q) \cdot \mathbf{d}\vec{\mathbf{x}} + \int_{C_4} (P, Q) \cdot \mathbf{d}\vec{\mathbf{x}} = \frac{\partial Q}{\partial x}(x^{**}, y^{**}).$$

6.3.3.

a) Look up the three forms of the fundamental theorem of calculus in $\mathbf{R}^3$ and identify how they fit into the general form of Stoke's theorem (3). Identify the regions and boundaries in each case and the form of the exterior derivative.

b) Starting with $F(x, y, z)$ verify that $d(dF) = \text{curl}(\nabla F) = 0$. Explain the corresponding interpretation of equation (5) in terms of the boundaries of the regions in question.

c) Now start with the vector function $(P(x, y, z), Q(x, y, z), R(x, y, z))$ and verify that $d(dF) = \text{div}(\text{curl}\,(P, Q, R)) = 0$ (where div stands for the divergence). Explain the corresponding interpretations of equation (5) for this case.

6.4 Green's Theorem

Green's theorem applies to vector fields $\vec{\mathbf{F}} = (P, Q)$. It states that the line integral of $\vec{\mathbf{F}}$ around the boundary of a region is equal to the double integral of the expression $\frac{\partial Q}{\partial x} - \frac{\partial P}{\partial y}$ over the entire region R. This is a fascinating result because it implies that observing vectors on the boundary of a region can tell us something about what is happening inside the region. In a sense Green's theorem describes how certain boundary behavior is related to the interior of a region, and conversely how interior properties propagate to the boundaries. In the physical world when a vector field has an interpretation such as some force or the velocity of a flow, the mathematical relationship in Green's theorem can have deep physical significance.

In Section 6.3, Green's theorem was presented as a manifestation of the general form of the fundamental theorem of calculus. However, historically the theorem was inspired by physical considerations. It first appeared in 1828 in a treatise by George Green on the mathematical description of electricity and magnetism. Later, in the course of developing equations for the behavior of fluids, Gabriel Stokes generalized

Green's theorem to three dimensions. Finally, Clerk Maxwell used the full power of these ideas to obtain a complete description of electromagnetic phenomena represented by the four Maxwell equations. Maxwell's beautiful and powerful vector field description has served as one of the most important models for how physicists mathematically describe and understand the world.

6.4.1 Green's Theorem for Simple Regions

We begin by proving Green's theorem for **simple regions**. These are regions such that no vertical or horizontal line intersects the boundary more than twice unless the line itself forms part of the boundary. (See Figure 6.4.1 for some examples.)

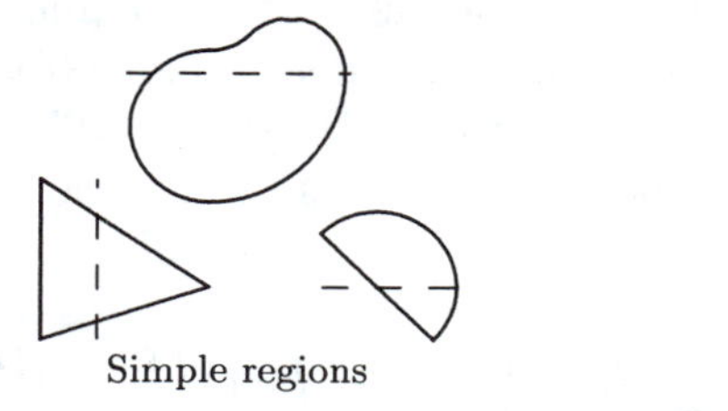

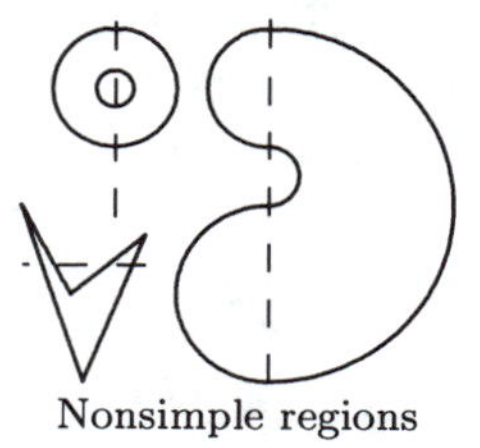

Figure 6.4.1

It follows that the boundaries of simple regions may be described analytically in two ways: either as the region between two curves $f_u(x)$ and $f_l(x)$ along the x-axis from a to b, or as the region between $g_u(y)$ and $g_l(y)$ along the y-axis from c to d (see Figure 6.4.2). In proving Green's theorem we compute a double integral over the region R twice, first along the x-axis and then along the y-axis. The dual descriptions of simple regions allow us to carry out these integrations easily. Later we will see that Green's theorem can be proven for more general regions formed by gluing together simple regions.

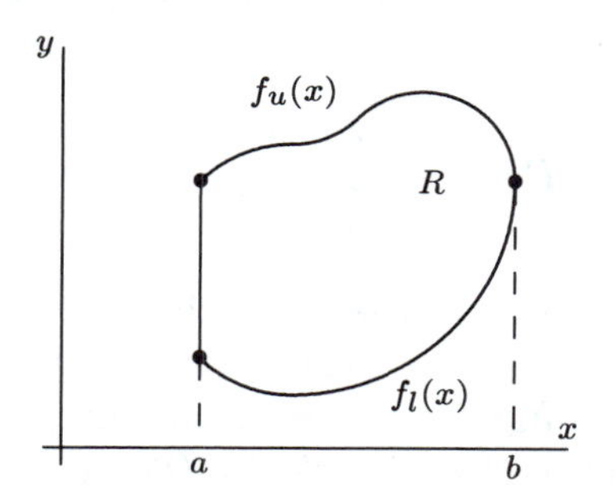

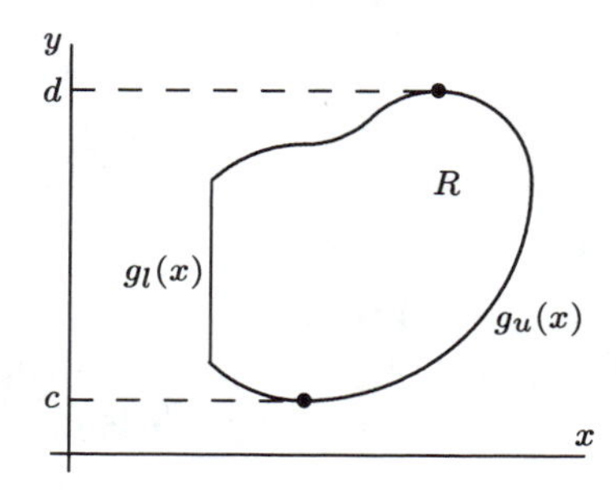

Figure 6.4.2

THEOREM 6.4.1 ***(Green's Theorem)*** *Let R be a simple region in the plane and let C be the boundary curve of R traced out in a counterclockwise direction. If $\vec{\mathbf{F}} = (P, Q)$ is a continuously differentiable vector field on R, then*

$$\iint_R \frac{\partial Q}{\partial x} - \frac{\partial P}{\partial y}\, dx\, dy = \oint_C P dx + Q dy. \tag{1}$$

From our work with line integrals we know that the orientation of C is important, since tracing a curve C in the opposite direction (clockwise) introduces a minus sign on the right side of (1).

PROOF The proof involves showing two equalities,

$$\iint_R -\frac{\partial P}{\partial y} dx\, dy = \oint_C P dx \quad \text{and} \quad \iint_R \frac{\partial Q}{\partial x} dx\, dy = \oint_C Q dy.$$

Then using the linearity properties for both line integrals and double integrals, these two equations are added to give Green's theorem. We will prove the first of these equalities; the second will be left as an exercise.

Using iterated integration on the simple region shown in Figure 6.4.3 we have

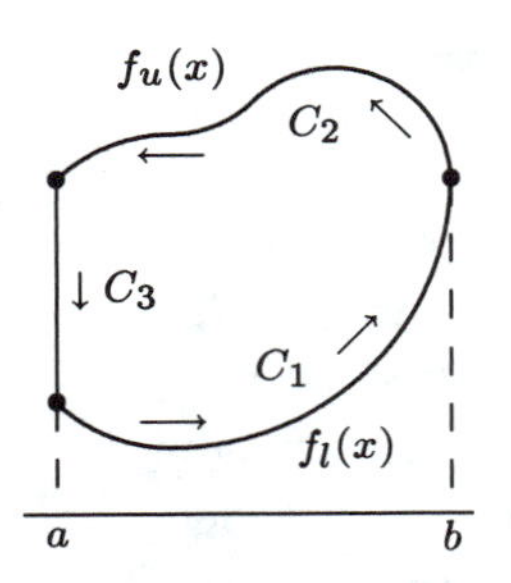

Figure 6.4.3

$$\iint_R -\frac{\partial P}{\partial y}\, dx\, dy = -\int_a^b \left[\int_{f_l(x)}^{f_u(x)} \frac{\partial P}{\partial y}\, dy \right] dx.$$

An antiderivative of $\frac{\partial P}{\partial y}$ with respect to y is just $P(x, y)$. Evaluating this at $y = f_u(x)$ and $y = f_l(x)$ yields

$$\begin{aligned}\iint_R -\frac{\partial P}{\partial y}\, dx\, dy &= -\int_a^b P(x, f_u(x)) - P(x, f_l(x))\, dx \\ &= -\int_a^b P(x, f_u(x))\, dx + \int_a^b P(x, f_l(x))\, dx. \quad (2)\end{aligned}$$

Now consider the line integral of P counterclockwise around R in Figure 6.4.3.

$$\oint_C P dx = \int_{C_1} P\, dx + \int_{C_2} P\, dx + \int_{C_3} P\, dx.$$

From problem 6.2.11, we know that the line integral $\int P dx$ is 0 along any vertical path, so the third term along the vertical segment C_3 is 0. The boundary C_1 is given by the function $f_l(x)$, which may be parameterized in the indicated direction by $x(t) = t,\ y(t) = f_l(t)$ for $t \in [a, b]$. Similarly, C_2 is given by $f_u(x)$, but the parameterization $x(t) = t,\ y(t) = f_u(t)$ for

$t \in [a, b]$ traces the curve in the opposite direction; therefore to use this parameterization we must change the sign of the integral. Subsituting these parameterizations for the line integrals over C_1 and C_2 we have

$$\oint_C P dx = \int_a^b P(t, f_l(t))\, dt - \int_a^b P(t, f_u(t))\, dt. \qquad (3)$$

Since the right-hand sides of (2) and (3) are the same integrals with different dummy variables, this proves that $\iint_R -\frac{\partial P}{\partial y}\, dx\, dy = \oint_C P\, dx$. The equation $\iint_R \frac{\partial Q}{\partial x}\, dx\, dy = \oint_C Q\, dy$ is proved similarly using the functions $g_l(y)$ and $g_u(y)$ in Figure 6.4.2. Adding these equations gives Green's theorem. ∎

The derivative expression appearing in Green's theorem warrants a special name owing to its importance.

DEFINITION 6.4.2 *If the partial derivatives of $\vec{\mathbf{F}}(x, y) = (P(x, y), Q(x, y))$ exist, then the expression $\frac{\partial Q}{\partial x} - \frac{\partial P}{\partial y}$ is called the **curl** of $\vec{\mathbf{F}}$ and is denoted* $\operatorname{curl} \vec{\mathbf{F}}$.

EXAMPLE 6.4.1 Let us verify Green's theorem for $\vec{\mathbf{F}}(x, y) = (xy, x + y)$ on the half disk shown in Figure 6.4.4 by computing both the line integral and the double integral. To compute the double integral, we first compute the curl of $\vec{\mathbf{F}}$.

$$\operatorname{curl} \vec{\mathbf{F}} = \frac{\partial Q}{\partial x} - \frac{\partial P}{\partial y} = 1 - x.$$

So

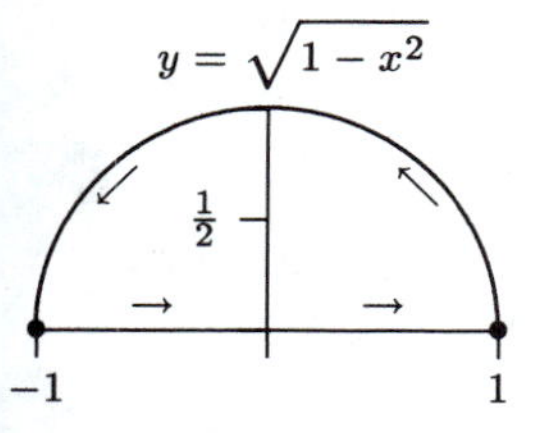

Figure 6.4.4

$$\begin{aligned}
\iint_D \frac{\partial Q}{\partial x} - \frac{\partial P}{\partial y}\, dy\, dx &= \int_{-1}^{1} \int_0^{\sqrt{1-x^2}} 1 - x\ dy\, dx \\
&= \int_{-1}^{1} \Big[(1 - x)y\Big]_{y=0}^{y=\sqrt{1-x^2}} dx \\
&= \int_{-1}^{1} (1 - x)\sqrt{1 - x^2}\, dx \\
&= \int_{-1}^{1} \sqrt{1 - x^2}\, dx + \int_{-1}^{1} -x\sqrt{1 - x^2}\, dx \\
&= \frac{\pi}{2} + \left.\frac{(1 - x^2)^{\frac{3}{2}}}{3}\right|_{-1}^{1} = \frac{\pi}{2}.
\end{aligned}$$

The line integral must be computed in a *counterclockwise* direction over the half circle and back along the x-axis. Parameterizing the half circle by $x(t) = \cos t,\ y(t) = \sin t$ for $t \in [0, \pi]$ gives as the line integral for $\vec{\mathbf{F}}$

$$\int_0^{\pi} (\cos t \sin t,\, \cos t + \sin t) \cdot (-\sin t, \cos t)\, dt.$$

Multiplying this out and separating the result into three integrals yields

$$\int_0^{\pi} -\cos t \sin^2 t\, dt + \int_0^{\pi} \cos^2 t\, dt + \int_0^{\pi} \cos t \sin t\, dt.$$

Integrating each gives

$$-\frac{\sin^3 t}{3}\Bigg|_0^{\pi} + \left(\frac{\sin 2t}{4} + \frac{t}{2}\right)\Bigg|_0^{\pi} + \frac{\sin^2 t}{2}\Bigg|_0^{\pi} = 0 + \frac{\pi}{2} + 0.$$

Finally, we must add the integral along the x-axis from -1 to 1, which is parameterized by $x(t) = t,\ y(t) = 0$ for $t \in [-1, 1]$. Integrating $\vec{\mathbf{F}}$ along this segment yields $\int_{-1}^{1} (0, t) \cdot (1, 0)\, dt = 0$. As asserted in Green's theorem, the results agree. ◻

In the preceding example, we reviewed all the details of computing line integrals and double integrals. But the beauty of Green's theorem is most appreciated when it enables us to avoid such details.

EXAMPLE 6.4.2 Compute the line integral for the vector field

$$\vec{\mathbf{F}}(x, y) = (ye^{xy} + \sin x, xe^{xy} + y^2)$$

taken counterclockwise around the contour C in Figure 6.4.4.

SOLUTION Using Green's theorem we first compute the curl of $\vec{\mathbf{F}}$:

$$\frac{\partial Q}{\partial x} = e^{xy} + xye^{xy} \qquad \text{and} \qquad -\frac{\partial P}{\partial y} = -(e^{xy} + xye^{xy}).$$

Since $\frac{\partial Q}{\partial x} - \frac{\partial P}{\partial y} = 0$, Green's theorem says that the integral around the contour must also be 0. Note that for this vector field we would obtain 0 for *any* contour about any simple region. ◻

EXAMPLE 6.4.3 Compute the line integral of

$$\vec{\mathbf{F}}(x, y) = (y + x \cos x, \sin y^2)$$

clockwise around the contour C^- in Figure 6.4.5.

SOLUTION In this problem Green's theorem is used to find the value for the counterclockwise parameterization; then the sign is changed to obtain the clockwise value.

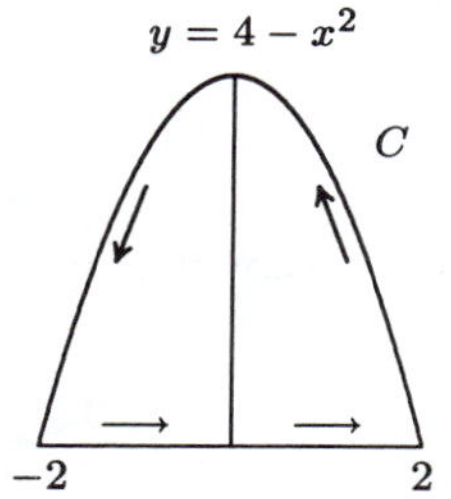

$$\frac{\partial Q}{\partial x} = 0 \qquad \text{and} \qquad -\frac{\partial P}{\partial y} = -1.$$

By Green's theorem

$$\int_{C-} \vec{\mathbf{F}} \cdot \mathbf{d}\vec{\mathbf{x}} = -\oint_C P\,dx + Q\,dy$$

$$= -\int_{-2}^{2}\int_{0}^{4-x^2} -1\,dy\,dx = \int_{-2}^{2} 4 - x^2\,dx = 32/3. \quad \square$$

Figure 6.4.5

Recall from problem 5.5.5 that $\iint_R dx\,dy$ gives us the area of the region R. If our vector field $\vec{\mathbf{F}}$ has curl equal to 1, Green's theorem enables us to compute this area as a *line integral of* $\vec{\mathbf{F}}$ *around the boundary.* This surprising result is one example of how knowing about a vector field on the boundary of a region gives information about the inside of the region. Many different vector fields have a curl of 1, but two of the simplest are $\vec{\mathbf{F}}(x, y) = (-y, 0)$ and $\vec{\mathbf{F}}(x, y) = (0, x)$. Thus the area of a region with boundary curve ∂R is given by

$$\text{area of } R = \oint_{\partial R} -y\,dx \;=\; \oint_{\partial R} x\,dy. \tag{4}$$

6.4.2 Interpreting the Curl

Green's theorem enables us to use the interpretation of the line integral to derive a geometric meaning for the curl derivative of a vector field. In the beginning of Section 6.2 we saw that the line integral of $\vec{\mathbf{F}}$ along a curve measured the magnitude of $\vec{\mathbf{F}}$ in the direction of C times the distance accumulated along the curve. If our vector field represents the velocity of some liquid in the plane, we may think of the line integral as the net flow of $\vec{\mathbf{F}}$ along C. Now if $C = \partial R$ is the closed boundary curve of a region R, the contour integral measures the flow of $\vec{\mathbf{F}}$ around the region, sometimes called the **circulation** of $\vec{\mathbf{F}}$ about R. The circulation may be thought of as the tendency of the vector field to revolve about the region. In Figure 6.4.6 the vector field has a net counterclockwise circulation on the unit circle.

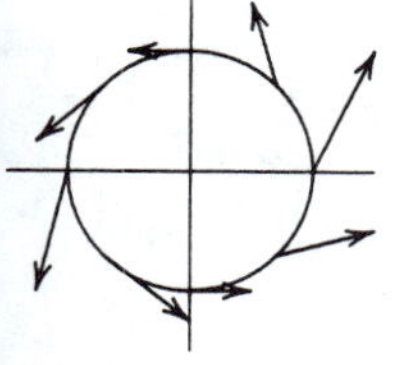

Figure 6.4.6

Using Green's theorem we can now show that $\operatorname{curl}\vec{\mathbf{F}}(a, b)$ *represents the circulation per unit area in the vector field at* (a, b). To see this assume that $\operatorname{curl}\vec{\mathbf{F}}$ is a continuous function and consider a small δ-disk

D about (a, b) such that for all (x, y) in the disk D

$$\operatorname{curl}\vec{\mathbf{F}}(x, y) - \epsilon < \operatorname{curl}\vec{\mathbf{F}}(a, b) < \operatorname{curl}\vec{\mathbf{F}}(x, y) + \epsilon.$$

Since this inequality holds for each (x, y) in D, we can integrate over the disk to obtain

$$\begin{aligned}\iint_D \operatorname{curl}\vec{\mathbf{F}}(x, y) - \epsilon \, dx\, dy &< \iint_D \operatorname{curl}\vec{\mathbf{F}}(a, b)\, dx\, dy \\ &< \iint_D \operatorname{curl}\vec{\mathbf{F}}(x, y) + \epsilon \, dx\, dy.\end{aligned}$$

Both ϵ and $\operatorname{curl}\vec{\mathbf{F}}(a, b)$ are constants; therefore integration of these terms is given by the area of the disk $(2\pi\delta^2)$ times the respective constant. Furthermore, by Green's theorem $\iint_D \operatorname{curl}\vec{\mathbf{F}}(x, y)\, dx\, dy$ is the circulation of $\vec{\mathbf{F}}$ about the δ-disk. Incorporating these observations in the preceding equation and dividing by the area of the δ-disk we have

$$\frac{\text{circulation}}{\text{area}} - \epsilon < \operatorname{curl}\vec{\mathbf{F}}(a, b) < \frac{\text{circulation}}{\text{area}} + \epsilon.$$

As ϵ becomes small, we see that $\operatorname{curl}\vec{\mathbf{F}}(a, b)$ is the circulation per unit area near the point. This is sometimes visualized as a measure of the twisting or torque per unit area in the vector field. For example, if a paddle wheel could be located at a point of positive curl, it would have a tendency to rotate counterclockwise. Vector fields that have this tendency to form vortices arise in electromagnetism.

EXAMPLE 6.4.4 **(Ampere's Law)** You have probably witnessed the experiment in which iron filings are scattered on a paper plate that has a wire through its center. If a current is run through the wire, a magnetic field is created around the wire and the filings become magnetically aligned in a circular pattern. With a stronger field the filings become more strongly aligned.

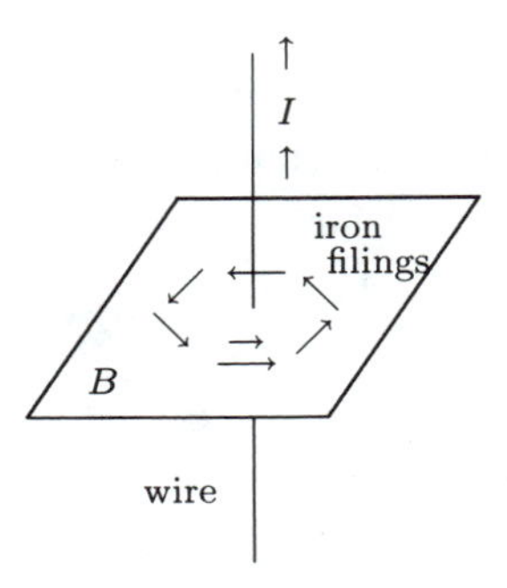

Figure 6.4.7

Ampere's law, which forms part of Maxwell's description of electricity and magnetism, describes this relationship between a current I and the induced magnetic field $\vec{\mathbf{B}}$. If L is a closed loop about a wire and I is the current running perpendicularly through the loop, then[1]

$$\oint_L \vec{\mathbf{B}} \cdot \mathbf{d}\vec{\mathbf{x}} = I. \tag{5}$$

Sometimes a current is not concentrated in a wire at the center but is diffused over a region. In this case it is described by a **current density** J over the plate. The total current I through the loop is then calculated as the double integral of the density J over the region R enclosed by the loop L. Applying this observation to the right side of (5) and using Green's theorem on the left side we may rewrite Ampere's law as

$$\oint_L \vec{\mathbf{B}} \cdot\ d\vec{\mathbf{x}} = \iint_R \operatorname{curl} \vec{\mathbf{B}}\, dx\, dy = \iint_R J\, dx\, dy = I.$$

Combining the integrals over R gives

$$\iint_R \operatorname{curl} \vec{\mathbf{B}} - J\, dx\, dy = 0.$$

Ampere's law is universal and applies to all regions R. Thus we are led to ask, what kind of function will integrate to 0 over *every* region R? The only continuous function with this property is $F = 0$. Therefore we must have

$$\operatorname{curl} \vec{\mathbf{B}} = J. \tag{6}$$

This equation states that current at a point causes a twisting in the magnetic field exactly equal to the density of the current in that region. Equation (5) is called the integral form of Ampere's law, and in (6) we have derived the equivalent differential form of Ampere's law. ☐

---

[1] Physics students will recognize that this version of Ampere's law is incomplete. Indeed, Maxwell's great insight was to add a term to this equation to obtain the correct description. If $\vec{\mathbf{E}}$ is the electric field,

$$\oint_L \vec{\mathbf{B}} \cdot \mathbf{d}\vec{\mathbf{x}} = I + \frac{d\vec{\mathbf{E}}}{dt}.$$

6.4.3 Green's Theorem for General Regions

Suppose that Green's theorem is applied to two contiguous simple regions (Figure 6.4.8). Then the integral of the curl of $\vec{\mathbf{F}}$ over R_1 is the counterclockwise line integral of $\vec{\mathbf{F}}$ around ∂R_1, and similarly for R_2. Adding these will give the integral of the curl over the whole region R.

$$\begin{aligned}\iint_R \operatorname{curl}\vec{\mathbf{F}}\,dx\,dy &= \iint_{R_1} \operatorname{curl}\vec{\mathbf{F}}\,dx\,dy + \iint_{R_2} \operatorname{curl}\vec{\mathbf{F}}\,dx\,dy\\ &= \oint_{\partial R_1} \vec{\mathbf{F}}\cdot d\vec{\mathbf{x}} + \oint_{\partial R_2} \vec{\mathbf{F}}\cdot d\vec{\mathbf{x}}.\end{aligned}$$

The internal boundary between R_1 and R_2 is traced once in each direction by the two contour integrals. Thus the line integrals on the interior boundary *cancel* each other out, leaving us with a line integral around the boundary of the entire region R traced once in the counterclockwise direction. This shows that Green's theorem holds for nonsimple regions R of the type in Figure 6.4.8.

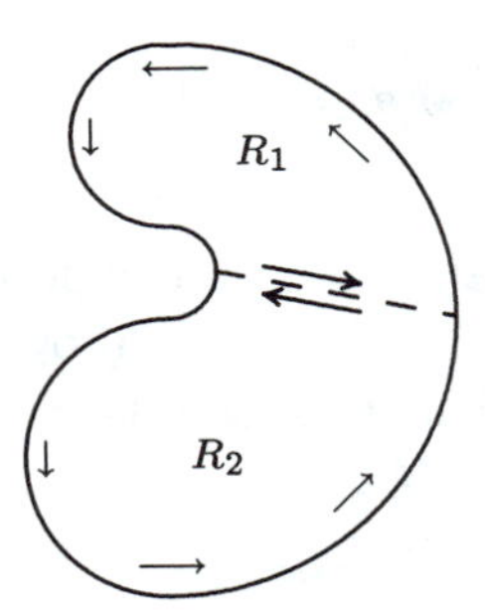

Figure 6.4.8

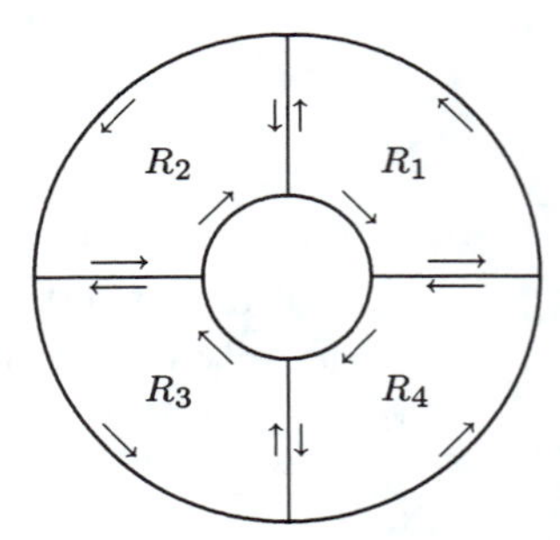

Figure 6.4.9

In a similar way we may show that Green's theorem holds for any region that can be decomposed into simple regions. For example, consider the annulus in Figure 6.4.9. It may be divided into four simple regions as shown. The double integrals of the curl on each of the four subregions sum to the integral of the curl over the entire annulus. The line integrals along the interior boundaries again cancel each other, leaving us with the outer circle traced once in the counterclockwise direction and the inner circle (which is also a boundary for the annulus) traced once *clockwise*. This apparent reversal of direction comes from the four simple regions composing the figure. Confusion in choosing the correct direction can be avoided by always keeping the region on the *left* as you traverse all boundaries. We state these observations in the following theorem:

THEOREM 6.4.3 ***(Green's Theorem)*** *Suppose that R is the union of finitely many simple regions with boundary ∂R. If $\vec{\mathbf{F}} = (P, Q)$ is continuously differentiable on R, then*

$$\iint_R \frac{\partial Q}{\partial x} - \frac{\partial P}{\partial y}\, dx\, dy = \oint_{\partial R} P dx + Q dy,$$

where all parts of ∂R are traced once in a direction that keeps the region always on the left.

EXAMPLE 6.4.5 Consider the vector field given by

$$\vec{\mathbf{F}}(x, y) = \left(\frac{-y}{x^2+y^2}, \frac{x}{x^2+y^2}\right).$$

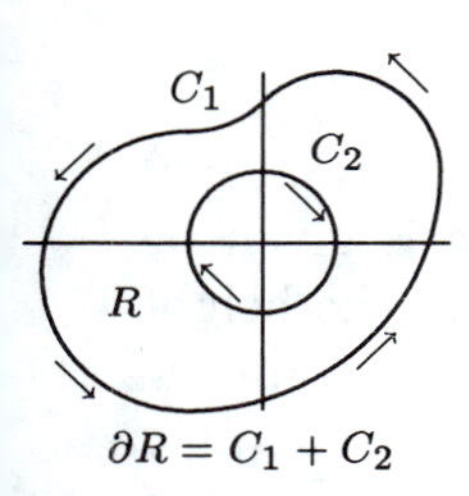

Figure 6.4.10

We cannot apply Green's theorem to any region containing the origin, since $\vec{\mathbf{F}}$ is undefined at that point. However, in the region shown in Figure 6.4.10, we have excluded the origin with a hole. We may apply Green's theorem to this region. Notice that

$$\operatorname{curl} \vec{\mathbf{F}} = \frac{\partial Q}{\partial x} - \frac{\partial P}{\partial y} = \frac{y^2 - x^2}{(x^2+y^2)^2} - \frac{y^2 - x^2}{(x^2+y^2)^2} = 0.$$

What does this say about the curves C_1 and C_2? Assuming that both curves are traced in a counterclockwise direction, Green's theorem can be expressed as follows, where the negative sign accounts for the reversal in the direction of C_2:

$$\iint_R \frac{\partial Q}{\partial x} - \frac{\partial P}{\partial y}\, dx\, dy = \oint_{C_1} \vec{\mathbf{F}} \cdot \mathbf{d}\vec{\mathbf{x}} - \oint_{C_2} \vec{\mathbf{F}} \cdot \mathbf{d}\vec{\mathbf{x}} = 0,$$

or

$$\oint_{C_1} \vec{\mathbf{F}} \cdot \mathbf{d}\vec{\mathbf{x}} = \oint_{C_2} \vec{\mathbf{F}} \cdot \mathbf{d}\vec{\mathbf{x}}.$$

Since the curl is 0, the particular curves in the example played no role in our computation. In fact, the line integral counterclockwise around *any two curves that surround the origin will be equal!* This value is easily computed to be 2π using the unit circle. Therefore the integral around any counterclockwise curve surrounding the origin will also be 2π. □

6.4.4 The Divergence in $\mathbf{R}^2$

In the beginning of Section 6.2 we presented an interpretation of the line integral as the flow of the vector field $\vec{\mathbf{F}} = (P, Q)$ *along* a curve C. A computation using dot products will allow us to determine the flow of $\vec{\mathbf{F}}$ *across* a curve C.

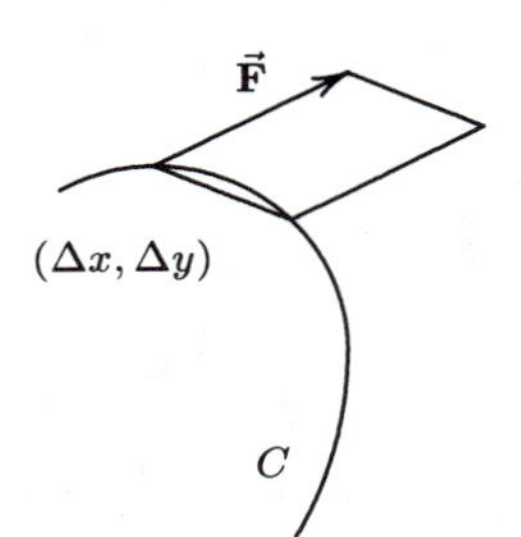

The flow of $\vec{\mathbf{F}}$ across a section of C.

Figure 6.4.11

In Figure 6.4.11 the flow across a little section of the curve $(\Delta x, \Delta y)$ is represented by the area of the parallelogram. This area is given by base $\times$ height, where the length of the base is $|(\Delta x, \Delta y)|$ and the height is the magnitude of $\vec{\mathbf{F}}$ perpendicular to the vector $(\Delta x, \Delta y)$. To compute this height we need to take the dot product of $\vec{\mathbf{F}}$ with a vector perpendicular to $(\Delta x, \Delta y)$. The vector $(\Delta y, -\Delta x)$ obtained by rotating $(\Delta x, \Delta y)$ clockwise by $90°$ is one such vector. Furthermore, its length is exactly equal to the length of $(\Delta x, \Delta y)$. Therefore the area of the paralleogram is given by the dot product:

$$\text{area of parallelogram} = (P, Q) \cdot (\Delta y, -\Delta x). \tag{7}$$

Summing these dot products over all the little segments $(\Delta x_i, \Delta y_i)$ along the curve will approximate the total flow of $\vec{\mathbf{F}}$ across the curve. But if we now transfer this twist in the vector $(\Delta y, -\Delta x)$ to the pair (P, Q), as follows, we obtain the Riemann sum for a line integral of a new auxiliary vector field $\vec{\mathbf{F}}^* = (-Q, P)$:

$$\sum_{i=1}^{n} (P, Q) \cdot (\Delta y, -\Delta x) = \sum_{i=1}^{n} (-Q, P) \cdot (\Delta x, \Delta y).$$

Thus by using the auxiliary vector field $\vec{\mathbf{F}}^* = (-Q, P)$ we may compute the flow of $\vec{\mathbf{F}}$ across a curve C as

$$\int_C \vec{\mathbf{F}}^* \cdot \mathbf{d\vec{x}} = \int_C (-Q, P) \cdot \mathbf{d\vec{x}} = \int_C -Q\, dx + P\, dy.$$

Although for purposes of computation we use $\vec{\mathbf{F}}^*$, our interpretation of the line integral applies to $\vec{\mathbf{F}}$ and represents *the sum along C of the coordinate of $\vec{\mathbf{F}}$ normal to the curve times the length of the curve.* Let $\vec{\mathbf{n}}$ denote the unit normal to the curve (taken 90° clockwise from the direction of the curve), then $\vec{\mathbf{F}} \cdot \vec{\mathbf{n}}$ is the coordinate of $\vec{\mathbf{F}}$ normal to the curve. Therefore, if ds represents an element of arc length, the flow of $\vec{\mathbf{F}}$ across C may be written in a shorthand notation as

$$\int_C \vec{\mathbf{F}}^* \cdot \mathbf{d}\vec{\mathbf{x}} = \int_C \vec{\mathbf{F}} \cdot \vec{\mathbf{n}}\, ds.$$

This quantity is called the flux of $\vec{\mathbf{F}}$ across C.

DEFINITION 6.4.4 *If $\vec{\mathbf{F}} = (P, Q)$ is a vector field defined on a curve C, then the* **flux** *of $\vec{\mathbf{F}}$ across C is*

$$\text{flux of } \vec{\mathbf{F}} \text{ across } C = \int_C \vec{\mathbf{F}} \cdot \vec{\mathbf{n}}\, ds = \int \vec{\mathbf{F}}^* \cdot \mathbf{d}\vec{\mathbf{x}} = \int_C -Q\, dx + P\, dy. \quad (8)$$

The flux represents the flow of $\vec{\mathbf{F}}$ perpendicularly in a clockwise sense across the curve C.

EXAMPLE 6.4.6 Let $\vec{\mathbf{V}}(x, y) = (1, y)$ represent the velocity field of a fluid in the plane. Compute the flux of $\vec{\mathbf{V}}$ across the parabolic segment C in Figure 6.4.12.

SOLUTION The curve C is parameterized by $x(t) = t,\ y(t) = t^2/2$, and the vector field $\vec{\mathbf{V}}^* = (-y, 1)$. Thus

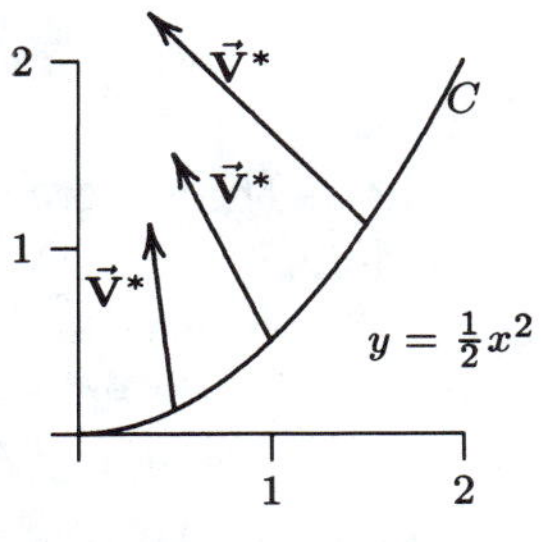

Figure 6.4.12

$$\begin{aligned}
\text{flux} = \int_C \vec{\mathbf{V}} \cdot \vec{\mathbf{n}}\, ds &= \int_C \vec{\mathbf{V}}^* \cdot \mathbf{d}\vec{\mathbf{x}} \\
&= \int_C (-y, 1) \cdot \mathbf{d}\vec{\mathbf{x}} \\
&= \int_0^2 (-t^2/2, 1) \cdot (1, t)\, dt \\
&= \int_0^2 -t^2/2 + t\, dt \\
&= -t^3/6 + t^2/2 \Big|_0^2 = 2/3.
\end{aligned}$$

This computes the volume of fluid crossing the parabola per unit time. ∎

Suppose that our curve C is the boundary of a region R. Then $\oint_C \vec{\mathbf{F}} \cdot \vec{\mathbf{n}}\, ds$ represents the net flux of $\vec{\mathbf{F}}$ across the boundary of R. If C traces this boundary in a counterclockwise direction, the normal to the curve taken clockwise from this direction will point out of the region. Therefore, $\oint_C \vec{\mathbf{F}} \cdot \vec{\mathbf{n}}\, ds$ measures the flow of $\vec{\mathbf{F}}$ *out* of the region R. Green's theorem enables us to compute this quantity using $\vec{\mathbf{F}}^*$.

THEOREM 6.4.5 ***(The Divergence Theorem in*** $\mathbf{R}^2$***)*** *Let* $\vec{\mathbf{F}} = (P, Q)$ *be a continuously differentiable vector field on a region* R *that is a finite union of simple regions. Then*

$$\oint_{\partial R} \vec{\mathbf{F}} \cdot \vec{\mathbf{n}}\, ds = \iint_R \frac{\partial P}{\partial x} + \frac{\partial Q}{\partial y}\, dx\, dy,$$

where ∂R *is traced in a counterclockwise direction.*

PROOF By definition $\oint_{\partial R} \vec{\mathbf{F}} \cdot \vec{\mathbf{n}}\, ds = \oint_{\partial R} \vec{\mathbf{F}}^* \cdot \mathbf{d}\vec{\mathbf{x}}$. Applying Green's theorem to the second integral, we obtain

$$\oint_{\partial R} \vec{\mathbf{F}}^* \cdot \mathbf{d}\vec{\mathbf{x}} = \oint_{\partial R} (-Q, P) \cdot \mathbf{d}\vec{\mathbf{x}} = \iint_R \operatorname{curl}(-Q, P)\, dx\, dy.$$

But the curl of $(-Q, P)$ is $\frac{\partial P}{\partial x} + \frac{\partial Q}{\partial y}$, which proves the theorem.[2] ∎

DEFINITION 6.4.6 *Let* $\vec{\mathbf{F}} = (P, Q)$ *be a vector field. The* **divergence** *of* $\vec{\mathbf{F}}$*, denoted by* $\operatorname{div} \vec{\mathbf{F}}$*, is*

$$\operatorname{div} \vec{\mathbf{F}} = \frac{\partial P}{\partial x} + \frac{\partial Q}{\partial y}.$$

In a region where $\operatorname{div} \vec{\mathbf{F}} > 0$ the divergence theorem shows that there will be a net positive flux of $\vec{\mathbf{F}}$ out of the region. Such regions are called **sources**. If $\operatorname{div} \vec{\mathbf{F}} < 0$, there is a negative flux, which is interpreted as a flow into the region. We call these regions **sinks** of the vector field $\vec{\mathbf{F}}$. Finally, if $\operatorname{div} \vec{\mathbf{F}} = 0$ everywhere, there is no net flow into or out of any region and the field is called **incompressible**. This descriptive terminology points toward some of the important physical implications arising from the divergence theorem.

---

[2] The divergence theorem usually refers to the three-dimensional version also known as Gauss's theorem. In that case the divergence is the analogous expression: $\frac{\partial P}{\partial x} + \frac{\partial Q}{\partial y} + \frac{\partial R}{\partial z}$, where $\vec{\mathbf{F}} = (P, Q, R)$.

EXAMPLE 6.4.7 **(The Continuity Equation)** Let $\rho_t(x, y)$ represent the density of a fluid at the point (x, y) at time t and let $\vec{\mathbf{V}} = (V_1, V_2)$ be the velocity field of the fluid. The total mass of fluid in a region R at time t can be computed by integrating this density over the region:

$$\text{mass in } R = \iint_R \rho_t(x, y)\, dx\, dy.$$

We define a vector quantity $\vec{\mathbf{J}} = \rho\vec{\mathbf{V}} = (\rho V_1, \rho V_2)$, which represents the velocity of the flow weighted by the density of the fluid, that is, the velocity flow of the mass. The rate at which mass is escaping from R is given by the flux of $\vec{\mathbf{J}}$ across the boundary of R:

$$\text{rate of mass escaping } R = \oint_{\partial R} \vec{\mathbf{J}} \cdot \vec{\mathbf{n}}\, ds.$$

If the mass is conserved and can neither be created or destroyed in the region, the change in mass in R must be due entirely to the flow across its boundaries. Thus the rate of change in the mass of the region must equal the negative of the flux (since if the flux is positive the mass is decreasing):

$$\frac{d}{dt} \iint_R \rho_t(x, y)\, dx\, dy = -\oint_{\partial R} \vec{\mathbf{J}} \cdot \vec{\mathbf{n}}\, ds. \tag{8}$$

Assuming that $\frac{\partial \rho}{\partial t}$ is continuous, we may use Leibniz's rule (Corollary 5.5.11) to bring the derivative on the left side of this equation under the integral. Applying this to the left side and using the divergence theorem on the right side gives

$$\iint_R \frac{\partial}{\partial t}\rho_t(x, y)\, dx\, dy = -\iint_R \operatorname{div} \vec{\mathbf{J}}\, dx\, dy.$$

We combine these to obtain

$$\iint_R \frac{\partial}{\partial t}\rho_t(x, y) + \operatorname{div} \vec{\mathbf{J}}\, dx\, dy = 0.$$

If mass is conserved everywhere this relationship holds for every region R. What kind of function integrates to 0 over every region? Exactly as in Example 6.4.4, the only continuous function with this property is 0. Therefore we conclude

$$\frac{\partial}{\partial t}\rho_t(x, y) + \operatorname{div}\vec{\mathbf{J}} = 0. \tag{9}$$

This statement, called the *equation of continuity*, is one of the basic equations for the behavior of fluids. We may regard it as the differential expression for the conservation of mass in contrast to the integral expression in (8). It states that decreasing density near a point is due to a net flow of fluid away from a point. □

We see how valuable the mathematics in Green's theorem is for the description of physical phenomena. This theorem is the source of much lovely mathematics filled with strong physical-geometric meanings.

Problems for Section 6.4

6.4.1. Complete the proof of Theorem 6.4.1 by showing that $\iint_R \frac{\partial Q}{\partial x}\,dx\,dy = \oint_C Q\,dy$.

6.4.2. Verify Green's theorem in each of the following cases by computing both the line integral and the double integral over the enclosed region.

a) $\vec{\mathbf{F}}(x,y) = (1-y, x)$ and the region is the disk of radius 2 centered at the origin.

b) $\vec{\mathbf{F}}(x,y) = (xy, -xy)$ over the region bounded by the curves $y = x^2$ and $y = 1$.

c) $\vec{\mathbf{F}}(x,y) = (xy, 2x^2)$ in the region bounded by $y = \frac{3}{x}$ and $y = -x + 4$.

6.4.3. For each of the following cases compute the line integrals in whichever way seems simplest (directly or by Green's theorem).

a) $\oint_C e^x \cos y\,dx + e^x \sin y\,dy$, where C is the contour tracing the triangle with vertices (0,0), (1,0), and $(1,\frac{\pi}{2})$ counterclockwise.

b) Same as in **(a)** but with C being the square with vertices (0,0), (1,0), (1,1), and (0,1) traced counterclockwise.

c) $\oint_C x^2 - y^2\,dx + x^2 + y^2\,dy$, where C is the unit circle traced clockwise.

6.4.4. **a)** Use Green's theorem to compute the line integral of the vector field given by $\vec{\mathbf{F}}(x,y) = (ye^{xy} - x^2, 3x + xe^{xy})$ counterclockwise about the region bounded by the x-axis and the parabola $y = 4 - x^2$.

b) Use this result to determine the line integral of $\vec{\mathbf{F}}$ along the parabolic arc $y = 4 - x^2$ from $(2,0)$ to $(-2,0)$. (Hint: Subtract the line integral along the x-axis from the result in **(a)**.)

6.4.5. Verify that the result of Green's theorem is not valid for $\vec{\mathbf{F}} = (\frac{-y}{x^2+y^2}, \frac{x}{x^2+y^2})$ over the disk of radius 1 centered at the origin. Why does the theorem not apply?

6.4.6. **a)** Show that the area of a simple region is given by

$$\text{area}\,(R) = \frac{1}{2}\oint_{\partial R} -y\,dx + x\,dy$$

Verify that this formula holds for the triangle with vertices $(0,0)$, $(b,0)$, and (a,h), where a, b, and h are positive constants.

b) Compute the area of an ellipse given by $\frac{x^2}{a^2} + \frac{y^2}{b^2} = 1$.

6.4.7. Use Green's theorem to compute the area of the loop enclosed by the curve $\vec{\mathbf{F}}(t) = (t - t^3, t^2 - 1)$ for $t \in [-1, 1]$ by integrating an appropriate vector field about the boundary (see Figure 5.2.1).

6.4.8. Compute the line integral of $\vec{\mathbf{G}} = (3x^2y - 2x, 2x^3 + y)$ along one arch of the cycloid parameterized by $(t - \sin t, 1 - \cos t)$ for $t \in [0, 2\pi]$. (This would be a tedious integral to do directly. Instead begin by using Green's theorem to find the integral about the entire region, from $(2\pi, 0)$ to (0,0) along the arch and back along the x-axis to $(2\pi, 0)$. Then subtract off the line integral along the x-axis from this result to find the line integral along the arch.)

6.4.9. Let R be a plane region with area $A(R)$. The **centroid** of the region R is the point $(\bar{x}, \bar{y})$, where $\bar{x} = \frac{1}{A(R)} \int\int_R x\,dx\,dy$ and $\bar{y} = \frac{1}{A(R)} \int\int_R y\,dx\,dy$.

a) Describe the centroid of a region using line integrals about the boundary of the region.

b) Find a formula for the centroid of the triangle with vertices (a_1, b_1), (a_2, b_2), and (a_3, b_3) listed counterclockwise.

6.4.10. Compute the flux of each of these vector fields across the indicated curve.

a) $\vec{\mathbf{F}} = (y, xy)$ across the line from $(1, 0)$ to $(0, 1)$.

b) $\vec{\mathbf{F}} = (x, y)$ across the semicircle centered at $(0, 0)$ from $(2, 0)$ to $(-2, 0)$.

6.4.11. Compute the flux of the vector field out of each these regions using Theorem 6.4.5.

a) $\vec{\mathbf{F}}(x, y) = (x + y^2, \sqrt{x} - 2y)$ out of the triangle with vertices $(0, 0)$, $(4, 0)$ and $(3, 5)$.

b) $\vec{\mathbf{F}}(x, y) = (2xy, x - y^2)$ out of the circle with center $(0, 0)$ and radius r.

c) $\vec{\mathbf{F}}(x, y) = (xy, x - y)$ out of the rectangle with vertices $(1, 1)$, $(3, 1)$, $(1, 4)$, and $(3, 4)$.

6.4.12. Suppose that $\vec{\mathbf{F}}$ is a continuously differentiable vector field on a region containing R, and that $\vec{\mathbf{F}}$ is constant on the boundary of R (i.e., on the boundary we have $\vec{\mathbf{F}} = (k_1, k_2)$ for some constants k_1 and k_2). Show that

$$\int\int_R \operatorname{curl} \vec{\mathbf{F}}\,dx\,dy = 0.$$

(Be careful! You cannot say that $\operatorname{curl} \vec{\mathbf{F}} = 0$, since you don't know the values of $\vec{\mathbf{F}}$ except on the boundary. Instead let $(x(t), y(t))$ for $t \in [a, b]$ be the curve tracing the boundary of R and show that $\oint_{\partial R} \vec{\mathbf{F}} \cdot \mathbf{d}\vec{\mathbf{x}} = 0$ and then invoke Green's theorem.)

6.4.13. Assume that f and g map $\mathbf{R}^2$ to $\mathbf{R}$ and are continuously differentiable everywhere. Show that if C is any closed contour tracing out the boundary of a region, then

$$\oint_C f \nabla g \cdot d\vec{\mathbf{x}} = - \oint_C g \nabla f \cdot d\vec{\mathbf{x}}.$$

(Hint: Combine the two line integrals and use Green's theorem. Alternatively, compute the gradient of fg.)

6.4.14. The **Laplacian** of a scalar function f is $\nabla^2 f = \frac{\partial^2 f}{\partial x^2} + \frac{\partial^2 f}{\partial y^2}$. Prove that for any region R

$$\oint_{\partial R} \nabla f \cdot \vec{\mathbf{n}}\, ds = \int\int_R \nabla^2 f\, dx\, dy.$$

6.4.15. **a)** Compute the flux $\int_C \vec{\mathbf{F}} \cdot \vec{\mathbf{n}}\, ds$ of the vector field $\vec{\mathbf{F}}(x,y) = (\frac{x}{x^2+y^2}, \frac{y}{x^2+y^2})$ across the circle C of radius r centered at the origin.

b) Compute the divergence of $\vec{\mathbf{F}}$.

c) Does the divergence theorem apply to this problem? Why or why not?

6.5 Path Independence and Potential Functions

We have seen the important role of potential functions in simplifying the computation of line integrals (see Example 6.2.6). At the end of Section 6.2 we saw how a potential function for a force field allowed us to define a *potential energy* for the field, which in turn led to the concept of the conservation of energy. Clearly, potential functions are very useful. Unfortunately, a vector field does not always have a potential function, as was illustrated in Example 6.2.7. In this section we explore some conditions on a vector field $\vec{\mathbf{F}}$ that are equivalent to the existence of a potential function (i.e., a function f such that $\nabla f = \vec{\mathbf{F}}$). The geometric nature of these conditions underscores again the connections between geometry and analysis that arise from Green's theorem.

6.5.1 Path Indpendence

The concept of path independence was first noted in connection with the fundamental theorem of calculus for line integrals.

Definition 6.5.1 *Let $\vec{\mathbf{F}} = (P, Q)$ be a vector field on a connected region R. Then $\vec{\mathbf{F}}$ is said to be* ***independent of path*** *in R if the line integral of $\vec{\mathbf{F}}$ between any two points $\vec{\mathbf{a}}$ and $\vec{\mathbf{b}}$ in the region is independent of the path connecting them. That is, if C_1 and C_2 are two contours in R between $\vec{\mathbf{a}}$ and $\vec{\mathbf{b}}$, then*

$$\int_{C_1} \vec{\mathbf{F}} \cdot \mathbf{d}\vec{\mathbf{x}} = \int_{C_2} \vec{\mathbf{F}} \cdot \mathbf{d}\vec{\mathbf{x}}.$$

By the fundamental theorem for line integrals (Theorem 6.2.6), if the vector field $\vec{\mathbf{F}}$ has a potential function f defined on an open set R and C is any curve in R from $\vec{\mathbf{a}}$ to $\vec{\mathbf{b}}$, then the line integral is given by

$$\int_C \vec{\mathbf{F}} \cdot \mathbf{d}\vec{\mathbf{x}} = \int_C \nabla f \cdot \mathbf{d}\vec{\mathbf{x}} = f(\vec{\mathbf{b}}) - f(\vec{\mathbf{a}}).$$

Thus gradients of a potential function give path independent vector fields. Are there other types of fields that are path independent? The answer is no. That is, *if a vector field is independent of path in a region, it must have a potential function.* In fact, potential functions can be constructed directly from line integrals, just as antiderivatives are constructed directly from integrals. (See Theorem 3.6.5.)

THEOREM 6.5.2 *Let $\vec{\mathbf{F}}(x, y) = (P(x, y), Q(x, y))$ be a continuous vector field defined on an open, connected region, R, in the plane. If $\vec{\mathbf{F}}$ is independent of path, then $\vec{\mathbf{F}}$ has a potential function $f(x, y)$ defined by the line integral*

$$f(x, y) = \int_C \vec{\mathbf{F}} \cdot \mathbf{d}\vec{\mathbf{x}} = \int_{(a,b)}^{(x,y)} \vec{\mathbf{F}} \cdot \mathbf{d}\vec{\mathbf{x}},$$

where (a, b) is an arbitrary point in R, and C is any contour in R between (a, b) and (x, y).

PROOF Note first that since R is *connected*, there is some path from (a, b) to (x, y); consequently f is defined at all points in R. Furthermore, the hypothesis that $\vec{\mathbf{F}}$ is independent of path assures us that the function f will have the same value at any point (x, y) no matter which curve in R is used to compute it! This guarantees that f is a well-defined function. Next we must show that f is actually a potential function for $\vec{\mathbf{F}}$, that is,

$$\nabla f = \left(\frac{\partial f}{\partial x}, \frac{\partial f}{\partial y}\right) = (P(x, y), Q(x, y)) = \vec{\mathbf{F}}(x, y).$$

We will prove that $\frac{\partial f}{\partial x} = P(x, y)$ by computing $\frac{\partial f}{\partial x}$ directly from the definition of partial derivative. The computation of $\frac{\partial f}{\partial y}$ is similar and is left as an exercise.

From the definition of partial derivative we have

$$\begin{aligned} \frac{\partial f}{\partial x} &= \lim_{\Delta x \to 0} \frac{1}{\Delta x} \left[\int_{(a,b)}^{(x+\Delta x,y)} \vec{\mathbf{F}} \cdot \mathbf{d}\vec{\mathbf{x}} - \int_{(a,b)}^{(x,y)} \vec{\mathbf{F}} \cdot \mathbf{d}\vec{\mathbf{x}}\right] \\ &= \lim_{\Delta x \to 0} \frac{1}{\Delta x} \int_{(x,y)}^{(x+\Delta x,y)} \vec{\mathbf{F}} \cdot \mathbf{d}\vec{\mathbf{x}}, \qquad (1) \end{aligned}$$

where the last step follows from the additive property of line integrals.

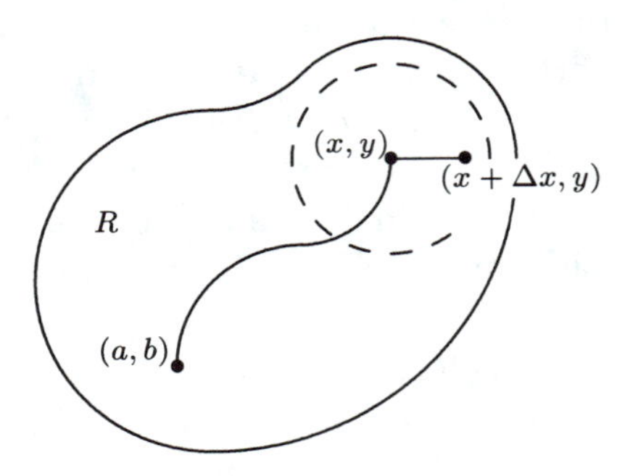

Figure 6.5.1

Since R is an *open* set, for a sufficiently small Δx the disk of radius Δx will be entirely contained in R. In particular, the straight line between (x, y) and $(x + \Delta x, y)$ will lie entirely in R (see Figure 6.5.1). Since $\vec{\mathbf{F}}$ is independent of path we may use this straight line to compute $\int_{(x,y)}^{(x+\Delta x,y)} \vec{\mathbf{F}} \cdot \mathbf{d}\vec{\mathbf{x}}$. Noting that y does not vary in this integration, we can parameterize the line by $x(t) = t$ and $y(t) = y$ for $t \in [x, x + \Delta x]$. This gives us

$$\int_{(x,y)}^{(x+\Delta x,y)} \vec{\mathbf{F}} \cdot \mathbf{d}\vec{\mathbf{x}} = \int_{x}^{x+\Delta x} (P(t, y), Q(t, y)) \cdot (1, 0)\, dt = \int_{x}^{x+\Delta x} P(t, y)\, dt.$$

P is continuous, so applying the integral mean value theorem says that for some $x^* \in [x, x + dx]$,

$$\int_{x}^{x+\Delta x} P(t, y)\, dt = P(x^*, y)\Delta x.$$

Incorporating this into (1), we have

$$\begin{aligned} \lim_{\Delta x \to 0} \frac{1}{\Delta x} \int_{(x,y)}^{(x+\Delta x,y)} \vec{\mathbf{F}} \cdot \mathbf{d}\vec{\mathbf{x}} &= \lim_{\Delta x \to 0} \frac{1}{\Delta x} P(x^*, y)\Delta x \\ &= \lim_{\Delta x \to 0} P(x^*, y) = P(x, y), \qquad (2) \end{aligned}$$

where in the last step we have again used the fact that P is continuous. ∎

Example 6.5.1 Assuming that the vector field $\vec{\mathbf{F}}(x, y) = (y^2, y + 2xy)$ is independent of path in all of $\mathbf{R}^2$, Theorem 6.5.2 may be used to construct a potential function for $\vec{\mathbf{F}}$. To do this select an arbitrary base point, say, $\vec{\mathbf{a}} = (0, 0)$. Next compute the line integral from $(0, 0)$ to (x, y) along some path in the region. We illustrate this using two possible paths (see Figure 6.5.2).

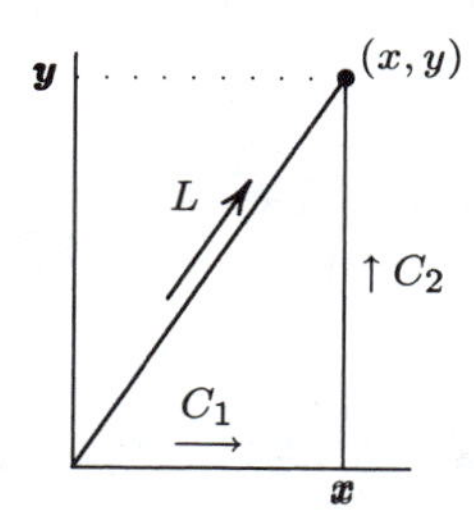

Figure 6.5.2

First we use the rectangular path $C = C_1 + C_2$, which gives us

$$f(x, y) = \int_C \vec{\mathbf{F}} \cdot \mathbf{d}\vec{\mathbf{x}} = \int_{C_1} (y^2, y + 2xy) \cdot \mathbf{d}\vec{\mathbf{x}} + \int_{C_2} (y^2, y + 2xy) \cdot \mathbf{d}\vec{\mathbf{x}}.$$

The horizontal segment C_1 is parameterized by $x(t) = t$, $y(t) = 0$ for $t \in [0, x]$, and the vertical segment C_2 by $x(t) = x$, $y(t) = t$ for $t \in [0, y]$. This yields

$$\int_0^x (0, 0) \cdot (1, 0)\, dt + \int_0^y (t^2, t + 2xt) \cdot (0, 1)\, dt = \tfrac{t^2}{2} + xt^2 \Big|_0^y = \tfrac{y^2}{2} + xy^2.$$

Using the straight line L between $(0, 0)$ and (x, y) parameterized by $x(t) = t$, $y(t) = (\frac{y}{x})t$ for $t \in [0, x]$, we have

$$\begin{aligned} \int \vec{\mathbf{F}} \cdot \mathbf{d}\vec{\mathbf{x}} &= \int_0^x \left(\frac{y^2 t^2}{x^2}, \frac{yt}{x} + \frac{2yt^2}{x} \right) \cdot \left(1, \frac{y}{x}\right) dt \\ &= \frac{y^2 t^3}{3x^2} + \frac{y^2 t^2}{2x^2} + \frac{2y^2 t^3}{3x^2} \Big|_0^x = \frac{y^2}{2} + xy^2. \quad \square \end{aligned}$$

In this example we *assumed* that $\vec{\mathbf{F}}$ was independent of path and then used Theorem 6.5.2 to construct the potential function. Our preceding computations for the two paths seem to corroborate this assumption. But how do we know that path independence holds for all the other possible routes? It is not feasible to check all paths. However, Theorem 6.5.2 and Theorem 6.2.6 together show that path independence is equivalent to the existence of a potential function. Therefore once we have a candidate for a potential function we simply verify path independence by checking that $\nabla f = \vec{\mathbf{F}}$.

6.5.2 Equivalent Conditions for a Potential

In the next theorem we present one more condition that is equivalent to the existence of a potential function.

THEOREM 6.5.3 *Let $\vec{\mathbf{F}}$ be a continuous vector field in an open connected region R of the plane. Then the following three conditions are equivalent:*

a) *$\vec{\mathbf{F}}$ is the gradient field for some potential function f.*

b) *$\vec{\mathbf{F}}$ is path independent.*

c) *$\int_C \vec{\mathbf{F}} \cdot \mathbf{d}\vec{\mathbf{x}} = 0$ for every closed curve C in R.*

PROOF Theorems 6.2.6 and 6.5.2 have shown that (**a**) $\iff$ (**b**). To complete the proof we must show that (**b**) $\Rightarrow$ (**c**) and (**c**) $\Rightarrow$ (**b**). We show the first implication and leave the second as an exercise.

Suppose that $\vec{\mathbf{F}}$ is path independent in R and consider any closed curve C in R. Choose two arbitrary points $\vec{\mathbf{a}}$ and $\vec{\mathbf{b}}$ on the curve and divide the curve into two segments C_1 and C_2 (Figure 6.5.3). From the additive property of line integrals we know that

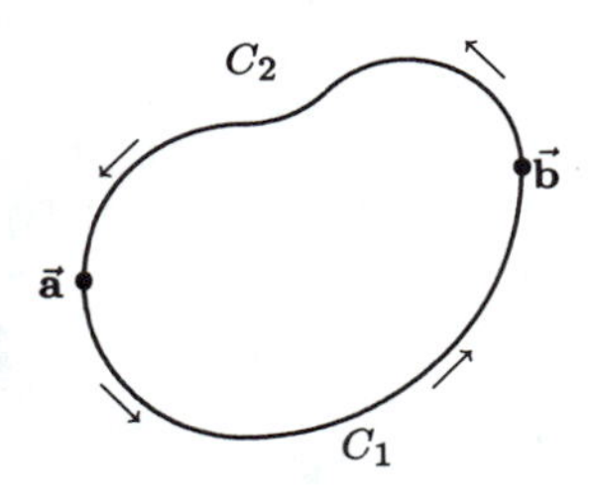

Figure 6.5.3

$$\int_C \vec{\mathbf{F}} \cdot \mathbf{d}\vec{\mathbf{x}} = \int_{C_1} \vec{\mathbf{F}} \cdot \mathbf{d}\vec{\mathbf{x}} + \int_{C_2} \vec{\mathbf{F}} \cdot \mathbf{d}\vec{\mathbf{x}}. \tag{3}$$

Denote by C_2^- the path C_2 traced in the opposite direction. Both C_1 and C_2^- connect $\vec{\mathbf{a}}$ to $\vec{\mathbf{b}}$. By path independence we know that

$$\int_{C_2^-} \vec{\mathbf{F}} \cdot \mathbf{d}\vec{\mathbf{x}} = \int_{C_1} \vec{\mathbf{F}} \cdot \mathbf{d}\vec{\mathbf{x}}.$$

But from Theorem 6.2.5 we have

$$\int_{C_2^-} \vec{\mathbf{F}} \cdot \mathbf{d}\vec{\mathbf{x}} = -\int_{C_2} \vec{\mathbf{F}} \cdot \mathbf{d}\vec{\mathbf{x}}.$$

Substituting these into (3) gives $\int_C \vec{\mathbf{F}} \cdot \mathbf{d}\vec{\mathbf{x}} = 0$. ∎

Theorem 6.2.7 showed that if a potential function exists for a continuously differentiable field $\vec{\mathbf{F}} = (P, Q)$, it necessarily satisfies $\frac{\partial Q}{\partial x} = \frac{\partial P}{\partial y}$. Our last theorem in this section provides a converse to this when we restrict ourselves to a nicely filled region with no holes. Once again the geometry of the regions is playing a role in the analysis.

THEOREM 6.5.4 *If $\vec{\mathbf{F}} = (P, Q)$ is a continuously differentiable vector field on an open disk D and*

$$\frac{\partial Q}{\partial x} = \frac{\partial P}{\partial y}$$

in D, then $\vec{\mathbf{F}}$ has a potential function of D.

PROOF It is tempting to try to prove this by using Green's theorem to show that all integrals around closed curves in D are 0 and then invoking Theorem 6.5.3. After all, from Green's theorem we know $\oint_{\partial R} \vec{\mathbf{F}} \cdot \mathbf{d\vec{x}} = \iint_R \frac{\partial Q}{\partial x} - \frac{\partial P}{\partial y}\, dx\, dy$. Since $\frac{\partial Q}{\partial x} = \frac{\partial P}{\partial y}$, this integral is 0. But this argument only shows $\oint_C \vec{\mathbf{F}} \cdot \mathbf{d\vec{x}} = 0$ for curves C that are *boundaries* of regions. To invoke Theorem 6.2.3 we must show that the line integral is 0 for *all* closed curves, including self-intersecting curves (such as a figure eight curve) that are not boundaries of simple regions. Instead of using Green's Theorem, we must construct our potential function bare-handed.

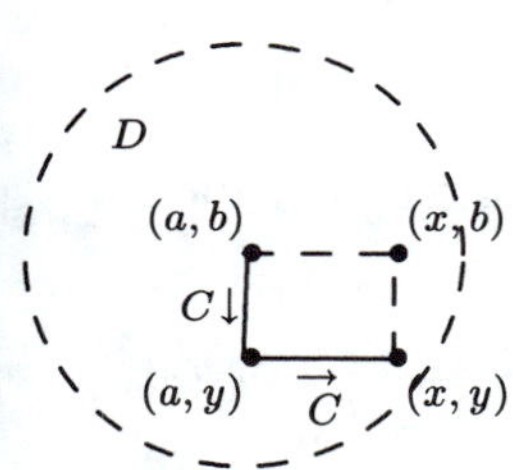

Figure 6.5.4

Let (a, b) be the center of the disk D. We will define a function $f(x, y)$ to be the line integral from the center of D to (x, y) along a specific path. Let

$$f(x, y) = \int_C \vec{\mathbf{F}} \cdot \mathbf{d\vec{x}},$$

where C is the rectangular path that goes from (a, b) to (a, y) and then to (x, y), as in Figure 6.5.4. Since (a, b) is the center of the disk, this path lies in D for every point $(x, y) \in D$. We claim that f is a potential function for $\vec{\mathbf{F}}$. To show this we compute $\frac{\partial f}{\partial x}$ as in equation (2) of the proof of Theorem 6.5.2:

$$\begin{aligned}\frac{\partial f}{\partial x} &= \lim_{\Delta x \to 0} \frac{1}{\Delta x}\left[\int_{(a,b)}^{(x+\Delta x, y)} \vec{\mathbf{F}} \cdot \mathbf{d\vec{x}} - \int_{(a,b)}^{(x+\Delta x, y)} \vec{\mathbf{F}} \cdot \mathbf{d\vec{x}}\right] \\ &= \lim_{\Delta x \to 0} \frac{1}{\Delta x} \int_{(x,y)}^{(x+\Delta x, y)} \vec{\mathbf{F}} \cdot \mathbf{d\vec{x}} \\ &= P(x, y).\end{aligned}$$

To show that $\frac{\partial f}{\partial y} = Q(x, y)$, we use Green's theorem to obtain an equivalent definition of $f(x, y)$ that will enable us to compute $\frac{\partial f}{\partial y}$ exactly as we computed $\frac{\partial f}{\partial x}$. Let Γ be the rectangular path from (a, b) to (x, b) and then to (x, y). (See Figure 6.5.5.) Now we will apply Green's theorem to the rectangle R with boundary formed by C^- and Γ.

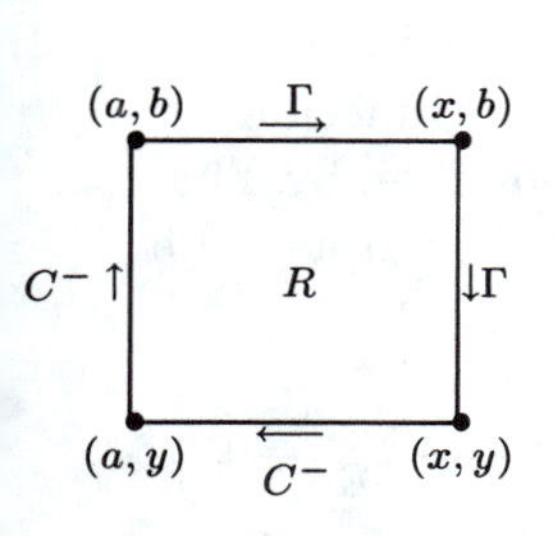

Figure 6.5.5

By assumption $\frac{\partial Q}{\partial x} = \frac{\partial P}{\partial y}$; consequently Green's theorem gives us

$$\int_\Gamma \vec{\mathbf{F}} \cdot \mathbf{d}\vec{\mathbf{x}} + \int_{C^-} \vec{\mathbf{F}} \cdot \mathbf{d}\vec{\mathbf{x}} = \iint_R \frac{\partial Q}{\partial x} - \frac{\partial P}{\partial y} = 0.$$

Since $\int_{C^-} \vec{\mathbf{F}} \cdot \mathbf{d}\vec{\mathbf{x}} = -\int_C \vec{\mathbf{F}} \cdot \mathbf{d}\vec{\mathbf{x}}$, we obtain

$$\int_\Gamma \vec{\mathbf{F}} \cdot \mathbf{d}\vec{\mathbf{x}} = \int_C \vec{\mathbf{F}} \cdot \mathbf{d}\vec{\mathbf{x}} = f(x, y).$$

Thus at (x, y), f is also given by $f(x, y) = \int_\Gamma \vec{\mathbf{F}} \cdot \mathbf{d}\vec{\mathbf{x}}$. We may again use the argument employed in the derivation of (2) to show the following (you are asked to supply the details in problem 6):

$$\frac{\partial f}{\partial y} = \lim_{\Delta y \to 0} \frac{1}{\Delta y} \left[\int_{(a,b)}^{(x,y+\Delta y)} \vec{\mathbf{F}} \cdot \mathbf{d}\vec{\mathbf{x}} - \int_{(a,b)}^{(x+,y+\Delta y)} \vec{\mathbf{F}} \cdot \mathbf{d}\vec{\mathbf{x}} \right] = Q(x, y),$$

where these integrals are computed along Γ. ∎

We end this chapter with a famous example showing that the restriction in the preceding theorem to a disk (or something like a disk) is necessary. In general the geometrical nature of the region on which $\frac{\partial Q}{\partial x} = \frac{\partial P}{\partial y}$ plays an important role.

EXAMPLE 6.5.2 Let $\vec{\mathbf{F}} = \left(\frac{-y}{x^2+y^2}, \frac{x}{x^2+y^2}\right)$. Example 6.4.5 showed that $\frac{\partial Q}{\partial x} = \frac{\partial P}{\partial y}$ for all points (x, y) except the origin. Therefore by Theorem 6.5.4 $\vec{\mathbf{F}}$ will have a potential function in any circular region that does not contain the origin. For example, if our disk doesn't touch the x-axis, the following is a potential function:

$$f(x, y) = \tan^{-1}\left(\frac{y}{x}\right).$$

If the circle doesn't touch the y-axis, then

$$f(x, y) = \tan^{-1}\left(\frac{-x}{y}\right)$$

is a potential function. There are many others. However, it is impossible to splice the functions together to obtain a single potential function for all points except the origin in the plane. For if this were possible, by Theorem 6.5.3 $\vec{\mathbf{F}}$ would be 0 around every closed curve. But Example 6.4.5 showed that this integral is 2π. □

If we take any region in which $\vec{\mathbf{F}} = (P, Q)$ is defined and $\frac{\partial Q}{\partial x} = \frac{\partial P}{\partial y}$, then by Green's theorem integrating $\vec{\mathbf{F}}$ around the entire boundary must give 0. The problem in the preceding example is that the region where $\vec{\mathbf{F}}$ is defined has a hole in it at the origin. This permits us to have *closed curves that are not boundaries of any region.* For instance, the geometric boundary of the unit disk with the origin removed is the unit circle *plus* the point at the origin. This intimate interplay between the geometry of regions and the analysis of functions defined on the regions is characteristic of Green's theorem and higher-dimensional versions of the fundamental theorem of calculus, and plays an important role in complex analysis in the next chapter.

Problems for Section 6.5

•6.5.1. Prove the second half of Theorem 6.5.2 by showing that $\frac{\partial f}{\partial y} = Q(x, y)$.

6.5.2. **a)** Let $\vec{\mathbf{F}}(x, y) = (y + 2xy, y + x^2 + x)$. As in Example 6.5.1, find a potential function for $\vec{\mathbf{F}}$ by integrating $\vec{\mathbf{F}}$ between $(0, 0)$ and (x, y) along a straight line.

b) Now determine the potential function by integrating along the vertical path from $(0, 0)$ to $(0, y)$ and then along the horizontal path $(0, y)$ to (x, y).

6.5.3. Prove that $\vec{\mathbf{G}}(x, y) = (y, 2x)$ is not independent of path between $(0, 0)$ and (x, y).

6.5.4. **a)** Integrate the vector field $\vec{\mathbf{F}}(x, y) = (p(x), q(y))$ from $(0, 0)$ to $(2, 4)$ by integrating first along the x-axis from $(0, 0)$ to $(2, 0)$ and then vertically up to $(2, 4)$. (Your answer will necessarily involve the respective antiderivatives P and Q of the functions p and q.)

b) Prove that $\int_{(0,0)}^{(2,4)} \vec{\mathbf{F}} \cdot \mathbf{d}\vec{\mathbf{x}}$ along the parabolic curve (t, t^2) for $t \in [0, 2]$ has the same value as in **(a)**. (You will need to use integration by parts.)

•6.5.5. Finish the proof of Theorem 6.5.3 by showing that if $\int_C \vec{\mathbf{F}} \cdot \mathbf{d}\vec{\mathbf{x}} = 0$ for every closed curve C, then $\vec{\mathbf{F}}$ is independent of path.

•6.5.6. Complete the proof of Theorem 6.5.4. by supplying the details to show that $\frac{\partial f}{\partial y} = Q(x, y)$

6.5.7. Suppose that the vector field $\vec{\mathbf{F}}(x, y) = (P(x, y), Q(x, y))$ satisfies $\frac{\partial Q}{\partial x} = \frac{\partial P}{\partial y}$ except at two points a and b (see Figure 6.5.6).

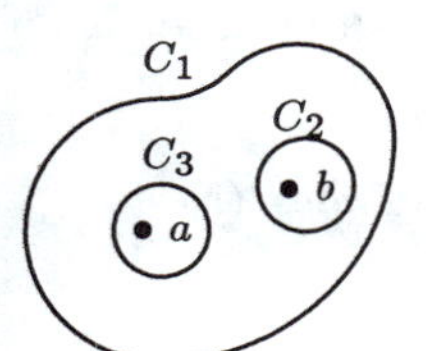

Figure 6.5.6

a) Describe how the line integrals $\int_{C_1} \vec{\mathbf{F}} \cdot \mathbf{d}\vec{\mathbf{x}}$, $\int_{C_2} \vec{\mathbf{F}} \cdot \mathbf{d}\vec{\mathbf{x}}$, and $\int_{C_3} \vec{\mathbf{F}} \cdot \mathbf{d}\vec{\mathbf{x}}$ are related. (Hint: Review Example 6.4.5.)

b) For which curves C will it be true that $\int_C \vec{\mathbf{F}} \cdot \mathbf{d}\vec{\mathbf{x}} = 0$? For which curves C will $\int_C \vec{\mathbf{F}} \cdot \mathbf{d}\vec{\mathbf{x}} = \int_{C_1} \vec{\mathbf{F}} \cdot \mathbf{d}\vec{\mathbf{x}}$?

7

Complex Analysis

In Chapter 1 we saw that mathematicians were led to construct the real number system because certain elementary problems had "solutions" that were not rational numbers. For example, the length x of the diagonal of a unit square cannot be expressed as a rational number, since by the Pythagorean theorem, we must have $x^2 - 2 = 0$. But Theorem 1.1.1 says that $\sqrt{2}$ is not rational. Such quantities, which arose in the most elementary physical way, could not be ignored, even if they could not be adequately described or defined. Thus some level of acceptance of irrational numbers existed even before Dedekind's work in the nineteenth century.

Even more fundamental for European mathematicians was the problem of negative numbers. It was not until the seventeenth century that negative numbers became known in Europe through Arab texts, though the Hindus had used them as early as the seventh century. Why were negative numbers problematic? Consider this argument given by Antoine Arnauld (1612–1694). We wish to claim that

$$-1 : 1 = 1 : -1 .$$

Of course -1 is less than 1. How can a smaller number be related to a larger one in the same way that a larger number is related to a smaller one?[1]

[1] Morris Kline, *Mathematical Thought from Ancient to Modern Times* (New York: Oxford University Press, 1972), p. 252.

Without quite understanding negative or irrational numbers, mathematicians began to struggle with what we now call complex or imaginary numbers. Again the motivation was a search for solutions. The quadratic equation $x^2 - 2 = 0$ at least has irrational solutions, but the equally simple equation $x^2 + 1 = 0$ has no real solutions, rational or irrational. This fact follows directly from the axiom system for real numbers. From problem 1.3.12, we know that for any real number x, we have $x^2 \geq 0$. But then by axiom 13 for real numbers, $x^2 + 1 > 0$. That is, no real number x satisfies $x^2 + 1 = 0$.

As early as the sixteenth century mathematicians began to explore the properties of nonreal solutions to such quadratic equations. The symbols $\sqrt{-1}$ or i were used to denote the "imaginary" solutions to the equation $x^2 + 1 = 0$. Acting as if the usual algebraic laws were valid here, mathematicians such as Jerome Cardan (1501–1576) began to use square roots of negative numbers to solve all sorts of "impossible" problems.

In his *Ars Magna* of 1545 Cardan poses the following problem. Find two numbers whose sum is 10 and whose product is 40. Of course, this requires us to find solutions to the quadratic equation

$$x(10 - x) = 40,$$

or equivalently,

$$x^2 - 10x + 40 = 0.$$

As every student of elementary calculus can show, the quadratic $x^2 - 10x + 40$ has an absolute minimum of 15 at $x = 5$, as long as x is constrained to be a real number. Therefore no real solutions to this problem exist. Cardan was not deterred, however, and he obtained the nonreal "solutions" that we today would denote by $5 \pm \sqrt{-15}$ or $5 \pm \sqrt{15}i$. We can verify that Cardan's answers are correct, acting on faith that the usual laws of algebra apply to such quantities: If $x = 5 + \sqrt{15}i$, then

$$\begin{aligned} x(10 - x) &= (5 + \sqrt{15}i)\big(10 - (5 + \sqrt{15}i)\big) \\ &= (5 + \sqrt{15}i)(5 - \sqrt{15}i) \\ &= 25 - 5\sqrt{15}i + \sqrt{15}i - 15i^2. \end{aligned}$$

But i is supposed to represent a square root of -1, so $i^2 = -1$; therefore

$$x(10 - x) = 25 - 15i^2 = 25 + 15 = 40,$$

solving Cardan's problem. A similar argument works for $x = 5 - \sqrt{15}i$. Yet even Cardan regarded these solutions as at best subtle and at worst useless.[2]

[2] David Eugene Smith, *A Source Book in Mathematics* (New York: Dover Publications, Inc., 1984), p. 202.

Cardan and others at this time operated without any careful or formal theory of complex numbers by blindly applying the ordinary rules of algebra as needed. But some mathematicians completely rejected the notion of such "imaginary" solutions to equations:

> *Imaginary numbers made their way into arithmetic calculation without the approval, and even against the desires of individual mathematicians, and obtained wider circulation only gradually and to the extent they showed themselves useful.* Imaginary numbers long retained a somewhat *mystic* coloring.... As evidence, I mention a very significant utterance by *Leibniz* in the year 1702, "Imaginary numbers are a fine and wonderful refuge of the divine spirit, almost amphibian between being and non-being."[3]

Our immediate goal is to make precise the nature of the complex number system and its rules of algebra. This will be followed by developing a calculus that employs the complex numbers. The chapter will conclude with a proof of the fundamental theorem of algebra, which states that every nonconstant polynomial with either real or complex coefficients has at least one solution over the complex numbers, thus ending the search for solutions.

7.1 The Complex Numbers

So far we have operated on an intuitive level applying the "usual" algebraic operations to i or $\sqrt{-1}$ without any justification. To clarify these operations, we now define the complex numbers and their sums and products.

7.1.1 Algebra

DEFINITION 7.1.1 *A **complex number** is an ordered pair of real numbers $z = (a, b)$. The set of all complex numbers will be denoted by* $\mathbf{C}$.

One should not fail to notice that a complex number (a, b) has the same representation as a point or vector in the real plane. We will exploit this fact shortly when it comes to giving a geometric representation of complex numbers.

[3] Felix Klein, *Elementary Mathematics from an Advanced Standpoint: Arithmetic, Algebra, Analysis* (New York: Dover Publications, 1945), p. 56.

DEFINITION 7.1.2 *Two complex numbers $z = (a, b)$ and $w = (c, d)$ are* **equal** *if and only if their corresponding components are equal, that is, if and only if $a = c$ and $b = d$.*

Equality for complex numbers reduces to equality of components, just as with vectors in the plane.

DEFINITION 7.1.3 *Given two complex numbers (a, b) and (c, d), their* **sum** *is defined by*

$$(a, b) + (c, d) = (a + c, b + d).$$

Note that the "+" on the left side of the equation is the operation being defined, while the "+" on the right side of the equation refers to addition of real numbers. Complex addition is "component addition," just like addition of vectors in the plane. Complex multiplication is somewhat trickier. It is *not* related to the ordinary dot product of two vectors in the plane.

DEFINITION 7.1.4 *Given two complex numbers (a, b) and (c, d), their* **product** *is defined by*

$$(a, b) \cdot (c, d) = (ac - bd, ad + bc).$$

The first three definitions make it clear that we can think of $\mathbf{C}$ as having the familiar vector space (additive) structure of $\mathbf{R}^2$. The crucial fourth definition shows that $\mathbf{C}$ has a multiplicative structure that $\mathbf{R}^2$ does not share. We will exploit this multiplicative structure, as well as the familiar vector space structure, in our development of complex calculus.

EXAMPLE 7.1.1 Using Definition 7.1.4, the product of $z = (-2, 4)$ and $w = (3, 1)$ is

$$zw = (-2, 4) \cdot (3, 1) = (-2 \cdot 3 - 4 \cdot 1, -2 \cdot 1 + 4 \cdot 3) = (-10, 10). \quad \square$$

A real number is not an ordered pair, so technically by Definition 7.1.1 a real number is not a complex number. But historically the real numbers were considered a subset (subfield) of the complex numbers, and this proved quite useful. By employing the following identification we too will think of $\mathbf{R}$ as a subset of $\mathbf{C}$.

The real number a is associated with the complex number $(a, 0)$ and conversely. In this way a one-to-one correspondence is set up between $\mathbf{R}$ and the elements of $\mathbf{C}$ of the form $(a, 0)$ by

$$a \longleftrightarrow (a, 0).$$

Most important, this correspondence preserves real addition and multiplication. That is, using the definition of complex addition and multiplication on such pairs yields

$$(a, 0) + (c, 0) = (a + c, 0),$$

and

$$(a, 0) \cdot (c, 0) = (ac - 0, 0 + 0) = (ac, 0).$$

The complex numbers of the form $(a, 0)$ under complex addition and complex multiplication behave exactly like their associated real numbers under real addition and real multiplication, so the identification that we have used is called an isomorphism.

This correspondence justifies the following convention: we denote the complex number $(a, 0)$ simply by a. Further, we use the standard convention of denoting $(0, 1)$ by i. Then using this notation and Definition 7.1.4 we have

$$i^2 = (0, 1) \cdot (0, 1) = (-1, 0) = -1.$$

Thus the number i that we have singled out *is* a square root of -1, just as in the introductory section. Further notice that

$$(b, 0) \cdot (0, 1) = (0, b).$$

Therefore it makes sense to extend our notational conventions and write

$$(b, 0) \cdot (0, 1) = bi.$$

Finally, for any complex number (a, b), using these conventions we have

$$(a, b) = (a, 0) + (0, b) = a + bi.$$

We will make frequent use of the "$a+bi$" notation for complex numbers because it simplifies multiplication. If we remember that $i^2 = -1$ and use the distributive property, the definition of complex multiplication will follow automatically.

DEFINITION 7.1.5 *If $z = a + bi$, then a is called the* ***real part*** *of z and b is called the* ***imaginary part*** *of z. These are denoted by* $\operatorname{Re} z$ *and* $\operatorname{Im} z$*, respectively.*

EXAMPLE 7.1.2 Let $z = a + bi$. Show that $\operatorname{Re} iz = -\operatorname{Im} z$ and that $\operatorname{Re} z^2 = a^2 - b^2$.

SOLUTION Using our notational conventions and Definition 7.1.4 it follows that

$$iz = i(a + bi) = -b + ai.$$

Therefore

$$\operatorname{Re} iz = -b = -\operatorname{Im} z.$$

Next,

$$z^2 = (a + bi) \cdot (a + bi) = a^2 - b^2 + (ab + ba)i = a^2 - b^2 + 2abi.$$

So

$$\operatorname{Re} z^2 = a^2 - b^2. \quad \square$$

Next we show that the "usual" algebraic laws hold for the complex numbers.

THEOREM 7.1.6 *The complex numbers,* **C**, *form a field.*

PROOF We must show that Field Axioms 1 through 11 (Chapter 1, Section 3) hold for the elements of **C** under complex addition and multiplication. Most of the axioms are easily verified and often depend, in part, on the corresponding property for real numbers. To illustrate these ideas we will check several of the axioms.

Axiom 4 (Existence of an Additive Identity): Any complex number (x, y) that *acts* like an additive identity must satisfy

$$(a, b) = (a, b) + (x, y) = (a + x, b + y).$$

Therfore $a = a + x$, which implies $x = 0$, and $b = b + y$, which implies $y = 0$. It follows that the additive identity is

$$(x, y) = (0, 0) = 0.$$

That is, 0 is the additive identity for **C** as well as for **R**.

Axiom 7 (Commutativity of Multiplication):

$$\begin{aligned}
(a, b) \cdot (c, d) &= (ac - bd, ad + bc) \\
&= (ca - db, da + cb) && \text{(Axiom 7 for } \mathbf{R}) \\
&= (ca - db, cb + da) && \text{(Axiom 2 for } \mathbf{R}) \\
&= (c, d) \cdot (a, b).
\end{aligned}$$

Axiom 9 (Existence of a Multiplicative Identity): If (x, y) is to be the multiplicative identity for the complex numbers, it must satisfy

$$(a, b) = (a, b) \cdot (x, y) = (ax - by, ay + bx) \tag{1}$$

for *any* complex number (a, b). In particular, if $b = 0$, this reduces to the case that for any real number a,

$$(a, 0) = (a, 0) \cdot (x, y) = (ax, ay).$$

It immediately follows that $x = 1$ and $y = 0$. A quick check shows that these values for x and y also satisfy (1). Thus

$$(x, y) = (1, 0) = 1$$

is the unique multiplicative identity for $\mathbf{C}$ as well as for $\mathbf{R}$.

Axiom 10 (Existence of Multiplicative Inverses): Given $(a, b) \neq 0$, we will determine its inverse (x, y). Since $1 = (1, 0)$ is the multiplicative identity for $\mathbf{C}$, we must have

$$(a, b) \cdot (x, y) = (1, 0),$$

or equivalently,

$$(ax - by, ay + bx) = (1, 0).$$

This yields a system of two linear equations in two unknowns:

$$\begin{aligned} ax - by &= 1 \\ bx + ay &= 0 \,. \end{aligned}$$

The system has a unique solution, which is (as you should check)

$$x = \frac{a}{a^2 + b^2}; \quad y = \frac{-b}{a^2 + b^2}.$$

Thus the multiplicative inverse of $z = (a, b)$ is

$$z^{-1} = \left(\frac{a}{a^2 + b^2}, \frac{-b}{a^2 + b^2} \right).$$

Checking the remaining axioms is a good exercise for becoming familiar with the algebra of $\mathbf{C}$. ∎

Division by a nonzero complex number w is now defined in the usual way as multiplication by w^{-1}. In particular, if $z = a + bi$ and $w = c + di$, then

$$\frac{z}{w} = zw^{-1} = (a + bi) \left(\frac{c}{c^2 + d^2} - \frac{d}{c^2 + d^2} i \right) = \frac{ac + bd}{c^2 + d^2} + \frac{bc - ad}{c^2 + d^2} i.$$

The definition of a complex number as an ordered pair of real numbers leads to a natural geometric representation of the complex numbers as points in the real plane. In particular, the point $P = (a, b)$ of the plane corresponds to the complex number $(a, b) = a + bi$ (Figure 7.1.1). It is clear that this is a one-to-one correspondence. In this context we will often refer to the plane as the **complex plane**.

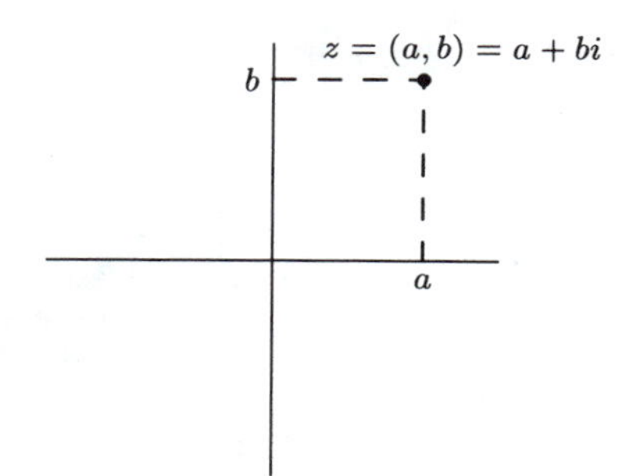

Complex numbers can be thought of as points in the plane.

Figure 7.1.1

The horizontal or x-axis is called the **real axis** because it consists of all the points of the form $(a, 0) = a$. Similarly the vertical or y-axis is called the **imaginary axis** because it consists of the purely imaginary points $(0, b) = bi$. This geometric interpretation of the complex numbers leads to two further important notions.

DEFINITION 7.1.7 *If $z = a + bi$, then the* **modulus** *of z, denoted $|z|$, is defined by $|z| = \sqrt{a^2 + b^2}$.*

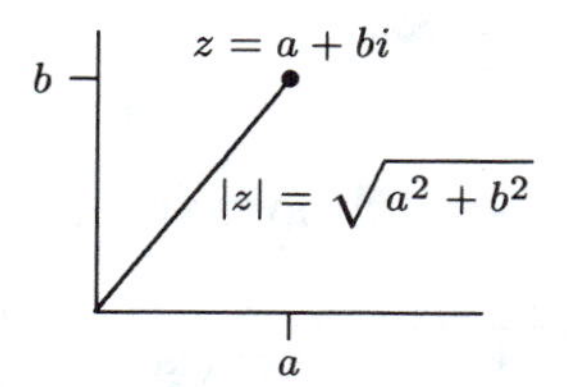

The modulus of z is the distance from z to the origin.

Figure 7.1.2

Geometrically the modulus is just the distance of $z = a + bi$ from the origin (see Figure 7.1.2). Notice that the vertical bar notation for the modulus is consistent with the notation for the norm of a vector in $\mathbf{R}^2$. That is, $|(a, b)| = a^2 + b^2$ whether we think of (a, b) as a complex number $z \in \mathbf{C}$ or as a vector $\vec{\mathbf{x}} \in \mathbf{R}^2$. This observation combined with the fact that addition in $\mathbf{C}$ is exactly vector addition means that the triangle inequality continues to hold.

THEOREM 7.1.8 ***(Triangle Inequality)*** *For any complex numbers z and w,*

$$|z + w| \leq |z| + |w|.$$

The modulus of z is sometimes called the **absolute value** of z. The use of the term absolute value here is consistent with its use for real numbers. If $z = (a, 0)$ is purely real, then $|z| = |(a, 0)| = \sqrt{a^2} = |a|$.

DEFINITION 7.1.9 *If $z = a + bi$, then the **conjugate** $\overline{z}$ is defined by $\overline{z} = a - bi$.*

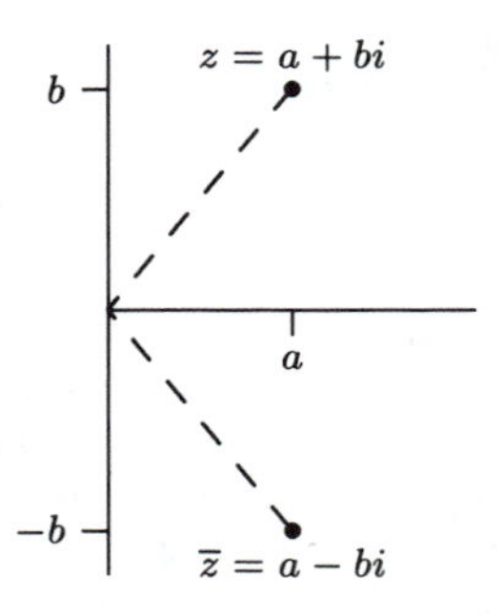

The conjugate of z is the reflection of z in the real axis.

Figure 7.1.3

Some of the elementary properties that the modulus and the conjugate possess are noted in the following theorem. Other properties are given in the problems at the end of this section.

THEOREM 7.1.10 *If $z = a + bi$ and $w = c + di$, then*

a) $\overline{z} = z$;

b) $\overline{z + w} = \overline{z} + \overline{w}$;

c) $\overline{z \cdot w} = \overline{z} \cdot \overline{w}$;

d) $|z|^2 = z \cdot \overline{z}$.

PROOF All parts can be verified by straightforward calculations. We check **(c)**.

$$\begin{aligned}\overline{z \cdot w} = \overline{((a + bi) \cdot (c + di))} &= \overline{(ac - bd) + (ad + bd)i} \\ &= ac - bd - (ad + bc)i \\ &= (a - bi) \cdot (c - di) \\ &= \overline{z} \cdot \overline{w}. \quad \blacksquare\end{aligned}$$

Addition of complex numbers can be interpreted geometrically. Let $z = (a, b)$ and $w = (c, d)$, and think of z and w as vectors in the plane. Since addition takes place componentwise, geometrically the sum $z + w = (a + c, b + d)$ is obtained using the parallelogram law for the addition of vectors (Figure 7.1.4). The **distance** between two complex numbers z and w is the length of the vector from w to z, which is just $|z - w|$ (Figure 7.1.5).

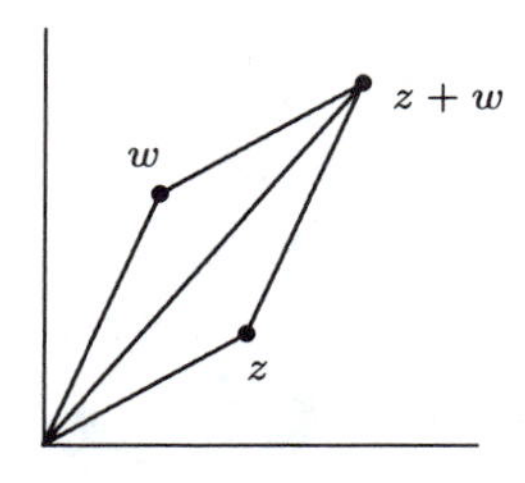

Figure 7.1.4

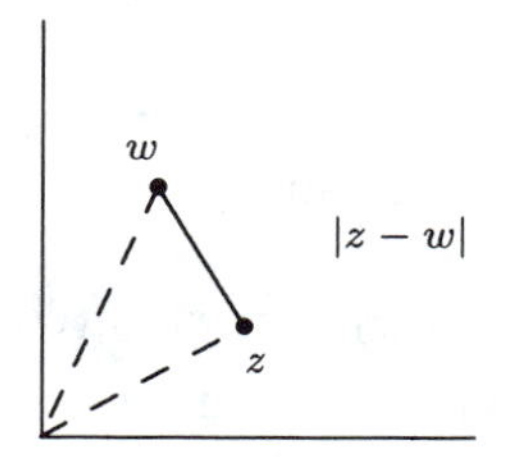

Figure 7.1.5

7.1.2 Polar Representation of Complex Numbers

To obtain a geometric interpretation for multiplication, we will use a **polar coordinate** representation to describe complex numbers. First, we think of the complex number $z = a + bi$ as a vector in the plane. Let r denote the length or modulus of z. As long as $z \neq 0$, its vector representation makes an angle θ with the positive real axis, where $-\pi < \theta \leq \pi$ (see Figure 7.1.6). Using basic trigonometry, it follows that $\tan\theta = b/a$,

$$a = r\cos\theta, \qquad b = r\sin\theta, \tag{2}$$

and

$$r = |z| = \sqrt{a^2 + b^2}.$$

The polar angle θ is called an **argument** of z.

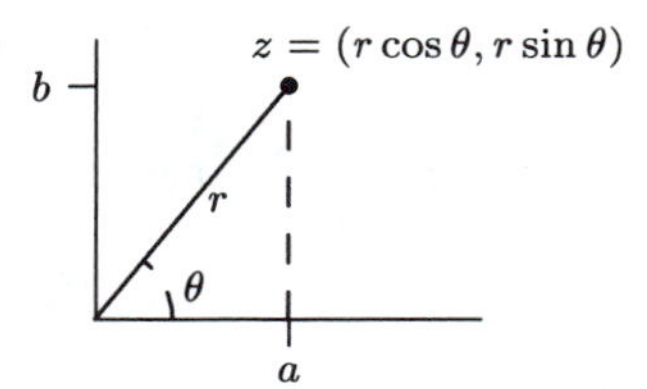

The polar representation of a complex number.

Figure 7.1.6

In general, we will allow values of θ that lie outside the interval $(-\pi, \pi]$. In fact, if θ satisfies (2), so does the angle $\theta + 2k\pi$ for any integer k, since the cosine and sine functions have period 2π. From this discussion we see that we can write

$$z = a + bi = r\cos\theta + ir\sin\theta = r(\cos\theta + i\sin\theta).$$

Let us show how such polar representations provide a geometric interpretation for complex multiplication. Let

$$z_1 = r_1(\cos\theta_1 + i\sin\theta_1),$$

and

$$z_2 = r_2(\cos\theta_2 + i\sin\theta_2).$$

Then

$$\begin{aligned} z_1 z_2 &= r_1 r_2[(\cos\theta_1\cos\theta_2 - \sin\theta_1\sin\theta_2) + i(\cos\theta_1\sin\theta_2 + \cos\theta_2\sin\theta_1)] \\ &= r_1 r_2[\cos(\theta_1+\theta_2) + i\sin(\theta_1+\theta_2)], \qquad (3) \end{aligned}$$

where the last step follows by using the addition formulas for the cosine and sine functions. Notice that (3) gives an expression for $z_1 z_2$ that is in polar form. It shows that the product of two complex numbers is the complex number that has length (modulus) equal to the product of the lengths of the two complex numbers and has an argument that is the sum of the arguments for those complex numbers. (See Figure 7.1.7.) This is stated formally in the following result:

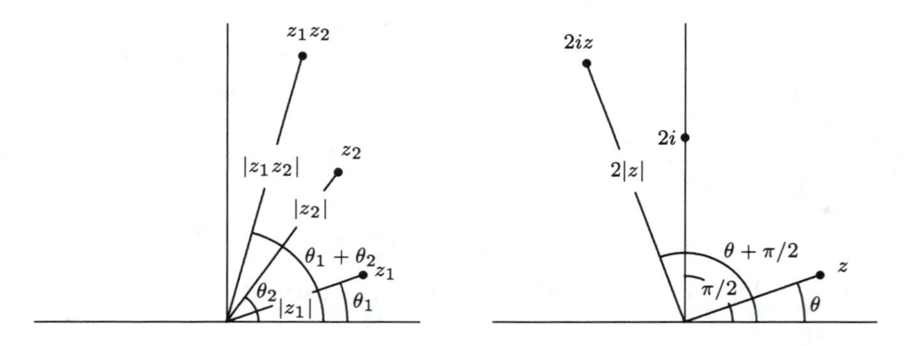

Figure 7.1.7 **Figure 7.1.8**

THEOREM 7.1.11 *For any complex numbers z_1 and z_2 with arguments θ_1 and θ_2, respectively,*

a) $|z_1 z_2| = |z_1||z_2|$;

b) $(\theta_1 + \theta_2)$ *is an argument for* $z_1 z_2$.

Suppose that we fix a complex number z_1. Then Theorem 7.1.11 says that for any other $z \in \mathbf{C}$, multiplication by z_1 has the effect on z of stretching its length by the factor $|z_1|$ and rotating z counterclockwise through an angle θ_1, where θ_1 is an argument of z. For example, multiplying z by $2i$ doubles the length of z and rotates z through $\pi/2$ radians counterclockwise. (See Figure 7.1.8.)

EXAMPLE 7.1.3 If $z = r(\cos\theta + i\sin\theta) \neq 0$, find z^{-1} in polar form.

SOLUTION The product $z^{-1}z = 1$ has modulus 1 and polar angle 0. Therefore the effect of multiplying $z = r(\cos\theta + i\sin\theta)$ by z^{-1} must be to stretch (shrink) the modulus of z by a factor of r^{-1} and to rotate z backward through the angle θ to 0. This means that z^{-1} must have modulus r^{-1} and $-\theta$ as an argument. That is,

$$z^{-1} = r^{-1}(\cos -\theta + i \sin -\theta). \quad \square$$

Using the polar form of complex numbers makes it simple to compute powers of a complex number. For example, if $z = r(\cos\theta + i\sin\theta)$, then by (3)

$$\begin{aligned} z^2 &= r^2(\cos\theta + i\sin\theta)^2 \\ &= r^2(\cos 2\theta + i \sin 2\theta). \end{aligned}$$

In fact, by a simple induction argument, it can be shown that for any positive integer n,

$$z^n = r^n(\cos n\theta + i \sin n\theta).$$

This result is called **DeMoivre's theorem.**

EXAMPLE 7.1.4 Calculate $(1-i)^4$ using polar coordinates.

SOLUTION $|1-i| = \sqrt{2}$ and an argument for $1-i$ is $-\pi/4$, so in polar form $1 - i = \sqrt{2}(\cos(-\pi/4) + i\sin(-\pi/4))$. Therefore

$$(1-i)^4 = (\sqrt{2})^4(\cos(-\pi) + i\sin(-\pi)) = 4(-1) = -4. \quad \square$$

As in the real case, we say that a complex number w is an nth **root** of z if $w^n = z$. Because integral powers of complex numbers are easy to compute, roots of complex numbers are easily calculated.

THEOREM 7.1.12 *Let n be a positive integer. Then any complex number $z \neq 0$ has exactly n distinct complex nth roots.*

This theorem is clearly not true for the real numbers. This is a first indication of the algebraic completeness of the complex numbers which will have its ultimate expression in the fundamental theorem of algebra.

PROOF Let $w = r(\cos\theta + i\sin\theta)$. If $z = s(\cos\phi + i\sin\phi)$, then z is an nth root of w if and only if $z^n = w$. By DeMoivre's theorem, $z^n = s^n(\cos n\phi + i \sin n\phi)$. Comparing the polar forms of w and z^n, we must have $|w| = r = s^n$ and $\theta + 2k\pi = n\phi$. Solving for z by solving for s and ϕ yields

$$z = \sqrt[n]{r}\left(\cos\left(\frac{\theta}{n} + 2\pi\frac{k}{n}\right) + i\sin\left(\frac{\theta}{n} + 2\pi\frac{k}{n}\right)\right). \tag{4}$$

Notice that each value of $k = 0, 1, \ldots, n-1$ yields a different value of z in this last equation. However, any other integer value of k merely repeats one of these first n values, since the cosine and sine functions have period 2π. ∎

Since all of the nth roots of z have the same modulus, they all lie on a circle of radius $\sqrt[n]{r}$ with the origin as its center. From (4) the roots occur at equally spaced angles of $\frac{2k\pi}{n}$ along this circle, starting at angle $\frac{\theta}{n}$.

EXAMPLE 7.1.5 Find the three cube roots of $z = -8i$.

SOLUTION $|z| = 8$ and an argument for z is $3\pi/2$. So in polar form,

$$z = 8\left(\cos\left(\frac{3\pi}{2}\right) + i\sin\left(\frac{3\pi}{2}\right)\right).$$

By (4) the cube roots of z have the form

$$w_k = \sqrt[3]{8}\left(\cos\left(\frac{\pi}{2} + \frac{2k\pi}{3}\right) + i\sin\left(\frac{\pi}{2} + \frac{2k\pi}{3}\right)\right),$$

for $k = 0$, 1, and, 2. Explicitly, the three cube roots of $-8i$ are

$$w_0 = 2\left(\cos\left(\frac{\pi}{2}\right) + i\sin\left(\frac{\pi}{2}\right)\right) = 2i,$$

$$w_1 = 2\left(\cos\left(\frac{7\pi}{6}\right) + i\sin\left(\frac{7\pi}{6}\right)\right) = -\sqrt{3} - i,$$

$$w_2 = 2\left(\cos\left(\frac{11\pi}{6}\right) + i\sin\left(\frac{11\pi}{6}\right)\right) = \sqrt{3} - i.$$

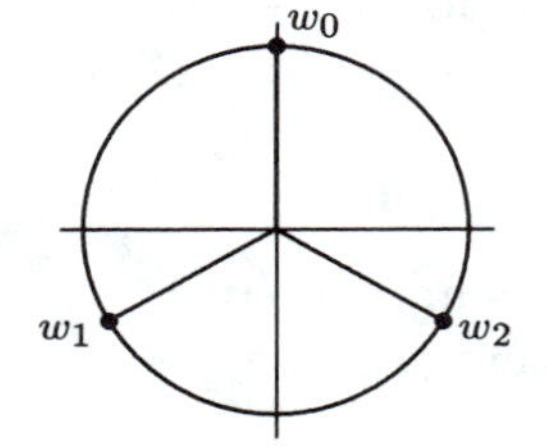

Figure 7.1.9

Problems for Section 7.1

7.1.1. Verify Field Axioms 1, 2, 3, 5, 6, 8, and 11 to complete the proof of Theorem 7.1.1.

7.1.2. Show that Field Axiom 12 (the order axiom) does not hold for **C**.

•7.1.3. Complete the proof of Theorem 7.1.10.

•7.1.4. Prove the complex version of problem 5.1.4. **The Complex Box Lemma:** *If $z \in \mathbf{C}$, then* ***(a)*** $|\operatorname{Re} z| \leq |z| \leq |\operatorname{Re} z| + |\operatorname{Im} z|$ *and* ***(b)*** $|\operatorname{Im} z| \leq |z| \leq |\operatorname{Re} z| + |\operatorname{Im} z|$.

7.1.5. **a)** Show by induction that if $z = r(\cos\theta + i\sin\theta)$ and n is a positive integer n, then $z^n = r^n(\cos n\theta + i\sin n\theta)$.

b) Show that if $z \neq 0$ and n is a positive integer n, then $z^{-n} = r^{-n}(\cos n\theta - i\sin n\theta)$.

7.1.6. Show that $\overline{z}$ and z^{-1} have the same arguments.

7.1.7. Show that if $|z| = 1$, then $z^{-1} = \overline{z}$.

7.1.8. Write the following complex numbers in polar form. In each case choose a polar angle θ such that $-\pi < \theta \leq \pi$.

a) $4 - 4i$ **b)** $\sqrt{3} - 3i$ **c)** $\dfrac{1-i}{1+i}$

d) $2 + i$ **e)** $(1+i)(1+i\sqrt{3})$ **f)** $(-3+i\sqrt{3})^{-1}$

7.1.9. Find and graph the solution sets for each of the following. (Hint: For the last three parts, think "geometrically" about what each equation is saying in terms of distance.)

a) $z = \overline{z}$ **b)** $z^2 = \overline{z}^2$ **c)** $|z^{-1}| < \delta$ $(\delta > 0)$

d) $|z - i| = 1$ **e)** $|z - 1| = |z + i|$ **f)** $|z - i| + |z + i| = 4$

7.1.10. **a)** Prove that 0, $\overline{z}$, and $1/z$ are always collinear $(z \neq 0)$.

b) If $z \neq 0$, when are z, $\overline{z}$, and $1/z$ collinear?

7.1.11. Find an argument θ with $-\pi < \theta \leq \pi$ for $z = -(2+i)^{-1}$.

7.1.12. **a)** Find the three cube roots of 27.

b) Find the two square roots of $-1 + \sqrt{3}i$.

c) Find the four fourth roots of $-2 - 2\sqrt{3}i$.

7.1.13. Let $a, b, c \in \mathbf{R}$, with $a \neq 0$. Show that the quadratic polynomial $az^2 + bz + c$ has two roots in $\mathbf{C}$ (counting multiplicities).

7.1.14. Show that the five roots of $(z+1)^5 + z^5 = 0$ all lie on a line parallel to the imaginary axis. (Hint: Divide by z^5 and observe that the equation becomes $\left(\frac{z+1}{z}\right)^5 = -1$. Solve for $\frac{z+1}{z}$ and then for z.)

7.1.15. Show that if n is a positive integer greater than 1, then the n nth roots of 1 sum to 0.

7.2 Complex Functions and Limits

Using the algebra and geometry of the previous section we are now prepared to describe complex functions and their limits. As we have done twice before, we will use these ideas to develop the concepts of differentiation and integration, this time in the complex setting. The pleasant surprise is that given all of your previous experience, this development is quite easy.

7.2.1 Functions

Definition 2.1.1 of a function still applies. But in the complex setting, the collection of ordered pairs that define the function are ordered pairs of complex numbers. The notions of domain and range (now subsets of **C**) carry over precisely as in the real case. (Since we have identified the real numbers as a subset of the complex numbers, any function we described in Chapter 2 can be thought of as a very special complex function.) We begin by singling out a few types of functions for particular attention.

A function f is called a **real-valued function** if $f(z)$ is a real number for all z in the domain of f, that is, the range of f is a subset of **R**. We have already seen three important real-valued functions. For any complex number $z = x + iy$ the real and imaginary parts of z are the real-valued functions defined by

$$\operatorname{Re} z = \operatorname{Re}(x + iy) = x$$

and

$$\operatorname{Im} z = \operatorname{Im}(x + iy) = y,$$

while the modulus of z is the real-valued function

$$|z| = |x + iy| = x^2 + y^2.$$

Another real-valued function of some importance has to do with the arguments of a complex number. Assume that $z \neq 0$. Recall that any real number θ is *an* argument for z if $z = r(\cos\theta + i\sin\theta)$, where $r = |z|$. We have noted that there are are an infinite number of arguments for z and that they all differ by multiples of 2π. However, exactly one of these arguments θ satisfies the condition $-\pi < \theta \leq \pi$. This particular value of θ is called the **principal value** of the argument of z (or more simply *the* argument of z). This real-valued function is denoted by $\operatorname{Arg} z$.

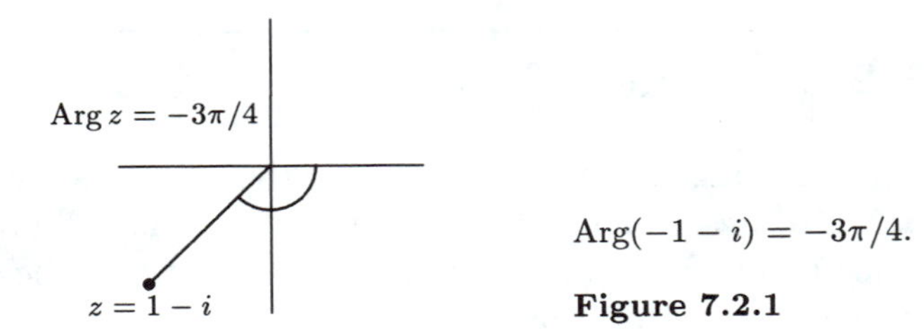

$\operatorname{Arg}(-1 - i) = -3\pi/4.$

Figure 7.2.1

When we want to emphasize the fact that a function might not be real-valued, we will say that it is a **complex-valued** function. The conjugate function is such an example:

$$\overline{z} = x - iy = \operatorname{Re} z - i \operatorname{Im} z.$$

Polynomials are another elementary type of complex-valued function. By analogy with the real case, a complex polynomial p is a fucntion that can be written in the form

$$p(z) = a_n z^n + a_{n-1} z^{n-1} + \ldots + a_1 z + a_0,$$

where the a_i are complex numbers.

Any complex-valued function f can be decomposed into the form

$$f = u + iv,$$

where u and v are real-valued functions of z. To obtain u and v we simply take the composition of f with the "real-part function" and the "imaginary-part function," respectively. That is,

$$u(z) = \operatorname{Re} f(z)$$

and

$$v(z) = \operatorname{Im} f(z).$$

If we again think of z as an ordered pair of real numbers (x, y), the functions $u(z) = u(x, y)$ and $v(z) = v(x, y)$ can be thought of as maps from $\mathbf{R}^2$ to $\mathbf{R}$. We already know a great deal about limits, continuity, and differentiability of such real-valued functions from our work in Chapters 5 and 6, and we will make use of this knowledge in the next few sections.

EXAMPLE 7.2.1 Find the real and imaginary parts of the polynomial $f(z) = z^2 - z$.

SOLUTION If we set $z = x + iy$, then

$$f(z) = z^2 - z = (x + iy)^2 - (x + iy) = x^2 - y^2 + 2ixy - x - iy.$$

Therefore

$$u(x, y) = \operatorname{Re} f(z) = x^2 - y^2 - x$$

and

$$v(x, y) = \operatorname{Im} f(z) = 2xy - y. \quad \square$$

7.2.2 The Complex Exponential Function

In Chapter 4 we saw that the real exponential function is represented on all of $\mathbf{R}$ by its Taylor series

$$e^x = \sum_{n=0}^{\infty} \frac{x^n}{n!} = 1 + \frac{x}{1!} + \frac{x^2}{2!} + \frac{x^3}{3!} + \frac{x^4}{4!} \cdots. \tag{1}$$

We would like to define a complex exponential function $f(z)$ that agrees with e^x when z is real. How might this be done?

Since we are trying to define an exponential function, we certainly want the basic exponent laws to hold. In particular, if $z = x + iy$, we want $e^z = e^{x+iy}$ to be equal to $e^x e^{iy}$. Since we know what e^x is, once we define e^{iy} we can define e^z as $e^x e^{iy}$.

To define e^{iy}, by analogy with (1), we might try

$$e^{iy} = \sum_{n=0}^{\infty} \frac{(iy)^n}{n!} = 1 + \frac{iy}{1!} + \frac{(iy)^2}{2!} + \frac{(iy)^3}{3!} + \frac{(iy)^4}{4!} \cdots. \tag{2}$$

This is not quite legitimate, since we have not discussed convergence of series in $\mathbf{C}$. However, since $|iy| = |y|$ for any y, it follows that this complex series is absolutely convergent, since

$$\sum_{n=0}^{\infty} \left| \frac{(iy)^n}{n!} \right| = \sum_{n=0}^{\infty} \frac{|y|^n}{n!} = e^{|y|}.$$

Although we will not systematically explore complex-valued series in this text, it is the case that if a complex-valued series converges absolutely, it converges. That is, the complex analogue of Theorem 4.4.2 holds. Therefore we will feel free to use the series expansion for e^{iy}, at least for motivational purposes.

In particular, notice that by rearranging the terms in (2) we can divide the sum into its real and imaginary parts. This yields

$$\begin{aligned} e^{iy} &= 1 + \frac{iy}{1!} + \frac{(iy)^2}{2!} + \frac{(iy)^3}{3!} + \frac{(iy)^4}{4!} + \frac{(iy)^5}{5!} + \cdots \\ &= 1 + \frac{iy}{1!} - \frac{y^2}{2!} - \frac{iy^3}{3!} + \frac{y^4}{4!} + \frac{iy^5}{5!} - \cdots \\ &= \left(1 - \frac{y^2}{2!} + \frac{y^4}{4!} - \cdots\right) + i\left(\frac{y}{1!} - \frac{y^3}{3!} + \frac{y^5}{5!} - \cdots\right) \\ &= \cos y + i \sin y, \end{aligned}$$

where the last step follows from recognizing the familiar Taylor series for the real cosine and sine functions. Thus we now *define* e^{iy} to be $\cos y + i \sin y$. Since we want e^z to be $e^x e^{iy}$, we are led to the following definition.

DEFINITION 7.2.1 *If $z = x + iy$, then e^z is defined by $e^x(\cos y + i \sin y)$.*

Notice that e^z has been carefully defined at this point without reference to complex series. This has the advantage of putting our definition on a firm foundation. However, in no way have we justified thinking of e^z as "e raised to the complex power z," since we have not even discussed complex powers. At this stage, e^z is simply a shorthand way of writing $e^x(\cos y + i \sin y)$. As an exercise, you are asked to prove some of the basic properties of the exponential function, such as $e^{z+w} = e^z e^w$ for all complex numbers z and w.

Since $e^{iy} = \cos y + i \sin y$, in exponential form the polar representation of a complex number $z = r(\cos\theta + i\sin\theta)$ is simplified to

$$z = re^{i\theta},$$

where $r = |z|$ and θ is any argument of z. In particular, we have $z = |z|e^{i\,\mathrm{Arg}\,z}$.

Finally, the exponential function can be used to parameterize circles in a simple fashion. If we fix a particular value of $r > 0$ and allow θ to vary from 0 to 2π, the graph of $re^{i\theta} = r(\cos\theta + i\sin\theta)$ traces out a circle of radius r centered at 0 in a counterclockwise fashion starting at $(r, 0)$. We will make extensive use of this parameterization when we discuss complex line integrals in the latter sections of the chapter.

7.2.3 Limits

In working with functions of two real variables in Chapters 5 and 6, we used open sets and neighborhoods as natural generalizations of open intervals of the real line. These notions are readily transported to the complex plane by using the distance function on $\mathbf{C}$ that is based on the modulus function. Further, as in the real case, the triangle inequality is valid (Theorem 7.1.8), that is, $|z + w| \leq |z| + |w|$. This is crucial when working with complex limits.

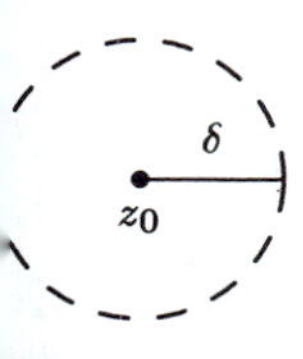

Figure 7.2.2

Because we have a distance function (metric) on $\mathbf{C}$, we can define open sets just as we did in $\mathbf{R}^2$. Let δ be a positve real number. Then the set satisfying the inequality $|z| < \delta$ consists of all those points interior to the circle of radius δ centered at the origin. Similarly the points satisfying $|z - z_0| < \delta$ form the interior of a circle of radius δ centered at z_0. We will call such a set of points a **neighborhood of** z_0 or an **open** δ**-ball** (Figure 7.2.2).

Now let S be an arbtrary set of points in the complex plane. A point $z_0 \in S$ is called an **interior point** of the set if there is some neighborhood of z_0 that contains only points of S. That is, we can find a $\delta > 0$ such that the set $|z - z_0| < \delta$ lies entirely within S (Figure 7.2.3).

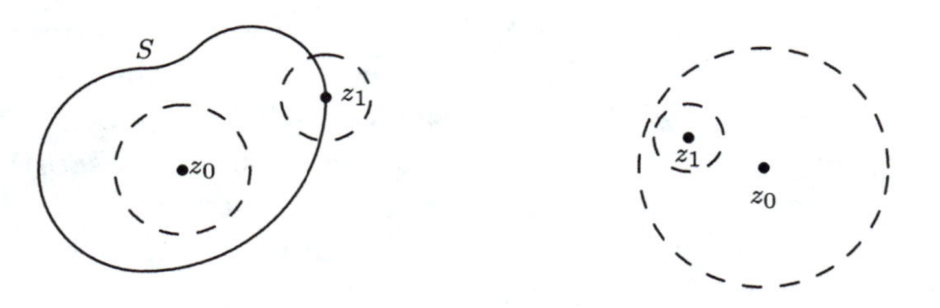

Figure 7.2.3 **Figure 7.2.4**

If every point of S is an interior point of the set, S is said to be an **open set**. The simplest open sets are the entire complex plane, **C** and the empty set. Any neighborhood of a point z_0 is also an open set, as is illustrated in Figure 7.2.4. Notice that the set S in Figure 7.2.3 is not open, since every neighborhood of the point z_1 contains points not in S.

Now we are able to give a definition for the limit of a complex-valued function of a complex variable that is analogous to the definition in the real case.

DEFINITION 7.2.2 *Let f be a function defined at each point in some neighborhood of the complex number z_0. Then the complex number l is the* ***limit of f at z_0*** *if for every real number $\epsilon > 0$ there exists a real number $\delta > 0$ such that*

$$\text{if } \ 0 < |z - z_0| < \delta, \ \text{ then } \ |f(z) - l| < \epsilon.$$

This is denoted by $\lim_{z \to z_0} f(z) = l$.

As in the real case, if $\lim_{z \to z_0} f(z)$ exists, this limit is unique. The proof of this can be constructed by appropriately modifying the proof of Theorem 2.3.1.

In Definition 7.2.21, the points z must satisfy the condition $0 < |z - z_0|$, so $z \neq z_0$. Therefore the value of f at z_0 itself (if f is even defined there) is irrelevant.

EXAMPLE 7.2.2 Show that $\lim_{z \to i} \frac{2z^2+2}{z-i} = 4i$.

SOLUTION For $z \neq i$,

$$f(z) = \frac{2z^2 + 2}{z - i} = \frac{2(z+i)(z-i)}{z - i} = 2(z + i).$$

So for $z \neq i$,

$$|f(z)-4i| < \epsilon \iff |2(z+i)-4i| < \epsilon \iff |2z-2i| < \epsilon \iff |z-i| < \frac{\epsilon}{2}.$$

We may choose $\delta = \epsilon/2$ to satisfy the limit definition. □

The familiar basic properties of limits remain valid in the complex setting and are easily obtained.

THEOREM 7.2.3 *If $\lim_{z\to z_0} f(z) = L$ and $\lim_{z\to z_0} g(z) = M$, then*

a) $\lim_{z\to z_0} cf(z) = cL$ *for any complex constant c;*

b) $\lim_{z\to z_0} (f+g)(z) = L + M;$

c) $\lim_{z\to z_0} cf(z) = cL$ *for any complex constant c;*

d) $\lim_{z\to z_0} (f\cdot g)(z) = L\cdot M;$

e) $\lim_{z\to z_0} \left(\frac{f}{g}\right)(z) = \frac{L}{M}$ (if $M \neq 0$).

PROOF The proofs of these are entirely analogous to the proofs of the corresponding results for limits of real-valued functions in Chapter 2, Section 3. We sketch the proof of **(c)**.

The triangle inequality provides the key to the proof. Notice that

$$\begin{aligned}|(f\cdot g)(z) - LM| &= |(f(z)g(z) - f(z)M) + (f(z)M - LM)|\\ &\leq |f(z)|\cdot|g(z) - M| + |M|\cdot|f(z) - L|.\end{aligned}$$

Since $\lim_{z\to z_0} f(z) = L$, there exists a $\delta_1 > 0$ such that if $0 < |z - z_0| < \delta_1$, then $|f(z) - L| < 1$. It follows (see problem 1.3.20) that if $0 < |z - z_0| < \delta_1$, then $|f(z)| < |L| + 1$. Let $m = \max\{|L| + 1, |M|\}$. If $0 < |z - z_0| < \delta_1$, then

$$|(f\cdot g)(z) - LM| \leq m|g(z) - M| + m|f(z) - L|.$$

But given any $\epsilon > 0$, there exists a $\delta_2 > 0$ and a $\delta_3 > 0$ such that

$$0 < |z - z_0| < \delta_2 \Longrightarrow |f(z) - L| < \epsilon/2m$$

and

$$0 < |z - z_0| < \delta_3 \Longrightarrow |g(z) - M| < \epsilon/2m.$$

Let $\delta = \min\{\delta_1, \delta_2, \delta_3\}$. If $0 < |z - z_0| < \delta$, then

$$|f(z)g(z) - LM| < m|g(z) - M| + m|f(z) - L| < m\cdot\frac{\epsilon}{2m} + m\cdot\frac{\epsilon}{2m} = \epsilon. \blacksquare$$

A number of elementary facts about limits can be derived from the preceding result. For example, it is clear that $\lim_{z\to z_0} z = z_0$. So $\lim_{z\to z_0} z^2 = \lim_{z\to z_0} z\cdot z = {z_0}^2$. More generally, by induction we can show that for any positive integer n,

$$\lim_{z\to z_0} z^n = {z_0}^n. \tag{3}$$

Next consider the polynomial

$$p(z) = a_n z^n + a_{n-1}z^{n-1} + \cdots + a_1 z + a_0.$$

It follows from Theorem 7.2.3 and (3) that $\lim_{z\to z_0} p(z) = p(z_0)$. This means that to calculate the limit of a polynomial p at any point z_0, we need only to evaluate p at z_0. As in the real case, such functions are called continuous.

DEFINITION 7.2.4 *The function f is said to be* **continuous at** z_0 *if* $\lim_{z\to z_0} f(z) = f(z_0)$. *If f is continuous at every point in its domain, then f is said to be* **continuous**.

You are asked to show that $f = u + iv$ is continuous if and only if $u(x,y)$ and $v(x,y)$ are continuous functions of two real variables (see problems 2 and 3). Our remarks about polynomials indicate that they are continuous. In the problem section you also are asked to show that several other familiar functions, such as $|z|$, $\overline{z}$, $\operatorname{Re} z$, and $\operatorname{Im} z$, are continuous.

There is a notable omission from this list: $\operatorname{Arg} z$ is *not* continuous along the nonpositive real axis. To see this, let x_0 be any point on the negative real axis of $\mathbf{C}$. As we approach x_0 from the lower half-plane, $\operatorname{Arg} z \to -\pi$ (Figure 7.2.5). But $\operatorname{Arg} x_0 = \pi$. Consequently $\operatorname{Arg} z$ cannot be continuous at x_0. (What happens if we approach x_0 from the upper half-plane?) Similar arguments can be made to show that $\operatorname{Arg} z$ is not continuous at 0.

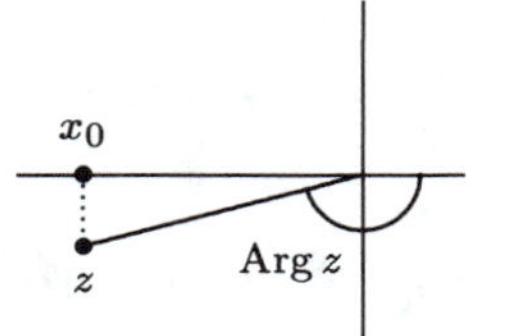

Arg z is not continuous along the nonpositive real axis.

Figure 7.2.5

Problems for Section 7.2

7.2.1. **a)** Use the definition of limit to show that $\lim_{z\to z_0} c = c$, where c is a constant, and that $\lim_{z\to z_0} z = z_0$.

b) Prove Theorem 7.2.3 **(a)**, **(b)**, and **(d)**.

7.2.2. **a)** Prove the complex analogue to Theorem 5.2.2 by using the complex box lemma of problem 7.1.4. Let $f(z) = u(z) + iv(z)$, where u and v are real-valued functions. Let $L = a + bi$, $z = x + iy$, and $z_0 = x_0 + iy_0$. Then

$$\lim_{z\to z_0} f(z) = L \iff \begin{cases} \lim_{(x,y)\to(x_0,y_0)} u(x,y) = a, \\ \lim_{(x,y)\to(x_0,y_0)} v(x,y) = b. \end{cases}$$

b) Use this result to give alternative proofs of **(a)** to **(c)** of Theorem 7.2.3.

7.2.3. **a)** Show that $f(z)$ is continuous if and only if $\operatorname{Re} f$ and $\operatorname{Im} f$ are continuous functions of x and y.

b) Use this result to show that e^z is continuous for all z.

•7.2.4. Show that $\operatorname{Re} z$, $\operatorname{Im} z$, $\overline{z}$, and $|z|$ are continuous.

7.2.5. **a)** Show that $\lim_{z\to 0} \operatorname{Arg} z$ does not exist. (Hint: Approach 0 along different rays.)

b) Describe the largest domain in which $\operatorname{Arg} z$ is continuous.)

7.2.6. **a)** Show that if $\lim_{z\to z_0} f(z) = w_0$, then $\lim_{z\to z_0} \overline{f(z)} = \overline{w_0}$.

b) Does the converse hold? Why?

7.2.7. **a)** Show that if $\lim_{z\to z_0} f(z) = w_0$, then $\lim_{z\to z_0} |f(z)| = |w_0|$.

b) Does the converse hold? Why?

c) Show that if $\lim_{z\to z_0} |f(z)| = 0$, then $\lim_{z\to z_0} f(z) = 0$.

•7.2.8. Show that $\lim_{z\to z_0} \frac{f(z)}{|z|} = 0$ if and only if $\lim_{z\to z_0} \frac{f(z)}{z} = 0$.

7.2.9. Show that if $\lim_{z\to z_0} f(z)$ exists, it is unique.

7.2.10. Assume that $f(z)$ is continuous and nonzero at a. Show that there is a $\delta > 0$ such that if $|z - a| < \delta$, then $f(z) \neq 0$.

7.2.11. **a)** Show that $e^{z+w} = e^z e^w$ for all complex numbers z and w.

b) Show that $e^{i\pi} + 1 = 0$. What a wonderful identity! Look at how four fundamental constants are related.

7.2.12. We wish to extend the definition of the natural logarithm function to the complex plane. Continue to let $\ln x$ denote the natural logarithm of a real number x. Let $z = re^{i \operatorname{Arg} z}$ be a nonzero complex number. Let $\log z$ denote the **principal value of the logarithm** of z which, is defined by

$$\log z = \ln r + i \operatorname{Arg} z = \ln |z| + i \operatorname{Arg} z.$$

a) Why didn't we say that if $z = re^{i\theta}$, then $\log z = \ln r + i\theta$?

b) Show that if a is a positive real number, then $\log a = \ln a$.

c) What is $\log -1$? What is $\log i$?

d) Does $\log zw = \log z + \log w$? (Hint: Consider $z = i$ and $w = -1$.)

e) Where is $\log z$ continuous?

7.3 Differentiability of Complex Functions

At this point in our treatment of complex analysis, many questions arise:

> What does it mean for a complex function to be differentiable? Are such differentiable functions continuous? Are there continuous functions that are not differentiable? Is there a complex version of the mean value theorem?

We can start answering these questions by generalizing the definition of the derivative of a real function. The idea is the same: take the limit of the difference quotient.

DEFINITION 7.3.1 *Let f be a function defined in a neighborhood of z_0. The function f is **differentiable at** z_0 if*

$$\lim_{z \to z_0} \frac{f(z) - f(z_0)}{z - z_0}$$

exists. When the limit does exist, it is denoted by $f'(z_0)$.

Though a complex number z is defined as an ordered pair (x, y) of real numbers, we have defined division by complex numbers. Therefore the definition of derivative of a complex function can use the difference quotient of complex numbers, in analogy with the definition of the derivative of a real-valued function of a single real variable. Such a direct analogy was not possible in the definition of differentiability of a real-valued function of a vector variable in Chapter 5, Section 4. There we saw that division by a vector in the difference quotient made no sense.

EXAMPLE 7.3.1 If $f(z) = z^2$, show that $f'(z_0) = 2z_0$ for any $z_0 \in \mathbf{C}$.

SOLUTION The calculation is as easy as in the real case.

$$\lim_{z \to z_0} \frac{f(z) - f(z_0)}{z - z_0} = \lim_{z \to z_0} \frac{z^2 - {z_0}^2}{z - z_0} = \lim_{z \to z_0} z + z_0 = 2z_0. \quad \square$$

The simplicity of such a calculation and its similarity to a real limit should not cause us to forget that the limits involved are complex ones; z is a complex variable and may approach z_0 in a variety of ways. Because of this, some rather simple functions turn out not to be differentiable.

EXAMPLE 7.3.2 Show that $f(z) = \overline{z}$ is not differentiable at any point.

SOLUTION Limits, and hence derivatives, are unique if they exist. Therefore to show that $\overline{z}$ is not differentiable at a point $z_0 \in \mathbf{C}$, we will show that as $z \to z_0$ in two different ways, the difference quotient in the definition of the derivative approaches two different values.

Let $z_0 = x_0 + iy_0$ and let $z = x + iy$. First we will approach z_0 along the line through z_0 parrallel to the real axis. This is accomplished by letting $y = y_0$, that is, $z = x + iy_0$ (Figure 7.3.1). Then for any $z \neq z_0$, the difference quotient is

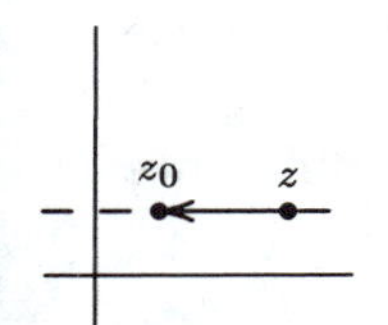

Figure 7.3.1

$$\frac{f(z) - f(z_0)}{z - z_0} = \frac{\overline{z} - \overline{z_0}}{z - z_0} = \frac{\overline{x + iy_0} - \overline{x_0 + iy_0}}{x - x_0} = \frac{x - x_0}{x - x_0} = 1.$$

On the other hand, we can approach z_0 along the line through z_0 that is parallel to the imaginary axis by letting $z = x_0 + iy$. This time for the difference quotient we obtain

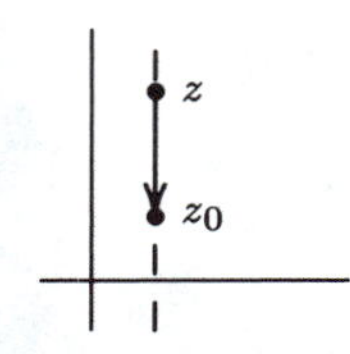

Figure 7.3.2

$$\frac{\overline{z} - \overline{z_0}}{z - z_0} = \frac{\overline{x_0 + iy} - \overline{x_0 + iy_0}}{i(y - y_0)} = \frac{-iy + iy_0}{iy - iy_0} = -1.$$

The difference quotient approaches at least two different values, 1 and -1, as $z \to z_0$, so the derivative of the conjugate function does not exist at any point z_0. $\square$

Example 7.3.2 is surprising because $\overline{z}$ is continuous (problem 7.2.4). Thus $\overline{z}$ is a continuous, nowhere differentiable function. There are real functions of a single real variable that exhibit this kind of behavior, but they are quite complicated to construct.[1] The next example also shows how special differentiability is in the complex case.

---

[1] For an example, see Walter Rudin, *Principles of Mathematical Analysis*, 3rd ed. (New York: McGraw-Hill Book Company, 1976), p. 154.

EXAMPLE 7.3.3 Show that $f(z) = |z|^2$ is differentiable only at 0.

SOLUTION The difference quotient is

$$\frac{f(z) - f(z_0)}{z - z_0} = \frac{|z|^2 - |z_0|^2}{z - z_0}. \tag{1}$$

Let $z = x + iy$. To show that f is differentiable at 0, we substitute $z_0 = 0$ into (1) and obtain

$$\frac{|z|^2 - |0|^2}{z - 0} = \frac{|z|^2}{z} = \frac{z\overline{z}}{z} = \overline{z}.$$

But $\overline{z}$ is continuous (problem 7.2.4), so

$$f'(0) = \lim_{z \to 0} \frac{f(z) - f(0)}{z - 0} = \lim_{z \to 0} \overline{z} = 0.$$

To show that f is not differentiable at any point other than 0, we proceed as in Example 7.3.2. First we approach z_0 along the line through z_0 parallel to the x-axis by letting $z = x + iy_0$. Then the difference quotient (1) becomes

$$\frac{|z|^2 - |z_0|^2}{z - z_0} = \frac{(x^2 + {y_0}^2) - ({x_0}^2 + {y_0}^2)}{(x + iy) - (x_0 + iy_0)} = \frac{x^2 - {x_0}^2}{x - x_0} = x + x_0.$$

As we approach z_0 along the line parallel to the x-axis, the difference quotient approaches $2x_0$.

Now approach z_0 along the line through z_0 parallel to the y-axis by letting $z = x_0 + iy$. Then

$$\frac{|z|^2 - |z_0|^2}{z - z_0} = \frac{{x_0}^2 + y^2 - ({x_0}^2 + {y_0}^2)}{(x + iy) - (x_0 + iy_0)} = \frac{y^2 - {y_0}^2}{i(y - y_0)} = \frac{y + y_0}{i}.$$

Thus as we approach z_0 along the line parallel to the y-axis, the difference quotient approaches $2y_0/i = -2iy_0$, where we have used the fact that $i^{-1} = -i$.

If $f'(z_0)$ exists, these difference quotients must approach the same value, that is,

$$2x_0 = -2iy_0.$$

But one quantity is real and the other is purely imaginary. The two are never equal unless $x_0 = y_0 = 0$, that is, unless $z_0 = 0$. Thus $|z|^2$ is not differentiable when $z_0 \neq 0$. ☐

The preceding example contrasts sharply with the behavior of the real absolute value function. We have previously seen that $|x|$ is differentiable for all x *except* 0. Further, since $|x|^2 = x^2$, $|x|^2$ is differentiable for *all* x. By comparison, we have just seen that $|z|^2$ is differentiable *only* at 0.

An obvious question at this stage is which of the elementary differentiation formulas for real functions carry over to the complex case. The next theorem tells us that all do. Indeed, the proofs are exactly the same as in the real case in Chapter 3, Section 2 and are left as exercises.

THEOREM 7.3.2 *Let f and g be complex functions differentiable at $z = z_0$ and let c be a complex constant.*

a) *cf is differentiable at $z = z_0$ and $(cf)'(z_0) = cf'(z_0)$.*

b) *$f + g$ is differentiable at $z = z_0$ and $(f + g)'(z_0) = f'(z_0) + g'(z_0)$.*

c) *$f \cdot g$ is differentiable at $z = z_0$ and*

$$(f \cdot g)'(z_0) = f'(z_0) \cdot g(z_0) + f(z_0) \cdot g'(z_0).$$

d) *If $g(z_0) \neq 0$, then f/g is differentiable at $z = z_0$ and*

$$\left(\frac{f}{g}\right)'(z_0) = \frac{f'(z_0)g(z_0) - f(z_0)g'(z_0)}{(g(z_0))^2}.$$

e) *Further, if f is differentiable at $g(z_0)$ and g is differentiable at z_0, then $f \circ g$ is differentiable at $z = z_0$ and*

$$(f \circ g)'(z_0) = f'(g(z_0)) \cdot g'(z_0).$$

The functions that we will study are those that have derivatives at every point in some open connected set. (A set is **connected** if any two points in the set can be joined by using a sequence of line segments with no self-intersections.) A function f is said to be **analytic** on a set D (the term **holomorphic** is also used) if f is differentiable at every point in some open connected set containing D. A function f is **analytic at a point** z_0 if f is analytic in some neighborhood of z_0. Example 7.3.3 shows that $|z|^2$ is differentiable only at 0. Thus no point has a neighborhood throughout which $|z|^2$ is differentiable. Therefore $|z|^2$ is not analytic anywhere. In Example 7.3.1 we saw that z^2 is analytic at every point of $\mathbf{C}$.

Problems for Section 7.3

7.3.1. Prove Theorem 7.3.2. (You may wish to review the proofs of the analogous theorems for real functions in Section 3.2).

7.3.2. **a)** Show that z is an function analytic on $\mathbf{C}$.

b) Use induction to show that $\frac{d}{dz} z^n = nz^{n-1}$ for any positive integer n.

c) Extend this result to the negative integers.

7.3.3. **a)** Show that any polynomial $p(z) = a_n z^n + \ldots + a_1 z + a_0$ is analytic on $\mathbf{C}$ and that $p'(z) = na_n z^{n-1} + \ldots + 2a_2 z + a_1$.

b) Show that any rational function $r(z) = p(z)/q(z)$, where $p(z)$ and $q(z)$ are polynomials, is analytic at every point in its domain.

7.3.4. **a)** Use the technique employed in Example 7.3.2 to determine where the function $\operatorname{Re} z$ is differentiable.

b) Now use Theorem 7.3.2 to determine where $\operatorname{Im} z$ is differentiable.

7.3.5. **a)** Show that $|z|$ is not differentiable for any $z_0 \neq 0$. (Hint: If it were differentiable, what could you say about $|z|^2$ and what would this mean for Example 7.3.3?)

b) Now show that $|z|$ is not differentiable at 0.

7.3.6. Let $f(z) = z^3 + z^2$. Let $z_0 = 0$ and let $z_1 = i$. Show that there is no point w on the straight line segment connecting z_0 to z_1 such that

$$f'(w) = \frac{f(z_1) - f(z_0)}{z_1 - z_0}.$$

This shows that the mean value theorem of real calculus does not extend to the complex setting.

•7.3.7. Show that if f is differentiable at z_0, then f is continuous at z_0. (Hint: See Theorem 3.1.2.)

•7.3.8. **a)** Show that if f is differentiable at z_0, then

$$\lim_{z \to z_0} \frac{f(z) - f(z_0) - f'(z_0)(z - z_0)}{|z - z_0|} = 0.$$

(Hint: Problem 7.2.8 may be helpful. Review Section 5.4.)

b) Conversely, show that if there is a complex constant α such that

$$\lim_{z \to z_0} \frac{f(z) - f(z_0) - \alpha(z - z_0)}{|z - z_0|} = 0,$$

then f is differentiable at z_0 and $f'(z_0) = \alpha$.

7.3.9. Prove the weak form of L'Hôpital's rule. Assume that f and g are analytic at z_0 and that $\lim_{z\to z_0} f(z) = 0$ and $\lim_{z\to z_0} g(z) = 0$. If $g'(z_0) \neq 0$, then

$$\lim_{z\to z_0} \frac{f(z)}{g(z)} = \frac{f'(z_0)}{g'(z_0)}.$$

(Hint: Start with the definition of $f'(z_0)/g'(z_0)$ and work backward.)

7.3.10. Use the preceding version of L'Hôpital's rule to evaluate the following limits:

a) $\displaystyle\lim_{z\to i} \frac{z^8+1}{z^6+1}$ **b)** $\displaystyle\lim_{z\to 1+i} \frac{z^4-4}{z^3+2z}$ **c)** $\displaystyle\lim_{z\to -1-i} \frac{z^2-2i}{z^5+4z^3}$

7.4 The Cauchy-Riemann Equations

In discussing the differentiability of $\overline{z}$ and $|z|^2$ in Examples 7.3.2 and 7.3.3 we calculated the difference quotient

$$\frac{f(z)-f(z_0)}{z-z_0}$$

in two very special ways. First we let $z = x + iy_0$, so only the real or x component of z varied, while the imaginary component was fixed at y_0. Next we let $z = x_0 + iy$, so that the y component of z varied while the x component was fixed at x_0. This is entirely analoguous to the process of partial differentiation. The Cauchy-Riemann equations, which we now develop, make an explicit connection between complex differentiation and partial differentiation.

To begin, let f be a function of a complex variable $z = (x, y)$ defined on a domain D. Write f in terms of its real and imaginary parts as

$$f(z) = f(x,y) = u(x,y) + iv(x,y).$$

If we now *assume that f is differentiable at $z_0 \in D$*, then by Theorem 7.3.2,

$$\begin{aligned} f'(z_0) &= \lim_{z\to z_0} \frac{f(z)-f(z_0)}{z-z_0} \\ &= \lim_{z\to z_0} \frac{u(z)+iv(z)-(u(z_0)+iv(z_0))}{z-z_0} \\ &= \lim_{(x,y)\to(x_0,y_0)} \frac{u(x,y)-u(x_0,y_0)}{(x+iy)-(x_0+iy_0)} + i \lim_{(x,y)\to(x_0,y_0)} \frac{v(x,y)-v(x_0,y_0)}{(x+iy)-(x_0+iy_0)}. \end{aligned} \tag{1}$$

No matter how z approaches z_0, we will always obtain $f'(z_0)$ as the limit of the difference quotient in (1). In particular, if we let $y = y_0$, so that $z = x + iy_0$, (1) becomes

$$\begin{aligned} f'(z_0) &= \lim_{x \to x_0} \frac{u(x, y_0) - u(x_0, y_0)}{x - x_0} + i \lim_{x \to x_0} \frac{v(x, y_0) - v(x_0, y_0)}{x - x_0} \\ &= u_x(x_0, y_0) + i v_x(x_0, y_0). \end{aligned} \tag{2}$$

By problem 7.2.2, the two real limits (partial derivatives) must exist because by assumption the complex limit (derivative) exists here. Notice that (2) says that the derivative of f can be expressed in terms of the partial derivatives of u and v with respect to x:

$$\operatorname{Re} f'(z_0) = u_x(x_0, y_0) \quad \text{and} \quad \operatorname{Im} f'(z_0) = v_x(x_0, y_0). \tag{3}$$

Instead, if we let $x = x_0$, so that $z = x_0 + iy$, (1) becomes

$$\begin{aligned} f'(z_0) &= \lim_{y \to y_0} \frac{u(x_0, y) - u(x_0, y_0)}{iy - iy_0} + i \lim_{y \to y_0} \frac{v(x_0, y) - v(x_0, y_0)}{iy - iy_0} \\ &= \frac{u_y(x_0, y_0)}{i} + v_y(x_0, y_0) \\ &= v_y(x_0, y_0) - i u_y(x_0, y_0). \end{aligned}$$

Thus $f'(z_0)$ can also be expressed in terms of the partial derivatives of u and v with respect to y:

$$\operatorname{Re} f'(z_0) = v_y(x_0, y_0) \quad \text{and} \quad \operatorname{Im} f'(z_0) = -u_y(x_0, y_0). \tag{4}$$

Comparing (3) and (4), we obtain the following result.

THEOREM 7.4.1 *If the function $f = u + iv$ is differentiable at $z_0 = (x_0, y_0)$, then at (x_0, y_0)*

$$u_x = v_y \qquad \text{and} \qquad v_x = -u_y.$$

These equations are known as the ***Cauchy-Riemann equations****.*

Using Theorem 7.4.1 we can quickly redo Example 7.3.2 and show that $\overline{z}$ is nowhere differentiable. Here $u(x, y) = \operatorname{Re} \overline{z} = x$ and $v(x, y) = \operatorname{Im} \overline{z} = -y$.Therefore $u_x = 1$ and $v_y = -1$, so $u_x \neq v_y$. The Cauchy-Riemann equations are never satisfied, so $\overline{z}$ is not differentiable at any point.

Theorem 7.4.1 gives *necessary* conditions for f to be differentiable, but we have not yet shown that satisfying the Cauchy-Riemann equations is a *sufficient* condition for f to be differentiable. For example, if $f(z) = e^z = e^x(\cos y + i \sin y)$, then $u_x = e^x \cos y = v_y$ and

$v_x = e^x \cos y = -u_y$, so the Cauchy-Riemann equations are satisfied. But is f differentiable? (Try to answer this by using Definition 7.3.1 to see how difficult such a question can be.) The next theorem states that satisfying the Cauchy-Riemann equations *is* sufficient for differentiability under some mild continuity assumptions.

THEOREM 7.4.2 *Assume that the real-valued functions $u(x,y)$ and $v(x,y)$ together with their partial derivatives of the first-order are continuous in a neighborhood N of (x_0, y_0) and that these partials satisfy the Cauchy-Riemann equations in N. Then $f = u+iv$ is differentiable at the point $z_0 = (x_0, y_0)$ and*

$$f'(z_0) = u_x(x_0, y_0) + iv_x(x_0, y_0) = v_y(x_0, y_0) - iu_y(x_0, y_0).$$

PROOF Let

$$\alpha = u_x(x_0, y_0) + iv_x(x_0, y_0) = v_y(x_0, y_0) - iu_y(x_0, y_0).$$

To prove the result, we use the alternative formulation of the derivative discussed in Chapter 5, Section 4 which was translated into complex terms in problem 7.3.8: $f'(z_0)$ exists and equals α if and only if

$$\lim_{z \to z_0} \frac{f(z) - f(z_0) - \alpha(z - z_0)}{|z - z_0|} = 0.$$

This will be the case if and only if the limits of the real and imaginary parts are both 0. To determine the limit of the real part, notice that

$$\begin{aligned} \operatorname{Re} \alpha(z - z_0) &= \operatorname{Re}\left((u_x(x_0, y_0) + iv_x(x_0, y_0))((x - x_0) + i(y - y_0))\right) \\ &= u_x(x_0, y_0)(x - x_0) - v_x(x_0, y_0)(y - y_0). \end{aligned}$$

But since the Cauchy-Riemann equations hold throughout N, we have $-v_x = u_y$, so

$$\operatorname{Re} \alpha(z - z_0) = u_x(x_0, y_0)(x - x_0) + u_y(x_0, y_0)(y - y_0).$$

Consequently

$$\operatorname{Re}\left(\frac{f(z) - f(z_0) - \alpha(z - z_0)}{|z - z_0|}\right)$$

may be written as

$$\frac{u(x, y) - u(x_0, y_0) - u_x(x_0, y_0)(x - x_0) - u_y(x_0, y_0)(y - y_0)}{|(x, y) - (x_0, y_0)|}.$$

But the partials of u are continuous throughout N, so by Theorem 5.4.6 u is differentiable at (x_0, y_0). Therefore by Theorem 5.4.4,

$$\lim_{(x,y)\to(x_0,y_0)} \frac{u(x,y) - u(x_0,y_0) - u_x(x_0,y_0)(x-x_0) - u_y(x_0,y_0)(y-y_0)}{|(x,y)-(x_0,y_0)|} = 0.$$

It is left as an exercise to show, in a similar fashion, that

$$\lim_{z\to z_0} \operatorname{Im} \frac{f(z) - f(z_0) - \alpha(z-z_0)}{|z-z_0|} = 0. \quad \blacksquare$$

EXAMPLE 7.4.1 We have noted that $e^z = e^x(\cos y + i \sin y)$ satisfies the Cauchy-Riemann equations: $u_x = e^x \cos y = v_y$ and $v_x = e^x \cos y = -u_y$. Since e^z and its partials are continuous everywhere, it follows from Theorem 7.4.2 that e^z is analytic throughout **C**. Further, as we might hope,

$$\frac{d}{dz}e^z = u_x + iv_x = e^x(\cos y + i \sin y) = e^z. \quad \square$$

EXAMPLE 7.4.2 Determine where $f(z) = z^{-1}$ is analytic.

SOLUTION Write

$$f(z) = z^{-1} = \frac{1}{z} = \frac{\overline{z}}{|z|^2} = \frac{x - iy}{x^2 + y^2}.$$

Thus

$$u(x,y) = \frac{x}{x^2+y^2} \qquad \text{and} \qquad v(x,y) = \frac{-y}{x^2+y^2}.$$

The first-order partial derivatives of u and v are

$$\frac{\partial u}{\partial x} = \frac{y^2 - x^2}{(x^2+y^2)^2} = \frac{\partial v}{\partial y}$$

and

$$\frac{\partial v}{\partial x} = \frac{2xy}{(x^2+y^2)^2} = -\frac{\partial u}{\partial y}.$$

Except at 0, these partial derivatives are continuous and satisfy the Cauchy-Riemann equations. Thus z^{-1} is analytic at every point in its domain. $\square$

Problems for Section 7.4

•7.4.1. Complete the proof of Theorem 7.4.2 by showing that

$$\lim_{z \to z_0} \operatorname{Im} \left(\frac{f(z) - f(z_0) - \alpha(z - z_0)}{|z - z_0|} \right) = 0.$$

7.4.2. **a)** Show that $\operatorname{Re} z$ and $\operatorname{Im} z$ are nowhere differentiable functions.

b) Determine where $z \operatorname{Re} z$ is analytic.

7.4.3. Determine where the following functions are analytic by using the Cauchy-Riemann equations.

a) $e^y(\cos x + i \sin x)$ **b)** $z \operatorname{Im} z^2$ **c)** $e^x(\sin y + i \cos y)$

d) $e^x(\sin y - i \cos y)$ **e)** $|z|^2 - \overline{z}^2$ **f)** $\operatorname{Re} \overline{z}^2 + \operatorname{Im} z^2$

g) $e^{x^2 - y^2}(\cos 2xy + i \sin 2xy)$ **h)** $\operatorname{Im} z + i \operatorname{Re} z$ **i)** $xy + i(y^2 - x^2)$

7.4.4. Show that if $f'(z) = 0$ for all $z \in \mathbf{C}$, then $f(z)$ is a constant function. (Hint: Let $f = u + iv$. Show that both first-order partials of u and of v are 0. Then show that both u and v must be constant functions.)

7.4.5. Use the Cauchy-Riemann equations and the preceding problem to prove the following results:

a) If $f(z)$ is analytic and f is *real valued*, then f is constant.

b) If $f(z)$ is analytic and f is *purely imaginary*, then f is constant.

c) If $f(z)$ is analytic and $\operatorname{Re} f$ is *constant*, then f is constant.

d) If $f(z)$ and $\overline{f(z)}$ are *both analytic*, then f is constant.

e) If $f(z)$ is analytic and $|f(z)|$ is *constant*, then f is constant. (Hint: Split the proof into two cases: $|f| = 0$ and $|f| \neq 0$ and use **(d)**.)

7.4.6. Let

$$f(z) = f(x, y) = \begin{cases} \frac{xy}{x^2 + y^2}, & \text{if } z = (x, y) \neq 0 \\ 0, & \text{if } z = (x, y) = 0. \end{cases}$$

Show that f satisfies the Cauchy-Riemann equations at 0, *but that f is still not differentiable at 0.* Why doesn't this contradict Theorem 7.4.2?

7.4.7. Suppose that $f(z) = u(x, y) + iv(x, y)$ is analytic. Which of the following functions are analytic?

a) $g(z) = v(x, y) + iu(x, y)$ **b)** $h(z) = v(x, y) - iu(x, y)$

c) $k(z) = u(y, x) - iv(y, x)$ **d)** $m(z) = v(x, y) - iv(x, y)$

7.4.8. We say that a function $\phi(x, y)$ is **harmonic** in a domain D if all its second-order partials are continuous in D and its Laplacian $\nabla^2\phi = 0$, that is

$$\frac{\partial^2\phi}{\partial x^2} + \frac{\partial^2\phi}{\partial y^2} = 0.$$

a) Use the Cauchy-Riemann equations to show that if $f(z)$ is analytic, then the imaginary part of f must be harmonic.

b) Is the real part of an analytic function harmonic?

c) If $ax^2 + bxy + cy^2$ is harmonic, what can you say about the coefficients a, b, c?

d) Show that $\ln|z|$ is harmonic ($z \neq 0$).

e) Suppose that $f(z)$ is analytic and nonzero in a domain D. Show that $\ln|f(z)|$ is harmonic in D.

7.4.9. If $u(x, y)$ is a harmonic function in D, then a harmonic function $v(x, y)$ in D is called a **harmonic conjugate** of u if the function $u + iv$ is analytic in D.

a) Find a harmonic conjugate of $u(x, y) = x^3 - 3xy^2$.

b) Find a harmonic conjugate of $u(x, y) = \frac{1}{2}\ln(x^2 + y^2)$.

c) Find a harmonic conjugate of $u(x, y) = y/(x^2 + y^2)$.

7.4.10. Suppose that $f = u + iv$ were analytic and that $u(x, y) = xy - x + y$. What must the function $v(x, y)$ be?

7.4.11. Suppose that $p(x)$ is a polynomial in x only and $q(y)$ is a polynomial in y only.

a) Prove that $f(z) = p(x) + iq(y)$ and its first order partials are continuous.

b) At which points is $f(z) = p(x) + iq(y)$ analytic?

7.4.12. Define $f(z) = \ln\left(\sqrt{x^2 + y^2}\right) + i\tan^{-1}(\frac{y}{x})$ on the set $x > 0$.

a) Show that f is analytic.

b) Find the real and imaginary parts of $\frac{1}{z}$ and show that $f'(z) = \frac{1}{z}$.

c) How are $f(z)$ and $\log z$ related? (See problem 7.2.12.)

7.4.13. Show that $f(z) = \sin x \cosh x + i\cos x \sinh y$ is analytic throughout $\mathbf{C}$.

7.5 Integration

At this stage in the development of complex analysis we can reap the rewards of all the previous hard work done on integration and line integrals. There is a rich interplay among the earlier ideas in the text and the properties of analytic functions that allows us to prove a number of beautiful and important theorems with a minimum of effort.

7.5.1 Line Integrals of Complex Functions

A simple way to generalize integration to the complex setting is to 'integrate' a complex function along a curve in the complex plane using line integrals of the type developed in Chapter 6. Assume that we are given a curve C in the complex plane. As in Chapter 6, we will assume that curves are parameterized, smooth, and of finite length. Let $f = u + iv$ be a function defined along C. We start by partitioning the curve C by using points $z_0, z_1, \ldots, z_n$, where z_0 is the initial point of C and z_n is the terminal point. We denote the difference $z_i - z_{i-1}$ by Δz_i.

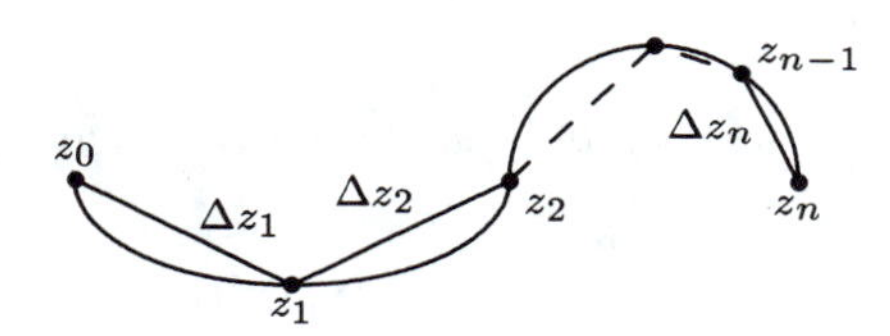

Figure 7.5.1

On each arc from z_{i-1} to z_i choose an intermediate point z_i^* at which to evaluate f and form the product $f(z_i^*)\Delta z_i$. Then the basic Riemann sum is

$$\sum_{i=1}^{n} f(z_i^*)\Delta z_i. \tag{1}$$

This sum of products has exactly the same form (and associated difficulties) as our initial attempt to define the integral of a real function of a single real variable in Chapter 3. However, the sum in (1) is actually taking place along a curve in the plane (not along an axis), so there ought to be a connection to line integrals here. However, the products being summed in a line integral are dot products of a vector field with tangent vectors to the curve. In (1) the product is multiplication of two complex numbers.

$$\begin{aligned} f(z_i^*)\Delta z_i &= (u(x_i^*, y_i^*) + iv(x_i^*, y_i^*))(\Delta x_i + i\Delta y_i) \\ &= u(x_i^*, y_i^*)\Delta x_i - v(x_i^*, y_i^*)\Delta y_i + i(v(x_i^*, y_i^*)\Delta x_i + u(x_i^*, y_i^*)\Delta y_i). \end{aligned}$$

The connection with line integrals becomes apparent when we use the preceding equation to split the Riemann sum in (1) into its real and imaginary parts.

$$\begin{aligned} \sum_{i=1}^{n} f(z_i^*)\Delta z_i = \sum_{i=1}^{n} & u(x_i^*, y_i^*)\Delta x_i - v(x_i^*, y_i^*)\Delta y_i \\ & + i\sum_{i=1}^{n} v(x_i^*, y_i^*)\Delta x_i + u(x_i^*, y_i^*)\Delta y_i. \end{aligned} \tag{2}$$

From Chapter 6, Section 2, we recognize the sums on the right-hand side of (2) as approximations to the sum of the pair of line integrals along C:

$$\int_C u\,dx - v\,dy + i\int_C v\,dx + u\,dy.$$

This suggests the following definition.

DEFINITION 7.5.1 *Let C be a smooth curve parameterized by $(x(t), y(t)) = x(t) + iy(t)$, for $t \in [a, b]$. Let $f = u + iv$ be a complex-valued function defined along C. Then the* **line integral of f along C** *is*

$$\begin{aligned}\int_C f(z)\,dz &= \int_C u\,dx - v\,dy \;+\; i\int_C v\,dx + u\,dy \\ &= \int_a^b u\big(x(t), y(t)\big)x'(t) - v\big(x(t), y(t)\big)y'(t)\,dt \\ &\qquad + i\int_a^b v\big(x(t), y(t)\big)x'(t) + u\big(x(t), y(t)\big)y'(t)\,dt.\end{aligned}$$

We have defined the integral of a complex-valued function along a curve C in terms of two *real* line integrals. If we let $z(t) = x(t) + iy(t)$ be the parameterization of C, then $z'(t) = x'(t) + iy'(t)$. Of course the values of f along C are given by the composite function $f(z(t))$. Therefore

$$\begin{aligned}f(z(t))z'(t) &= \Big(u(z(t)) + iv(z(t))\Big)z'(t) \\ &= \Big(u(x(t), y(t)) + iv(x(t), y(t))\Big)\big(x'(t) + iy'(t)\big) \\ &= u\big(x(t), y(t)\big)x'(t) - v\big(x(t), y(t)\big)y'(t) \\ &\qquad + i\Big[v\big(x(t), y(t)\big)x'(t) + u\big(x(t), y(t)\big)y'(t)\Big]. \qquad (3)\end{aligned}$$

The real and imaginary parts of $f(z(t))z'(t)$ are just the integrands in the definition of the line integral. Consequently, using (3),

$$\int_C f(z)\,dz = \int_a^b \mathrm{Re}\Big(f(z(t))z'(t)\Big)dt \;+\; i\int_a^b \mathrm{Im}\Big(f(z(t))z'(t)\Big)dt. \qquad (4)$$

This formulation of the definition of $\int_C f(z)\,dz$ is often quite useful. For example, from (4) it is clear that

$$\mathrm{Re}\int_C f(z)\,dz = \int_a^b \mathrm{Re}\Big(f(z(t))z'(t)\Big)\,dt \qquad (5)$$

and

$$\operatorname{Im} \int_C f(z)\, dz = \int_a^b \operatorname{Im}\Big(f(z(t))z'(t)\Big)\, dt. \tag{6}$$

We will often write $\int_C f(z)\, dz$ as $\int_a^b f(z(t))z'(t)\, dt$ with the understanding that it is the expansion in (4) that we have in mind.

As in the real case, if f is continuous along C, then $\int_C f(z)\, dz$ exists. To see this observe that if f is continuous at each point of C, so are u and v. Further since C is a smooth curve, $x'(t)$ and $y'(t)$ are both continuous. Thus the two real integrals that define $\int_C f(z)\, dz$ exist because their integrands are continuous.

EXAMPLE 7.5.1 Let $f(z) = z^{-1}$ and let C be the upper half of the unit circle parameterized by $z(\theta) = e^{i\theta} = \cos\theta + i\sin\theta$ for $\theta \in [0, \pi]$. Evaluate $\int_C f\, dz$. (See Figure 7.5.2.) Notice that $f(z) = z^{-1} = e^{-i\theta}$ on C. Further, $z'(\theta) = -\sin\theta + i\cos\theta = ie^{i\theta}$. So $f(z(\theta))z'(\theta) = e^{-i\theta}ie^{i\theta} = i$. Therefore

$$\int_C z^{-1}\, dz = \int_0^\pi f(z(\theta))z'(\theta)\, d\theta = \int_0^\pi i\, d\theta = \pi i. \quad \square$$

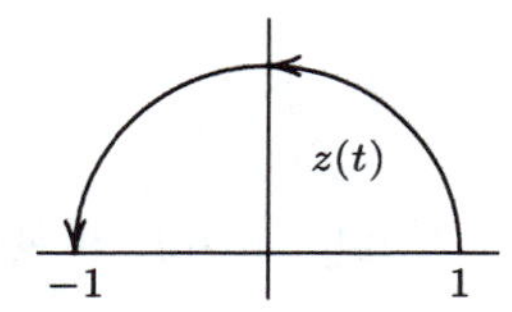

Figure 7.5.2

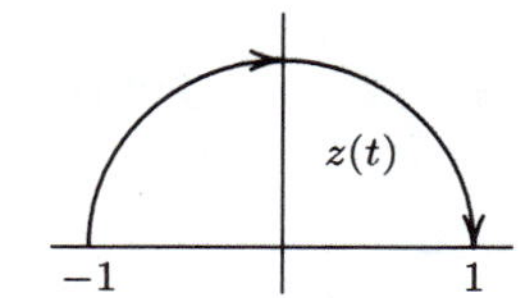

Figure 7.5.3

The definition of a complex line integral appears to depend on the particular parameterization of the curve C. Problem 8 asks you to show that, just as in the real case, complex line integrals are independent of parameterization.

THEOREM 7.5.2 *Let C and C' be two equivalent curves. If the orientations of C and C' are the same, then*

$$\int_{C'} f\, dz = \int_C f\, dz.$$

If the orientation of C' is opposite to that of C, then

$$\int_{C'} f\, dz = -\int_C f\, dz.$$

EXAMPLE 7.5.2 Let C' be the clockwise upper unit semicircle. Evaluate $\int_{C'} z^{-1}\,dz$.

SOLUTION Since C' is traced in a clockwise fashion, its orientation is opposite to the curve C in Example 7.5.1. (See Figure 7.5.3.) From Theorem 7.5.1 we conclude that

$$\int_{C'} z^{-1}\,dz = -\int_C z^{-1}\,dz = -\pi i.$$

Indeed, the actual calculation bears this out. C' can be parameterized using $z(\theta) = e^{i(-\theta)}$ for $\theta \in [\pi, 2\pi]$. So $z^{-1} = e^{i\theta}$ on C' and $z'(\theta) = -ie^{-i\theta}$. Therefore

$$\int_{C'} z^{-1}\,dz = \int_{\pi}^{2\pi} e^{i\theta}(-ie^{-i\theta})\,d\theta = \int_{\pi}^{2\pi} -i\,d\theta = -\pi i. \ \square$$

As in the real case, the definition of a complex line integral has a natural extension to a general contour C in the complex plane. Recall that a contour consists of a finite number n of smooth curves C_i that satisfy the condition that for $i = 1, \ldots, n-1$ the terminal point of C_i is the initial point of C_{i+1}. The **integral of f along the contour C** is defined by

$$\int_C f(z)\,dz = \int_{C_1} f(z)\,dz + \int_{C_2} f(z)\,dz + \cdots + \int_{C_n} f(z)\,dz\,.$$

EXAMPLE 7.5.3 Let $f(z) = z^2$. Let C be the contour consisting of the segments C_1 and C_2, where C_1 is parameterized by $z(t) = t + i$, for $t \in [0,1]$ and C_2 is parameterized by $z(t) = 1 + it$ for $t \in [1,2]$. Evaluate $\int_C f(z)\,dz$.

SOLUTION On C_1, $(z(t))^2 = (t+i)^2 = t^2 - 1 + 2it$ and $z'(t) = 1$. Therefore

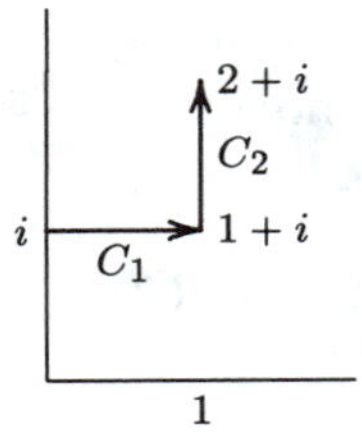

$$\int_{C_1} z^2\,dz = \int_0^1 t^2 - 1 + 2it\,dt = \frac{t^3}{3} - t\Big|_0^1 + it^2\Big|_0^1 = -\frac{2}{3} + i.$$

Figure 7.5.4

On C_2, $(z(t))^2 = 1 - t^2 + 2it$ and $z'(t) = i$. Therefore

$$\int_{C_2} z^2\,dz = \int_1^2 -2t + i(1-t^2)\,dt = -t^2\Big|_1^2 + i\left(t - \frac{t^3}{3}\right)\Big|_1^2 = -3 - \frac{4i}{3}.$$

Adding the values of the integrals over C_1 and C_2 together yields

$$\int_C z^2\,dz = \int_{C_1} z^2\,dz + \int_{C_2} z^2\,dz = \frac{-11 - i}{3}. \ \square$$

The next theorem gives a bound on this modulus of a line integral and is similar to the boundedness result for real line integrals in Theorem 6.2.3.

THEOREM 7.5.3 *Let C be a curve such that $|f(z)| \leq M$ for all $z \in C$. If L is the length of C, then*

$$\left| \int_C f(z)\, dz \right| \leq ML.$$

PROOF Assume that C is parameterized by $z(t) = (x(t), y(t))$ for $t \in [a, b]$. When $\int_C f(z)\, dz = 0$, the result clearly holds. Thus we may assume that $\int_C f(z)\, dz \neq 0$.

Using polar form, let $r = \left|\int_C f(z)\, dz\right|$ and $\theta = \operatorname{Arg}(\int_C f(z)\, dz)$. Then

$$re^{i\theta} = \int_C f(z)\, dz. \tag{7}$$

In this notation, we must show that $r = |\int_c f(z)\, dz| \leq ML$. But by multiplying (7) by $e^{-i\theta}$,

$$r = e^{-i\theta} \int_C f(z)\, dz = \int_C e^{-i\theta} f(z)\, dz = \int_a^b e^{-i\theta} f(z(t))z'(t)\, dt.$$

Since r is real, so is $\int_a^b e^{-i\theta} f(z(t))z'(t)\, dt$. Therefore by (5), we have

$$r = \operatorname{Re} \int_a^b e^{-i\theta} f(z(t))z'(t)\, dt = \int_a^b \operatorname{Re}\left(e^{-i\theta} f(z(t))z'(t)\right) dt.$$

But the absolute value of the real part of any complex number never exceeds its modulus (problem 7.1.4), thus

$$|\operatorname{Re}(e^{-i\theta} f(z(t)))| \leq |e^{-i\theta} f(z(t))| = |f(z(t))| \leq M.$$

Since $\int_a^b \operatorname{Re}\left(e^{-i\theta} f(z(t))z'(t)\right) dt$ is a real line integral, it follows from Theorem 6.2.3 that:

$$\begin{aligned}
r &= \int_a^b \operatorname{Re}\left(e^{-i\theta} f(z(t))z'(t)\right)\, dt \\
&\leq \left| \int_a^b \operatorname{Re}\left(e^{-i\theta} f(z(t))z'(t)\right)\, dt \right| \\
&\leq \int_a^b |\operatorname{Re}\left(e^{-i\theta} f(z(t))z'(t)\right)|\, dt \leq \int_a^b M|z'(t)|\, dt \leq ML. \quad \blacksquare
\end{aligned}$$

7.5.2 Extending the Fundamental Theorem of Calculus

In Chapter 6, Section 2 we were able to extend the fundamental theorem of calculus to line integrals. An important elementary result in complex analysis is to extend the fundamental theorem of calculus to contour integrals. Such a result is critical because it shows that under the proper circumstances the integral of a function is independent of the particular path used and depends only on the initial and terminal points of the contour. In the real case, the simplest situation in which line integrals are independent of path is when the integrand is the gradient of a function (Theorem 6.2.6). We will use this result to prove a similar fact in the complex setting.

THEOREM 7.5.4 *Assume that $F(z)$ is analytic with continuous derivative $F'(z) = f(z)$ in a domain D (i.e., F is an antiderivative of f in D). If C is any contour within D whose initial point is z_0 and whose terminal point is z_1, then*

$$\int_C f(z)\,dz = F(z)\Big|_{z_0}^{z_1} = F(z_1) - F(z_0).$$

PROOF Let $F(z) = U(x,y) + iV(x,y)$ and $f(z) = u(x,y) + iv(x,y)$. Since F is analytic on D the Cauchy-Riemann equations hold. Using this and the fact that $F'(z) = f(z)$ yields the following equalities:

$$\frac{\partial U}{\partial x} = \frac{\partial V}{\partial y} = u$$

and

$$\frac{\partial V}{\partial x} = -\frac{\partial U}{\partial y} = v.$$

In particular, then, we can express the gradients of U and V as

$$\nabla U = \left(\frac{\partial U}{\partial x}, \frac{\partial U}{\partial y}\right) = (u, -v)$$

and

$$\nabla V = \left(\frac{\partial V}{\partial x}, \frac{\partial V}{\partial y}\right) = (v, u). \tag{8}$$

The proof is finished by applying Theorem 6.2.6.

$$\int_C f(z)\,dz = \int_C u\,dx - v\,dy + i\int_C v\,dx + u\,dy$$

$$= \int_C \nabla U \cdot \mathbf{d\vec{x}} + i \int_C \nabla V \cdot \mathbf{d\vec{x}} \qquad \text{(using (8))}$$
$$= U(x_1, y_1) - U(x_0, y_0) + i(V(x_1, y_1) - V(x_0, y_0))$$
$$= U(x_1, y_1) + iV(x_1, y_1) - (U(x_0, y_0) + iV(x_0, y_0))$$
$$= F(z_1) - F(z_0). \ \blacksquare$$

Theorem 7.5.4 says that the integral of the derivative of a continuously differentiable function F depends only on the endpoints of the contour and is independent of the particular contour used. This makes integrating such a function around a closed contour trivial.

COROLLARY 7.5.5 *If $F(z)$ is analytic with continuous derivative $F'(z) = f(z)$ in a domain D and if C is any closed contour within D, then*

$$\int_C f(z)\,dz = 0\,.$$

PROOF The initial and terminal points of C are the same and the result follows by applying Theorem 7.5.4. $\blacksquare$

EXAMPLE 7.5.4 Notice that the contour integral of Example 7.5.3 can now be evaluated quickly. The initial point of the contour C is i and the terminal point is $1 + 2i$. Since $z^3/3$ is an antiderivative for z^2 throughout the entire complex plane,

$$\int_C z^2\,dz = \left.\frac{z^3}{3}\right|_i^{1+2i} = \frac{1}{3}((1+2i)^3 - i^3) = \frac{-11-i}{3}. \ \square$$

Not all contour integrals are independent of path, which makes contour integrals interesting. In particular the integral in Example 7.5.2 is *not* independent of the contour used, as the next example shows.

EXAMPLE 7.5.5 Let C'' be the lower unit semicircle traversed in the counterclockwise direction using the parameterization $z(\theta) = e^{i\theta}$ for $\theta \in [\pi, 2\pi]$. (See Figure 7.5.5.) Then, as in Example 7.5.1,

$$\int_{C''} z^{-1}\,dz = \int_\pi^{2\pi} i\,dt = \pi i.$$

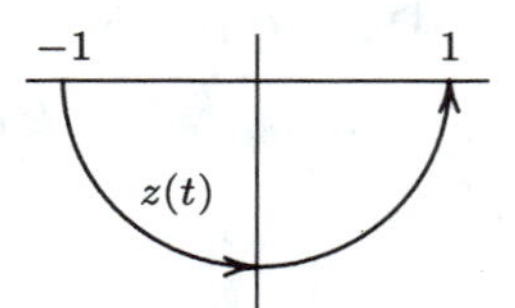

Figure 7.5.5

Even though curve C' of Example 7.5.2 begins at -1 and ends at 1 as C'' does,

$$\int_{C''} z^{-1}\,dz \neq \int_{C'} z^{-1}\,dz.$$

The integral of z^{-1} from -1 to 1 is not independent of path. (Since Theorem 7.5.4 fails here, what can you say about the analyticity of z^{-1} on the interior of the unit circle?) □

Theorem 7.5.4 is helpful in integrating certain elementary functions over contours, as Example 7.5.4 shows. The theorem would be much more useful if we knew precisely which complex functions had antiderivatives. The next section is devoted to showing that analytic functions in appropriately simple domains always have antiderivatives.

Problems for Section 7.5

7.5.1. Compute $\int_C f(z)\,dz$ for the each of the following functions along the indicated curve C.

a) $f(z) = \bar{z}$ along the curve C defined by $z(t) = t + i(t^2 + 1)$ for $t \in [0, 2]$.

b) $f(z) = x^2 + iy^2$ along the straight line C from i to $2 + 5i$.

c) $f(z) = z^2 - 4z + 3$ around the circle C of radius r centered at the origin. (Does the radius matter? Do the initial and terminal points matter? Does the orientation matter?)

7.5.2. Let C be the unit circle traversed counterclockwise starting and ending at 1.

a) Evaluate $\int_C \bar{z}\,dz$.

b) Evaluate $\int_C |z|\,dz$.

c) Evaluate $\int_C z^{-1}\,dz$.

d) Evaluate $\int_C p(z)\,dz$, where $p(z)$ is any polynomial.

7.5.3. Let $f(z) = f(x, y) = x + y - ix^2$.

a) Let C be the line segment from 0 to $1 + i$. Evaluate $\int_C f(z)\,dz$.

b) Let C be the contuour consisting of the following two line segments: from 0 to $1 + i$ and from $1 + i$ to $2i$. Evaluate $\int_C f(z)\,dz$.

7.5.4. Let $f(z) = (1+z)^{-1}$.

a) Let C be the line segment from 0 to $1+i$. Evaluate $\int_C f(z)\,dz$.

b) Let C be the semicircle defined by $z(\theta) = 2e^{i\theta}$, where $0 \le \theta \le \pi$. Evaluate $\int_C f(z)\,dz$.

c) Let C be the semicircle parameterized by $z(\theta) = 2e^{i\theta}$, where $\pi \le \theta \le 2\pi$. Evaluate $\int_C f(z)\,dz$.

7.5.5. Let $f(z) = \dfrac{1}{1-\overline{z}}$. Let C be the semicircle parameterized by $z(\theta) = 1 + e^{i\theta}$, where $0 \le \theta \le \pi$. Evaluate $\int_C f(z)\,dz$.

7.5.6. Let C be the unit circle parameterized by $z(\theta) = e^{i\theta}$, where $0 \le \theta \le 2\pi$.

a) Show that if n is an integer, then

$$\int_C z^n\,dz = \begin{cases} 0, & \text{if } n \ne -1 \\ 2\pi i, & \text{if } n = -1. \end{cases}$$

b) Show that if m and n are integers, then

$$\int_C z^m(\overline{z})^n\,dz = \begin{cases} 0, & \text{if } n \ne m+1 \\ 2\pi i, & \text{if } n = m+1. \end{cases}$$

7.5.7. Let C_ρ be the circle of radius ρ centered at z_0 parameterzied by $z(\theta) = z_0 + \rho e^{i\theta}$, where $0 \le \theta \le 2\pi$. Show that if f is any function defined at z_0, then

$$f(z_0) = \frac{1}{2\pi i}\int_{C_\rho} \frac{f(z_0)}{z - z_0}\,dz.$$

7.5.8. Prove Theorem 7.5.2 by modifying the proof of Theorem 6.2.5.

7.5.9. **a)** Let C be the (counterclockwise) square whose vertices are 0, 1, $1+i$, and i. Using Theorem 7.5.3 show that $\left|\int_C z^4 dz\right| \le 16$.

b) Better still, using Theorem 7.5.3 on each side of the square show that $\left|\int_C z^4 dz\right| \le 10$.

7.5.10. Let C be the unit circle and n an integer.

a) Show that $\left|\int_c z^n - 1\,dz\right| \le 4\pi$.

b) Show that $\left|\int_c (z^n - 2)^{-1}\,dz\right| \le 2\pi$.

7.5.11. Let C be the (counterclockwise) square whose vertices are 1, i, -1, and $-i$. Using Theorem 7.5.3 show that $\left|\int_C z^{-2}\,dz\right| \le 8\sqrt{2}$.

7.5.12. Assume that f is continuous at z_0. Show that given any $\epsilon > 0$, there is a $\rho > 0$ such that if C_ρ is the circle of radius ρ centered at z_0, then

$$\left| \int_{C_\rho} \frac{f(z) - f(z_0)}{z - z_0} \, dz \right| \leq 2\pi\epsilon.$$

(Hint: Make $|f(z) - f(z_0)| \leq \epsilon$.)

7.5.13. Let $f(z) = z$ and let C be the unit circle with parameterization $e^{i\theta}$, where $0 \leq \theta \leq 2\pi$. Show that for all points c between 0 and 2π,

$$f(z(c)) \neq \int_C f(z) \, dz.$$

This shows that the integral mean value theorem need not hold for complex-valued functions.

7.5.14. Suppose that $|f'(z)| < M$ for all z in an open disk D. Show that $|f(z_1) - f(z_0)| \leq M|z_1 - z_0|$. (Hint: In the real variable case this is proven using the mean value theorem, but that theorem doesn't hold for complex functions. Instead look at $\int f'(z) \, dz$ between the points z_0 and z_1 and apply Theorem 7.5.3.)

⋆7.5.15. Suppose that $f(z)$ is analytic on a disk D. Use Green's theorem and the Cauchy-Riemann equations to show that $\oint_{\partial D} f(z) \, dz = 0$. (Note: You have just proven Cauchy's theorem, a central theorem in complex analysis.)

⋆7.5.16. Begin by reviewing Theorems 6.5.2 and 6.5.3, which relate the path independence of $\vec{\mathbf{F}}$ to $\vec{\mathbf{F}}$ being a gradient vector field. Assume that $f(z)$ is a continuous function in an open disk D. Show that the following conditions are equivalent:

i) $\int_a^b f(z) \, dz$ is independent of path C in D.

ii) $\int_C f(z) \, dz = 0$ for any closed contour C in D.

iii) f has an antiderivative in D.

7.6 The Cauchy Integral Theorem

This section is devoted to proving some important results about contour integrals. The first of these results says that the integral of an analytic function f is independent of path as long as the contour lies within a simply connected domain. (We will prove a restricted version of this theorem.) Using this first result, we will show that an analytic function has an antiderivative in an appropriate domain. This result generalizes the second fundamental theorem of calculus. Both of these results are the complex versions of theorems developed for real line integrals in Chapter 6, Section 5. The final result of this section is the Cauchy integral formula, which states that the value of an analytic function f at a point z_0 is determined, under certain conditions, by the values of f on a simple closed contour that encloses z_0. We will use this result in the next section to prove the fundametal theorem of algebra. Several more consequences of these integral theorems are described in the last section of the text.

7.6.1 The Integral Theorems

THEOREM 7.6.1 ***(The Cauchy Integral Theorem)*** *Let $f = u + iv$ be an analytic function in a simply connected domain D such that the first-order partial derivatives of u and v are continuous. Let C be any closed curve within D. Then*

$$\int_C f(z)\,dz = 0\,.$$

PROOF The proof of this version of the Cauchy Integral theorem is almost trivial using Green's theorem.

Let R be the region bounded by C. Since the partials of u and v are continuous, by Green's Theorem

$$\int_C v\,dx + u\,dy = \iint_R u_x - v_y\,dx\,dy$$

and

$$\int_C u\,dx - v\,dy = \iint_R -v_x - u_y\,dx\,dy.$$

But f is analytic, so the Cauchy-Riemann equations are valid. This means that the integrands of both double integrals are 0. Therefore

$$\int_C f\,dz = \int_C u\,dx - v\,dy \; + \; i\int_C v\,dx + u\,dy = 0\,. \; \blacksquare$$

EXAMPLE 7.6.1 Let C be any closed contour and $p(z)$ be any polynomial. Then $p(z)$ is analytic. The real and imaginary parts of $p(z)$ are just polynomials in x and y so their partials exist and are continuous. The Cauchy integral theorem applies, so

$$\int_{C_r} p(z)\,dz = 0. \; \square$$

EXAMPLE 7.6.2 By contrast, the function $f(z) = (z - z_0)^{-1}$ is not analytic at z_0. Let C_ρ be the circle of radius ρ centered at $z_0 = x_0 + iy_0$ that is parameterized by $z(\theta) = z_0 + \rho e^{i\theta}$ for $\theta \in [0, 2\pi]$. For a general point $z(\theta)$ on C_ρ, $z(\theta) - z_0 = \rho e^{i\theta}$ and $z'(\theta) = i\rho e^{i\theta}$. Thus

$$\int_{C_\rho} (z - z_0)^{-1} dz = \int_0^{2\pi} (\rho e^{i\theta})^{-1} i\rho e^{i\theta}\, d\theta = \int_0^{2\pi} i\, d\theta = 2\pi i. \quad \square$$

EXAMPLE 7.6.3 Even if a function is *not* analytic, an integral around a closed curve may *still* be 0. Let $f(z) = z^{-2}$, which is not analytic at the origin. Let C_r be the circle of radius r centered at the origin. Though $f(z)$ is not analytic at the origin, it is analytic in the domain $\mathbf{C} - \{0\}$, which contains C_r. Since $F(z) = -z^{-1}$ is an antiderivative of $f(z) = z^{-1}$ in this domain, Corollary 7.5.5 applies and we have $\int_{C_r} z^{-2}\, dz = 0$. $\square$

Cauchy proved a version of the integral theorem in 1822 for rectangular contours. In 1825 he generalized the result, and he did not assume the continuity of $f'(z)$, though he used it in his proof.[1] In 1900 Edouard Goursat was able to give a proof of the Cauchy integral theorem that did not make use of the continuity assumption on $f'(z)$.

THEOREM 7.6.2 ***(The Cauchy-Goursat Integral Theorem)*** *If $f(z)$ is analytic in a simply connected domain D, then for any closed contour C within D*

$$\int_C f(z)\, dz = 0\,.$$

The proof is beyond the scope of this text, but the interested reader will find it in most elementary texts on complex analysis. The Cauchy-Goursat theorem generalizes the Cauchy integral theorem in two ways. First, as was mentioned previously, the continuity condition on $f'(z)$ has been dropped. Second, the closed contour C need not be simple. The importance of the relaxation of both of these restrictions will be evident as we explore a few of the consequences of the Cauchy-Goursat theorem. In light of Theorem 6.5.3 we should anticipate the following result:

---

[1] Kline, *Mathematical Thought*, p. 637. Kline also notes that during Cauchy's time continuous functions were thought to be differentiable and that the derivative could be discontinuous only where the function itself was discontinuous. So for Cauchy, assuming that f was continuous implied that f' was continuous.

COROLLARY 7.6.3 *Assume that $f(z)$ is analytic in a simply connected domain D. If C_1 and C_2 are two contours within D that have the same initial point and the same terminal point, then*

$$\int_{C_1} f(z)\,dz = \int_{C_2} f(z)\,dz.$$

PROOF Let z_0 denote the initial point of C_1 and C_2 and let z_1 denote the terminal point. Let C be the *closed* contour obtained by going from z_0 to z_1 via C_1 and then back to z_0 along C_2^-. By the Cauchy-Goursat theorem

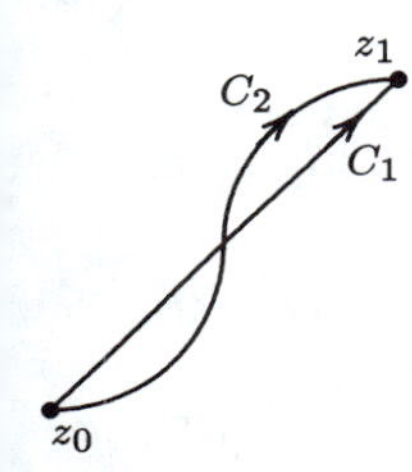

Figure 7.6.1

$$\begin{aligned} 0 = \int_C f(z)\,dz = \int_{C_1+C_2^-} f(z)\,dz &= \int_{C_1} f(z)\,dz + \int_{C_2^-} f(z)\,dz \\ &= \int_{C_1} f(z)\,dz - \int_{C_2} f(z)\,dz. \end{aligned}$$

This implies that $\int_{C_1} f(z)\,dz = \int_{C_2} f(z)\,dz$. ∎

It is clear that the closed contour C constructed in the proof of the corollary need *not* be simple, since C_1 and C_2 may cross. Thus the Cauchy integral theorem is not sufficient to prove this last result.

REMARK When an integral is independent of path C from z_0 to z_1 we will sometimes write $\int_{z_0}^{z_1} f(z)\,dz$ instead of $\int_C f(z)\,dz$.

COROLLARY 7.6.4 *If f is analytic in a simply connected domain D, then f has an antiderivative in D.*

PROOF The proof generalizes the ideas of the second fundamental theorem of calculus and is very similar to the proof of Theorem 6.5.2 (which you might wish to review at this point). Let z_0 be any point in D. Define a function F on D as follows. For a point z_1 in D, let C be any contour within D that has z_0 as its initial point and z_1 as its terminal point. Set

$$F(z_1) = \int_C f(w)\,dw.$$

Note that w is the variable of integration. F is well defined because by Corollary 7.6.3 the integral from z_0 to z_1 is independent of path (in D).

We now show that F is an antiderivative of f on D. Showing that $F'(z_1) = f(z_1)$ is equivalent to showing that

$$\lim_{z\to z_1} \left| \frac{F(z)-F(z_1)}{z-z_1} - f(z_1) \right| = 0\,. \qquad (1)$$

Therefore choose $\epsilon > 0$. Because $z_1 \in D$ and D is open, if we choose z sufficiently close to z_1, the straight line segment from z_1 to z will lie entirely within D. Denote this segment by S.

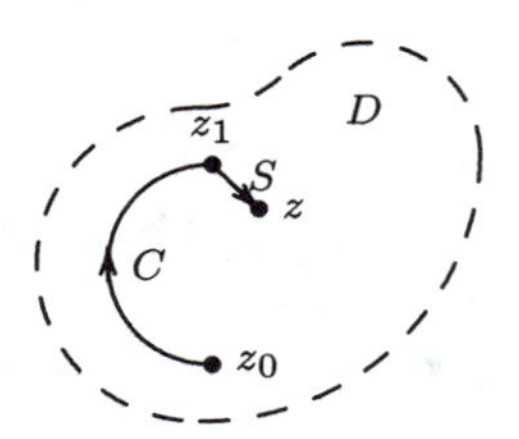

S lies entirely within D.

Figure 7.6.2

From the definition of F, we can evaluate $F(z)$ by using the contour $C + S$ that has initial point z_0, goes to z_1 via C, and then to z along S.

$$F(z) - F(z_1) = \int_{C+S} f(w)\, dw - \int_C f(w)\, dw = \int_S f(w)\, dw = \int_{z_1}^{z} f(w)\, dw. \tag{2}$$

We now express $f(z_1)$ in terms of a similar integral. Noting that $f(z_1)$ is a constant with respect to integration by w and that $z - z_1 = \int_{z_1}^{z} dw$ we have

$$f(z_1) = \frac{f(z_1)}{z - z_1} \int_{z_1}^{z} dw = \frac{1}{z - z_1} \int_{z_1}^{z} f(z_1)\, dw. \tag{3}$$

The expression in (1) can now be translated in to a single integral as follows using (2) and (3):

$$\begin{aligned} \frac{F(z) - F(z_1)}{z - z_1} - f(z_1) &= \frac{1}{z - z_1} \int_{z_1}^{z} f(w)\, dw - \frac{1}{z - z_1} \int_{z_1}^{z} f(z_1)\, dw \\ &= \frac{1}{z - z_1} \int_{z_1}^{z} \big(f(w) - f(z_1)\big)\, dw. \end{aligned} \tag{4}$$

Since f is continuous (analytic) in D, there is a $\delta > 0$ such that if $|w - z_1| < \delta$, then $|f(w) - f(z_1)| < \epsilon$. Consequently, if $|z - z_1| < \delta$, then for any point w on the segment S from z_1 to z we have $|f(w) - f(z_1)| < \epsilon$. The length of the segment S is $|z - z_1|$, so Theorem 7.5.3 and (4) imply that

$$\begin{aligned} \left| \frac{F(z) - F(z_1)}{z - z_1} - f(z_1) \right| &= \left| \frac{1}{z - z_1} \int_{z_1}^{z} \big(f(w) - f(z_1)\big)\, dw \right| \\ &< \frac{1}{|z - z_1|} \cdot \epsilon \cdot |z - z_1| = \epsilon. \end{aligned}$$

Since ϵ was arbitrary, we have

$$\lim_{z \to z_1} \left| \frac{F(z) - F(z_1)}{z - z_1} - f(z_1) \right| = 0,$$

so $F'(z_1) = f(z_1)$ at each point $z_1 \in D$. ∎

The Cauchy-Goursat theorem can be extended to regions that are **multiply connected**, that is, regions that are not simply connected and have holes in them (see Section 6.4.3). In this brief introduction to complex analysis, consideration of general multiply connected regions will not play a major role. However, the following simple example illustrates the ideas involved in the more general situation.

EXAMPLE 7.6.4 Suppose that g is analytic in a simply connected domain D *except* at the point z_0. Let C be any simple closed contour within D that encloses z_0. Let C_r be a circle of radius r centered at z_0 with r chosen sufficiently small so that C_r lies entirely within C. If both contours have counterclockwise orientations,

$$\int_C g(z)\,dz = \int_{C_r} g(z)\,dz.$$

SOLUTION Notice that g is analytic in the region between C_r and C, but this region is not simply connected (see Figure 7.6.3). To apply the Cauchy-Goursat theorem we must rectify this. To do this, introduce an additional boundary curve S from C_r to C of the type shown in Figure 7.6.4. Now form the *closed* contour $\Gamma = C + S^- + C_r^- + S$. Note the orientation of the individual curves (see Figure 7.6.5).

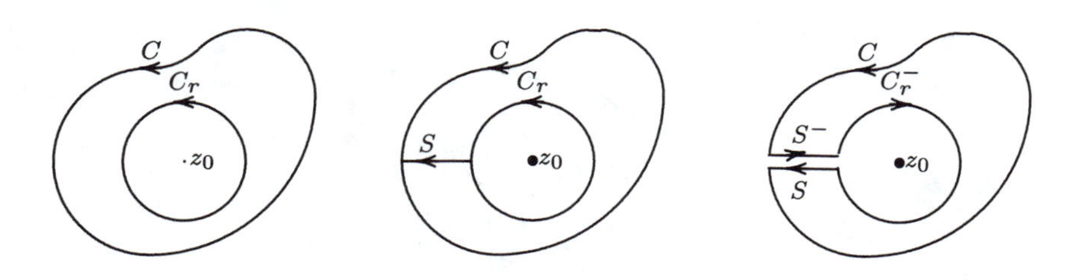

Figure 7.6.3 **Figure 7.6.4** **Figure 7.6.5**

The region determined by Γ and its interior is simply connected, so the Cauchy-Goursat theorem applies,

$$\begin{aligned} 0 = \int_\Gamma g(z)\,dz &= \int_C g(z)\,dz - \int_S g(z)\,dz - \int_{C_r} g(z)\,dz + \int_S g(z)\,dz \\ &= \int_C g(z)\,dz - \int_{C_r} g(z)\,dz. \end{aligned}$$

From this equation it follows that

$$\int_{C_r} g(z)\,dz = \int_C g(z)\,dz. \quad \square$$

7.6.2 The Cauchy Integral Formula

One of the most important uses of the Cauchy-Goursat theorem is to evaluate an analytic function at a point by using an appropriate contour integral.

THEOREM 7.6.5 ***(The Cauchy Integral Formula)*** *Let f be analytic in a simply connected domain D. Let C be a simple closed contour within D with counterclockwise orientation. If z_0 is any point within C, then*

$$f(z_0) = \frac{1}{2\pi i}\int_C \frac{f(z)}{z - z_0}\,dz.$$

PROOF Since f is analytic within D, the integrand is analytic within D *except* at $z = z_0$. We use the same technique as in Example 7.6.4 to create a closed contour within which $f(z)/(z - z_0)$ is analytic so that we may apply the Cauchy-Goursat theorem. Choose a sufficiently small positive number ρ so that the circle of radius ρ centered at z_0 with counterclockwise orientation lies within C. Then $f(z)/(z - z_0)$ is analytic in the region between C and C_ρ.

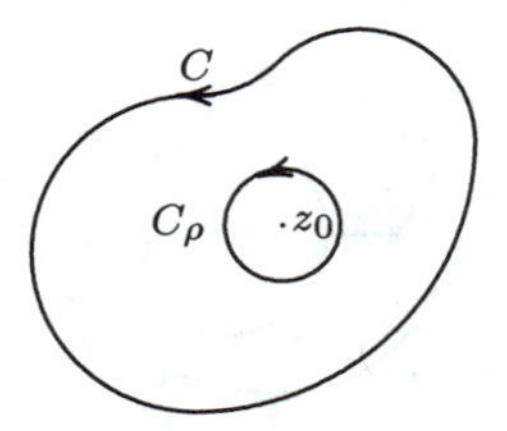

Figure 7.6.6

From Example 7.6.4 we know that

$$\int_C \frac{f(z)}{z - z_0}\,dz = \int_{C_\rho} \frac{f(z)}{z - z_0}\,dz. \tag{5}$$

Next, from Example 7.6.2, we know that $\int_{C_\rho} \frac{1}{z-z_0}\,dz = 2\pi i$, so we can write the number $f(z_0)$ as

$$f(z_0) = \frac{1}{2\pi i}\int_{C_\rho} \frac{f(z_0)}{z - z_0}\,dz. \tag{6}$$

Dividing (5) by $2\pi i$ and subtracting (6) gives

$$\begin{aligned}\frac{1}{2\pi i}\int_C \frac{f(z)}{z-z_0}\,dz - f(z_0) &= \frac{1}{2\pi i}\int_{C_\rho}\frac{f(z)}{z-z_0}\,dz - \frac{1}{2\pi i}\int_{C_\rho}\frac{f(z_0)}{z-z_0}\,dz \\ &= \frac{1}{2\pi i}\int_{C_\rho}\frac{f(z)-f(z_0)}{z-z_0}\,dz. \qquad (7)\end{aligned}$$

Notice that the left side of (7) is independent of ρ, so the right side must also be independent of ρ. That is, the value of the last integral is constant for all sufficiently small ρ. If we can show that this constant is 0, then (7) will complete the proof.

Choose any $\epsilon > 0$. Since f is continuous, there exists a $\delta > 0$ such that if $|z - z_0| < \delta$, then $|f(z) - f(z_0)| < \epsilon$. Take $\rho < \delta$. If $z \in C_\rho$, then $|z - z_0| = \rho < \delta$, so $|f(z) - f(z_0)| < \epsilon$. Since the length of C_ρ is $2\pi\rho$, Theorem 7.5.3 implies that

$$\left|\frac{1}{2\pi i}\int_{C_\rho}\frac{f(z)-f(z_0)}{z-z_0}\,dz\right| \le \frac{1}{2\pi}\cdot\frac{\epsilon}{\rho}\cdot 2\pi\rho = \epsilon.$$

But ϵ is an arbitrary positive number, and we have already remarked that this integral is independent of ρ. Therefore we must in fact have

$$\frac{1}{2\pi i}\int_{C_\rho}\frac{f(z)-f(z_0)}{z-z_0}\,dz = 0\,.$$

This last observation combined with (7) gives

$$f(z_0) = \frac{1}{2\pi i}\int_C \frac{f(z)}{z-z_0}\,dz. \quad \blacksquare$$

The Cauchy integral formula further illustrates how special analytic functions are. If f is analytic within and on a circle C, the Cauchy integral formula says that the values of f at any point *within* the circle are completely determined by the values of f *on* the circle. (An analogous statement for functions of a single real variable would be that knowing the values of f at the endpoints of a closed interval $[a, b]$ completely determines the function within the interval!)

The Cauchy integral formula can be used to evaluate certain complicated integrals, as the next example illustrates.

EXAMPLE 7.6.5 Let C be the circle of radius 2 centered at i with counterclockwise orientation. We evaluate

$$\int_C \frac{2z^4 - z^3 + 5z + 6i}{z^2+4}\,dz\,.$$

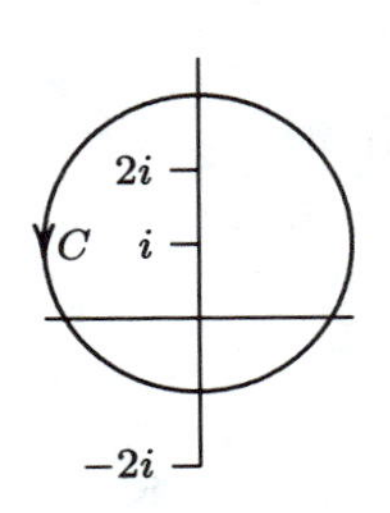

Figure 7.6.7

The integrand is not analytic at $z = \pm 2i$. However, $-2i$ lies outside C. Thus the function $f(z) = \frac{z^4 - z^3 + 5z + 6i}{z + 2i}$ is analytic within and on C. Therefore, by the Cauchy integral theorem

$$\int_C \frac{2z^4 - z^3 + 5z + 6i}{z^2 + 4}\, dz = \int_C \frac{\frac{2z^4 - z^3 + 5z + 6i}{z + 2i}}{z - 2i}\, dz$$
$$= \int_C \frac{f(z)}{z - 2i}\, dz = f(2i) = 6 - 4i. \quad \square$$

Problems for Section 7.6

7.6.1. Let C be the unit circle centered at the origin. Show that $\int_C (\overline{z})^2\, dz \neq 0$. Why doesn't the Cauchy integral theorem apply?

7.6.2. Let C_1 be the line segment from 0 to $2 + 3i$. Let C_2 be the path consisting of the segment from 0 to 2 followed by the segment from 2 to $2 + 3i$. Show that

$$\int_{C_1} \overline{z}\, dz \neq \int_{C_2} \overline{z}\, dz.$$

7.6.3. Let C be the path from 0 to $3+2i$ parameterized by $z(t) = 3t^3 + i(1 - \cos \pi t)$ for $0 \leq t \leq 1$. Evaluate $\int_C 3z^2 + 2z + i\, dz$ first by using the given parameterization, then do it another way.

7.6.4. Let C be the circle of radius 2 centered at the origin. Evaluate each of the following integrals:

$$\text{a)} \int_C \frac{z^2 + 6z}{z - i}\, dz \qquad \text{b)} \int_C \frac{z^3 + 5z + 9}{z^2 - 4z + 3}\, dz \qquad \text{c)} \int_C \frac{z^3 - 7z + 11}{z^2 - 5iz - 4}\, dz$$

7.6.5. Assume that f is analytic inside and on the simple closed contour C. If z_0 lies outside C, evaluate

$$\int_C \frac{f(z)}{z - z_0}\, dz.$$

7.6.6. Let C be any square centered at z_0. Show that if n is any integer, then

$$\int_C z^n\, dz = \begin{cases} 0, & \text{if } n \neq -1 \\ 2\pi i, & \text{if } n = -1. \end{cases}$$

(Hint: Use the Cauchy-Goursat theorem and problem 7.5.6.)

7.6.7. Let C be the positively oriented triangle whose vertices are $1 + i$, $2 + 3i$, and $5i$. Evaluate $\int_C (z - z_0)^{-1}\, dz$, where z_0 is any point interior to C.

7.6.8. **Gauss's Mean Value Theorem:** *Let f be analytic within and on a circle C with radius r and center z_0. Parameterize C by $z(\theta) = re^{i\rho e^{i\theta}}$. Then*

$$f(z_0) = \frac{1}{2\pi}\int_0^{2\pi} f(z(\theta))\, d\theta.$$

Prove this result by using the Cauchy integral formula.

7.6.9. **Morera's theorem** is a type of converse to the Cauchy Integral Theorem: *Let $f = u + iv$ be a function in a simply connected domain D such that the first-order partial derivatives of u and v exist and are continuous. Assume that for every closed curve C within D that $\int_C f(z)\, dz = 0$. Then f is analytic within D.*

a) Use a proof by contradiction. If f is not analytic show that there is some point a within D such that $u_x - v_y \neq 0$ or $v_x + u_y \neq 0$.

b) Show, using continuity, that there is a $\delta > 0$ such that $u_x - v_y \neq 0$ or $v_x + u_y \neq 0$ for any point z such that $|z - a| \leq \delta$.

c) If C is the circle $|z - a| = \delta$ oriented positively, use Green's theorem to show that $\int_C f(z)\, dz \neq 0$. (Hint: $\int_C f\, dz = \int_C u\, dx - v\, dy \ + \ i \int_C v\, dx + u\, dy$. Make sure that you show that at least one of these integrands does not change sign on C.)

7.6.10. Every nonzero complex number has two distinct square roots. This leads to what are called the two **branches** of the square root function $z^{1/2}$. This is often the starting point of the study of **Riemann surfaces** in a course devoted entirely to complex analysis. What is clear at this point is that if $z \neq 0$ and $r = |z|$ and $\theta = \operatorname{Arg} z$, then $z = r(\cos\theta + i\sin\theta)$, so $\sqrt{r}e^{i\theta/2}$ is one of the square roots of z. Using this branch of the square root, show that $z^{1/2}$ is not analytic at every point inside the unit circle C. Do this by showing that $\int_C z^{1/2}\, dz \neq 0$.

7.7 The Fundamental Theorem of Algebra

Much effort in elementary algebra courses is spent on finding roots of polynomials. The quadratic formula completely solves the problem for degree 2 polynomials. You may even be vaguely familiar with techniques that help in calculating the roots of cubic polynomials. Beyond degree 3 the problem becomes increasingly difficult, in complete disproportion to the usual view that polynomials are quite simple.

In this section we will show that every nonconstant polynomial with complex coefficients (and consequently any polynomial with real coefficients) has at least one complex root. It will then follow that any nth degree complex (or real) polynomial has n *complex* roots, being careful to count multiplicities. We are aware that this does not hold if we confine ourselves to *real* roots of *real* polynomials. Even quite simple quadratic polynomials with real coefficients often have *no* real roots. For example, we saw that one of the real polynomials studied by Cardan, $p(x) = x^2 - 10x - 40$, has no real roots. However, the fundamental theorem of algebra guarantees that $p(x)$ has a complex root. In fact, we found that $z = 5 + \sqrt{15}i$ and $\overline{z} = 5 - \sqrt{15}i$ are roots of p. (The fact that the complex roots to this real polynomial appeared in a conjugate pair is no accident; see problem 1.)

In 1799 Gauss gave the first fairly complete proof of the fundamental theorem of algebra. Mathematicians had struggled with the problem for over three centuries. The work that led to proving the fundamental theorem of algebra had a profound influence on many fields in mathematics. For us the fundamental theorem of algebra represents a culmination of a number of ideas. The text began with the problem of the "incompleteness" of the rational numbers, and in Chapter 1 the reals were constructed as a way to topologically complete the rational numbers. Now we are confronted with the algebraic incompleteness of the real number field. That is, certain polynomials with real coefficients that no roots in the real number system. This algebraic incompleteness is remedied by extending the real number system to the complex number system through the "adjoining of i"; any complex number is written as $a + bi$ where a and b are real. The fundamental theorem of algebra says that not only can we now solve any quadratic equation, but we can find all n roots of a general nth degree polynomial with complex coefficients. That is, the complex numbers are **algebraically closed**, whereas the reals are not. The price for this algebraic completeness is a loss of ordering; **C** is not ordered, whereas **R** is (see problem 7.1.2).

There are a number of different proofs of the fundamental theorem of algebra.[1] The following proof makes use of the Cauchy integral forumula:

THEOREM 7.7.1 ***(The Fundamental Theorem of Algebra)*** *Let $p(z)$ be a nonconstant polynomial*

$$p(z) = a_n z^n + \cdots + a_1 z + a_0,$$

where $n \geq 1$ and $a_n \neq 0$. Then there is at least one complex number z_0 that is a root of $p(z)$, that is, $p(z_0) = 0$.

PROOF The proof is by contradiction. Assume that $p(z)$ is never 0. The polynomial $p(z)$ is analytic with continuous partials. Since $p(z)$ is never 0, the rational function $1/p(z)$ is well defined and analytic with continuous partials. Thus $1/p(z)$ satisfies the conditions of the Cauchy integral formula on all of **C**. The remainder of the proof is devoted to showing that

[1] There are at least a half dozen proofs given in the text by Levinson and Redheffer listed in the Bibliography.

$1/p(z)$ is constant and hence that $p(z)$ is constant, which contradicts the hypothesis.

The first step is to show that $|1/p(z)|$ is small when $|z|$ is large. By using the triangle inequality twice,

$$\begin{aligned}|p(z)| &= |a_n z^n + a_{n-1}z^{n-1} + \cdots + a_1 z + a_0| \\ &\geq |a_n z^n| - |a_{n-1}z^{n-1} + \cdots + a_1 z + a_0| \\ &\geq |a_n z^n| - (|a_{n-1}z^{n-1}| + \cdots + |a_1 z| + |a_0|) \\ &= |a_n z^n| - |a_{n-1}z^{n-1}| - \cdots - |a_1 z| - |a_0|.\end{aligned}$$

In particular, on the circle C_r defined by $|z| = r$,

$$|p(z)| \geq |a_n|r^n - |a_{n-1}|r^{n-1} - \cdots - |a_1|r - |a_0|.$$

The right side of this inequality is a real polynomial in r. By problem 2.4.9, there is a real number R such that if $r > R$, then $|p(z)| > |a_n|/2$. Consequently

$$\left|\frac{1}{p(z)}\right| < \frac{2}{|a_n|} \qquad (r > R). \tag{1}$$

Now we use the Cauchy integral theorem to show that $1/p(z)$ is constant. Fix an arbitrary point z_0. If r is sufficiently large so that C_r encloses z_0, then

$$\begin{aligned}\frac{1}{p(z_0)} - \frac{1}{p(0)} &= \frac{1}{2\pi i}\int_{C_r} \frac{1/p(z)}{z - z_0}\,dz - \frac{1}{2\pi i}\int_{C_r} \frac{1/p(z)}{z - 0}\,dz \\ &= \frac{1}{2\pi i}\int_{C_r} \left(\frac{1}{z - z_0} - \frac{1}{z}\right)\frac{1}{p(z)}\,dz.\end{aligned}$$

We need to show that this last integral is small, so we now determine a bound for the integrand. Using the triangle inequality and the fact that $|z| = r$ on C_r,

$$\left|\frac{1}{z - z_0} - \frac{1}{z}\right| = \frac{|z_0|}{|z(z - z_0)|} \geq \frac{|z_0|}{|z|(|z| - |z_0|)} = \frac{|z_0|}{r(r - |z_0|)}.$$

From (1), if $r > R$, then $|1/p(z)| < 2/|a_n|$. Consequently by Theorem 7.5.3,

$$\begin{aligned}\left|\frac{1}{p(z_0)} - \frac{1}{p(0)}\right| &= \left|\frac{1}{2\pi i}\int_{C_r} \left(\frac{1}{z - z_0} - \frac{1}{z}\right)\frac{1}{p(z)}\,dz\right| \\ &\leq \frac{1}{2\pi} \cdot \frac{|z_0|}{r(r - |z_0|)} \cdot \frac{2}{|a_n|} \cdot 2\pi r \leq \frac{2|z_0|}{|a_n|(r - |z_0|)}.\end{aligned}$$

Thus for a fixed z_0, the difference $\left|\frac{1}{p(z_0)} - \frac{1}{p(0)}\right|$ is bounded above by the quantity $\frac{2|z_0|}{|a_n|(r-|z_0|)}$, for any $r > R$. Therefore

$$0 \leq \left|\frac{1}{p(z_0)} - \frac{1}{p(0)}\right| \leq \lim_{r\to\infty} \frac{2|z_0|}{|a_n|(r-|z_0|)} = 0.$$

It follows that $p(z_0) = p(0)$. Since z_0 was arbitrary, $p(z)$ must be a constant polynomial. But this contradicts the hypotheses. ∎

Nowhere in the proof of the fundamental theorem of algebra is any indication given of where to look for the roots of $p(z)$. We have not developed an algorithm for finding the roots of p (like the quadratic formula). The proof merely guarantees the existence of a root.

> Gauss's approach to the fundamental theorem of algebra inaugurated a new approach to the entire question of mathematical existence. The Greeks had wisely recognized that the existence of mathematical entities must be established before theorems about them can be entertained. Their criterion of existence was constructibility. In the more formal work of the succeeding centuries, existence was established by actually obtaining or exhibiting the quantity in question. For example, the existence of the solutions of a quadratic equation is established by exhibiting quantities that satisfy the equation. But in the case of equations of degree higher than four, this method is not available. Of course a proof of existence such as Gauss's may be of no help at all in computing the object whose existence is being established.[2]

Such existence proofs have become increasingly common in mathematics in the last hundred years.

Problems for Section 7.7

7.7.1. Let $P(z) = a_n z^n + \cdots + a_1 z + a_0$ be a degree n complex polynomial. Let $\overline{P}(z)$ denote the polynomial whose coefficients are the conjugates of the coefficients of P, that is

$$\overline{P}(z) = \overline{a}_n z^n + \cdots + \overline{a}_1 z + \overline{a}_0.$$

a) Show that $\overline{P(z)} = \overline{P}(\overline{z})$.

b) Show that α is a root of $P(z) \iff \overline{\alpha}$ is a root of $\overline{P}(z)$.

c) Suppose now that the coeficients of P are *real numbers*. Show that α is a root of $P(z) \iff \overline{\alpha}$ is a root of $P(z)$.

---

[2] Kline, *Mathematical Thought*, p. 599.

d) Conclude that if $P(x) = a_n x^n + \cdots + a_1 x + a_0$ is a real polynomial and α is a root of P, then either (i) α is a real number; or (ii) α and $\overline{\alpha}$ are both roots of P, that is, complex roots of real polynomials come in conjugate pairs.

7.7.2. **a)** Show that if $p(z)$ is a polynomial of degree n and $p(\alpha) = 0$, then we may write $p(z) = (z - \alpha)q(z)$, where $q(z)$ is a polynomial of degree $n - 1$. (Hint: Rewrite $p(z)$ using the fact that $p(z) = p(z) - p(\alpha)$ and that for all $k \in \mathbf{N}$

$$z^k - \alpha^k = (z - \alpha)(z^{k-1} + z^{k-2}\alpha + \cdots + z\alpha^{k-2} + \alpha^{k-1}).)$$

b) Show that if $p(z)$ is a polynomial of degree n and $\alpha_1, \alpha_2, \ldots, \alpha_n$ are n distinct points such that $p(\alpha_i) = 0$, then

$$p(z) = c(z - \alpha_1)(z - \alpha_2)\cdots(z - \alpha_n)$$

for some constant c.

c) Show that if $p(z)$ and $q(z)$ are polynomials of degree at most n that agree at $n + 1$ distinct points, then $p(z) = q(z)$.

7.7.3. Show by using the previous problem, the fundamental theorem of algebra, and induction that any complex polynomial of degree n factors completely into n linear factors over $\mathbf{C}$.

7.7.4. Let $p(z)$ be a polynomial. Then α is called a **simple root** of $p(z)$ if $(z - \alpha)$ divides (is a factor of) $p(z)$ but $(z - \alpha)^2$ does not divide $p(z)$. More generally, if $p(z)$ is divisible by $(z - \alpha)^k$ but not by $(z - \alpha)^{k+1}$, then α is a **root of multiplicity** k. When $k > 1$, α is called a **multiple root** of $p(z)$.

a) Show that if α is a simple root of $p(z)$, then α is not a root of $p'(z)$. (Hint: Write $p(z)$ as $(z - \alpha)q(z)$, where α is not a root of the polynomial $q(z)$. Now use the product rule for derivatives.)

b) Show that if α is a multiple root of $p(z)$, then α is a root of $p'(z)$. (In fact, show that if the multiplicty of α for $p(z)$ is k, then the multiplicity for α in $p'(z)$ is $k - 1$. Part **(a)** can then be subsumed in this more general statement.)

7.7.5. Show that every real polynomial can be factored using real numbers into a product of linear factors or irreducible real quadratic factors, or both. For example,

$$x^4 - 1 = (x - 1)(x + 1)(x^2 + 1).$$

Here, the quadratic factor $x^2 + 1$ is irreducible over $\mathbf{R}$. (Hint: Use Problems 1d and 2b.)

7.8 Consequences of the Cauchy Integral Formula

The Cauchy integral formula is one of the most basic tools in complex function theory. We have already seen that the fundamental theorem of algebra is one of its consequences. In this section we briefly describe a number of results that are within reach using the integral formula.[1]

To begin assume that $f(z)$ is analytic within and on a simple closed curve C with counterclockwise orientation. Let w be any point within C. Then by the Cauchy integral formula

$$f(w) = \frac{1}{2\pi i}\int_C \frac{f(z)}{z-w}\,dz. \tag{1}$$

Since f is analytic within C, $f'(w)$ exists. We can obtain a formula for $f'(w)$ by simply "differentiating under the integral sign" with respect to w in (1). This yields

$$f'(w) = \frac{1}{2\pi i}\int_C \frac{f(z)}{(z-w)^2}\,dz. \tag{2}$$

In general, one cannot simply differentiate an integral, as we have just done, since it involves interchanging the order of certain limit processes. Yet one can show in this case that (2) is, indeed, valid. In fact, this idea may be used again. Differentiating (2) with respect to w yields

$$f''(w) = \frac{2!}{2\pi i}\int_C \frac{f(z)}{(z-w)^3}\,dz, \tag{3}$$

so that f'' exists within C. We now see that *if f is analytic within and on C (i.e., if we assume f has a first derivative within and on C), then f'' exists (i.e., f has a second derivative there).* Now we can apply this result to f'. We know that it has a first derivative (f'' exists), so f' has a second derivative within C (f''' exists). Continuing in this fashion we obtain the following result.

Theorem 7.8.1 *If f is analytic within and on the simple closed contour C, then f has derivatives of all orders. Further at any point z_0 within C,*

$$f^{(n)}(z_0) = \frac{n!}{2\pi i}\int_C \frac{f(z)}{(z-z_0)^{n+1}}\,dz. \tag{4}$$

[1] The complete proofs of any of these results may be found in most texts devoted to complex analysis.

Next, let C_r denote the circle $|z - z_0| = r$ centered at z_0 with counterclockwise orientation. Again by the Cauchy integral formula,

$$f(z_0) = \frac{1}{2\pi i} \int_{C_r} \frac{f(z)}{z - z_0}\, dz. \tag{5}$$

Let M be the maximum value of $|f(z)|$ on C_r. Since $|z - z_0| = r$ on C_r, Theorem 7.5.3 implies

$$|f(z_0)| = \left| \frac{1}{2\pi i} \int_{C_r} \frac{f(z)}{z - z_0}\, dz \right| \leq \frac{1}{2\pi} \cdot \frac{M}{r} \cdot 2\pi r = M\,. \tag{6}$$

Now (6) says that the modulus of f *within* C_r cannot exceed the maximum modulus *on* C_r itself. This result can be improved, using different techniques, to make the inequality strict when f is nonconstant.

THEOREM 7.8.2 ***(The Maximum Principle)*** *Assume that $f(z)$ is analytic within and on a simple closed contour C. If M is the maximum value of $|f(z)|$ on C, then $M > |f(z)|$ for all points within C unless f is a constant function (in which case $|f(z)| = M$).*

In short, a nonconstant analytic function has its maximum modulus *on* the boundary curve C, never within the curve.

Again let C_r denote the counterclockwise oriented circle $|z - z_0| = r$ and assume that f is analytic within and on C. Then from Theorem 7.8.1

$$f^{(n)}(z_0) = \frac{n!}{2\pi i} \int_{C_r} \frac{f(z)}{(z - z_0)^{n+1}}\, dz. \tag{7}$$

If M is the maximum value of $|f(z)|$ on C_r, arguing as in (6) we obtain from (7) that

$$|f^{(n)}(z_0)| \leq \frac{n!M}{r^n}\,. \tag{8}$$

These are called the **Cauchy estimates** of $|f^{(n)}(z_0)|$. One of their most important consequences is the following result:

THEOREM 7.8.3 ***(Liouville's Theorem)*** *Assume that $f(z)$ is analytic on the entire complex plane. If $|f(z)|$ is bounded for all z, then f is a constant function.*

PROOF Let M be a bound for $|f(z)|$. Letting $n = 1$ in (8) we obtain

$$|f'(z_0)| \leq \frac{M}{r}\,. \tag{9}$$

Since f is analytic throughout the entire complex plane, (9) is valid for any point z_0 and any radius r. Letting $r \to \infty$ we see that $|f'(z_0)| = 0$, so $f'(z_0) = 0$ at every point in the plane. From this it follows that $f(z)$ is a constant function (problem 7.4.4). ∎

Example 7.8.1 Finally, we use the Cauchy integral formula to illustrate how certain real integrals can be evaluated using contour integrals. Let's evaluate the improper real integral

$$\int_0^{\infty} \frac{1}{x^2+1}\, dx.$$

SOLUTION Because the integrand is an even function, we can rewrite it as

$$\int_0^{\infty} \frac{1}{x^2+1}\, dx = \frac{1}{2}\int_{-\infty}^{+\infty} \frac{1}{x^2+1}\, dx = \frac{1}{2}\lim_{r\to\infty}\int_{-r}^{r} \frac{1}{x^2+1}\, dx. \tag{10}$$

We can think of the interval $[-r, r]$ as a straight line path I_r from $-r$ to r along the real axis in the complex plane. To use the Cauchy integral formula, we will need a closed contour. If we let S_r denote the upper half of the circle $|z| = r$ traversed in the counterclockwise sense, then $\Gamma_r = I_r + S_r$ is a simple closed contour with counterclockwise orientation.

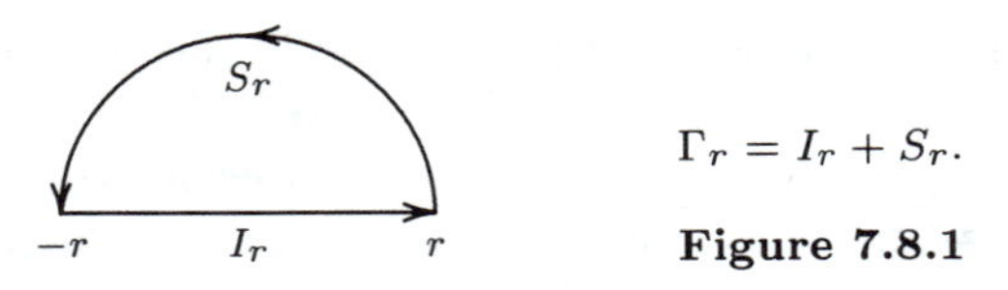

Figure 7.8.1

Now consider the complex valued function $g(z) = \frac{1}{z^2+1}$. The final integral in (10) may be rewritten as

$$\begin{aligned}\int_{-r}^{r} \frac{1}{x^2+1}\, dx &= \int_{I_r} g(x)\, dx = \int_{I_r+S_r} g(z)\, dz - \int_{S_r} g(z)\, dz \\ &= \int_{\Gamma_r} g(z)\, dz - \int_{S_r} g(z)\, dz. \end{aligned} \tag{11}$$

The last two integrals in (11) are easily evaluated. Since we will eventually want to let $r \to \infty$, we may assume that $r > 1$. Notice that

$$g(z) = \frac{1}{z^2+1} = \frac{1}{(z-i)(z+i)} = \frac{1/(z+i)}{z-i}.$$

The function $f(z) = \frac{1}{(z+i)}$ is analytic within Γ_r. Since i is inside Γ_r, we may apply the Cauchy integral formula:

$$\int_{\Gamma_r} g(z)\, dz = \int_{\Gamma_r} \frac{f(z)}{z-i}\, dz = 2\pi i f(i) = \frac{2\pi i}{2i} = \pi.$$

Next evaluate $\int_{S_r} g(z)\,dz$. On S_r we have $|z^2+1| = |z-i||z+i| \geq r(r-1)$, so

$$\frac{1}{|z^2+1|} \leq \frac{1}{r(r-1)}\,.$$

Since the arc length of S_r is πr, by Theorem 7.5.3,

$$\left|\int_{S_r} g(z)\,dz\right| \leq \frac{1}{r(r-1)}\pi r = \frac{\pi}{r-1}\,.$$

Consequently $\lim_{r\to\infty} \left|\int_{S_r} g(z)\,dz\right| = 0$, and

$$\lim_{r\to\infty} \int_{S_r} g(z)\,dz = 0\,. \tag{12}$$

Combining all of our results, we have

$$\begin{aligned}
\int_0^\infty \frac{1}{x^2+1}\,dx &= \frac{1}{2}\lim_{r\to\infty}\int_{-r}^{r} \frac{1}{x^2+1}\,dx \\
&= \frac{1}{2}\lim_{r\to\infty}\left(\int_{\Gamma_r} g(z)\,dz - \int_{S_r} g(z)\,dz\right) \\
&= \frac{1}{2}(\pi - 0) \\
&= \pi/2\,. \quad \square
\end{aligned}$$

Of course this particular integral could have been evaluated using $\tan^{-1} x$, but for other integrals of this type there may be no simple real antiderivative available. The general technique just described then becomes quite useful because nowhere in the process did we actually use an antiderivative to evaluate any of the particular contour integrals!

Problems for Section 7.8

7.8.1. Show that if f is analytic within and on the simple closed contour C, then for any point z_0 inside C

$$\int_C \frac{f'(z)}{z-z_0}\,dz = \int_C \frac{f(z)}{(z-z_0)^2}\,dz.$$

7.8.2. Prove the fundamental theorem of algebra using Liouville's theorem. (Hint: If the polynomial P has no roots, apply Liouville's theorem to $1/P(z)$.)

7.8.3. **a)** Assume that $f(z)$ is nonconstant and analytic within and on a simple closed contour C. Assume further that $f(z) \neq 0$ within or on C. Let m be the minimum value of $|f(z)|$ on C. Show that $m < |f(z)|$ for all points within C. (Hint: Consider the function $F(z) = 1/f(z)$.)

b) Show that the restriction $f(z) \neq 0$ in **(a)** is necessary by considering the function $f(z) = z$ and an appropriately chosen contour. (That is, show that $|f(z)|$ can achieve a minimum interior to C when that minimum is 0.)

c) Assume that $f(z)$ is nonconstant and analytic within and on a simple closed contour C. Show that if $|f(z)|$ is constant on C, then f must have at least one zero within C.

7.8.4. Let $f(z) = (z + i)^2$ within and on the triangle whose vertices are 0, 3, and $1+2i$. Find the maximum and minimum values of $|f(z)|$ on this region. (Hint: Think of $|f(z)|$ as the square of the distance between z and $-i$.)

7.8.5. Suppose that f and g are analytic functions within and on the contour C. Show, using the maximum principle, that if $f(z) = g(z)$ at every point *on* C, then $f(z) = g(z)$ at every point *within* C.

7.8.6. Find all functions analytic inside the disk $|z| < r$ such that $f(0) = -i$ and $|f(z)| \leq 1$.

7.8.7. Show that if f is analytic *within* the unit circle $|z| < 1$ and if $|f(z)| \leq 1 - |z|$, then $f \equiv 0$.

7.8.8. Assume that f is analytic and nonconstant within and on the contour C and that $|f(z) - 1| \leq 1$. Show that f has no zeros within C.

7.8.9. Evaluate the following improper integrals.

a) $\displaystyle\int_{-\infty}^{+\infty} \frac{1}{x^2 + 2x + 2}\, dx$ **b)** $\displaystyle\int_{-\infty}^{0} \frac{1}{x^2 + 4}\, dx$

c) $\displaystyle\int_{-\infty}^{+\infty} \frac{1}{x^2 + 2x + 2}\, dx$

Bibliography

Ahlfors, Lars V., *Complex Analysis*, 2nd ed. New York: McGraw-Hill Book Company, 1966.

Bottazzini, Umberto, *The Higher Calculus: A History of Real and Complex Analysis from Euler to Weierstrass.* New York: Springer-Verlag, 1986.

Boyer, Carl B., *The History of the Calculus and Its Conceptual Development.* New York: Dover Publications, 1949.

Boyer, Carl B., *A History of Mathematics*, Princeton, NJ: Princeton University Press, 1985.

Buck, R. Creighton, *Advanced Calculus*, 2nd ed. New York: McGraw-Hill Book Company, 1965.

Cajori, Florian, *A Hisory of the Conceptions of Limits and Fluxions in Great Britain, from Newton to Woodhouse.* Chicago: The Open Court Publishing Company, 1919.

Cauchy, Augustin-Louis, *Resumé des Leçon a l'Ecole Royale Polytechnique sur le Calcul Infinitesimal, Œuvres*, Ser. 2, Vol. 4. Paris: Gauthier-Villars, 1899.

Churchill, Ruel V. and James Ward Brown, *Complex Variables and Applications*, 4th ed. New York: McGraw-Hill Book Company, 1984.

Clark, Colin, *Elementary Mathematical Analysis*, 2nd ed. Belmont, CA: Wadsworth Publishers of Canada, Ltd., 1982.

Crowell, Richard H. and William E. Slesnick, *Calculus with Analytic Geometry.* New York: W. W. Norton & Company, Inc., 1968.

Edwards, Jr., C. H., *The Historical Development of the Calculus.* New York: Springer-Verlag, 1979.

Gaughn, Edward D., *Introduction to Analysis*, 3rd ed. Monterey, CA: Brooks/Cole Publishing Company, 1987.

Grabiner, Judith V., *The Origins of Cauchy's Rigorous Calculus.* Cambridge, MA: MIT Press, 1981.

Grattan-Guinness, I., editor, *From the Calculus to Set Theory, 1630–1910, An Introductory History.* London: Gerald Duckworth & Co. Ltd., 1980.

Genther, Robert M., "A Simple Estimate of the Error in Linear Approximation," *The American Mathematical Monthly*, 96, no. 6 (June-July 1989), 522–523.

Klein, Felix, *Elementary Mathematics from an Advanced Standpoint: Arithmetic, Algebra, Analysis.* New York: Dover Pulications, 1945.

Kline, Morris, *Mathematical Thought from Ancient to Modern Times.* New York, Oxford University Press, 1972.

Lay, Steven R., *Analysis with an Introduction to Proof*, 2nd ed. Englewood Cliffs, NJ: Prentice-Hall, 1990.

Levinson, Norman and Raymond M. Redheffer, *Complex Variables.* San Francisco, CA: Holden-Day, Inc., 1970.

Marsden, Jerold E. and Anthony J. Tromba, *Vector Calculus*, 2nd ed. San Francisco, CA: W. H. Freeman and Company, 1981.

Ray, William O., *Real Analysis.* Englewood Cliffs, NJ: Prentice-Hall, 1988.

Rudin, Walter, *Principles of Mathematical Analysis*, 3rd ed. New York: McGraw-Hill Book Company,1976.

Saaf, E. B., and A. D. Snider, *Fundamentals of Complex Analysis for Mathematics, Science, and Engineering.* Englewood Cliffs, NJ: Prentice-Hall, 1976.

Smith, David Eugene, *A Source Book in Mathematics.* New York: Dover Publications, 1984.

Smith, Keenan T., *A Primer of Modern Analysis.* Tarrytown-on-Hudson, NY: Bogden & Quigley Inc., Publishers, 1971.

Sondheimer, Ernst and Alan Rogerson, *Numbers and Infinity: A Historical Account of Mathematical Concepts.* Cambridge: Cambridge University Press, 1981.

Struik, D. J., editor, *A Source Book in Mathematics 1200–1800.* Princeton, NJ: Princeton University Press, 1986.

Spivak, Michael, *Calculus*, 2nd ed. Berkeley, CA: Publish or Perish, Inc., 1980.

Williamson, Richard E., Richard H. Crowell, and Hale F. Trotter, *Calculus of Vector Functions*, 3rd ed. Englewood Cliffs, NJ: Prentice-Hall, 1972.

Williamson, Richard E. and Hale F. Trotter, *Multivariable Mathematics: Linear Algebra, Calculus, Differential Equations*, 2nd ed. Englewood Cliffs, NJ: Prentice-Hall, 1979.

Index

A

B

C

D

E

F

G

H

I

K

L

M

N

O

Index of Symbols Used